CUSTOMIZING AutoCAD®

*A Complete Guide To Integrating AutoLISP®
Menus, Macros And More*

Joseph Smith and Rusty Gesner

With Technical Assistance from Patrick Haessly

 New Riders Publishing, Thousand Oaks, California

CUSTOMIZING AutoCAD®

A Complete Guide To Integrating AutoLISP®
Menus, Macros And More

By Joseph Smith and Rusty Gesner

Published By:

New Riders Publishing
Post Office Box 4846
Thousand Oaks, CA 91360, U.S.A.

All rights reserved. No part of this book may be reproduced or transmitted in any form or by any means, electronic or mechanical, including photocopying, recording or by any information storage and retrieval system, without written permission from the publisher, except for the inclusion of brief quotations in a review.

Copyright © 1987, 1988 Joseph J. Smith and B. Rustin Gesner

First Edition, 1988

Printed in the United States of America

Library of Congress Card Catalog Data

Smith, Joseph J. & Gesner, B. Rustin

CUSTOMIZING AutoCAD®
A Complete Guide To Integrating AutoLISP® Menus, Macros And More
Library of Congress Card Catalog Number: **88-60083**
ISBN 0-934035-19-0 Softcover

Cover Drawing By Charles Huckeba

Illustrations By Patrick Haessly

Printed by Day and Night Graphics, Santa Barbara, California

WARNING AND DISCLAIMER

This book is designed to provide information about AutoCAD® and AutoLISP®. Every effort has been made to make this book complete and as accurate as possible. But no warranty or fitness is implied.

The information is provided on an "as is" basis. The authors and New Riders Publishing shall have neither liability nor responsibility to any person or entity with respect to any loss or damages arising from the information contained in this book.

If you do not agree to the above, you may return this book for a full refund.

TRADEMARKS

AutoCAD and AutoLISP are registered trademarks of Autodesk, Inc.

dBASE III is a registered trademark of the Ashton-Tate Company.

IBM/PC/XT/AT and IBM PS/2 are registered trademarks of the International Business Machines Corporation.

MS/DOS and OS/2 are trademarks of Microsoft Corporation.

Lotus 1-2-3 is a registered trademark of the Lotus Development Corporation.

UNIX is a registered trademark of the AT&T Company.

Additional software and hardware products are listed by company in the Acknowledgments and Appendices.

ABOUT THE AUTHORS

Joseph Smith

Joseph J. Smith is president of ACUWARE, Inc., in Portland, OR. His company provides an AutoCAD based application program called AUTOPE for structural engineering drafting. He has used AutoCAD since Version 1.4 and writes about AutoCAD customization from the standpoint of a user and a developer, producing commercial application packages.

Before moving to the west coast, Mr. Smith was a product manager with MiCAD Systems, Inc. in New York. He was in charge of AutoCAD systems development and provided corporate training for customized AutoCAD systems. Mr. Smith is trained as a civil engineer with a B.S. degree from Villanova University, Villanova, PA. Prior to joining MiCAD Systems, he worked as an engineer with James T. Smith & Co., Philadelphia, PA. and Piasecki Aircraft Corp., Philadelphia, PA.

Rusty Gesner

B. Rustin Gesner is Technical Editor of New Riders Publishing and heads New Rider's technical support office in Portland, OR. He is responsible for technical review and support for New Rider's books and software products. He has used AutoCAD since Version 1.1 and writes about AutoCAD from the standpoint of a long-time user and from customizing AutoCAD for architectural applications.

Prior to joining New Riders, Mr. Gesner was president of CAD Northwest, Inc., Portland, OR. He was responsible for the sale, installation, and support of AutoCAD systems. Mr. Gesner is a registered architect. Before forming CAD Northwest, he was a practicing architect in Portland. He attended the College of Design, Art and Architecture at the University of Cincinnati, and Antioch College.

Patrick Haessly

Patrick Haessly is Senior Technical Systems Developer at New Riders Publishing in Portland, OR. He provided technical support to the authors in developing the book and contributed to many of the AutoLISP routines used in the book.

Mr. Haessly has used AutoCAD since Version 1.3. Prior to joining New Riders, he was the AutoCAD Systems Manager for A to Z Steel Systems, Inc., a structural steel detailing company in Wilsonville, OR. He was responsible for developing and maintaining customized AutoCAD systems.

CONTENTS

INTRODUCTION

Scope of Personal Applications	I-1
Checklist for Customization	I-1
How CUSTOMIZING AutoCAD Is Organized	I-2
How To Use CUSTOMIZING AutoCAD	I-3
Pick and Choose	I-3
Sample How to Skills Checklist	I-4
Sample Menus, AutoLISP Tools and Programs	I-4
Building a Customized Menu	I-5
Learn By Doing	I-5
How Exercises Are Shown	I-6
How Screens Are Shown	I-7
Prerequisites for CUSTOMIZING AutoCAD	I-7
AutoCAD Experience	I-7
DOS Experience and Hardware	I-7
The CUSTOMIZING AutoCAD DISK SET	I-8
Things to Watch in the Exercises	I-9
Moving On with a Little LISP	I-9

PART I. SYSTEM ORGANIZATION

CHAPTER 1 Getting Started 1-1
A LITTLE LISP AND ORGANIZING DOS

The Benefits of a Well Organized System	1-1
How To Skills Checklist	1-1
DOS and AutoLISP Tools	1-2
A Little LISP	1-2
DOS Directories	1-7
Necessary Directories	1-7
AutoCAD Program and Support Files	1-8
Setting Up AutoCAD's Configuration Files	1-8
ADI Drivers	1-9
Installing the CUSTOMIZING AutoCAD DISKS	1-9
Selecting Text Editors	1-10
The Bootup Environment	1-11
CONFIG.SYS	1-11
AUTOEXEC.BAT	1-12
More on DOS	1-13
Summary Tips and Techniques	1-13

CHAPTER 2 Your AutoCAD System 2-1
ORGANIZING AutoCAD AND CRACKING THE SHELL

The Benefits of Organizing AutoCAD's Environment	2-1
How To Skills Checklist	2-1

	Programs and Files	2-1
	A Common Base for Jumping into the Book	2-2
	PGP: Cracking the SHELL	2-2
	The PGP File	2-3
	Avoiding ACAD.LSP Conflict	2-6
	Prototype Drawing	2-9
	Layering and Style Conventions	2-10
	Establishing a Layering Convention	2-11
	Configuring and Testing CA-PROTO	2-12
	Tips and Techniques	2-12
CHAPTER 3	**Scaling and Block Management**	**3-1**
	SYMBOLS AND PARTS	
	The Benefits of Well Organized Blocks	3-1
	How To Skills Checklist	3-1
	Simple Is Better	3-1
	Symbols, Parts and Blocks	3-2
	Block Commands and Block Definitions	3-3
	How To Create Blocks	3-4
	Scaling Parts and Symbols	3-4
	Layers/Colors/Linetypes and Blocks	3-7
	Multiple and Nested Blocks	3-11
	Minserts of Blocks	3-13
	Side Effects of Exploding Blocks	3-14
	Blocks vs. Shapes	3-15
	Block Management Summary Tips and Techniques	3-16

PART II. BASIC TOOLS

CHAPTER 4	**Menu Macros**	**4-1**
	The Standard AutoCAD Menu	4-1
	Basic Macro Skills	4-1
	The Benefits of Using Menu Macros	4-1
	How To Skills Checklist	4-2
	Menus and Macros	4-2
	How to Make Menu Files with the CA DISK	4-2
	How to Make Menu Files without the CA DISK	4-3
	Special Characters in Menus	4-4
	Backslash, SPACE and Special Characters	4-6
	Semicolons Are a Special Case	4-6
	Putting a Semicolon in Menu Text	4-7
	Labeling Macro Commands	4-7
	Using Special Characters and Path Names	4-8
	Making Long Macros	4-12
	Layers and Macros	4-14
	Repeating Macros and Multiple Commands	4-15
	Indefinitely Repeating Macros and Commands	4-16
	Macros that Edit Using Selection Sets	4-17
	Selection Set Modes	4-18
	Cleanup	4-19
	Practice with Macros	4-20
	Tips And Techniques	4-20

CHAPTER 5	**The Anatomy Of A Menu**	**5-1**
	Pages of Macros	5-1
	The Benefits of a Well Organized Menu	5-1
	How To Skills Checklist	5-1
	Macros, AutoLISP Tools and Programs	5-3
	Menu Layouts	5-3
	Adding Some Symbols	5-4
	Naming Menu Pages	5-9
	Structuring the Menu	5-11
	Menu Devices	5-11
	Assigning Pages to Devices	5-11
	Tablet Menu	5-12
	A ***BUTTONS Menu	5-14
	Menu Page Order	5-16
	Switching Between Menu Pages	5-16
	Suspended Commands and Menu Macros	5-17
	Macros and AutoCAD's Transparent Commands	5-19
	PopUp Menus AutoCAD's Special Interface	5-21
	ICON Menus	5-23
	Summary Tips and Techniques for Menu Devices	5-26
CHAPTER 6	**A Menu System**	**6-1**
	Surveying Your Menu Needs	6-1
	The Benefits of Developing a Menu System	6-1
	How To Skills Checklist	6-2
	Macros, AutoLISP Tools and Programs	6-2
	Automating a Setup Menu	6-2
	The Setup Menu CA-SETUP.MNU	6-3
	Controlling Access to Your Menu Pages	6-10
	Application Menus	6-11
	How To Work in TEST.MNU Menus	6-17
	Making a Standard Text Screen Menu	6-17
	Appending the Chapter Menus	6-18
	Integrating the Text Menu	6-18
	Tips and Techniques	6-19
CHAPTER 7	**A Little More LISP**	**7-1**
	What Is AutoLISP?	7-1
	The Benefits of Using AutoLISP in Macros	7-1
	How To Skills Checklist	7-1
	Macros, AutoLISP Tools and Programs	7-2
	Variables	7-2
	AutoCAD's System Variables	7-3
	Variable Types	7-3
	AutoLISP Variables	7-6
	Flavoring Your Menu with AutoLISP	7-8
	Simple AutoLISP Expressions and Evaluation	7-8
	Math with AutoLISP	7-9
	Using GET Functions for User Input	7-10
	Getting Distances, Angles and Points	7-11
	Base Point Arguments and the GET Functions	7-13
	GET Functions in Menu Commands	7-13

Pausing Problems with GET Functions in Macros	7-15
LISP Lists	7-16
A Little Entity Stuff	7-18
A Little Table Searching	7-20
Integrating the Example Menu	7-22
Tips and Techniques	7-23

PART III. SUPPORT LIBRARIES, FONTS AND HATCHES

CHAPTER 8 Text Styles and Fonts — 8-1

AutoCAD Text Fonts	8-1
The Benefits of Customizing AutoCAD Text Fonts	8-1
How To Skills Checklist	8-2
Macros, AutoLISP Tools and Programs	8-2
AutoCAD Styles	8-2
Making a Centerline Character	8-4
Developing Special Fraction Characters	8-8
BigFonts	8-9
BIGFONTS Files	8-9
Using Bigfonts with AutoCAD	8-13
Integrating the Specials Menu	8-16
Summary Tips and Techniques	8-17

CHAPTER 9 Linetypes, Hatches and Fills — 9-1

The Benefits of Customized Hatches and Fills	9-1
How To Skills Checklist	9-1
Macros, AutoLISP Tools and Programs	9-1
Patterns of Dots and Dashes	9-1
Linetypes	9-2
Scaling Linetype Patterns	9-3
Hatch Patterns	9-5
Making a Hatch Pattern	9-7
Partial and Irregular Fills	9-14
Integrating the Hatch Menu	9-16
Tips and Techniques	9-17

PART IV. AutoLISP: AutoCAD's PROGRAMMING LANGUAGE

CHAPTER 10 AutoLISP: Theory and Program Structure — 10-1

LISP and AutoLISP	10-1
The Benefit of Using AutoLISP	10-1
How To Skills Checklist	10-2
Macros, AutoLISP Tools and Programs	10-2
AutoLISP Program Vs. Data	10-2
The Structure of Functions in AutoLISP	10-4
Defining a Subroutine with DEFUN	10-5
The COMMAND Pipeline	10-7
Adding Command Names to AutoCAD	10-8
External Function Storage and AutoLISP's File Format	10-9
Loading a Function	10-11

NIL and NON-NIL Conditional Program Branching	10-11
Logical and Relational Operators	10-12
Understanding Logical and Relational Operators	10-12
Relational Operators	10-13
The IF Structure	10-13
The PROGN Structure	10-14
The COND Structure: A Multiple IF	10-15
Program Looping Structures	10-16
The REPEAT Structure	10-16
The WHILE Program Structure	10-17
Processing Lists with the FOREACH Function	10-19
Integrating the Example Menu	10-20
Tips and Techniques	10-20

CHAPTER 11 AutoLISP Data Processing — 11-1

The Benefits of AutoLISP Data Processing	11-1
How To Skills Checklist	11-1
Macros, AutoLISP Tools and Programs	11-2
Determining Data Types: What are You?	11-2
Processing STRINGs	11-5
Character Functions	11-6
Formatting STRINGs	11-7
Displaying and Printing Strings	11-9
Converting Numbers to STRINGs and STRINGs to Numbers	11-10
Formatting for Linear Distances	11-11
AutoCAD Angles	11-12
Making A Dimensioning Command	11-18
Formatting Prompt Strings	11-20
Controlling User Input with INITGET	11-22
More UFUNS Functions	11-25
Integrating the Text Menu	11-28
Summary Tips and Techniques	11-28

CHAPTER 12 The Drawing Database — 12-1

The Benefits of Accessing AutoCAD's Database	12-1
How To Skills Checklist	12-1
Macros, AutoLISP Tools and Programs	12-2
Entity Names	12-2
Entity Selection Sets	12-6
More on Selection Sets	12-9
Developing a Selection Set Toolkit SSTOOLS	12-10
Traversing The Database	12-13
Retrieving Entity Data	12-16
Entity Association Lists	12-17
DXF Group Codes and AutoLISP	12-18
Entity Properties and Default Values	12-22
Modifying and Updating Entity Data	12-26
Complex Entities	12-28
Modifying Polylines via Entity Access	12-29
Integrating the Example Menu	12-32
Tips and Techniques	12-34

CHAPTER 13 AutoLISP Table Access and More! — 13-1
- The Benefits of AutoLISP Table Access — 13-1
- How To Skills Checklist — 13-1
- Macros, AutoLISP Tools and Programs — 13-1
- Symbol Tables — 13-2
- Named Blocks — 13-2
- Named Views, Layers and Styles — 13-11
- Named Properties: Linetypes and More — 13-13
- Converting Data Types to Strings ETOS — 13-17
- Defining and Protecting Your PVARS — 13-18
- The PVAR Functions — 13-20
- Integrating the Example Menu — 13-28
- Summary Tips and Techniques — 13-29

CHAPTER 14 AutoLISP Input/Output — 14-1
- The Benefits of AutoLISP I/O — 14-1
- How To Skills Checklist — 14-1
- Macros, AutoLISP Tools and Programs — 14-2
- Device Input/Output — 14-2
- Reading and Writing — 14-3
- General File Handling — 14-4
- Writing to the Printer and Other Devices — 14-6
- Testing for Files and Paths — 14-7
- AutoLISP and DOS File Handling — 14-11
- Automatic Hatch Pattern Generator — 14-13
- Using ANSI Formatting Codes — 14-19
- File Formats: The Basis of External Data Handling — 14-23
- The REFDWG Function — 14-26
- Integrating the Example Menu — 14-31
- Tips and Techniques — 14-32

CHAPTER 15 AutoLISP Device Access — 15-1
- The Benefits of Using AutoLISP's Device Access — 15-1
- How To Skills Checklist — 15-1
- Macros, AutoLISP Tools and Programs — 15-2
- Dynamic Screen Labeling Using GRTEXT — 15-2
- GRDRAW: Drawing on the Screen — 15-5
- Making a Dynamic View — 15-7
- Alternatives for Displaying Screen Information — 15-9
- GRREAD: Device Access for Input — 15-12
- ETEXT An AutoCAD Text Editor — 15-13
- Continuous Coordinate Tracking — 15-16
- Integrating the Example Menu — 15-21
- Tips and Techniques — 15-21

PART V. ADVANCED APPLICATIONS

CHAPTER 16 Developing an AutoLISP Toolkit — 16-1
- The Benefits of Building a Standard Set of Subroutines — 16-1
- How To Skills Checklist — 16-2
- Macros, AutoLISP Tools and Programs — 16-2
- The ACAD.LSP File — 16-2

Memory Management	16-3
The CLEAN Alternative	16-5
Memory Management Practices	16-5
AutoLISP Setup and Initialization	16-6
General Utility Functions	16-7
User Interface Functions	16-10
ERROR Error Handling When Functions Crash	16-10
An Error Trap System	16-11
Working with Other Application Programs	16-19
Integrating the Example Menu	16-22
Summary Tips and Techniques	16-23

CHAPTER 17 3-D Manipulations with AutoLISP — 17-1

The Benefits of Making 3-D Tools	17-1
How to Skills Checklist	17-1
Macros, AutoLISP Tools and Programs	17-1
2-D vs. 3-D ENTITIES	17-2
3-D Distances	17-4
3-D Polar Functions	17-6
Polar Rotations in Space	17-9
Other Polar and Curve Formulas	17-10
Mesh Generation	17-12
Converting 2-D Plans to Elevations	17-16
Integration From 3-D to 2-D	17-24
Integrating the 3D Menu	17-25
Tips and Techniques.	17-26

CHAPTER 18 An ISO Dimensioning System — 18-1

The Benefits of Building an Iso-dimensioning System	18-1
How To Skills Checklist	18-1
Menus, AutoLISP Tools and Programs	18-2
An Overview of Iso-Dimensioning	18-2
Isometric Dimension Text Styles and Symbols	18-3
Making an Iso Screen Page	18-3
The ISODIM Functions	18-8
Putting IsoDims on Your Tablet	18-12
Integrating the Iso Menu	18-15
Using Associative Dimensioning with Iso-Dimensioning	18-16
Summary Tips and Techniques	18-16

CHAPTER 19 Attributes as Data Tools — 19-1

The Benefits of Using Attributes	19-1
How to Skills Checklist	19-2
Macros, AutoLISP Tools and Programs	19-2
The Title Block System.	19-3
Attributes in Menu Macros	19-6
Automatic Title Sheet Maintenance	19-10
Automating ATTEDIT	19-10
The Drawing Revision System	19-13
Tracking Drawing Revisions	19-15
AutoCAD's Treatment of Attribute Data	19-17
Drawing Revision Updates Function	19-19

AutoBreaking Blocks	19-24
A Word about Block Redefinition and Lost Attributes	19-28
Integrating the TitleBlk Menu	19-29
Tips and Techniques	19-29

CHAPTER 20 Parametrics and Material Tagging — 20-1

An Overview of Parametric Systems	20-1
The Benefits of Parametrics	20-1
How To Skills Checklist	20-2
Macros, AutoLISP Tools and Programs	20-2
The Parametric Screen Menu	20-3
The External File Format	20-5
A Little Help	20-7
Retrieving External Data	20-8
Generating the Image	20-11
Drawing a Side View of a 90 degree Elbow	20-12
Multiple Views: The Beauty of Parametrics	20-15
A Double Line Pipe Command	20-19
Labeling Components with Material Tags	20-21
Integrating the Pipefits Menu	20-25
Tip and Techniques	20-26

CHAPTER 21 AutoLISP, Script and DXF Batch Processing — 21-1

The Benefits of Batch Processing	21-1
How To Skills Checklist	21-1
Macros, AutoLISP Tools and Programs	21-2
Script vs. Menu vs. AutoLISP vs. External Software	21-3
Scripts	21-3
Using Long Menu Items	21-4
DXF Importation	21-4
Reading in Data with AutoLISP	21-5
What, When and Where for Scripts, AutoLISP and DXF?	21-8
Script Batch Builder	21-8
Slide Libraries	21-15
Scripts and Plotting	21-16
Starting, Stopping and Resuming Scripts	21-16
Coordinated Scripts and AutoLISP	21-18
XINSERT External Block Extraction	21-22
AutoCAD Drawing File Formats	21-22
When to Use the DXF	21-23
DXF Group Codes and Data Elements	21-24
DXF File Format, Data Types and Codes	21-25
Header Information	21-25
Reference Table Information	21-26
Block References	21-27
The Entities Section	21-27
A Version Conversion Reversion Utility	21-28
Translating from Other CAD Programs	21-29
Integrating the Example Menu	21-30
Summary Tips and Techniques	21-31

CHAPTER 22 LOTUS and dBASE — 22-1
IMPORTING AND EXPORTING DATA

The Reporting Scenario	22-1
The Benefits of Importing and Exporting AutoCAD Data	22-1
How to Skills Checklist	22-2
Macros, AutoLISP Tools and Programs	22-2
The Template File	22-2
Importing Data Into Lotus	22-6
Bringing It Back In	22-18
Using dBASE with AutoCAD	22-19
Preparing the Input Record	22-21
Tracking CAD Drawings	22-24
The PRO_TRAK Database Structure	22-25
Importing the Project Data	22-26
Reporting Project Data	22-28
Time Log Reports	22-29
Instructions for Using PRO_TRAK	22-30
Integrating the Reports Menu	22-31
Tips and Techniques	22-31

CHAPTER 23 Summing Up with System Controls — 23-1

The Benefits of System Controls	23-1
How To Skills Checklist	23-1
Macros, AutoLISP Tools and Programs	23-2
Controlling Your System	23-2
Directory and File Control	23-2
DOS SUBST Directory Path Control	23-4
Drawing Control	23-5
New or Existing Drawing?	23-5
Reset Controls	23-6
Command Redefinition	23-6
Keyboard Control	23-8
ANSI.SYS Key Redefinition	23-9
Limited Power to the User	23-12
Encryption and Security	23-14
Customization Security	23-14
Cleaning .LSP Files with LSPSTRIP	23-15
Documenting and Presenting an Application	23-16
Finalizing ACAD.LSP	23-18
Finalizing Menus	23-19
Authors' Farewell	23-23

APPENDIX A Menu Macros and AutoLISP Functions — A-1

Menus and Macros	A-1
AutoLISP Functions and Programs	A-2

APPENDIX B Setup, Memory and Errors — B-1

Setup Problems with CONFIG.SYS	B-1
Problems with AUTOEXEC.BAT	B-1
Problems with DOS Environment Space	B-3
Using DOS 2	B-3
Memory Settings and Problems	B-4

	Using CUSTOMIZING AutoCAD with a RAMDISK	B-6
	Finding Support Files	B-7
	Current Directory Errors	B-7
	SHELL Errors	B-8
	Common AutoLISP Errors	B-8
	Miscellaneous Problems	B-9
	Insufficient Files Errors	B-9
	Tracing and Curing Errors	B-9
APPENDIX C	**Reference Tables**	**C-1**
	AutoCAD System Variables	C-2
	DXF Group Codes	C-4
APPENDIX D	**The Authors' Appendix**	**D-1**
	How Customizing AutoCAD Was Produced	D-1
	Our Hardware	D-1
	Our AUTOEXEC.BAT Files	D-1
	Our Text Editor	D-3
	Our Multitasking Interactive Environment	D-3
	Document Illustration, Formatting and Printing	D-4
	Tools, Sources and Support	D-4
	Advanced AutoCAD Classes	D-4
	A Brief Look at the CompuServe Autodesk Forum	D-4
	Bulletin Boards	D-6
	Users groups	D-6
	Magazines	D-6
	Books	D-7
	Commercial Utilities	D-7
	Hard Disk Management and Backup	D-7
	Keyboard Macros	D-8
	User Interface Shells	D-8
	General Utility Packages	D-8
	Freeware and Shareware	D-8
	Authors Last Word and Mail Box	D-9
INDEX		**X-1**

ACKNOWLEDGMENTS

Rusty would like to thank Kathy, Alicia and Roo for their patience, encouragement and good cheer when the going was tough, and to thank his parents for support in the endeavors which led to this book.

Joe wishes to thank his parents, Jim and Rita for their abundance of love and support. They have given me a chance at every passing opportunity, most recently this book.

Joe, Rusty and Pat wish to thank Harbert Rice for his able editing, and for making this project possible. Thanks to Carolyn Porter and Todd Meisler for helping with page layout and production.

Special thanks to Jon C. A. DeKeles for developing the dBASE and Clipper compiled drawing revision program, PRO_TRAK. Thanks also to Jon for helping to develop the Ventura style sheet for the book.

The authors wish to thank Tom Mahood, Mauri Laitinen, Duff Kurland, Eric Lyons, John Sergneri, Keith Marcelius, Dave Kalish, and many others from Autodesk, Inc. for their help and support over the years.

Special thanks to Andy Hood and Robert Palioca of KETIV Technologies, Inc. KETIV provided material used in developing the ANSI functions. Andy Hood reviewed the book. Robert Palioca provided invaluable help. KETIV Technologies, Inc. provided computing equipment used in developing the book.

Special thanks to Keith McIntyre, PE, for ideas, support and his review of the book, and to Dorothy Kent, Lambda Systems, Inc. for her review.

Thanks to Willamette Industries and Margo Bilson for providing a set of AutoCAD drawings. The drawings were a source of ideas and a starting point for many illustrations in the book.

Thanks also to Dan Stone, Ken Eichler, Dan Belmont, Eugene Jones, Bill Work and the rest of the people at MiCAD Systems, Inc. for their many ideas and techniques that contributed to material in the book. Thanks to the people at Smith Engineering for serving as the testbed of application experiments. Thanks to Phil Kreiker, Looking Glass Microcomputer Products, who provided the technique used in self-loading LISP command functions.

Autodesk, Inc. supplied AutoCAD, AutoShade and the Kelvinator. Xerox Corp. supplied Xerox Ventura Publisher. Microsoft Corp. supplied Microsoft Word. Michael Cuthbertson of Symsoft provided copies of Hotshot and Hotshot Plus. Thanks to Barry Simon and Richard Wilson, the developers of CTRLALT.

Verticom Inc. provided a 2Page Display System for Ventura Publisher displays. Verticom also provided H256e and M256e Video Display Systems for AutoCAD displays. Vermont Micro Systems provided a VMI1024 Video Display System for AutoCAD displays. CALCOMP provided Model 2300 and 2500 Digitizers.

INTRODUCTION

Welcome to the most creative part of using AutoCAD, CUSTOMIZING AutoCAD. The goal of this book is to bring you the information and tools that you need to make AutoCAD do what you want it to do. Whether you want to simply add a few menus and macros to AutoCAD, or write a complete "system" including AutoLISP commands to execute your application, CUSTOMIZING AutoCAD will give you knowledge you need to make AutoCAD work for you.

Scope of Personal Applications

Most AutoCAD users adapt AutoCAD to their own use. There are three stages to customization. Most users start customizing AutoCAD by writing macros which they add to their screen menus. Writing and using menu macros is a natural extension to using AutoCAD. Many users move on to a second stage of customizing AutoCAD, making their macros more efficient with AutoLISP expressions, or adding new drafting and calculation functions, commands and subroutines to their systems. They integrate these commands and functions into their menus. Finally, some users and groups move on to developing "full-blown" integrated application systems. These systems include everything from developing complete user interfaces for AutoCAD to integrating and running external programs.

This book addresses the key issues you encounter in planning, developing, programming and testing a professional AutoCAD application system. Whether you are a user entering the first stages of customization, or a developer putting together a complete custom application, CUSTOMIZING AutoCAD will help you assess your application needs and help you choose the best avenues for customizing AutoCAD. The book provides menu macros and AutoLISP routines to help you gain greater control over AutoCAD, and control access to your programs and data. The book provides menu and AutoLISP tools to help you develop a user interface, including how program prompts, questions and options will appear. The book addresses issues like integrating drawing standards, drawing scales, project and drawing schedules into an application.

Checklist for Customization

Here is a checklist of things that you need to do to get started customizing AutoCAD. The book will help you with each of these tasks.

- ❑ Set your application goals. You need to know what you want to do.
- ❑ Lay out your customization plan for modifying AutoCAD. You need to know how to do your applications in AutoCAD.
- ❑ Develop your macros. Get them tested and working.
- ❑ Develop your menus and integrate your macros.

- ❏ Develop the AutoLISP routines and programs that you need. Integrate these into your menu system.
- ❏ Develop and integrate external programs and program interfaces that you need. Integrate these external programs into your system. The book will help with some examples.
- ❏ Document your system so that you, and your users, will know how to use what you have developed. The book will help with some guidelines.

How CUSTOMIZING AutoCAD Is Organized

CUSTOMIZING AutoCAD is organized into five parts. Think of the book as five handy reference guides in one place. These five guides take you through customizing AutoCAD, from setup all the way through a set of applications.

PART I. SYSTEM ORGANIZATION gives you the basic organization of your MS-DOS and AutoCAD environments for customizing an AutoCAD application. Chapter 1 starts off by showing you a little AutoLISP, then gives the book's bootup and DOS directory structures. Chapter 2 leads you through making a startup batch file for CUSTOMIZING AutoCAD, and using AutoCAD's Shell to integrate your text editor to do program development. Chapter 3 sets out the book's structure of AutoCAD colors, layers, and linetypes. CUSTOMIZING AutoCAD's basic working environment is designed to give a common jumping off point for custom development, and to let you work through the book without interfering with your other AutoCAD applications.

PART II. BASIC TOOLS gives you a set of working menu tools to customize AutoCAD. Chapter 4 shows you how to build menu macros. Chapter 5, Anatomy of a Menu, shows you how to use AutoCAD's menu system, including how to use popup menus and icons. Chapter 6 shows you how to organize and integrate a menu system. Chapter 7, A Little More LISP, shows you how to build and integrate AutoLISP routines into your menu system. If you are an advanced user and are familiar with some of the material covered in these chapters, we recommend that you still review them. This ensures that we have a common base of understanding for the advanced AutoLISP applications that form the rest of the book.

PART III. SUPPORT LIBRARIES, FONTS and HATCHES. Many custom applications require support libraries. Chapters 8 and 9 show you how to build custom libraries for AutoCAD's text styles, fonts, hatches and fills. Chapter 9 shows you a process for making hatch patterns that can be automated. Later in the book, you will build an AutoLISP program that automates the hatch building process.

PART IV. AutoLISP: AutoCAD's PROGRAMMING LANGUAGE has six chapters. These chapters will give you both theory and the practical AutoLISP tools that you need to develop your own custom applications. Chapter 10 discusses LISP theory and AutoLISP's program structure. Chapter 11 shows you how to work with AutoLISP's data types. Chapter 12 shows you how to access and control AutoCAD's drawing database with AutoLISP. Chapter 13 shows you how AutoLISP's table access works. Chapter 14 shows you how to use AutoLISP's file I/O. Chapter 15 shows you how AutoLISP's device access works, including how to control devices for user input. After you have worked through these chapters on AutoLISP, you will have a thorough understanding of AutoLISP and a core set of AutoLISP programs that you can use in your own applications.

PART V. ADVANCED APPLICATIONS provides a set of advanced applications, showing you how to build and manage complete applications. Chapter 16 develops an AutoLISP Toolkit and shows you how to organize your ACAD.LSP library, and to how to manage AutoLISP memory. Chapter 17 provides a set of AutoLISP 3-D tools to manipulate 3-D drawings. Chapter 18 uses AutoLISP to develop a complete Isometric Dimensioning System. Chapter 19 shows how to use Attributes as Data Tools, including making a drawing update and revision system. Chapter 20 develops a parametric system with materials tagging. Chapter 21 shows you how to use and control batch processing and DXF file processing. Chapter 22 shows how to import and export data to Lotus 123 and dBASE III. The Lotus application extracts attribute data. The dBASE III application extracts and processes drawing revisions and schedules. Chapter 23 sums up and gives you some tools and guidelines for controlling your customized application.

APPENDICES. There are four Appendices. Appendix A gives a complete indexed list of the menus and AutoLISP programs used in the book. This list includes a brief description of each menu and AutoLISP program. Appendix B Problems, Errors and Debugging covers common problems and solutions encountered in setting up and using AutoCAD in a DOS environment. Appendix C provides two invaluable Reference Tables, a complete AutoCAD System Variable Table and a complete DXF Table. Appendix D, the Authors' Appendix, discusses the system setups and software used to produce CUSTOMIZING AutoCAD along with the authors' comments on software and information sources helpful to customizing AutoCAD.

How To Use CUSTOMIZING AutoCAD

There are three ways to use CUSTOMIZING AutoCAD. First, you can read it front to back. Second, after you have set up the CUSTOMIZING AutoCAD environment in chapters 1 and 2, you can pick and choose what menus, macros, and AutoLISP programs you want to learn and use. Third, you can pick and choose what customizing skills you wish to acquire from the chapters that interest you.

Pick and Choose

Obviously, CUSTOMIZING AutoCAD provides you with a wealth of information about how to customize AutoCAD. We expect that you will go through parts of the book several times, using it as a reference and as a source of macros and AutoLISP programs. The book tries to make your access to the material easy by providing lists of both tools and customization skills in each chapter. Each chapter starts with a How To Skills Checklist, like the one shown below. The example checklist is taken from chapter 10 on AutoLISP. Pick and choose the skills that you need by looking at the skills checklist.

Sample How to Skills Checklist

How you structure your AutoLISP program's flow determines how your programs will behave. The flow is set by how you organize your program statements using AutoLISP's logical functions and relational operators. One key skill to defining AutoLISP functions is to distinguish between global and local environment variables. Besides gaining an understanding of AutoLISP's global and local environments, you will learn how to:

- Create an executable AutoLISP program.
- Define functions with DEFUN.
- Store functions in external .LSP files, format the files, and load the functions into AutoCAD's memory.
- Use AutoLISP's COMMAND function to pipeline instructions to AutoCAD.
- Make your AutoLISP commands act like any other AutoCAD command by adding them to AutoCAD's command list.
- Use multiple test conditions, including the use of AND, OR and comparative functions like LESS THAN and EQUAL TO.
- Make looping programs to construct data lists, using AutoLISP's REPEAT function.
- Batch process data lists using AutoLISP's FOREACH function.

After the skills checklist, each chapter has a list of menus, macros, and AutoLISP programs used in the chapter. We call this list the "Toolkit." These toolkits are re-indexed in Appendix A. By looking at these toolkits in each chapter, you can pick and choose what menus and programs you want to learn. A typical toolkit looks like the one shown below. This toolkit is taken from chapter 18 on Isometric Dimensioning.

Sample Menus, AutoLISP Tools and Programs

MENUS
**ISO is the main isometric dimensioning page of the isometric menu system. It uses an integrated set of screen menu toggle pages.
**TISO is an isometric Tablet Area 1 menu that integrates with the screen menu. It also can be used alone.

AutoLISP TOOLS
C:DIMLINE is an AutoLISP command function that draws isometric dimensions and text.
C:EXTLINE is an AutoLISP command function that draws isometric extension lines.

PROGRAMS
ISODIM is a program that integrates the isometric menu page system working in conjunction with the ISODIM.LSP file. ISODIM.LSP contains the C:DIMLINE and the C:EXTLINE functions.

Building a Customized Menu

If you follow the book front to back, you will build a complete customized menu system. The CUSTOMIZING AutoCAD menu has a dozen major AutoCAD screen menu pages. After you set up your initial screen menus with the BASIC TOOLS in PART II., you add a menu page (or more) with most chapters. These menu pages are illustrated in each chapter. You can follow the overall menu plan used in the book by looking at these menu pages. A chapter screen menu page looks like the one shown below. This menu page is taken from chapter 19 on using Attributes as Data Tools.

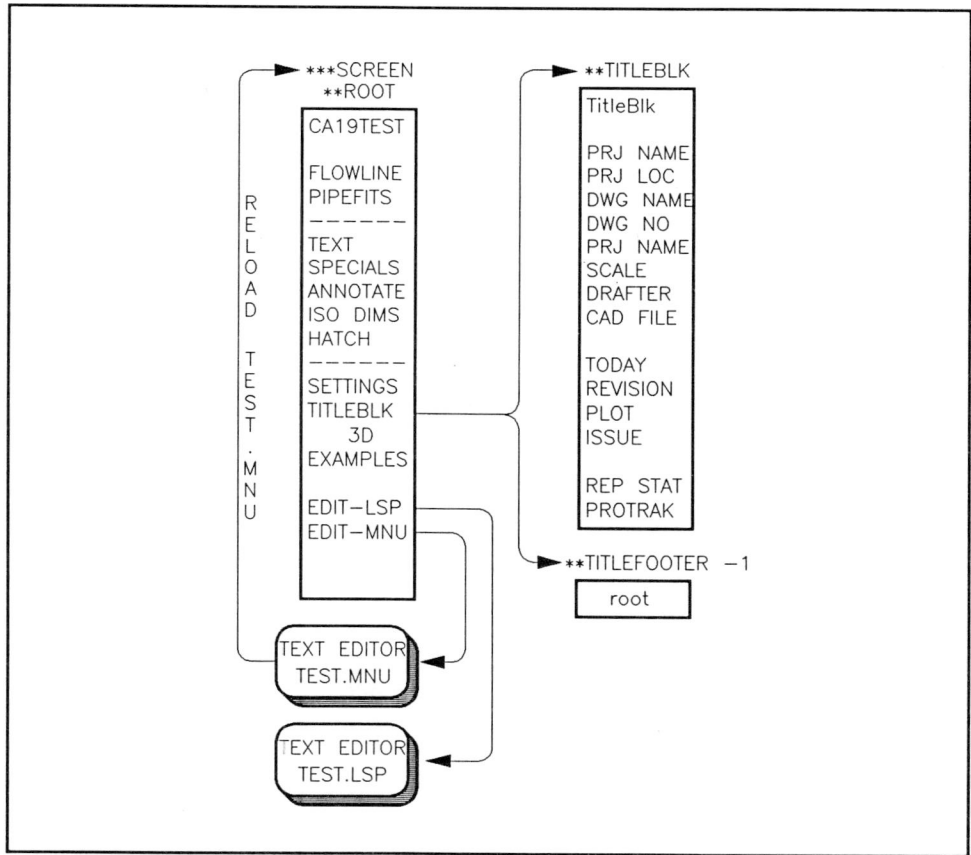

An Example Chapter Menu

Learn By Doing

Learning to use AutoLISP may look like a formidable task. It isn't. You also may harbor the suspicion that you have to be a programmer to use AutoLISP to develop a custom AutoCAD application. You don't. We are AutoCAD users and we have written CUSTOMIZING AutoCAD for AutoCAD users.

We learned by doing and built the book around hands-on exercises.

How Exercises Are Shown

In fact, let's do an "exercise" to show you how information is displayed in the book. This exercise will introduce the AutoCAD version requirements for the book, and will show you the book's format and syntax for representing AutoCAD commands.

To get started, all you need to know is how to start up AutoCAD. AutoLISP should be loaded. Make sure your system doesn't have any custom menu or custom ACAD.LSP file installed.

Checking Your Version of AutoLISP

```
Enter selection:                        Begin a NEW drawing named INTRO=
                                        The = forces all ACAD default settings.

Command: <Coords on>                    Toggle ON.
Command: SNAP                           Set ON at 0.1.
Command: GRID                           Set to 1.

Command: (ver)                          Type (ver) and hit <RETURN>.
Lisp returns: "AutoLISP Release 9.0"    Or your version number.
```

The book uses the full width of the page for command sequences. Commands and required steps are shown at the left. The "Command:" prompt is how the book shows all AutoCAD prompts, including those that you will create with AutoLISP. Prompts are shown as they will appear on your screen. The exercises show the most important commands and prompts, but they are not literal sequences. The example shows "**(ver)**" in bold to indicate that it is input that you need to type in. All input is in bold text.

The right-hand section provides "in-line" comments and instructions about the sequence. "Type (ver) and hit <RETURN>" is an instruction. All you need to do in an exercise is follow the Command: sequence, refer to any in-line instructions that are provided, and input any text shown in bold.

Lisp returns: is used to call your attention to AutoLISP's backtalk. You won't see the *Lisp returns:* on your screen, but what follows it should appear on your screen as shown in the book. If you got "AutoLISP Release 9.0", you're in great shape.

If you didn't get at least "AutoLISP version 2.6", then you need to look at your AutoCAD system. You must have AutoCAD 2.6 or later to follow the material in this book. If you got "AutoLISP version 2.3", or a lower number, you will need to update your AutoCAD. This is money as wisely spent as the money spent for this book. If you got an "Unknown command", or other error, check to see that you have AutoCAD 2.6, or Release 9, with ADE3. Then check to see that

AutoLISP is not disabled by insufficient memory or the Configuration Menu item "Configure operating parameters."

How Screens Are Shown

Pure uninterrupted lines of command sequences can become mesmerizing. To help you along with the hands-on exercise, the book provides illustrations and screen "shots" of what you should see on your screen at key points. Remember, we had to do the exercises too! We have captured our screens as we tested the prompt sequences and AutoLISP routines for the book. A typical screen shot is shown below.

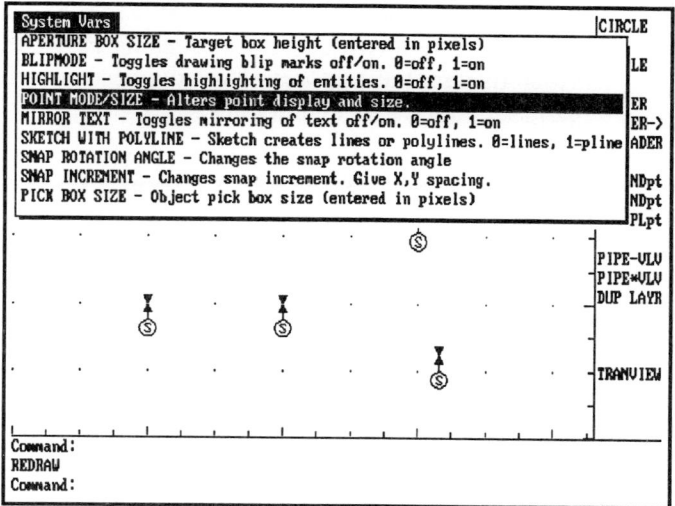

Sample Screen Shot

Prerequisites for CUSTOMIZING AutoCAD

The hardest part of CUSTOMIZING AutoCAD is typing accurately. If you have problems with your typing, we recommend that you get the CUSTOMIZING AutoCAD DISKS. This simplifies the typing by eliminating most of it.

AutoCAD Experience

To use this book, you should have enough experience to feel comfortable using AutoCAD. If you have worked through any of New Riders' AutoCAD books, like INSIDE AutoCAD or WORKING OUT WITH AutoCAD, you should have no problem picking up with this book. You should be familiar with most of AutoCAD's commands, but you do not need to be an expert. You should know how to configure your AutoCAD system. Beyond that, you just need a desire to make your AutoCAD system more effective and more productive.

DOS Experience and Hardware

You need PC or MS-DOS 3.0, or a later DOS version on your system. You should be familiar with the major DOS commands: CD, CHKDSK, COPY, DEL, DIR, DISKCOPY, FORMAT, MD, PATH, REN, RD, and TYPE. You should also know how to make batch files, including an AUTOEXEC.BAT file. If you have MS-DOS 2.0, you can still use the book, but refer to our Appendix C.

You don't need a fancy or particularly fast AutoCAD system. However, if your system is not running at IBM-AT speed, you may find some AutoLISP programs a little slower than you'd like. A hard disk, or equivalent, is required. You will need 640K of RAM with AutoCAD 2.6 to utilize AutoLISP. Extra EXTended or EXPanded RAM is nice, but not required. A mouse is OK, but a digitizer tablet is preferable, and a tablet is required to do the Tablet and Button menu examples.

As we have indicated, you need AutoCAD 2.6 or later to use the book. If you have AutoCAD Release 9, you may be able to take advantage of its advanced display features. The menu exercises covering POPUPS and ICON MENUS require a video driver that supports the advanced interface. If you have Release 9, you can test your video display now.

Testing for POPUPS and ICONS

```
Command: SETVAR
Variable name or ?: POPUPS
POPUPS = 0 (read only)
Command: QUIT
```

If you got a "1", you can use POPUPS and ICONS. If you got a "0", you need an ADI 3.0 video driver, or a later version ADI driver from your video board manufacturer, or you need a new video board. If you don't care about ICONS and POPUPS, don't worry. You can use all the other Release 9 features. If you got "Unknown variable name", you don't have Release 9. The book shows where Release 9 features are used in the exercises.

The CUSTOMIZING AutoCAD DISK SET

There is a disk set available for use with the book, The CUSTOMIZING AutoCAD DISKS (or CA DISK for short). The disks include all the menu .MNU files, AutoLISP .LSP files, Script .SCR files, batch .BAT files, symbol and example .DWG files, and other support files used in the book. An optional CA BONUS DISK also includes BONUS AutoLISP routines, utilities and source code files.

While not absolutely essential for most exercises, the CA DISK will save you time and energy. Using the CA DISK ensures that your customization menus, macros and AutoLISP routines are accurate. The CA DISK also releases you from tedious typing. This lets you focus your attention on the real material when you work through the book's exercises.

The back of the book includes an order form for the CUSTOMIZING AutoCAD DISK SET and the BONUS DISK. Instructions on installing and using the disks are given in the first chapter.

Things to Watch in the Exercises

The book's routines were tested by us and by our reviewers. Here are three things we suggest you watch for in the exercises:

The printing font used in the exercises doesn't distinguish clearly between zero and the letter O; and the number one and lower case l. You need to watch these closely:

```
0   This is a zero.
O   This is an upper case letter O.
1   This is the number one.
l   This is a lower case letter l.
```

The exercise sequences are not literal sequences. The exercises show the most important command prompts, but they don't show every command prompt.

When you are using the CA DISK menus with the exercises, the disks will show the completed menu as it will appear at the end of the chapter. You follow a chapter's menu macro edits by examining the menu with your text editor. Even if you have the CA DISK, there are a few menu tasks that you need to do in the book to use material that occurs in a later chapter.

To help you along with the hands-on exercises, the book provides some simple disk icons showing you when to use the CA DISK in the exercises. A pointing finger icon indicates material that is used in a later chapter. These icons are shown below.

 Do "this" if you have the CA DISK.

 Do "this" if you don't have the CA DISK.

 Do "this" whether or not you have the CA DISK. You need the material for later chapters.

Moving On with a Little LISP

Let's move on and get our feet wet by playing around with a few AutoLISP routines in chapter 1, then organize the DOS system environment for CUSTOMIZING AutoCAD.

1—0 Customizing AutoCAD

Customizing AutoCAD Setup

- AutoCAD Main Menu
- Drawings .DWG (CREATE/EDIT, CONVERT OLD DWG.)
- Working Drawing Database
- AutoLISP
- The AutoCAD Drawing Editor and Command Processor
- Text & Shapes .SHX
- Text .SHP
- TEXT — TEXTSCR — Screen
- STANDARD I/O
- Mouse or Digitizer
- TEXT EDITOR
- DOS Test File
- CONFIG.SYS
- AUTOEXEC.BAT

DIRECTORY STRUCTURE
- Root C:\
  - C:\ACAD
  - C:\CA-ACAD
  - C:\DOS

System Organization for Customizing AutoCAD

PART I. SYSTEM ORGANIZATION

CHAPTER 1

Getting Started

A LITTLE LISP AND ORGANIZING DOS

CUSTOMIZING AutoCAD is built around using AutoLISP, menus and organizing your DOS system environment to develop custom applications. This first chapter gives you a taste of AutoLISP, showing you how easy it is to create an AutoLISP command, and how AutoLISP accesses AutoCAD's drawing database. After showing AutoLISP access, the chapter shows you how to organize your DOS system environment to use CUSTOMIZING AutoCAD.

The Benefits of a Well Organized System

AutoLISP is easy to use. It is AutoCAD's most powerful tool to help you customize your drawing applications. You need a well organized system environment to develop AutoLISP applications. The benefits to creating a good DOS development and maintenance environment are:

- DOS performs system resource functions for AutoCAD. Configuring your DOS environment well will improve AutoCAD's performance.

- DOS subdirectories organize your files into groups. A key to working with DOS is using subdirectories to manage your files. A well managed system helps AutoCAD and other programs find your application files.

- Your text editor is your key developer's tool. You will use it to develop your menus, macros and AutoLISP programs. A well organized DOS environment makes it easy to use your text editor.

This chapter will help you test your text editor and help you set up your DOS environment for customizing AutoCAD.

How To Skills Checklist

Here are skills that you will acquire. You will learn how to:

- ❏ Write and test a simple AutoLISP routine.
- ❏ Use AutoLISP to access the AutoCAD drawing database.
- ❏ Test your text editor for customization use.

1—2 Customizing AutoCAD

❑ Configure your DOS environment using the CONFIG.SYS and AUTOEXEC.BAT files.
❑ Make a DOS subdirectory environment suitable for customizing AutoCAD.
❑ Set DOS PATH searches and make custom DOS prompt strings.
❑ Install the CA DISK.

DOS and AutoLISP Tools

DOS PROGRAMS AND FILES
CONFIG.SYS is the root directory file that sets the general DOS environment. AUTOEXEC.BAT is a batch program that initializes DOS environment settings when you boot the system.

 You need to create or verify your CONFIG.SYS and AUTOEXEC.BAT file to use the book.

AutoLISP TOOLS
C:SH-BOX is an AutoLISP command that draws a shadow box.

A Little LISP

Let's start by looking at a little AutoLISP. To do this exercise you need to start a new drawing. Type in what is shown below. The book assumes your hard disk is C: and AutoCAD is in the \ACAD directory. If your drive and directory names are different, use your own.

You are going to make a new AutoLISP command, called ZL. It executes a ZOOM Limits. Try it!

Don't panic if AutoLISP returns a 1>. It is just AutoLISP telling you that you need another right ")" parenthesis. Type a) and a <RETURN>, then hit a <^C> Control C and try again. The hardest part of AutoLISP is getting matching pairs of () parentheses and "" quotes!

Making ZL an AutoLISP Command

```
C:\ACAD> ACAD                  Start AutoCAD normally.
Enter selection:               Begin a NEW drawing named TEST=
                               The = forces all ACAD default settings.

Command: SNAP                  Set to 0.1.
Command: GRID                  Set to 1.
Command: <Coords on>

Command: ZOOM                  Zoom to Lower Right 1/4 corner.

Command: STYLE                 Name it COMPLEX. Use the COMPLEX font. Default all options.

Command: TEXT                  Set height to 0.5 and type some text.
```

```
Command: (defun C:ZL () (command "ZOOM" "W" (getvar "LIMMIN")(getvar "LIMMAX")))

Lisp returns: C:ZL                     That's all it takes. Now, let's test it.

Command: ZL                            The new command.
Command: ZOOM                          AutoLISP is in control now.

All/Center/Dynamic/Extents/Left/Previous/Window/<Scale(X)>: W
First corner: Other corner:            It Zooms to the limits.
```

That was easy. The new ZL "command" you made will remain available until you Quit, or End this drawing. Later you will store these "commands."

Now let's build a lengthier AutoLISP routine. It is a command to draw a shadowed box with text inside. You will find it handy for creating flow charts and labels.

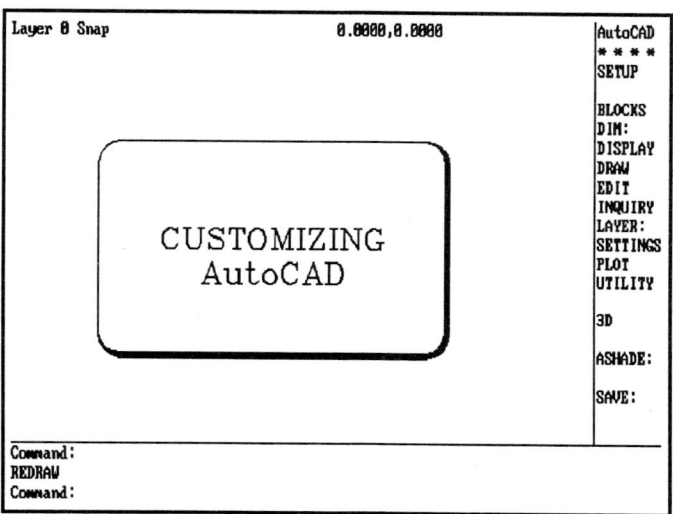

AutoLISP Automation with SH-BOX

The shadow box routine is broken into small pieces to make it easier to type in at the AutoCAD command prompt. In real life, you would create this program as a single function in a text editor, then run it in AutoCAD.

AutoLISP Automation with SH-BOX

The first part is a subroutine function to get the two corner points of a rubberbanded rectangle. Type these two lines:

```
Command: (defun cornrs () (setq ll (getpoint "\nPick LL corner: "))
                                  Hit a <RETURN> and continue typing.
1> (setq ur (getcorner ll "\nPick UR corner: ")))
Lisp returns: CORNRS

Command: (cornrs)                      Try out your function.
```

```
Pick LL corner:                     Pick a point, try 3,2.
Pick UR corner:                     Pick a point, try 9,6.
Lisp returns: (9.000000 6.000000)
```

AutoLISP took the points you picked and stored them under the variable names LL and UR. Check one. Then, write the next part to calculate the other two corner coordinates.

```
Command: !LL                        An ! in front of LL tells AutoCAD it's a variable and queries
                                    AutoLISP for its value.
Lisp returns:(3.000000 2.000000)    AutoLISP returns the point value to AutoCAD.

Command: (defun others () (setq ul (list (car ll) (cadr ur)))
1> (setq lr (list (car ur) (cadr ll))))
Lisp returns: OTHERS

Command: (others)                   Try it.
Lisp returns: (9.000000 2.000000)   It only returns the last of the two points.

Command: !ul                        Check the other one.
Lisp returns: (3.000000 6.000000)
```

You have all four corner points stored by AutoLISP as LL, LR, UR, and UL. Now, you need to get them back out to make AutoCAD do something.

```
Command: (defun drawbx () (command "PLINE" ll "W"
2> (/ (setq tx (getvar "TEXTSIZE")) 3) "" lr ur "W" "0" "" ul "C"))
Lisp returns: DRAWBX

Command: (drawbx)                   Try it out.
PLINE                               All this stuff scrolls on by...
From point:
Current line-width is 0.0000
Arc/Close/Halfwidth/Length/Undo/Width/<Endpoint of line>: W
Starting width <0.0000>: 0.166666666666671
Ending width <0.1667>:
Arc/Close/Halfwidth/Length/Undo/Width/<Endpoint of line>:
Arc/Close/Halfwidth/Length/Undo/Width/<Endpoint of line>:
Arc/Close/Halfwidth/Length/Undo/Width/<Endpoint of line>: W
Starting width <0.1667>: 0
Ending width <0.0000>:
Arc/Close/Halfwidth/Length/Undo/Width/<Endpoint of line>:
Arc/Close/Halfwidth/Length/Undo/Width/<Endpoint of line>: C
Command:
Lisp returns: nil                   And draws a box.
```

Your screen should look like Shadow Box.

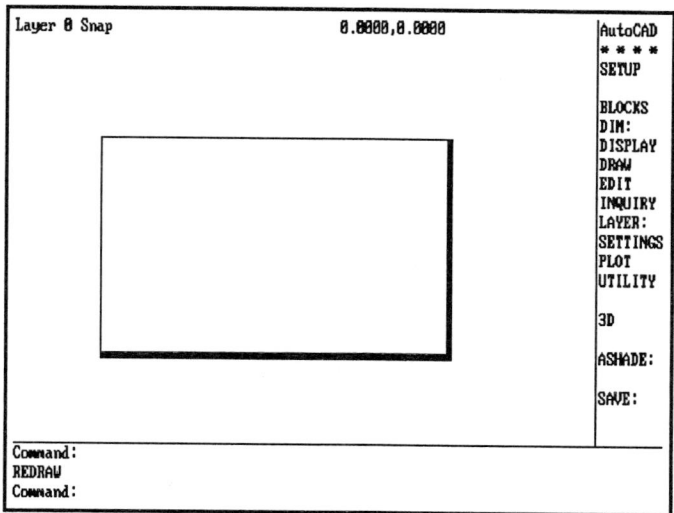

Shadow Box

Now, let's add Text. Enter these three lines to create DRAWTX. Then, test the routine.

```
Command: (defun drawtx ()
1> (setq pt (polar ll (angle ll ur) (/ (distance ll ur) 2)))
1> (command "TEXT" "C" pt "tx" "0" "CUSTOMIZING" "TEXT" "" "AutoCAD"))
Lisp returns: DRAWTX
```

Command: **(drawtx)**   Try it out.
TEXT Start point or Align/Center/Fit/Middle/Right/Style: C
Center point:
Height .5000:
Rotation angle : 0
Text: CUSTOMIZING
Command: TEXT Start point or Align/Center/Fit/Middle/Right/Style:
Text: AutoCAD

If DRAWTX worked OK, it put the text in the center of the box. You are ready to DEFUN the C: command function, SH-BOX. The AutoLISP definition (DEFUN) will put it all together. You can add a Fillet to dress up the routine.

Command: **ERASE**   Erase everything so you can test SH-BOX.

```
Command: (defun C:SH-BOX () (cornrs) (others) (drawbx)
1> (drawtx) (command "FILLET" "R" tx "FILLET" "P" ll))
Lisp returns: C:SH-BOX
```

Command: **SH-BOX**   Finally, try the whole thing.
Pick LL corner: **2,2**   Type, or pick the points.
Pick UR corner: **10,7**

All the commands, prompts and AutoLISP should scroll on by and you should be left with a screen like Shadow Box with Text.

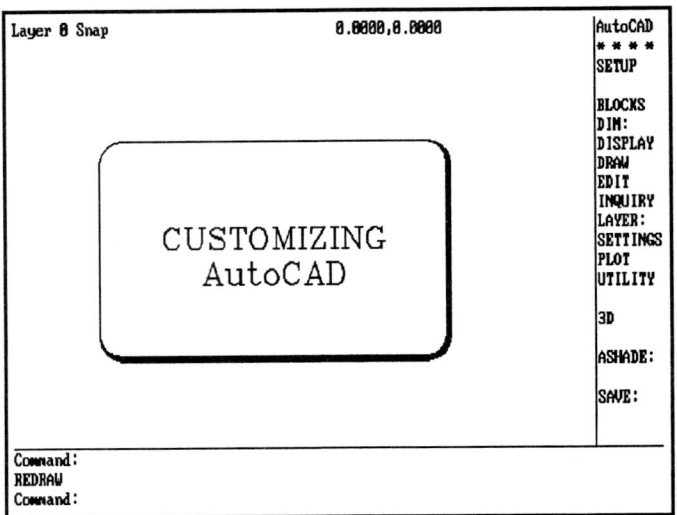

Shadow Box with Text

The shadow box is a simple example of the automation that AutoLISP provides. Some of AutoLISP's more powerful customization features come from its access to AutoCAD's drawing database. While this might sound complex, a quick little exercise will give you a peek at AutoCAD's database using AutoLISP's entity access.

A Peek at AutoLISP's Entity Access

Command: **LAYER**   Make a new TEST layer. Make sets it current.

Command: **LINE**   Draw a line from 1,1 to 2,2.

Command: **(entget (car (entsel)))**   Tells AutoLISP to select an entity and get its data.

Select object: Pick the Line. You get:

Lisp returns: ((-1 . <Entity name: 60000104>) (0 . "LINE") (8 . "TEST") (10 1.000000 1.000000) (11 2.000000 2.000000))

The entity name lets AutoLISP refer to and manipulate the entity. You can see that the type is listed as "LINE." The integer "0" is a code for the entity type. The "8" is a code for Layer, and the layer is listed as "TEST." The two endpoints also are listed. The "10" is the code for the start point, and the "11" is the code for the end point.

Generally, anything that is a default is not stored or listed. The leading integers in the parenthesis are used to identify the type of data so that AutoLISP can process it. You can grab and manipulate any part of any drawing entity's data.

The sky's the limit. That is why AutoLISP is such a powerful tool. It gives you direct access to the AutoCAD drawing database.

```
Command: QUIT                And exit AutoCAD to DOS.
C:\ACAD> CD\                 Return to your root directory.
C:\>
```

Now that you are back in good old DOS, let's start setting up the development environment.

DOS Directories

CUSTOMIZING AutoCAD assumes that your hard disk is C:, and that you have a directory structure like that shown in the Directory Structure illustration. If your drive letter or subdirectory names vary from those shown, you need to substitute your drive letter and names wherever you encounter the C: or directory names in the book.

Subdirectory Setup

Take a look at your directory names, and look at the book's directory names. Get into the root directory of drive C: and type:

```
C:\> DIR *.                            This lists directories and little else. The book's are:
Volume in drive C is DRIVE-C
Directory of  C:\

ACAD       <DIR>  12-01-87  11:27a     AutoCAD program, config. and standard support files.
DOS        <DIR>  12-01-87  11:27a     All of the DOS files.
CA-ACAD    <DIR>  12-01-87  11:27a     CUSTOMIZING AutoCAD config. and support files.
123        <DIR>  12-01-87  11:27a     Lotus 123 directory.
DBASE      <DIR>  12-01-87  11:27a     dBASE III files.
5 File(s)   8753472 bytes free         Your list will be different.
```

Necessary Directories

The CA-ACAD, DOS and ACAD directories are necessary for CUSTOMIZING AutoCAD's and for the CUSTOMIZING AutoCAD DISKs' application environment to work. You should be in the root directory of your hard disk. Make the directory CA-ACAD.

```
C:\> MD \CA-ACAD
C:\> DIR *.                            Your displayed list now should include:

ACAD       <DIR>  12-01-87  11:27a
DOS      , <DIR>  12-01-87  11:27a
CA-ACAD    <DIR>  12-01-87  11:27a
```

The CA-ACAD directory is the working and support directory for the book. The DOS directory is where CUSTOMIZING AutoCAD expects to find DOS command files. If your DOS files are not in a directory named DOS, they need to be in some other directory in the DOS PATH of your AUTOEXEC.BAT file.

If your AutoCAD directory is not named ACAD, you will have to make sure your AutoCAD directory name is in the book's startup batch file's PATH. The book's startup batch file is called CA.BAT. You will create this startup batch file, CA.BAT, in the next chapter.

The DOS, ACAD, and CA-ACAD directories, are the minimum directories necessary for the book.

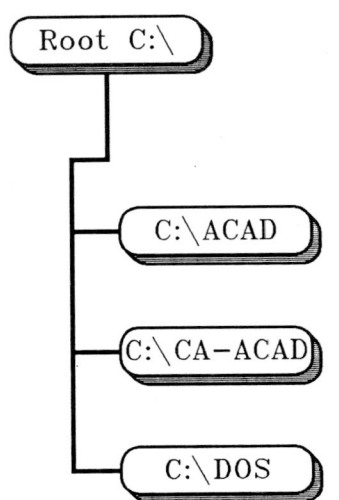

Customizing AutoCAD Directory Structure

AutoCAD Program and Support Files

AutoCAD is a big program, one of the biggest running on microcomputers. The ACAD.EXE file executes AutoCAD and loads core functions, but most of AutoCAD's program code is contained in several overlay files (ACADx.OVL). In addition, AutoCAD uses several support files, such as Text fonts, Linetypes, and Hatch patterns. If you are unfamiliar with AutoCAD's file structure, take a look at the program files list in the AutoCAD Installation and Performance Guide that accompanied your software.

Setting Up AutoCAD's Configuration Files

The book assumes that AutoCAD's support files are in the \ACAD subdirectory. Besides support files, AutoCAD also requires device driver files, *.DRV files, during configuration. These drivers are used by the Configuration Menu to create four ACAD*.OVL device configuration files. Configuration also creates an ACAD.CFG configuration file that stores the settings and configuration related System Variables used by AutoCAD. You need the *.OVL files and ACAD.CFG to run AutoCAD.

In order not to disturb established systems, we won't change your existing AutoCAD setup when you work through this book. So the book establishes a separate configuration for CUSTOMIZING AutoCAD by copying AutoCAD configuration files to the \CA-ACAD directory.

Copying AutoCAD Files to the CA-ACAD Directory

You need to copy your AutoCAD configuration files to the \CA-ACAD directory to establish the book's configuration.

```
C:\> CD \ACAD
C:\ACAD> COPY ACADP?.OVL \CA-ACAD\*.*
ACADPL.OVL                      Plotter overlay file.
ACADPP.OVL                      Printer/plotter overlay file.
2 File(s) copied
```

```
C:\ACAD> COPY ACADD?.OVL \CA-ACAD\*.*
ACADDS.OVL                                    Display (video) overlay file.
ACADDG.OVL                                    Digitizer (or mouse) overlay file.
2 File(s) copied

C:\ACAD> COPY ACAD.CFG \CA-ACAD\*.*
1 File(s) copied                              General AutoCAD configuration file.
```

ADI Drivers

ADI drivers are memory resident (Terminate and Stay Resident) device driver programs for plotters, printers, digitizers and video cards. If you are using an ADI (Autodesk Device Interface) driver, you must install this type of driver prior to starting AutoCAD. You can install an ADI driver in your CONFIG.SYS file, AUTOEXEC.BAT file, or your AutoCAD startup batch file. See Appendix B for more information on ADI drivers.

Installing the CUSTOMIZING AutoCAD DISKS

Now, you are ready to install the CUSTOMIZING AutoCAD DISKS. If you have the CUSTOMIZING AutoCAD DISKS, install them. If you don't have them yet, see the order form in the back of the book. We recommend getting the disks. They will save you a lot of typing and debugging time.

Installing the CUSTOMIZING AutoCAD DISKS

You need to copy the CA DISK files into the \CA-ACAD directory. To conserve disk space, the symbol Blocks are merged into the single CA-BLOCK.DWG file. The CA-BLOCK.SCR script writes these blocks out to disk as individual .DWG files.

```
Put CA DISK 1 in your diskette drive A:
C:\> CD \CA-ACAD
C:\CA-ACAD> COPY A:*.*
Put CA DISK 2 in your diskette drive A:
C:\CA-ACAD> COPY A:*.*

C:\CA-ACAD> \ACAD\ACAD                        Start up AutoCAD.
Enter selection:                              Edit the EXISTING drawing named CA-BLOCK.

Command: SCRIPT                               This script Wblocks the blocks to .DWG files in \CA-ACAD.
Script file <CA-BLOCKS> \CA-ACAD\CA-BLOCKS              It writes all needed blocks, including:

Command: WBLOCK File name: \CA-ACAD\ARROW
Block name: =

Command: WBLOCK File name: \CA-ACAD\VALVE
Block name: =
                                    ...and others scroll by, then:
Command: QUIT
Really want to discard all changes to drawing?          Y if OK, and exit to DOS.
```

1—10 Customizing AutoCAD

If you have created any of the symbol block drawing files and placed them in the \CA-ACAD directory, the script will not run to completion. Delete the offending file(s) and rerun it.

Selecting Text Editors

This chapter, and most of the book, depends on the use of a text editor. We used Norton's Editor to develop the book, but any good text editor will work. You need to be sure that your text editor will work for menu and AutoLISP development. Norton's Editor, Sidekick, PC Write (a "shareware" editor), Wordstar in non-document mode, the Word Perfect Library Program Editor, and the DOS program EDLIN all produce the standard ASCII file format that you will need. EDLIN is awkward and we recommend its use only as a last resort, or for temporary use until you settle on an editor.

Make a copy of your text editor in your CA-ACAD directory and permanently configure it for non-document mode so that you can use it without switching directories. If your text editor stores any of its format settings or modes, you need a separate non-document configured copy in your CA-ACAD directory. This avoids any conflicts in the text files that you will create for your development application.

Your editor must create standard ASCII files. We assume that you have a suitable editor at hand. Then, if you have doubts about its ability to produce ASCII files, test your editor with the following steps.

Text Editor Test

Install and configure a copy of your editor in the \CA-ACAD directory.
Load your text editor. Get into its edit mode and make a new file named TEXT.TXT.
Write a paragraph of text and block copy it to get a few screenfuls.
Save the file and exit to DOS. Then test it:

```
C:\> CD \CA-ACAD
C:\CA-ACAD> COPY TEXT.TXT CON         All the text you entered scrolls by if your editor produced
                                      a standard ASCII file.

C:\CA-ACAD> DEL TEXT.TXT
```

Your text editor is OK if "COPY TEXT.TXT CON" showed text identical to what you typed in your editor, no extra àÇäÆ characters or smiling faces! If you got any garbage, particularly at the top or bottom of the COPY, then your text editor is not suitable, or not configured correctly for use as a development editor. If you are unsure of your editor or want better alternatives, see Appendix D. **You need a suitable text editor to continue.**

The Bootup Environment

When DOS boots ups, it reads COMMAND.COM and two hidden files named IBMBIO.COM and IBMDOS.COM. Then, it looks around for some more information. It gets its DOS environment information from two important files: CONFIG.SYS and AUTOEXEC.BAT.

You need a CONFIG.SYS file and an AUTOEXEC.BAT file, like those shown below, as a minimum base for using the book.

CONFIG.SYS

CONFIG.SYS is the place to install "device drivers" that tell the computer how to talk to devices, like disk drives, RAMdisks and unusual video cards. It is also the place to put instructions that will improve your system performance and increase your DOS environment space for customization.

CONFIG.SYS must be located in the root directory. It is read automatically when your computer boots up. You may never even know it is there. If you have a CONFIG.SYS file, you need to examine it. The CA DISK includes a CONFIG.CA example file, shown below. We recommend that your file include the following lines. If you have a CONFIG.SYS file, display your file.

CONFIG.SYS and AUTOEXEC.BAT

```
C:\CA-ACAD> CD \
C:\> COPY CONFIG.SYS CON
```
This assumes your drive is C:
Copies the file to the CONsole, the screen.

The CONFIG.CA file contains:

```
BUFFERS=32
FILES=24
BREAK=ON
SHELL=C:\COMMAND.COM /P /E:512
DEVICE=C:\DOS\ANSI.SYS
```

Use a number from 20 to 48.

Allows <^C> and <BREAK> to break whenever possible.
/E:512 is for DOS 3.2 or 3.3. Use E:32 for DOS 3.0 or 3.1.
Allows use of extended character set.

 COPY \CA-ACAD\CONFIG.CA to \CONFIG.SYS and edit it, or edit your existing CONFIG.SYS file.

 Edit your CONFIG.SYS file. If you do not have a CONFIG.SYS file, create one.

Edit or create your CONFIG.SYS in your root directory, using your tested ASCII text editor. Include lines similar to those above. Use the discussion below to help you in any modifications.

The BUFFERS line allocates more RAM to hold your recently used data. If a program frequently accesses recently used data, buffers reduce disk accesses and increases speed. Each 2 buffer increment steals 1K from the DOS memory available to your programs. You may have to use a smaller number if AutoCAD runs short of memory.

The FILES line allocates more RAM to keep recently used files open. This reduces directory searching and increases data access speed. FILES uses very little memory, and a large value helps with AutoCAD and AutoLISP.

The SHELL line ensures adequate space for DOS environment variables. DOS allocates a small portion of RAM to store environment variable settings and information. AutoCAD and AutoLISP use several of these. You will need at least 256 bytes. We recommend allocating at least 512 bytes with DOS 3.2 and DOS 3.3. If you use DOS 2.x, or if you get an "Out of environment space" error, see Appendix B on DOS errors and problems.

The ANSI.SYS line is necessary for some of the routines that you will develop in the book. It provides the full 256 ANSI character set, including all characters like åÇäÆ£Ü and Ç. It enables other functions, like screen cursor control and key redefinition, to work. You need to have the DOS file ANSI.SYS in your DOS directory, or you need to change the path in the ANSI.SYS line to wherever your ANSI.SYS file is located.

AUTOEXEC.BAT

AUTOEXEC.BAT is a batch file like any other, with one important exception: It is automatically executed every time the system is booted. Like CONFIG.SYS, it must be in the root directory.

The AUTOEXEC.BAT file is the place to install your TSR (Terminate and Stay Resident) programs, like Prokey, Sidekick, and Superkey. It also is the place to install the other setup commands and DOS environment settings that you need to complete your application environment. Examine your AUTOEXEC.BAT file. We recommend that it include the following lines.

```
C:\>COPY AUTOEXEC.BAT CON
```
Be sure you are in the root C:\ directory.
Examine it.

The following are DOS environment modifiers.

```
PROMPT $P$G
PATH C:\;C:\DOS;
```

Other information may follow...

Edit your AUTOEXEC.BAT file. If you do not have one, create one.

Edit or create your AUTOEXEC.BAT in your root directory, using your tested ASCII text editor. Include lines similar to those above. Use the discussion below to help you with any modifications.

PROMPT PG is extremely valuable. It causes the DOS prompt to display your current directory path so you don't get lost.

PATH is essential for automatic directory access to programs and DOS commands. The C:\ root and C:\DOS paths are essential to the book's setup. If your DOS files are in a different directory, include the directory.

You should use whatever is relevant to your setup. It is likely that your path contains additional directories.

Reboot

The CONFIG.SYS and AUTOEXEC.BAT changes do not take effect until you reboot your computer. Reboot to test your configuration and batch file.

More on DOS

If you encounter problems in setting up your DOS environment, or you want information about setting up more complex environments, see our Appendix B on DOS.

If you wish to install a more advanced autoexec batch file, a more extensive example file is given in Appendix D, the Authors' Appendix.

This completes the initial setup for the book's DOS environment. Next, we move on to setting up AutoCAD. Here are a few tips and reminders about text editors and setting up DOS.

Summary Tips and Techniques

A good text editor is invaluable in customizing AutoCAD. Besides creating ASCII files, it helps if your editor is comfortable, compact, quick, and easy to use. But there are three essential things that your editor needs to do.

- ❏ It must create pure ASCII files, including the ability to create the ASCII <ESCAPE> character.
- ❏ It must be able to merge files.
- ❏ And you must be able to toggle word wrap off.

Here are two DOS organization tips and techniques:

- ❏ It helps to keep each program that you use, like AutoCAD, in its own subdirectory. File access is faster. You don't get files mixed up, and future program upgrades are easier to install.
- ❏ Update your DOS to at least Version 3.1. It is easy to remember to keep AutoCAD software up to date, but it is easy to forget to update PC-DOS, or MS-DOS. DOS Versions 3.1 and later DOS versions offer valuable features for a customizing environment, including a better ability to deal with DOS environment space limitations. We recommend updating your DOS.

2—0 Customizing AutoCAD

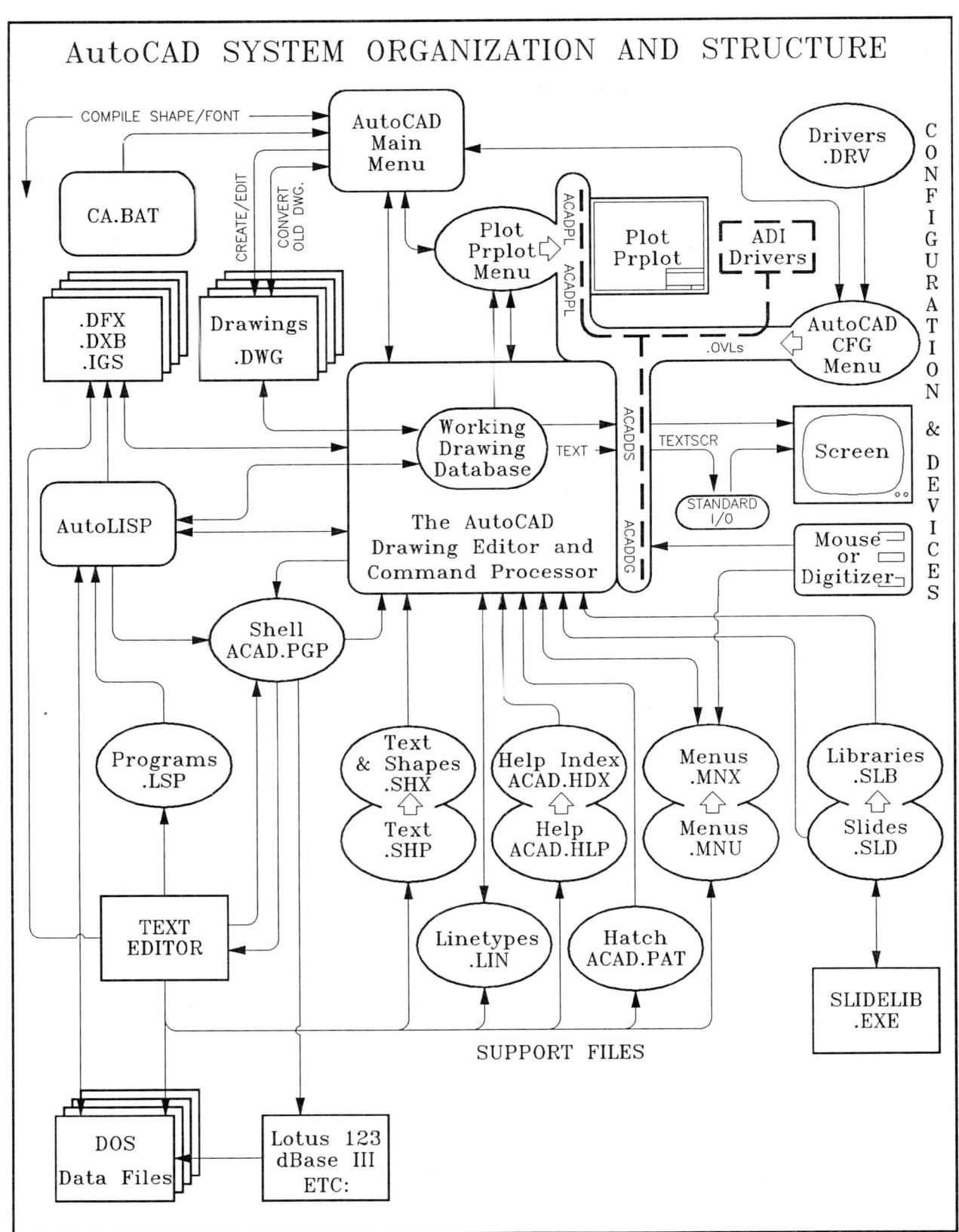

AutoCAD System Organization and Structure

CHAPTER 2

Your AutoCAD System

ORGANIZING AutoCAD AND CRACKING THE SHELL

Getting the right system environment is half the battle in customizing AutoCAD. Setting your AutoCAD DOS system environment with a batch file and interactively using your text editor via AutoCAD's Shell are two key elements in custom applications development. This chapter shows you how to set up AutoCAD, and use Shell to run other programs without leaving AutoCAD. You also will create a prototype drawing to use with the rest of the book.

The Benefits of Organizing AutoCAD's Environment

There are three main benefits to organizing your AutoCAD setup.

- Controlling AutoCAD's DOS environment lets you make more efficient use of subdirectories and run multiple applications and development configurations.

- Knowing how to use AutoCAD's Shell gives you a simple, productive environment that lets you flip back and forth between developing your custom routines with your text editor, and testing them in AutoCAD.

- Using a customized prototype drawing eliminates the need for making most AutoCAD drawing settings and standardizes new drawings.

How To Skills Checklist

Here are the skills that you will acquire. You will learn how to:

- ❑ SET the AutoCAD DOS environment variables with a startup batch file.
- ❑ Use SHELL to suspend AutoCAD in the background.
- ❑ Add new commands to AutoCAD using the ACAD.PGP file.
- ❑ Make a prototype drawing for Customizing AutoCAD.

Programs and Files

 You need to create or verify CA.BAT, ACAD.PGP, and configure AutoCAD with CA-PROTO.DWG for use with the book.

PROGRAMS
CA.BAT sets AutoCAD's DOS environment variables and starts up AutoCAD.

2—2 Customizing AutoCAD

FILES
ACAD.PGP defines the link between AutoCAD commands and DOS programs.
CA-PROTO.DWG defines the prototype drawing used in the book.

A Common Base for Jumping into the Book

You have already created part of your system environment with the AUTOEXEC.BAT and CONFIG.SYS files. Now you need to crack AutoCAD's SHELL.

PGP: Cracking the SHELL

You can run other programs, utilities or DOS commands without ending AutoCAD or reloading your drawing. A few predefined "External Commands" are included with AutoCAD to do this. They are set in a file, ACAD.PGP. PGP stands for ProGram Parameter file. To see how SHELL fits into AutoCAD, look at the System Organization illustration at the beginning of the chapter.

The next exercise will show you how to set up AutoCAD's Shell to jump directly into a test menu and AutoLISP files. This setup provides the development cycle that you can use throughout the rest of the book.

AutoCAD's standard SHELL features are set in the ACAD.PGP file, giving direct access to DOS commands. This access can be general purpose access or predefined access to jump out and run a specific DOS command, or program. If you have the CA DISK, you have a custom ACAD.PGP file in the \CA-ACAD directory. Otherwise, copy the ACAD.PGP file from your \ACAD directory.

PGP Going Outside AutoCAD.

Try SHELL with some standard DOS commands. Start with a simple directory command:

C:\> **CD \CA-ACAD**   Change to the book's directory.

  Do NOT copy ACAD.PGP.

  Copy \ACAD\ACAD.PGP like this:

C:\CA-ACAD> **COPY \ACAD\ACAD.PGP**   Copy your original PGP file.

C:\CA-ACAD> **\ACAD\ACAD**   Start up AutoCAD.
Enter selection: Begin a NEW drawing named EGGS.

Command: **SHELL**   This is the general SHELL DOS access command.
DOS Command: **DIR ACAD*.***

```
Volume in drive C is BRG-AT
Directory of  C:\CA-ACAD
```
The files you copied in the first chapter:
```
ACADPP   OVL     1314   7-06-87  11:53a
ACADPL   OVL     9555   7-16-87   8:22p
ACADDG   OVL     2461   9-04-87   4:07p
ACADDS   OVL    13842   9-14-87   5:21p
ACAD     CFG     1516   9-14-87   5:28p
ACAD     PGP      501  10-19-87   9:00a
```
And possibly other files.
```
6 File(s)   2838528 bytes free
```

The PGP File

AutoCAD freed some memory, permitting DOS to run the DIR command. Although the DIR command is not a true AutoCAD command, you can enter it as if it were. Its execution is enabled by its definition in the ACAD.PGP file.

Let's take a look at the contents of the PGP file with another PGP "command," using the DOS command TYPE:

```
Command: TYPE
File to list: ACAD.PGP

CATALOG,DIR /W,25000,*Files: ,0
DEL,DEL,25000,File to delete: ,0
DIR,DIR,25000,File specification: ,0
EDIT,EDLIN,40000,File to edit: ,0
SH,,25000,*DOS Command: ,0
SHELL,,125000,*DOS Command: ,0
TYPE,TYPE,25000,File to list: ,0
```

If your ACAD.PGP listing does not include these lines, you need to find and copy the ACAD.PGP file from your original AutoCAD diskettes.

Each line in the file defines one PGP command, and uses five fields of information. Each field is a piece of information separated by a comma. Look at the last line, which defines the TYPE command. Let's examine the parts of the PGP specification.

TYPE is the AutoCAD Command. The first field, tells AutoCAD what name you want to use for the command at the AutoCAD command prompt.

TYPE is the External Command. The second field is what AutoCAD feeds to the DOS command line prompt, but it is not displayed inside AutoCAD. The external command field may include spaces. If the external command field is blank, as the field is in the SH line and the SHELL line, it is not fed to DOS. DOS is fed only what you enter as input in response to the prompt.

25000 is the Memory Reserve. This number sets the amount of memory (in bytes) that you want AutoCAD to release for DOS use. Depending on your DOS version, it requires 25000 bytes, or more. Memory reserve is the total RAM available under AutoCAD, less a few Kbytes. Certain resident devices, like mice, may interfere with large settings. If you get errors, see Appendix B on DOS. UNIX systems ignore the memory reserve setting and can run larger programs. The ACAD default is 25000, the book's CA DISK PGP file sets memory reserve at 30000.

File to list: is The Prompt. AutoCAD uses the fourth field as a prompt for user input. It adds the input to the external command field and feeds the combined string to DOS. If an * asterisk precedes the prompt, AutoCAD accepts spaces in the input line, but it requires an actual <RETURN> to enter the input.

Return Code: The final item tells AutoCAD which screen to return to in a single screen system. A 0 returns to the text screen, while a 4 returns to whatever screen was displayed just before the PGP command was issued. For information on the other return codes, see the AutoCAD Reference Manual on DXB files.

So, when you entered **TYPE** and hit **<RETURN>,** AutoCAD put away 25000 bytes of its memory, jumped out to DOS and issued the text string **TYPE ACAD.PGP,** exactly as if you had typed it at the C:\CA-ACAD> DOS prompt. Then, AutoCAD returned directly to the AutoCAD text screen.

Now, let's add some new utility commands to the PGP file and integrate a text editor.

Interactive, easy use of a text editor is a key element in customizing AutoCAD. The following exercise sets up AutoCAD's SHELL to do this. While it is possible to set up a multi-tasking environment, the book uses AutoCAD's SHELL because it is clean, simple and reasonably quick. SHELL is available to everyone.

Integrating the Text Editor

While you are in the EGGS drawing, "Shell out" to DOS.

```
Command: SHELL
DOS Command:
```
Remember, SHELL sends DOS only what you enter on this prompt line. If you enter nothing with a <RETURN>...

```
Type EXIT to return to AutoCAD
The IBM Personal Computer DOS
Version 3.30 (C)Copyright International Business Machines Corp 1981, 1987
(C)Copyright Microsoft Corp 1981, 1986

C:\CA-ACAD>>
```
...it simply dumps you into DOS:

Most PGP commands jump into DOS and straight back to AutoCAD and you never see a DOS C:\> prompt. If you use the Shell command, SH and SHELL, and hit a <RETURN>, you simply get a DOS C:\>> with a subdirectory. The double >> indicates that AutoCAD still lurks in the background. You can run as

The PGP File 2—5

many DOS commands and programs as you want before you go back to AutoCAD with EXIT.

If you have the CA DISK, examine your custom ACAD.PGP file and read on. Otherwise, use your text editor to examine and modify the original ACAD.PGP file you copied, as shown below:

You should still be SHELLed out to DOS.

 Modify the ED, LSP and MNU lines to accommodate your text editor.

 Change the memory values from 25000 to 30000 in the PGP file. Enter the following lines. If you need to, modify the ED, LSP and MNU lines.

Load your editor with the file ACAD.PGP. For Norton's Editor, the book uses:

`C:\CA-ACAD>> `**`NE ACAD.PGP`**

Add these lines to the end of the copied ACAD.PGP file:

```
SHMAX,,480000,*DOS Command: ,0
REN,RENAME,30000,*Oldname Newname: ,0
DWGS,DIR *.DWG,30000,,0
DUP,COPY,30000,*SOURCEfile(s) TARGETfile(s): ,0
ED,NE,130000,File to edit: ,4
LSP,NE TEST.LSP,60000,,4
MNU,NE TEST.MNU,130000,,4
SHOW,MORE <,30000,File to list: ,0
```

Save the file and exit your editor.
`C:\CA-ACAD>> `**`EXIT`**   Takes you back to AutoCAD.

Here is what these additions to the PGP file do:

SHMAX runs a maximum size Shell. You may need to adjust the 480000 size to your available CHKDSK memory less about 10 Kbytes.

REN implements the standard DOS command REName, with a prompt.

DWGS lists all drawings in the current directory.

DUP is the DOS command COPY.

ED starts the text editor for use.

LSP starts the text editor and edits TEST.LSP, a testbed AutoLISP file that the book uses.

MNU starts the text editor and edits a testbed menu file.

SHOW is an improved "TYPE" command, which pauses for pages.

Make SURE you use the correct command to start YOUR text editor in the ED, LSP and MNU lines. Include any command line parameters your editor needs. Your editor may require additional memory. Increase the 130000 value in the ED line until your editor works. See Appendix B for more information.

The \ACAD\ACAD.PGP is the factory standard file. AutoCAD only recognizes the name ACAD.PGP, so your file needs to have the same name. Be careful not to mix your files. Keep backup copies of PGP files under unique, unusable names like ACAD-9.PGP for the Release 9 standard file, or CA-ACAD.PGP for the one you just created for the book.

AutoCAD loads the PGP file when it starts a drawing. You have to force it to reload the new PGP file to test it. Quit your drawing, then come back in and test the file. Type ED and a file name at the command line and your text editor will automatically start.

```
Command: QUIT                         Yes.
Enter selection:                      Begin a NEW drawing again named EGGS.

Command: ED
File to edit: TEST                    Starts your editor in TEST.
Quit your text editor and try the other PGP commands you added:

Command: SHMAX                        <RETURN> to go to DOS, then type EXIT back to AutoCAD.
Command: SHOW                         Enter ACAD.PGP and it should display it.
Command: DUP                          Test DUP. Back it up for safekeeping.

SOURCEfile(s) TARGETfile(s): ACAD.PGP CA-ACAD.PGP

Command: LSP                          Starts editor with new file TEST.LSP. Exits back to AutoCAD.
Command: MNU                          Edits a new file TEST.MNU. Exits to AutoCAD.
Command: DWGS                         Lists .DWG files, if any.
```

If your editor doesn't auto-load files when you use it, you may have to <RETURN> at the "File to edit:" prompt, and manually load files in your editor. If SHMAX fails, reduce its memory reserve slightly. See Appendix B on DOS, if you encounter other problems.

You are nearly finished setting up. Next, you will clear any possible conflict from any ACAD.LSP file your system may have; then, make a CA.BAT startup batch file and a CA-PROTO.DWG prototype drawing file.

Avoiding ACAD.LSP Conflict

Your system may have an ACAD.LSP file. If it does, AutoCAD automatically loads it at the start of each drawing edit session. This file may interfere with the book's setup. To play it safe, make a dummy ACAD.LSP file in your CA-ACAD directory.

Making a Dummy ACAD.LSP File

Command: **ED**	Enter the following line:
File to edit: **ACAD.LSP**	

`(prompt "CA empty ACAD.LSP...")`

Save the "empty" file and exit back to AutoCAD.

This empty dummy ACAD.LSP will load instead of any other ACAD.LSP file. Later on, you will create and use an ACAD.LSP file containing AutoLISP programs that you develop while using the book.

AutoCAD lets you preset several of its startup settings. These control its memory usage and support file search order. The book's CA.BAT startup batch file sets memory allocations, file search order, and loads AutoCAD automatically.

Creating a CA.BAT File

If you have the CA DISK, use or modify the CA.BAT file from the CA-ACAD directory. Otherwise create a new file. The CA.BAT file assumes your AutoCAD path is \ACAD. If not, substitute your path. You should still be in EGGS.DWG:

Command: **ED**	Start up your text editor.
File to edit: **CA.BAT**	Load, or create the file, CA.BAT.

Enter or modify these lines where applicable:

```
SET ACAD=\ACAD
SET ACADCFG=\CA-ACAD
SET ACADFREERAM=20
SET LISPHEAP=25000
SET LISPSTACK=10000
C:
CD \CA-ACAD
\ACAD\ACAD %1 %2
CD\
SET ACADCFG=
SET ACAD=
```

Save the CA.BAT file, exit your editor to AutoCAD.

Command: **DUP**	Copy CA.BAT to your root dir, or elsewhere on your PATH.
SOURCEfile(s) TARGETfile(s): **CA.BAT \CA.BAT**	
Command: **QUIT**	Then exit to DOS.

NOTE. The book assumes you will use the CA.BAT file. Type CA when you want to start AutoCAD. If you do not use the CA.BAT file, you may get incorrect DOS environment settings.

Here is a brief discussion of the settings in the CA.BAT file. See our Appendix B, your AutoCAD User Reference, and the AutoCAD Installation and Performance Guide for more details.

SET ACAD= tells AutoCAD where to look if it doesn't find a needed support file in the current directory. AutoCAD searches the current directory first, then this specified support directory, then the program directory. The program directory is the directory where ACAD.EXE was started.

SET ACADCFG= tells AutoCAD where to look for configuration files. Creating several configuration directories and startup batch files is useful if you need to support more than one environment, or to support more than one device, like different plotters.

SET ACADFREERAM= reserves RAM for AutoCAD's working storage. The default is 14K, the maximum depends on the system, usually about 24-26K. If you get "Out of RAM" or other errors, see Appendix B for more information on setting memory use.

SET LISPHEAP= allocates memory for AutoLISP functions and variables (nodes). If you use many programs, or you use large programs, you may need to increase this value. The book uses VMON (Virtual Memory ON) in AutoLISP to make AutoLISP page functions to disk or EXTended/EXPanded RAM, reusing HEAP space. More HEAP space increases AutoLISP speed by reducing the swapping (paging) of functions.

SET LISPSTACK= defines AutoLISP's temporary working data area during execution. Complex AutoLISP programs, using many arguments, recursive and/or nested routines, or large amounts of data may require more stack space.

HEAP and STACK space combined cannot exceed 45000 bytes. They reduce memory that otherwise is available to AutoCAD for Free RAM and I/O Page Space. If you encounter problems running large programs, you have to adjust these settings to achieve a working balance. Don't be alarmed. It is not hard. If it works, use it. If it's broken, change your settings until it works.

These SET environment settings do not affect memory outside of AutoCAD.

That is all there is to settings. The rest of the CA.BAT batch file is made up of straightforward startup DOS commands.

C: ensures that you are on the right drive. Substitute another letter if your drive isn't drive C:.

CD \CA-ACAD changes the working directory to \CA-ACAD.

\ACAD\ACAD %1 %2 executes ACAD. If \ACAD is on your PATH, you could use ACAD alone here, but specifying the directory avoids having DOS search the PATH. The %1 and %2 are replaceable parameters that you can enter when

Prototype Drawing   2—9

you run CA.BAT. For example, to run a Script with the name, NAME, you would enter CA X NAME and CA.BAT would execute this line as \ACAD\ACAD X NAME, running the Script. You will use these parameters later in the book.

When you exit AutoCAD, CD\ returns you to the root directory. SET ACADCFG= and SET ACAD= clear their settings.

Prototype Drawing

To follow the book, you need a consistent drawing setup. The easiest way to do this is to set a new standard prototype drawing for use with the book, and configure AutoCAD to use it. This standard prototype drawing, CA-PROTO.DWG, establishes a common drawing base and sets up AutoCAD to work with our customization exercises.

Cleaning the Slate

First, start with fresh defaults and set your units and limits along with some dimension settings. Then, set the layer and style conventions for CA-PROTO. If you have the CA DISK, you have CA-PROTO.DWG.

 Just read this exercise.

 Create CA-PROTO.

```
C:\CA-ACAD> CA                       Use CA.BAT to start up.
Enter selection:                     Begin a NEW drawing named CA-PROTO=.
                                     Make the following settings:

Command: UNITS                       Set Arch. (4), denom. 64, Angular frac. 2. Default the remainder.
Command: GRID                        Set to 96". This is 1" at a plot scale of 1:96.
Command: SNAP                        Set to 24. This is 1/4" on the plot.
Command: AXIS                        Set to 48. This is 1/2" on the plot. This is optional.
Command: LTSCALE                     Set to 36 (0.375 x 96).
Command: LIMITS                      Set to 0,0 and 288',192'. That's feet!
Command: ZOOM                        Zoom All.
Command: VIEW                        Save a view named A, for All.

Command: DIM                         Reset the following DIM Variables.
```

Type these in regardless of the defaults shown:

DIMSCALE	96	DIMASZ	3/16	DIMCEN	1/16	DIMEXO	3/32
DIMDLI	3/8	DIMEXE	3/16	DIMTXT	1/8		

```
Dim: EXIT
Command: MENU                        Enter period "." for none! You will set a menu later.
Command: <Coords on>                 Toggle them on.
```

DIMSCALE is set to the plot:drawing scale with DIMSCALE=96. DIMASZ will plot at 3/16-inch. You need to type in the DIM settings, regardless of the apparent defaults. The defaults are rounded off. For example, DIMASZ's default is 0.1800, but architectural units misleadingly rounds off the display to 3/16-inch. DIMTXT works the same way, but your fixed height text styles will override it.

CA-PROTO's default limits are set to get started. When you create different size drawings in the customization exercises, you can adjust these limits and their scalar settings. Later, you will create a Menu and an AutoLISP-based setup routine to automatically handle all these drawing setups.

Layering and Style Conventions

Let's establish the layers and text styles for the book. Unless you have the CA DISK, create the following Layers, Colors, and Linetypes.

Command: **LAYER**   Create the following:

```
Layer name           State    Color          Linetype
-----------------------------------------------------------
0                    On       7 (white)      CONTINUOUS
ANN02                On       4 (cyan)       CONTINUOUS
ANN03                On       6 (magenta)    CONTINUOUS
CEN31                On       2 (yellow)     CENTER
DSH11                On       3 (green)      DASHED
HID21                On       3 (green)      HIDDEN

OBJ01                On       3 (green)      CONTINUOUS
OBJ02                On       1 (red)        CONTINUOUS
OBJ12                On       1 (red)        DASHED
REF00                On       9 (varies)     CONTINUOUS
REF01                On       2 (yellow)     CONTINUOUS
TXT01                On       3 (green)      CONTINUOUS
TXT02                On       1 (red)        CONTINUOUS
```

Enter ? * to list and verify all above Layers.
Set the Current Layer to 0.

Command: **STYLE**   Create these Text Styles. All settings are defaults except Height.

```
Style name: STANDARD    Font files: TXT        Height: 0'-0"
Style name: STD3-16     Font files: SIMPLEX    Height: 1'-6"
Style name: STD1-4      Font files: SIMPLEX    Height: 2'-0"
Style name: STD3-32     Font files: SIMPLEX    Height: 0'-9"
Style name: STD1-16     Font files: SIMPLEX    Height: 0'-6"
Style name: STD1-8      Font files: SIMPLEX    Height: 1'-0"
```

Set STD1-8 last to leave it current.

 If you loaded a drawing, QUIT to the Main Menu.

 END to save CA-PROTO.

The books's macros and AutoLISP routines assume a fixed height text style. AutoCAD does not prompt for height in Text and related commands if you use a fixed height. The prototype drawing initially fixes text height for a 1:96 plot scale.

Establishing a Layering Convention

Customization requires a common layering scheme to automate drawing setups with menu macros. To do this, you need to establish your conventions in advance, so that macros can be written.

DOS uses wild card characters to filter file names. AutoCAD uses them to filter Layer names, using the same "?" and "*" syntax. The book's layer names are designed for filtering. The names distinguish text from material components, and dimensions from annotations, using names like TXT01 or DIM01 and CEN02. Here is the naming convention:

```
OBJECTS ON LAYER                              LAYER NAMES
TEXT                                          TXT01 thru TXT03
COMPONENTS, ASSEMBLIES and MATERIALS          OBJ01 thru OBJ93
DIMENSIONS                                    DIM01 thru DIM03
SYMBOLS & ANNOTATIONS                         ANN01 thru ANN93
TITLE SHEETS or FORMS                         REF01 thru REF93
LINETYPES - Dashed, Center, Hidden...         DSH01, CEN01, HID01
                                         thru DSH93, CEN93, HID93
DON'T PLOT reference layers                   REF00
```

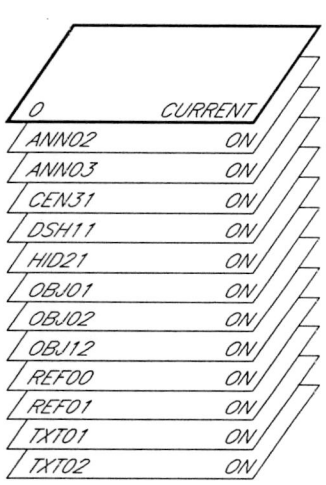

Prototype Layers

The first three characters tells you "what" you are dealing with. The two digit code describes the "appearance." The first "appearance" code keys to Layer linetype. The second digit keys to plotting line weight, 1, 2 or 3, using the color assigned to the Layer. The book groups the standard 7 colors and assigns them to 3 pens. Here are two tables, an Ltype Code and Pen Weight Code Table:

LTYPE CODE	AutoCAD LINETYPE	WEIGHT CODE	AutoCAD COLOR	PEN WEIGHT	PEN SIZE
0	Continuous				
1	Dashed	1	2 (yellow)	Fine	0.25mm
2	Hidden	1	3 (green)	Fine	0.25mm
3	Center	1	5 (blue)	Fine	0.25mm
4	Phantom				
5	Dot	2	1 (red)	Medium	0.35mm
6	Dashdot	2	4 (cyan)	Medium	0.35mm
7	Divide				
8	Border	3	6 (magenta)	Bold	0.60mm

This is a simple layering scheme. Feel free to adopt and modify it.

Configuring and Testing CA-PROTO

Your AutoCAD setup probably uses the default name ACAD.DWG as its automatic prototype drawing. New drawings will start up using this default prototype drawing unless you instruct AutoCAD to use a different prototype.

Configuring CA-PROTO as the Default Drawing

Since you set up a separate configuration for Customizing AutoCAD in the CA-ACAD directory, you can change the default without affecting your normal AutoCAD setup. Do this whether you have the CA DISK or not. You should be in the Main Menu.

```
Enter selection: 5                          Configure AutoCAD.
Press RETURN to continue:

Enter selection <0>: 8                      Configure operating parameters.
Enter selection <0>: 2                      Initial drawing setup.

Enter name of default prototype file for new drawings
or . for none <ACAD>: CA-PROTO

Enter selection <0>:                        <RETURN> 3 times to save and exit to the Main Menu.

Enter selection:                            Begin a NEW drawing named TEST.
                                            It should start up identically to the CA-PROTO you saved.

Command: QUIT                               And if OK, exit to DOS.
```

Now, you should be on common ground with the book's starting environment. We have a common bootup environment, a common DOS environment to use AutoCAD, and a common AutoCAD prototype drawing, CA-PROTO. It is time to move on to create some symbols and parts that you will use in customizing AutoCAD.

Tips and Techniques

Here are tips on organizing your DOS and AutoCAD environments.

- ❏ Customize your ACAD.PGP file by adding additional DOS commands, utilities, or programs that you would like to access from AutoCAD.

- ❏ Spaces cause errors in DOS variable names and their assigned values. SET ACADCFG=\CA-ACAD is OK, but SET ACADCFG = \CA-ACAD won't work.

- ❏ Use SET ACAD= and SET ACADCFG= to clear any SET ACAD=name and SET ACADCFG=name settings that you make in a startup batch file, like CA.BAT. If you do not clear your settings, your other AutoCAD applications will find the settings and be directed to the wrong configuration and support files.

- You can determine the maximum amount of RAM available in your system by running a CHKDSK. Make sure you're not in AutoCAD when you run CHKDSK.

- Run CHKDSK /F at the DOS prompt on a regular basis. It will verify your hard disk file structure and free up "lost clusters." Lost clusters are created when programs crash. Answer N when it asks if you want to convert the clusters to files. Do not run CHKDSK /F in AutoCAD.

- You can use an AutoCAD script file for plotting, if you use standard color and pen assignments in your plotting scheme.

3—0 Customizing AutoCAD

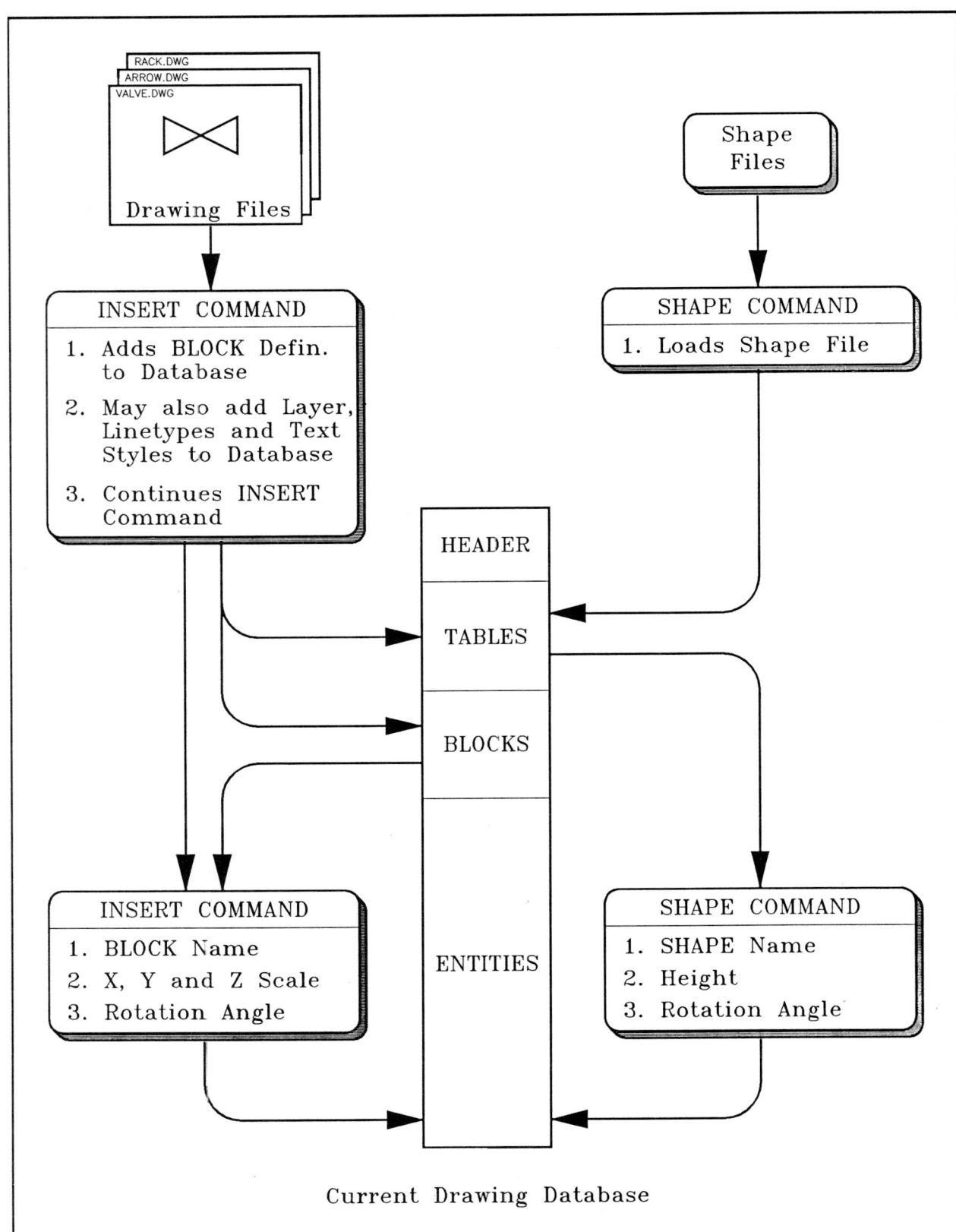

AutoCAD Blocks, Files, Inserts and Shapes

CHAPTER 3

Scaling and Block Management

SYMBOLS and PARTS

Most users build their customized drawing applications around a set of symbols and parts. You create and manage symbols and parts by using AutoCAD's block commands. Customization requires that you set up, standardize and control your block use in AutoCAD. This chapter shows you how AutoCAD's block structure works, and recommends simple layering schemes as the best method for managing your blocks.

The Benefits of Well Organized Blocks

There are three main benefits to effective block management.

- Good symbol management can reduce the time and effort that goes into creating a symbol library.

- You gain greater symbol flexibility when you exploit the block features of AutoCAD.

- Most important, you can lock visual effects and a standardized "look" into every drawing that you make with your customized application.

How To Skills Checklist

Here are the skills that you will acquire. You will learn how to:

- ❑ Make a block, understand where it is stored and how AutoCAD manages the block's definition in your application.
- ❑ Draw your symbols so that they work with any plotted scale.
- ❑ Control your symbol colors, linetypes and layer assignments to get both flexibility and control over your drawing.
- ❑ Nest block definitions.

Simple Is Better

For typical FIXED PURPOSE Blocks, like Parts and single purpose Symbols, we recommend using Bylayer entity settings on named layers, as the best means for controlling your symbols and parts. You can organize your data by layers and use layer control as a means of sorting data for evaluation. You can control what gets plotted by different weight pens by changing the colors of entire layers.

For MULTIPURPOSE SYMBOLS, we recommend that you use Bylayer on Layer 0, and control your blocks by the layer you insert them on. If you use "pseudo-entities," like unit scaled rectangles and ellipses, we recommend that you use Layer 0 Byblock definitions.

Symbols, Parts and Blocks

Let's define SYMBOLS and PARTS. Then, let's look at AutoCAD's Blocks and block commands to make sure that we have a common understanding of how AutoCAD treats Blocks.

SCALE	PARTS	SYMBOLS
1/2" = 1'-0"	ACTUAL SIZE	24 TIMES ORIGINAL
1/4" = 1'-0"	ACTUAL SIZE	48 TIMES ORIGINAL
1/8" = 1'-0"	ACTUAL SIZE	96 TIMES ORIGINAL

Scaling of Parts and Symbols

SYMBOLS represent intangible objects. Symbols are drawn at the plotted size measured on the plot. A Symbol-in-the-drawing requires a size adjustment based on the plotting scale. A "1/4 inch BUBBLE" would be 24 inches for a 1/8"=1'-0" plan.

PARTS represent real objects. A chair is a part. So is a tank, column, window or desk. Parts are drawn or inserted at "full scale," their actual dimensions in the real world.

UNIT PARTS are blocks drawn at a one-unit scale and stretched by their insertion scale to represent sizes of real objects. Examples include doors, hex bolt heads, or a simple 1x1 square that you stretch to represent any rectangular object.

You use an AutoCAD BLOCK to make SYMBOLs and PARTs.

Block Commands and Block Definitions

BLOCK. When you use the command Block, the selected entities are erased. Block does not create a visible entity, it creates a block definition in the AutoCAD drawing. There is no such thing as an "entity" called a "BLOCK." A Block Definition invisibly holds a record of the original entities. We'll call it a "Block Definition" or Block Def. You use the Insert command to see the Block's entities.

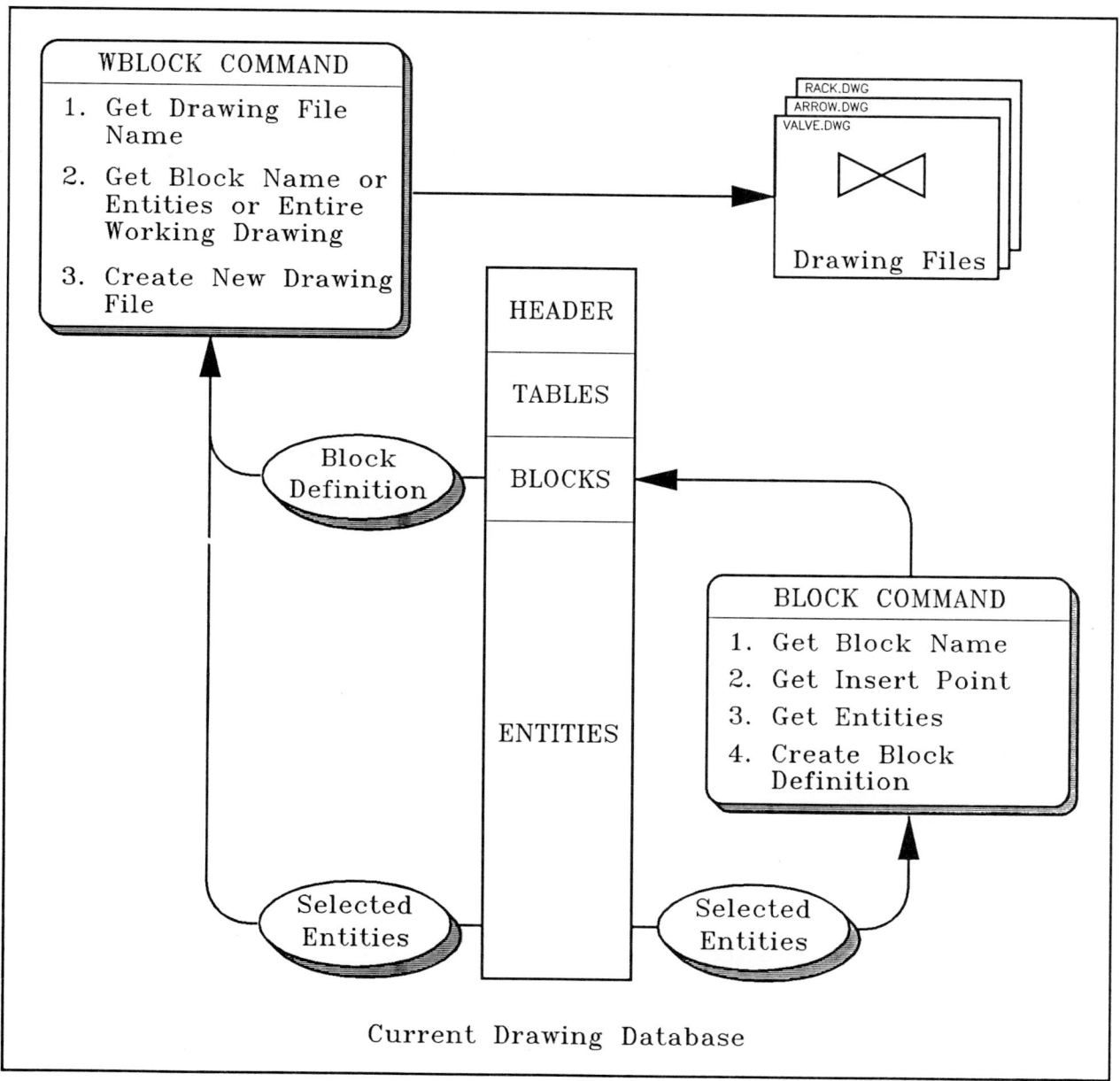

Block and Wblock Commands

INSERT. When you use Insert, you place an entity called an "INSERT" in the drawing. However, the List command confusingly lists it as a "BLOCK REFERENCE." Block Reference is a more apt description than INSERT, but AutoLISP and DXF use INSERT. We refer to the entity as a Block Insert, or just an Insert. A Block Insert is a very simple entity. It references, by name, the Block Def from a table of definitions. Each Insert is modified by the scale, rotation and location of its insertion. If you Insert a Block with *name, AutoCAD does not create an INSERT. It inserts a copy of the Block Def's entities.

WBLOCK. When you use WBlock, you create a new drawing file, or you overwrite an existing drawing file. There is no such "thing" as "a WBLOCK." The Wblock command does not create a Block Def, but you can insert any drawing file, including those created by WBlock, into a drawing. An inserted drawing becomes a Block Def in the current drawing. Any Block Def can be Blocked or WBlocked, and any drawing file can be Inserted to create a Block Def in the current drawing. If you insert a drawing with *name, it does not create a Block Def. It inserts copies of the drawing entities.

How To Create Blocks

It does not matter how you create a Block Def in AutoCAD. The Block Def may contain a single entity within a complex drawing. It may itself contain another Block Def. You can create many individual drawing files for use as Block Inserts, or you can create many Block Defs in a single drawing.

We recommend that you combine both methods in your customization. In the book, we create a group of related symbols in a single "library" drawing, then export each to separate files with WBlock for Insert use.

To actually use symbols and parts, you need to scale them in your drawing. Let's look at Scaling.

Scaling Parts and Symbols

Scaling PARTS is easy, you build and insert them at "full scale."
To see how symbol scaling works, consider the ARROW shown in the margin. It's designed for insertion at the end of a leader line. What size should it be?

You have two basic options in scaling symbols. You can base your symbols on an insertion factor of 1.0, and define one set of symbols for each plot scale that you use. Or you can define your symbols at the exact plotted size that you want on the final output. This scale is 1:1 to the "paper." You scale it by the drawing's plot scale when you insert it.

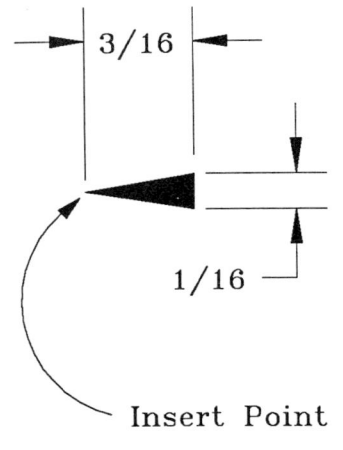

A Pointing Arrow

Both scale methods work, but working at 1:1 to the "paper" output lets you "think" and "program" to the paper size. We recommend creating symbols at 1:1 scale to the "paper." You have only one defined Block for each symbol, and you can adjust the size of an Inserted symbol with scale factors.

Symbol Scale and Plot Scale

Create the arrow. To help you visualize symbol creation at the actual plot size, make an 8-1/2 x 11 "plot sheet." Create the arrow with a Pline, Wblock the arrow, and insert it at a drawing scale of 96.

 You have the ARROW.DWG drawing.

 Create ARROW.DWG

```
C:\CA-ACAD> CA                                Use CA.BAT to start ACAD.
Enter selection:                              Begin a NEW drawing named BLOCKS.

Command: LIMITS                               Enter 0,0 to 88',68'. That's 88 feet: 8-1/2" x 11" at 1/8"=1'-0".
Command: SNAP                                 Set to 1/16.
Command: GRID                                 Set to 1.

Command: ZOOM                                 Zoom W 0,0 to 12,9 to draw a "plot sheet" border.
Command: PLINE                                Draw 11" wide x 8-1/2" rectangle with lower left corner at 0,0.
                                              This represents our plotted sheet size.

Command: VIEW                                 Save a VIEW named PLOTTED.

Command: PLINE                                Draw the ARROW in the lower left of the screen:
From point:                                   Pick point. Set starting Width 0 and ending width 1/16":

Arc/Close/Halfwidth/Length/Undo/Width/<Endpoint of line>: W
Arc/Close/Halfwidth/Length/Undo/Width/<Endpoint of line>: @3/16,0
And then exit the Pline command.
```

 Do not Wblock the arrow.

 Wblock the arrow:

```
Command: WBLOCK                               Wblock it to ARROW, snapping the Insertion base point
                                              to the arrow's point.

Command: OOPS                                 OOPS it back for comparison.

Command: INSERT                               Insert ARROW in the top left of the screen, scale 96,
                                              rotation 0.
```

3—6 Customizing AutoCAD

Original ARROW and ARROW at 96:1

A little big, is it? Zoom and draw a LEADER for comparison. If your DIMSCALE is correct, both arrows will be the same. DIMSCALE should be 96, the same as the plot scale.

Command: **ZOOM**	ZOOM Left <RETURN> to default the corner, height 3' (feet). See your original arrow, a tiny dot in the rectangle?
Command: **DIM1** DIM: **LEA**	Draw a Leader for comparison. Pick 1st point at far left above arrow, 2nd point at far right. <RETURN> to enter no text.
Command: **ZOOM**	Zoom All.

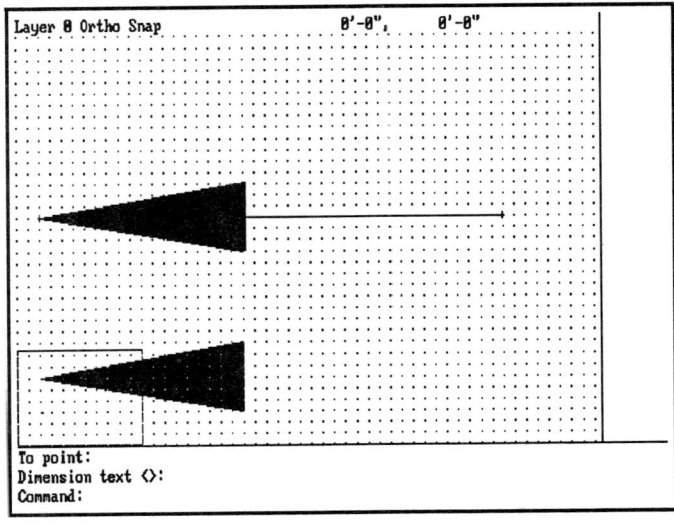

Arrow and Leader Comparison

Now you are Zoomed to see the entire full scale drawing. This drawing is 96 times bigger than a plot. It would fill an entire 11 x 8-1/2 sheet if plotted at 1:96. The Inserted and DIM arrows show on the screen as they actually would in the plot.

The following table sums up the scaling:

```
Original Arrow Size            1/16" x 3/16"
Insertion scale factor         96
Arrow Size in AutoCAD Drawing  6" x 18"
Plotting scale                 1:96
Final size on paper -          1/16" x 3/16"
```

You will use this arrow to make some leader macros in later exercises.

Now that you understand the book's convention for scaling blocks, let's look at some Layer/Color/Linetype issues that you will encounter in customization.

Layers/Colors/Linetypes and Blocks

Symbols and parts frequently contain entities with a variety of Layers, Colors and Linetypes. We assume you are familiar with AutoCAD's ways of controlling Layers, Colors and Linetypes. AutoCAD gives you flexibility with these settings in making your blocks. In fact, AutoCAD is so flexible that Layer, Color and Linetype settings in Blocks can become complex and confusing. If you need to refresh your memory, refer to INSIDE AutoCAD (New Riders Publishing) or your AutoCAD User Reference.

Let's review the controls AutoCAD uses for these settings, then look at some interactions with a few examples.

Every entity in an AutoCAD drawing is drawn in a color and linetype.

- You use the Color command to set an explicit color, like Red, Green or Byblock, or you can use the Color command to set color Bylayer, the default.

- You use the Linetype command to set an explicit linetype, like Dashed, Center, or Byblock, or you can use the Linetype command to set linetype Bylayer, the default.

- If the entity's color and/or linetype are set Bylayer, the Layer command settings control the entity's color and/or linetype. For example, the ARROW block was created on Layer 0, with both color and linetype settings set Bylayer. Each inserted ARROW shows up on the Layer that is current when it is Inserted, with the Color and Linetype current for that layer. If you Change the ARROW's layer, the color and linetype change to the settings of the new layer.

- If you explicitly set or assign entities an Entity Color and/or an Entity Linetype, they override the default settings for the Layer. The entities are "molded or modified" by the explicit settings. CHANGE is the only AutoCAD command that can modify them further.

- If you make a setting explicit, AutoCAD stores the setting with the entity. Byblock is treated as an explicit setting. Color Byblock is stored as color 0. Byblock means "not yet established." It is similar to the action of Layer 0.

- It helps to think of Layer 0 as "Layer Byblock." Blocks created on Layer 0 float to the layer current at the time of their insertion.

- Blocks with Color and Linetypes Byblock also adopt the current settings at the time of their insertion, whether Bylayer or explicit.

Layer 0 Blocks with settings Byblock act exactly like primitive entities. These make good general purpose blocks, like 1-unit squares and circles that are stretched to insert as rectangles and ellipses.

NOTE. The Color and Linetype shown by the List command for a Block Insertion are confusing. You can't see an Insert, just the entities within it. The List command shows the properties of the Insert, and not those of the inner entities.

Let's see how some of these block combinations work.

```
Command: ZOOM            Previous.
Command: ERASE           Erase the big arrows.
Command: VIEW            Restore View PLOTTED.

Command: CIRCLE          Draw a 1" radius circle. It's white and continuous.

Command: LAYER           Set HID21 current.
Command: CIRCLE          Draw a 2"radius circle. It's Green and HIDDEN.

Command: COLOR           Set Color to Red.
Command: LINETYPE        Set Linetype to DIVIDE.
Command: CIRCLE          3"radius. It's Red and DIVIDE, even though on Layer HID21.

Command: COLOR           Set Color BYBLOCK.
Command: LINETYPE        Set Linetype BYBLOCK.
Command: CIRCLE          4" radius. It's White and CONTINUOUS.
```

The 3-inch Red Divide Circle's explicit settings overrode the current Layer's Green Hidden layer defaults. The 4-inch BYBLOCK explicit settings also overruled the Layer for the new Circle. You can verify these with the Layer and Status commands.

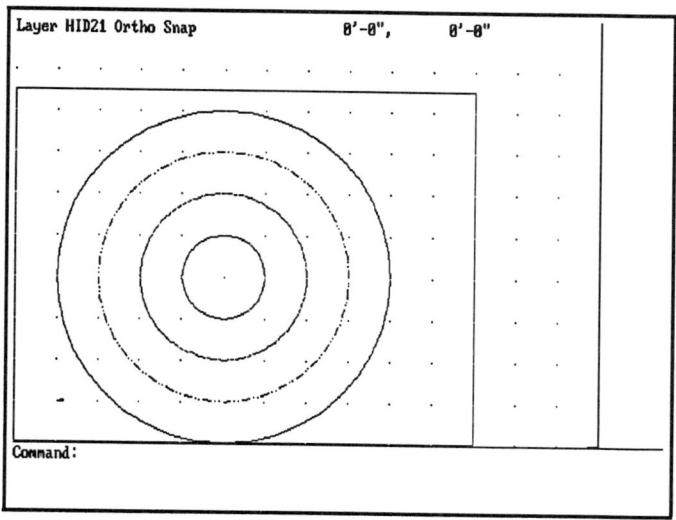

BYLAYER, Explicit and BYBLOCK Circles

Command: **STATUS**	Look at the settings.
Command: **LIST**	Select all circles and examine.
Command: **LAYER**	Enter ?. It shows Green and HIDDEN.
Command: **BLOCK**	Block all to CIRCLES, insertion base at center.
Command: **INSERT**	Insert CIRCLES. They're unchanged except the 1" Layer 0 circle floated to current layer HID21, Green and HIDDEN.
Command: **ERASE**	Erase it.
Command: **COLOR**	Set Color to Blue.
Command: **LINETYPE**	Set Linetype to DIVIDE.
Command: **INSERT**	Again. The same as above but the 4" Byblock circle adopted the current explicit settings. Keep it.
Command: **COLOR**	Set Color to BYLAYER.
Command: **LINETYPE**	Set Linetype to BYLAYER.
Command: **LAYER**	Set Layer REF01 current.
Command: **INSERT**	Again. Now the 1" Layer 0 Bylayer circle and the 4" Byblock circle adopted the current settings, the layer defaults.
Command: **CHANGE**	Change the previous Insert to Layer REF01. The 4" Byblock remains Blue DIVIDE, but the 1" Bylayer becomes Yellow CONTINUOUS.
Command: **END**	You will need the drawing later.

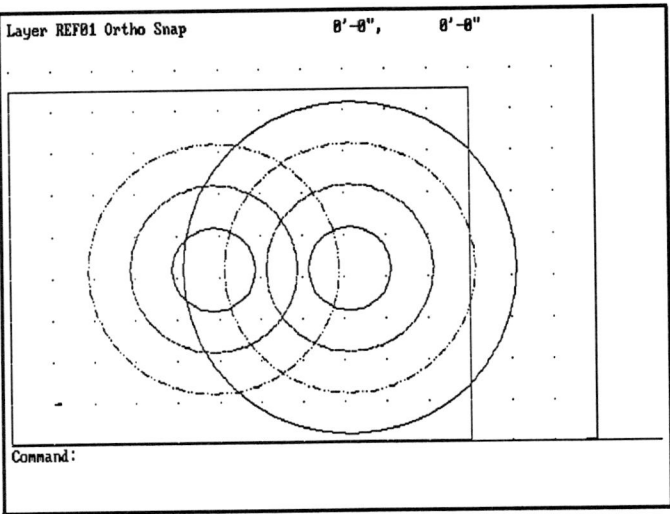

BYBLOCK vs. BYLAYER

If the 4-inch Byblock circle originally had been inserted with current settings Bylayer, it would have become Bylayer and Change would affect it.

An entity that looks red on the screen can be explicitly Red, or can be set Bylayer with a Layer Color of Red. The screen display is the same. However, these settings have an important effect on the organization and ease of updating blocks in customization.

If you organize your entities Bylayer, you can manipulate a group of entities using the Layer On/Off and Freezing/Thawing options. If a Layer's Color and/or Linetype is changed, it changes all entities on the layer unless they have overriding explicit Colors or Linetypes. If properties are explicit, say Red Dashed, you can still Freeze/Thaw and turn the layer On/Off, but you can only reset entity Color or Linetype with the Change command.

If entities within the Block Def do not have explicit Colors and Linetypes, they become chameleons and assume the Insert's properties. Look at ARROW, which was defined on Layer 0 with Color and Linetype Bylayer. Its Inserts vary because they all adopt the current-time-of-insertion defaults of the Layer.

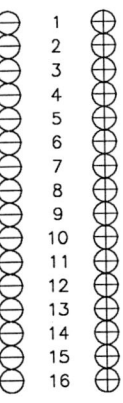

A Switch Board Terminal

Deciding to use Bylayer, Byblock, or explicit colors and linetypes is important to your Block construction. Once the Block is defined, explicit settings are locked in. This means someone can't easily change them. You can't change them unless you redefine the Block! If you set Color and Linetype Bylayer, you can change them easily by changing the default for the Layer. Organizing your drawing Bylayers imposes a structure and a means of control. You (and your users) can turn Layers On/Off and Freeze/Thaw them to control what you plot, or display. Simple schemes for blocks are better than complex schemes.

For typical Blocks, like Parts and single purpose Symbols, we recommend using Bylayer entity settings as the best means for controlling your symbols and parts. You can organize your data by layers and use layer control as a means of sorting data for evaluation. You can control what gets plotted by different weight pens by changing the colors of entire layers.

Multiple and Nested Blocks   3—11

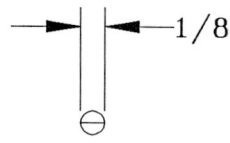

NEGATIVE

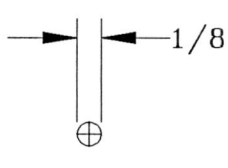

POSITIVE

Terminals

For multipurpose Symbols, we recommend that you use Bylayer on Layer 0, and control your blocks by layer insertion.

Multiple and Nested Blocks

So far the book has used simple single level Blocks, asking you to Insert them one-at-a-time. AutoCAD supports nesting Blocks inside other blocks. It also supports the Multiple INSERT (MINSERT) of Blocks. Let's look at nested blocks and multiple Inserts. We will use a switch board terminal as the exercise example. The terminal is shown in the Switch Board Terminal figure.

Create an electrical terminal rack. First, make a positive and a negative terminal connector, Wblocking each. The terminals are shown in the Terminals drawing. The electrical rack needs 16 rows of terminals, one POSitive and one NEGative per row. Insert the top row, add the Text number "1." Then, Array the three objects downwards.

Nesting Blocks

Enter selection:	Begin a NEW drawing named TERM-BLK=.
Command: **LIMITS**	Set to 0,0 to 11,8-1/2.
Command: **ZOOM**	Zoom W, 0,0 to 3,2-1/2.
Command: **SNAP**	Set to 1/16.
Command: **GRID**	Set to 1/2.
Command: **<Ortho on>**	Make sure ORTHO is ON.
Command: **LAYER**	Set Layer OBJ02 current.
Command: **SAVE**	Save the drawing settings for later. Do not Save again or End it later.
Command: **CIRCLE**	Draw a 1/16 Radius.
Command: **LINE**	Put a 1/8 horizontal LINE through the Circle's center.
Command: **BLOCK**	Block it to the name NEG, insertion base at its center.
Command: **OOPS**	Oops it back to reuse it.
Command: **LINE**	Draw a 1/8 vertical Line through its center.
Command: **BLOCK**	Block it to the name POS, insertion base at its center.
Command: **INSERT**	Insert NEG at 1:1 scale in the upper left screen corner.
Command: **INSERT**	Insert POS @1/2,0 to the right of NEG.
Command: **STYLE**	
Text style name (or ?) <STD1-8>: **STD3-32**	Reset for full scale.
Existing style.	
Font file <SIMPLEX>:	
Height <0'-9">: **3/32**	Set for 3/32 at 1:1, default the rest.
Command: **TEXT**	Style STD3-32 and a Middle point. Center the start point between the terminals. Enter a "1" as text.
Command: **ARRAY**	Select all three and use a Rectangular array. Use 16 rows, 1 column with a Unit distance between rows of -1/8 (minus 1/8).

Command: **CHANGE**	Select all the Text. Dance through Change, renumbering the terminals 1 thru 16.
Command: **WBLOCK**	Make this into one block called RACK. Pick the insertion base point just above top center and select all 48 entities.

```
Layer OBJ02 Ortho Snap           0'-0",0'-0"
       . 1     1
       Ⓞ  Ⓞ  Ⓞ  Ⓞ
       2     2
       Ⓞ  Ⓞ  Ⓞ  Ⓞ
       3     3
       ...
       16    16
Command:
REDRAW
Command:
```

Two RACK Inserts

Command: **STATUS**	Note the number of entities in the drawing, about 54, but erasures and errors count!
Command: **INSERT**	Hit a "?" to see the defined Blocks list, only POS and NEG.
Command: **QUIT** Enter selection:	Quit the drawing. You Saved it before. Edit the same drawing again.
Command: **STATUS**	No entities yet.
Command: **INSERT**	Hit a "?" to see that no Blocks are defined.
Command: **INSERT**	Insert the RACK in the top left of the drawing.
Command: **STATUS**	Compare the number of entities to before. 60 now.
Command: **INSERT**	Hit a "?" to see defined Blocks list: RACK, POS and NEG.
Command: **INSERT**	Insert another RACK in the top center of the drawing.
Command: **STATUS**	Compare the number of entities. Only 1 more than above

Remember, Blocks are not entities. When you Wblocked RACK the entities that made it up were erased, but they remain temporarily in the drawing database until you re-edit it. Deleted Blocks are purged from the drawing database when you reload an existing drawing. The Block RACK was not yet defined in the current drawing's Block Definitions.

When you re-edited the drawing, inserting RACK, AutoCAD put the definition, as well as the nested NEG and POS, in the Block Definitions. AutoCAD counts each of the items within the Block Def in the entity count, even though they don't occur in the drawing as individual entities.

Your nested block RACK contains 16 copies of the NEG Block Insert, 16 copies of the POS Block Insert, as well 16 Text entities. If RACK, POS and NEG were not Blocks, the AutoCAD drawing would store 32 Circles, 48 Lines and 16 Text Entities, a total of 96 total Entities.

Nesting blocks makes files smaller. The first Insertion of RACK adds 60 entities. Without nesting, it would have added 96 individual entities. When you Insert additional RACKs, AutoCAD adds only one more entity, the RACK Insert, and simply "looks up" the RACK data. This is why there is only one definition allowed per block in any one drawing. Since all definitions are stored in a table, only one description can exist for each named block.

Minserts of Blocks

For repetitive block insertions, AutoCAD provides a more compact way of storing the data, the MINSERT command. MINSERT is like a combined INSERT and ARRAY command. MINSERT Inserts the block, but it also prompts for the number of rows and columns.

Minserted RACKs

Reconstruct your RACK, using MINSERT for the top positive and negative symbols.

```
Command: EXPLODE                                    Explode one of the RACKs.
Command: ERASE                                      Erase the other RACK and each of the 16 POS and NEG
                                                    Inserts in the Exploded RACK, leaving only the 16 Text entities.
Command: STATUS                                     Note the number of entities in the drawing.

Command: MINSERT
Block name (or ?) <RACK>: NEG
Insertion point:                                    Pick point 1/4" to the left of the "1".
                                                    Default scale to 1 and rotation to 0.
Number of rows (---) <1>: 16
Number of columns (||||) <1>:
Unit cell or distance between rows (---): -1/8      That's minus 1/8.

Command: MINSERT                                    Repeat the Minsert with the POS Block and same parameters.
                                                    It should look identical to original RACK.

Command: STATUS                                     Note the number of entities. Only 2 more, not 32!
```

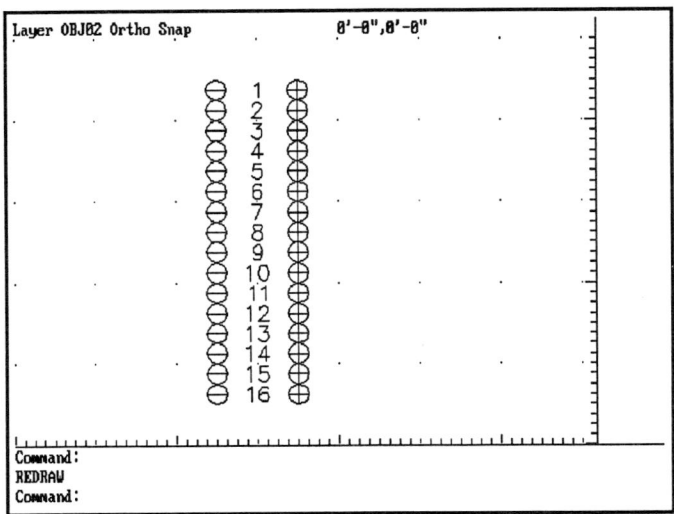

Minserted Rack

Command: **WBLOCK**	WBlock it to RACKM.
Command: **INSERT**	Insert RACKM at the left. The duplicate definitions are ignored.
Command: **STATUS**	Note the number of entities. 21 more, not 60 like RACK.
Command: **INSERT**	*RACK to make a comparison.
Command: **ERASE**	Erase the terminals.
Command: **WBLOCK**	All of the text to file name RACKTEXT.
Command: **DIR**	Do a DIRectory of RACK*.* and compare the files.
Command: **QUIT**	You can also delete files RACK*.DWG.

Compare the file sizes. RACKTEXT is the base file size. RACKM is about 880 bytes larger, and RACK about 1640 bytes larger. You can see from the file size of RACKM that MINSERT is about twice as efficient. RACK contains 32 Inserts entities. RACKM contains 2 Minserts. Both contain the same Text.

Side Effects of Exploding Blocks

There are drawbacks to using Blocks. You cannot edit INSERT entities as easily as individual entities. You can Scale, Rotate, Copy, Move or Change them to different Layers. But, if you have to change a part of the internal structure of the Insert, you're out of luck unless you can EXPLODE it.

Explode breaks down an Insert one level at a time. It replaces the Insert entity with the first level of the Block Definition's individual entities, including any nested Blocks. Exploding an Insert has one small, but often annoying side affect: the entities are put back to their original layer of definition.

Blocks vs. Shapes

While we are looking at blocks, let's look at shapes. AutoCAD has an older feature called Shapes. You can use Shapes to store symbols in a compact form. AutoCAD's text fonts are special shape files. Many AutoCAD users are curious about shapes, and wonder about using them in applications, but haven't played with them. The book has a shape file, called CA.SHP, on the CA-DISK. You can use this file to compare the difference in speed and file size in drawing with shapes and blocks.

A Typical Valve

AutoCAD draws vectors, not lines, arcs and circles. Entities are groups of data used to determine how to draw vectors. Shapes are hexadecimal coded vector lists that tell AutoCAD, how far and in which direction to draw. AutoCAD stores each Shape as a single reference in a drawing. The actual makeup of the shape is stored outside of the drawing file in a shape file, with "name.SHX." The .SHX file is a binary file which is compiled from the .SHP definition file before use. Since Shapes are already defined in vector format, AutoCAD handles and draws them faster.

Are You In Shape? A Performance Test

 The CA DISK includes a file named CA.SHP which is a typical valve symbol.

 If you don't have the CA DISK, just read along. Skip doing the exercise.

```
Enter selection:                                AutoCAD Main Menu option 7 to compile shape file CA.SHX.
Enter NAME of shape file (default term-blk): CA

Enter selection:                                Begin a NEW drawing named SHAPES.

Command: LOAD                                   Load Shape file CA.

Command: SHAPE
Shape name (or ?): VALVE                        Pick a point in the center.
                                                Drag its Height so it's big enough to see,
                                                and use 0 rotation. Notice that it scales equally X,Y.

Command: LINE                                   Trace another symbol over the Shape. It doesn't need to be
                                                exact for the test results to be valid.

Command: BLOCK                                  Block the Lines to the name VALVE.
Command: ARRAY                                  Do a large Array, 24x20 or more, of the Shape VALVE.
Command: ZOOM                                   All.

Command: REGEN                                  Time it.
Command: WBLOCK                                 Wblock all the Shape valves to VALVE-S.

Command: INSERT                                 Insert the Block VALVE.
Command: ARRAY                                  Do the same Array, 24x20 or more, of the Block VALVE.
```

```
Command: REGEN          Time it.
Command: WBLOCK         Wblock the array of Blocks to VALVE-B.

Command: DIR            Do a directory of VALVE*.* and compare file size.
Command: QUIT           You can also delete VALVE-?.DWG.
```

An Array of Valves

The test shows something like a 25% faster drawing Regen, and a 30% smaller file size. The file size savings is actually greater, since an empty drawing file is about 2700 bytes.

Shapes are more efficient than Blocks when the drawings you produce contain few insertions of complex symbols. Why doesn't everyone use shapes? Shapes have no flexibility in editing. You can't break, stretch, or assign multiple colors to Shapes. You can't explode them, and they only support one OSNAP mode, "Insertion." Finally, Shapes are difficult to create. Blocks are easy.

Block Management Summary Tips and Techniques

Here are our two summary recommendations for handling blocks:

- ❑ Use Bylayer and explicit layers when you define your blocks. Bylayer gives the best compromise of flexibility with control over color, linetype and visibility.

- ❑ If you need flexibility in layers for blocks, create all blocks on Layer 0.

Here are some added tips and techniques for handling blocks in your applications.

- ❑ Use entity explicit colors and linetypes sparingly for any visual effects you need locked in the block.
- ❑ Use Byblock if you need the colors and linetypes of entities to vary, like a primitive entity from one insertion to another.
- ❑ Use consistent insertion base points when defining your blocks. Don't keep your users guessing.
- ❑ Do not over nest your blocks.
- ❑ Use Minsert for very regular, fixed patterns that will not be edited.
- ❑ Consider using shapes, if you have standardized, complex symbols that are used sparingly in a drawing. You can automatically update shapes.

Your Block Defs can grow rapidly as you develop your application. We recommend keeping your blocks in separate subdirectories. Later, you can subdivide these subdirectories into other categories based on your application. If you have more than a hundred blocks in a subdirectory, it is a good idea to subdivide the blocks into more subdirectories.

This completes the basic setup tour of AutoCAD. It is time to start making some menu macros.

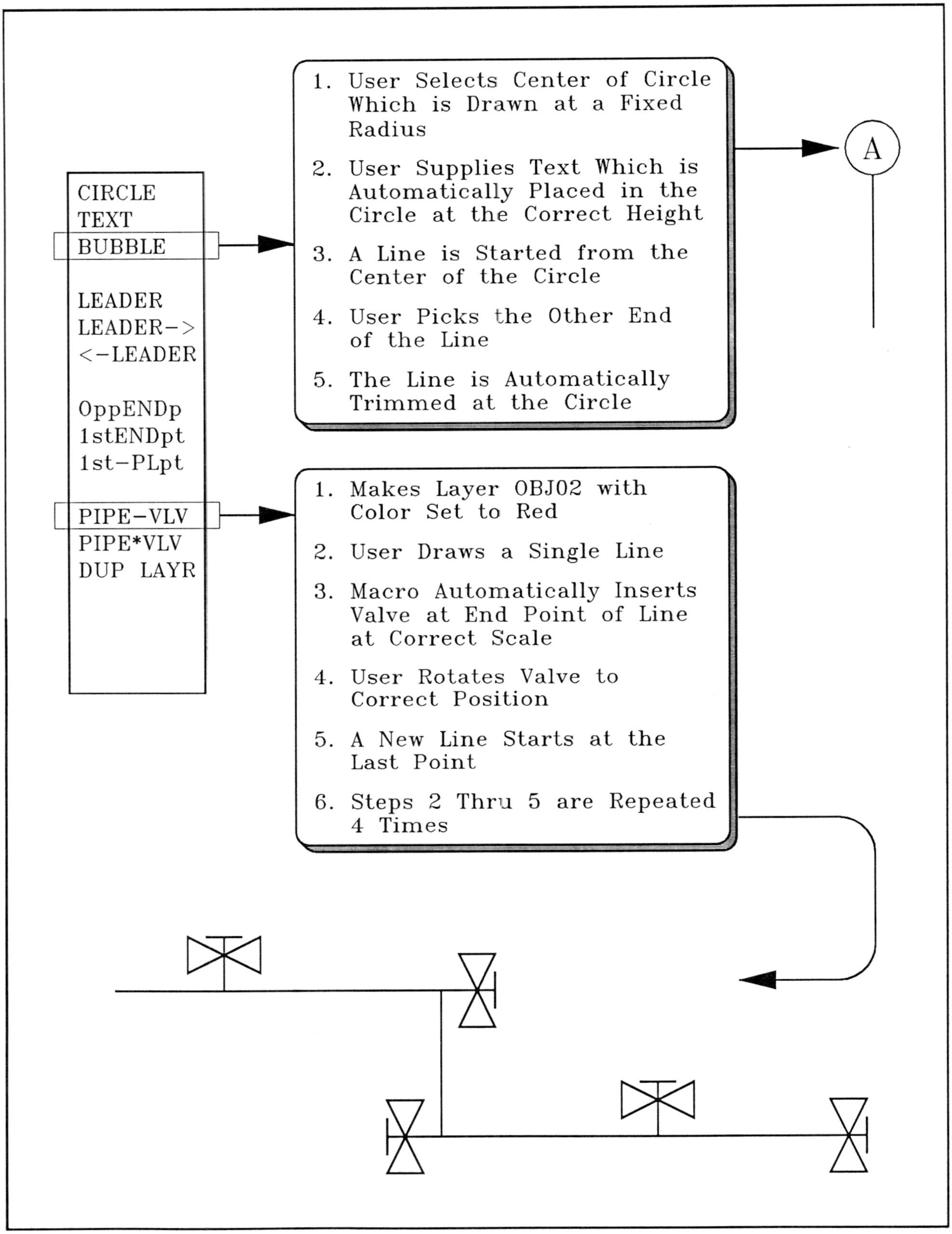

4—0 **Customizing AutoCAD**

```
CIRCLE
TEXT
BUBBLE

LEADER
LEADER->
<-LEADER

OppENDp
1stENDpt
1st-PLpt

PIPE-VLV
PIPE*VLV
DUP LAYR
```

BUBBLE →
1. User Selects Center of Circle Which is Drawn at a Fixed Radius
2. User Supplies Text Which is Automatically Placed in the Circle at the Correct Height
3. A Line is Started from the Center of the Circle
4. User Picks the Other End of the Line
5. The Line is Automatically Trimmed at the Circle

→ (A)

PIPE-VLV →
1. Makes Layer OBJ02 with Color Set to Red
2. User Draws a Single Line
3. Macro Automatically Inserts Valve at End Point of Line at Correct Scale
4. User Rotates Valve to Correct Position
5. A New Line Starts at the Last Point
6. Steps 2 Thru 5 are Repeated 4 Times

Screen Menu Macros

PART II. BASIC TOOLS

CHAPTER 4

Menu Macros

Like most computer programs, you can control AutoCAD's drawing editor by typing commands. Typing commands and input is tedious and prone to error. This chapter introduces you to the AutoCAD menu system. Using this customizable menu system is a key element in customizing AutoCAD.

The Standard AutoCAD Menu

Although AutoCAD's standard menu is fine for beginners, it becomes cumbersome to wade through as you become a more experienced user. You use 20% of the AutoCAD commands 80% of the time. Because you can create your own commands using menu macros, a custom menu system is an enormous time saver and an easy way to automate repetitive drawing procedures. We do not rely on AutoCAD's standard menu in this book. Instead, we will help you learn how to make your own menus, ones that are designed for your needs.

Basic Macro Skills

A MACRO is a series of AutoCAD commands and parameters put together to perform a task. Macros can call AutoCAD commands, pause for input from the user, and draw or edit objects on the screen. Macros also can automatically repeat.

Macros find their home in menu files. A MENU is an ASCII text file listing the commands and macros for each box of the screen, tablet, buttons, popups and icons menus. A typical screen menu displays a list of macro choices beside the drawing area. Although the real body of the menu item is never seen on the screen, it is similar to the sequence that you would type at the command prompt.

The Benefits of Using Menu Macros

A customized menu gives you important advantages over keyboard style input. Using a menu for input is faster. It is more error free in issuing complex instructions to AutoCAD. At its simplest, a menu system is a "super typist." There are three main benefits to using menu macros. These are:

- Macros help you increase productivity by reducing drawing steps.

- Macros help you set drafting standards.

- You can add new "commands" with drawing features that would be too tedious to do without menu customization.

How To Skills Checklist

Here are the skills that you will acquire. You will learn:

- ❏ How to put AutoCAD commands in a menu file.
- ❏ How to string commands together to form a basic macro.
- ❏ How to make macros pause for input from a user.
- ❏ How to use special characters in a macro to toggle Snap, Grid, Ortho and several other AutoCAD modes.
- ❏ How to use Object snaps in macros, including using the QUIck modifier to filter the points that you want.
- ❏ How to use labels for macros.
- ❏ How to use path names in macros.
- ❏ How to make a macro line repeat indefinitely.
- ❏ How to automatically repeat entire macros using an * modifier, and repeat individual commands using the Multiple command modifier.

Menus and Macros

MACROS
[BUBBLE] draws column line grids and bubbles. Puts in bubble text.
[LEADER] is a straight line annotation leader with an arrow.
[LEADER->] is a single entity leader that uses a polyline to create a straight segment and arrow end.
[<-LEADER] starts with an arrow and can have unlimited curved and/or straight leader segments.
[OppENDpt] gets the opposite endpoint of an entity.
[1stENDpt] gets the first endpoint of an entity, not necessarily the closest.
[1st-PLpt] gets the starting point of a selected pline.
[PIPE-VLV] initializes layer controls, draws pipe flow lines and puts in valve symbols.
[DUP LAYR] duplicates a set of entities on another layer.

How to Make Menu Files with the CA DISK

The book uses menu files from the CUSTOMIZING AutoCAD DISK SET (CA DISK). The disks contain the menus, example drawing files, AutoLISP routines and symbols libraries used in the hands-on exercises. Each chapter uses a working TEST.MNU file. At the beginning of each chapter we will ask you to copy the CAxx.MNU chapter file to a file called TEST.MNU. When the exercises ask you to "Add such and so to your TEST menu...," you can "edit" the TEST.MNU by simply "examining" its code.

At the end of each chapter, you delete your TEST.MNUs. This test method lets you experiment in the TEST.MNU file, but insures that you are on common ground with the book when you start each new chapter.

How to Make Menu Files without the CA DISK

If you don't have the CA DISK, we will ask to make a new TEST.MNU with each chapter. Later, we will make this task easier by creating and copying a "template" file called MAIN.MNU. Copy the MAIN.MNU file to create the TEST.MNU for the chapter. At the end of each chapter, you will rename TEST.MNU to names like MY04.MNU. The book will instruct you each step of the way. The exercise sequences are written as if you **don't** have the CA DISK.

Writing a Simple Menu Item

First, make a new menu file called TEST.MNU. Even if you have the CA DISK, type this one as shown. This first test menu makes sure your menu editing process is OK, and tests your MNU command from the .PGP file. Just follow our sequence.

```
C:\> CA                           Use the CA.BAT batch file to start ACAD.
Enter selection:                  Begin a NEW drawing named CA04.
                                  It should look like CA-PROTO.DWG, Grid and all. If not, check your
                                  CA.BAT, DOS directories, and the "Initial drawing setup" in your
                                  AutoCAD configuration.

Command: MNU                      This is the PGP command you made.
```

If everything is set up right, you're now in your text editor. If your text editor didn't automatically start a new TEST.MNU file in the CA-ACAD directory, please start one now. Some editors can't automatically load files. From now on we will assume that your editor loads the file automatically.

Type the next two lines into the TEST.MNU file.

```
CIRCLE
TEXT
```

If you have the CA DISK, your menu shows additional commands and parameters.
Save the file and exit your editor. Go back to AutoCAD.

```
Command: MENU                              Load the menu "TEST".
Menu file name or . for none <.>: TEST
Compiling menu TEST.mnu...                 [CIRCLE] and [TEXT] should appear at the top of the menu.
```

The book shows screen menu items in **[SQUARE BRACKETS]**.

Select **[CIRCLE]**	Works just as if typed at the "Command:" prompt. Draw a circle to test it:
Command: CIRCLE 3P/2P/TTR/<Center point>:	
Select **[TEXT]**	Test it also.
Command: **ERASE**	Wipe them out. Then make a bubble using the new menu:
Command: **ZOOM**	Zoom Center with a height of 16'.
Command: **SNAP**	Set to 2".
Command: **GRID**	Set to 1'.
Select **[CIRCLE]**	Pick a point centered on the screen:
Command: CIRCLE 3P/2P/TTR/<Center point>:	
Diameter/<Radius>:	24. That's 1/4" at 1:96 plot scale.
Select **[TEXT]**	And you get:
TEXT Start point or Align/Center/Fit/Middle/Right/Style: **S**	
Style name (or ?) <STANDARD>: **STD3-16**	
Start point or Align/Center/Fit/Middle/Right/Style: **M**	
Middle point: **@**	The @ lastpoint in the center.
Rotation angle <0>:	Default to 0.
Text: **A**	Just the letter "A".

NOTE. When you edit menus, type characters that you input exactly as the book shows them. Do not use tabs or invisible trailing <SPACES>. Do not use blank lines.

This first example menu was simple. The items send a single command to AutoCAD and terminate, letting the user do the rest. Text didn't prompt for height because the Styles preset height.

Special Characters in Menus

Macros need automation. The menu feature that makes automation possible is the **pause**. The macro pause stops for input from the user, then resumes execution with more commands or options. To string together multiple commands and user input requires special characters that the AutoCAD menu interpreter won't mistake for commands or input.

The "\" backslash is the most important of these special characters. It makes AutoCAD pause for input.

Repeat the "bubble" sequence. Write down each step. Use a blank <SPACE> to represent each <SPACE> or <RETURN> that you enter at the keyboard. Use a "\" backslash to mark where you pick a point or type input.

Command:	Repeat the bubble sequence.
	CIRCLE \24 TEXT S STD3-16 M @ 0 \
	Write the these steps down as you repeat the bubble.

Next, let's input this sequence to make a macro that creates circles with text! This macro starts the Circle command, pauses for location, draws a 24-inch Circle, and waits for Text before automatically putting the text into the center of the circle.

Command: **MNU**   Use the PGP command MNU to go edit the TEST.MNU file.

Add the following below the TEXT and CIRCLE lines:

`CIRCLE \24 TEXT S STD3-16 M @ 0`

Save and exit back to AutoCAD.

Command: **MENU**   Enter TEST. After editing you must reload the menu, or the old menu gets used. We'll automate this later.

Select **[CIRCLE \]**   Pick a point and type a character.

Your screen should shows a circle with text.

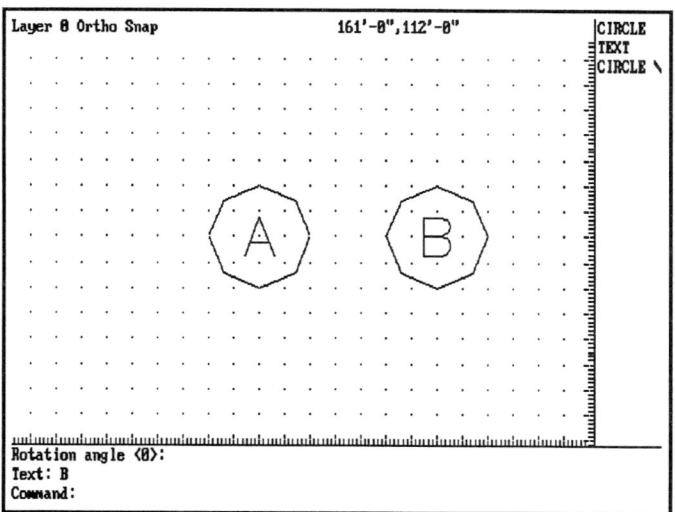

Circle with Text

A single backslash tells AutoCAD to wait for a single piece of input. In the circle-with-text macro, it pauses for the Circle's center point. Without a backslash, AutoCAD would continue taking its input from the macro and would pass the next item along to the command processor. It would read the "24" as the center point and get confused because it isn't a coordinate point. A redundant second backslash is omitted at the end of the macro since the Text command is waiting for input anyway.

You may wonder why we specify the Style when it was already set. You never know what was used last in macros. It is good practice to explicitly set modes and settings, like Style, if they are important for your macro to function.

Backslash, <SPACE> and Special Characters

Backslashes and <SPACES> are two of the control characters that you use in menus. They work just as if you typed them from the keyboard. AutoCAD automatically places a <SPACE> at the end of a menu line, unless the line ends with a special character. When AutoCAD encounters a "special character" at the end of a menu line, it **never** adds an automatic <SPACE> to the end of the line.

The following table lists the special menu characters. The @ (lastpoint) is not listed as a special character. The @ does not share in the special treatment by the AutoCAD menu interpreter. Here is the list of special characters:

SPECIAL MENU CHARACTERS

```
\    Pauses for input              ^B  Toggles SNAP        ^O  Toggles ORTHO
;    Issues RETURN                 ^D  Toggles COORDS      ^P  Toggles MENUECHO
+    Continues to next line        ^E  Toggles ISOPLANE    ^Q  Toggles Echo to Printer
[]   Encloses Label                ^G  Toggles GRID        ^T  Toggles Tablet
*    Autorepeats, or marks page    ^H  Issues BACKSPACE    ^M  Issues RETURN
     <SPACE>                       ^C  *Cancel*            ^X  *Delete* input in buffer
^Z   Does "nothing". Put at end of line to suppress automatic SPACE
```

Enter control characters, like <CTRL-B>, into a menu file as two characters. Use a carat "^" followed by an uppercase letter, like "^B."

Semicolons Are a Special Case

Let's look at the semicolon ";" by adding extension lines to the bubble macro. The semicolon is the special character for <RETURN>. The bubble macro uses more backslashes and the QUAdrant Osnap mode. The exercise shows **additions** to the existing menu lines by **bolding** the additions.

Command: **MNU**   Edit TEST.MNU. Add **bold** part to end of line shown:

CIRCLE \24 TEXT S STD3-16 M @ 0 **\LINE QUA \\;**

Save and exit to AutoCAD.

Command: **MENU**   Reload the menu.
Select **[CIRCLE \]**
Command: CIRCLE 3P/2P/TTR/<Center point>:   Pick point.
Diameter/<Radius>: **24**
Command: TEXT   The Text command scrolls by:
Text: **12**   Enter 2 characters.
Command: LINE From point: QUA of   Pick the QUA point anywhere on Circle.
To point: Pick point.
To point: Macro issues ; as a <RETURN> and ends.

The \\ paused for two input points after the Line command, then the last ";" terminated the command. You must supply one backslash for each point that you want for input.

The last character is a semicolon ";." You commonly can use semicolons and spaces interchangeably in macros, like <RETURN>s and <SPACE>s typed from the keyboard. In macros, spaces act like the <SPACE> from the keyboard and semicolons act like the <RETURN> key. When you terminate a macro, use a ";" that you can see. Don't use an "invisible" trailing <SPACE>.

```
Command: MNU                          Add these two lines. Note the "A1 LINE" vs the "A2;LINE":

CIRCLE \24 TEXT S STD3-16 M @ 0 A1 LINE QUA \\;
CIRCLE \24 TEXT S STD3-16 M @ 0 A2;LINE QUA \\;
```

Reload your menu, test both items. Which one works?

You would think that both lines would put their text label, A1 or A2, in their bubbles. The first line tries to use a <SPACE> after the "A1" while the second line uses a semicolon for a <RETURN>. Only the "A2" item works.

Since you must be able to type <SPACE> characters in the middle of text strings, AutoCAD's menu interpreter treats them as true spaces in your text, not <SPACE>s like <RETURN>s. If you want the macro to continue after the text input, you need some way to tell AutoCAD that the string of text is complete. You can't use the automatic <SPACE> at the end of the menu line for text. You use a ; semicolon.

Putting a Semicolon in Menu Text

Fortunately, all special characters except <SPACE>s act exactly the same whether they are in the middle of a text string or not. You can use a semicolon anywhere to issue a <RETURN>. The sacrifice you must make is that the only way to enter a real semicolon ";" in text created by a menu is to use ASCII codes. The AutoCAD menu interpreter doesn't recognize ASCII codes, but the text string does.

You are familiar with AutoCAD's underscores in text where the %%u gives an underscore in a text string. The %% is the "escape" character in fonts. ASCII 59 is the ASCII code for ";." If you imbed a %%59 in a text string, you get a semicolon. For example, "Word%%59 item, stuff" becomes "Word; item, stuff." This is the text method for special characters and text modes. The %%nn works for any character where nn is the ASCII code number.

Labeling Macro Commands

Now that you have several macros on your screen, it is hard to distinguish what each command does. You can control what is displayed on the screen by putting a macro label in [SQUARE BRACKETS]. Only eight characters display on the screen menu, but you can make labels longer for documentation. Labels can include letters, numbers, and any displayable character. Control and extended ASCII characters are simply ignored. The square [brackets] identify the label to AutoCAD's menu interpreter.

Let's add labels to the menu. Don't leave any blank <SPACE>s between your screen labels and the body of the macros. Everything that follows the label is treated by AutoCAD as an instruction, so a <SPACE> will cause an error.

Making a Menu Label

 Copy CA04.MNU to TEST.MNU.

 Continue editing TEST.MNU.

Command: **MNU** Add these labels and delete the other items:

```
[CIRCLE]CIRCLE
[TEXT  ]TEXT
[BUBBLE]CIRCLE \24 TEXT S STD3-16 M @ 0 \LINE QUA \\;
```
Save, exit and reload the menu. See the screen differences.

Your screen should display the screen menu labels shown in the screen shot.

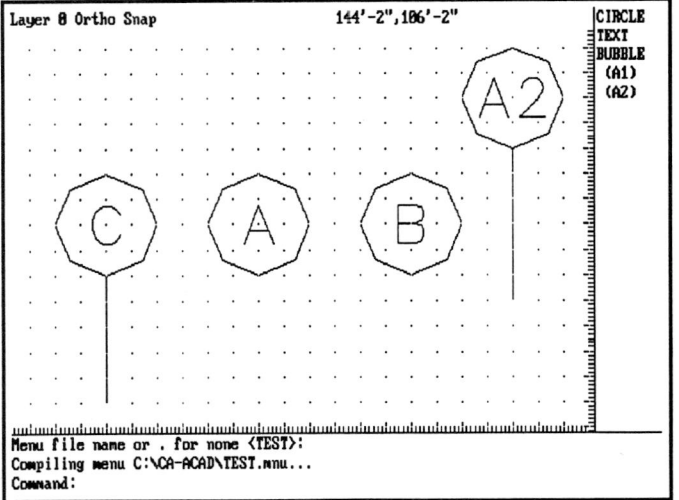

Screen Menu Labels

Pointing Arrow Block

Using Special Characters and Path Names

The next exercise shows you how to use more special control characters while you make a series of [LEADER] macros. The exercise also shows how to use directory path names in macros.

The first [LEADER] draws a two segment Pline and Inserts an arrow Block at the end. The macro uses the ARROW block that you created in the chapter on Blocks. The macro Rotates the arrow to the midpoint of the line, using Osnap MIDpoint.

Making a [LEADER] Macro

Command: **MNU** Edit TEST.MNU. Add the following:

`[LEADER    ]^C^C^CORTHO ON PLINE \\^O\ INSERT /CA-ACAD/ARROW @ 96 ;MID @ ^O`

Save, exit and reload the menu.

Select **[LEADER]** Try it. The Ortho, Pline and Block commands scroll by and it ends with <Ortho on> leaving a 2-pline leader and arrow.

Your screen should show the two-pline leader.

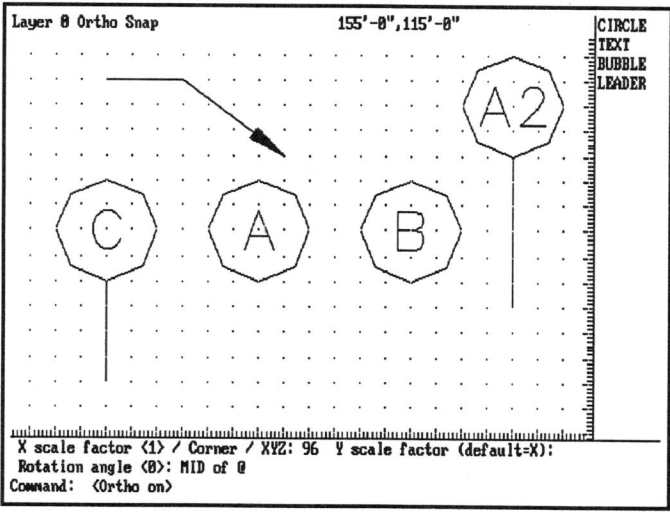

The LEADER Macro

Several items in this [LEADER] macro are new. You probably use the <F8> key to control Ortho, but you need to use the Ortho command, or control character toggles in menus. <^O> is the Ortho toggle. The [LEADER] macro uses the Ortho command to turn Ortho on. After the first segment of the Pline, the <^O> flips Ortho off. Later it flips Ortho back on.

The [LEADER] macro uses three <^C> control-Cs that act as CANCELs. We recommend using three because some deep dark recesses of AutoCAD and AutoLISP require three <^C>s to get back to the command prompt. It is good practice to begin command macros with <^C>s to cancel anything that is outstanding.

Look at the directory path name for the ARROW being inserted in the [LEADER] macro. The slashes are forward "/" slashes, not the backslashes "\" that you are used to in the DOS operating system. This macro used a path to show how to handle path names. The macro didn't need a path, since ARROW was in the current directory. But when macros need path names, you **must** use the "/" forward slash. There is no exception. Backslashes make menus pause!

The [LEADER] macro used the @ symbol to recall the lastpoint at the end of the Pline during the Block insertion. The macro then fed the X-scale value of 96 (plot scale). The ";" semicolon of "96 ;MID" defaults the Y scale factor to equal the X. When you accept default values, show them with a semicolon.

Last, the [LEADER] macro Osnapped the Block rotation with "MID @." Since the @ lastpoint is at the end of the Pline, the MIDpoint orients the ARROW symbol to the angle of the line. ARROW was created with this in mind, with its insertion point at the arrow point.

Making a Single Entity [LEADER->]

Now, try a few macro tricks. Create an entire leader, including its arrow, as a single PLINE. This [LEADER->] has the advantage that you can select it as a single entity. First draw interactively with AutoCAD, writing down the characters and picks, then write the macro.

```
Command: ORTHO
ON/OFF <Off>: ON
Command: PLINE
From point:                                              Pick a point
Current line-width is 0'-0"
Arc/Close/Halfwidth/Length/Undo/Width/<Endpoint of line>: W Make sure it is 0 width.
Starting width <0'-0">: 0
Ending width <0'-0">: 0
Arc/Close/Halfwidth/Length/Undo/Width/<Endpoint of line>:   Pick a point.
<Ortho off>                                              Toggle with ^O and pick another point.
Arc/Close/Halfwidth/Length/Undo/Width/<Endpoint of line>: W
Starting width <0'-0">: 0
Ending width <0'-0">: 6
Arc/Close/Halfwidth/Length/Undo/Width/<Endpoint of line>: L
Length of line: -18                                      Yes, that's minus 18.
Arc/Close/Halfwidth/Length/Undo/Width/<Endpoint of line>: W Reset width 0 0. <RETURN> to exit.
```

Collect all your typed input and picks. String them together to make the [LEADER->] macro.

Command: **MNU**   Edit the TEST.MNU and add this line:

[LEADER->]^C^C^CORTHO ON PLINE \W 0 0 \^O\W 0 6 L -18 W 0 0 ;

Save, exit and reload the menu.
Select **[LEADER->]**   Pick points when it pauses, in the same sequence as above.

Command: **LIST**   List it. It is a single Pline.

Your screen should show a single entity leader.

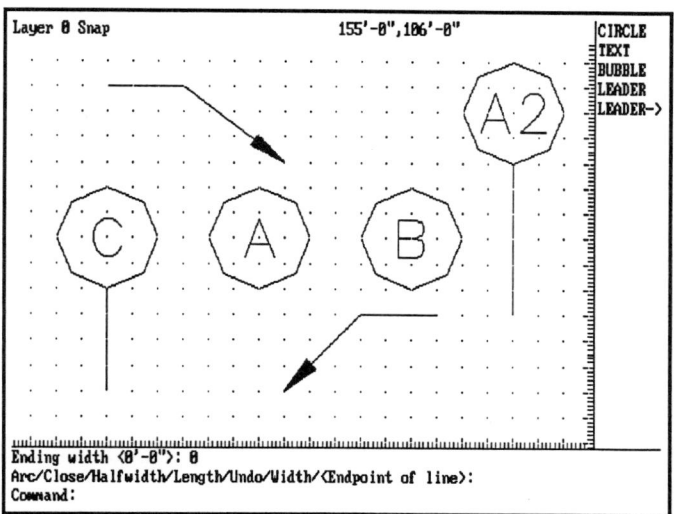

A Single Entity Leader

The label [LEADER->] "graphically" indicates the direction of the leader. Since you can't predict the incoming default Pline Width, your Pline macros always must set the width explicitly, as in "W 0 0 ." Remember that \ backslashes and control characters, like ^O, are executed without needing a <SPACE> or <RETURN>. "W 0 6 " sets the width to draw the arrowhead.

The "L" Length parameter draws the 18-inch long arrow segment by using a negative length Pline trick. It folds the pline back upon the previous Pline segment with "L -18 ."

Making a Multi-segment [<-LEADER] Macro

The last leader macro starts with the arrow. Its label is [<-LEADER]. The macro uses a Pedit trick to reset the @ lastpoint to the start point of a Pline. This lets the macro start with an arrow, but you must repick the second point. You can finish it up with as many line or arc Pline segments as you like.

Command: **MNU**   Edit TEST.MNU and add the following line:

[<-LEADER]^C^C^CPLINE \\;PEDIT L E M @ X ;ERASE L ;PLINE @ W 0 6 L 18 W 0 0 L 1

Save, exit, reload the menu.
Select [<-LEADER]
Command: PLINE   Pick two points. It starts with a Pline.
                                         Many Pline and Pedit prompts scroll by, then:
Arc/Close/Halfwidth/Length/Undo/Width/<Endpoint of line>:
                                         Re-pick the 2nd point, a 3rd point, and <RETURN> to end.

Your screen should show a curved pline leader.

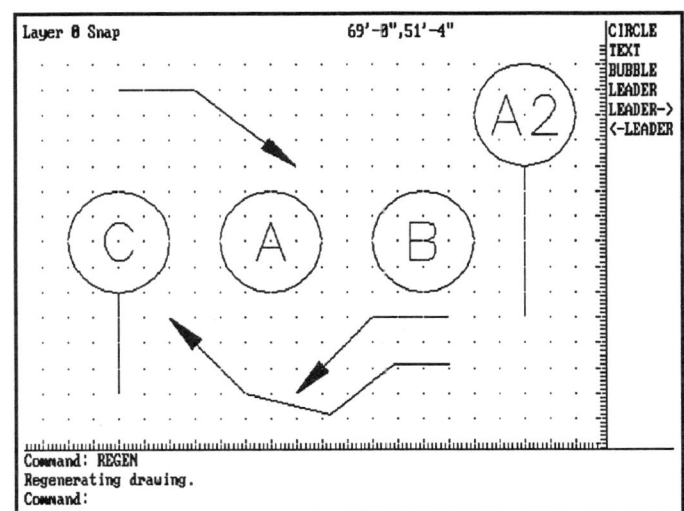

Multi-segment Pline Leader

The "PLINE \\;" draws a single segment that establishes a starting direction for the arrow. The "PEDIT Last Edit Move @ " is the key trick used in the macro. It doesn't actually move the vertex, the @ resets the lastpoint to the start of the Pline.

The "X ;" exits Pedit. Then "ERASE L ;" erases the temporary Pline. The macro starts the real Pline with an arrow at the lastpoint with "PLINE @ ." The "W 0 6 " sets the arrow width so you can draw it with the Length parameter using "L 18 ." The "W 0 0 " sets width back to 0. Add a small Length segment with "L 1 " to keep the next segment from distorting the arrow. If you don't, it mitres the arrow base relative to the angle of the next segment. The macro leaves you in the Pline command. Re-pick the second point and draw line segments, hit an A to draw arcs, or <RETURN> to end.

Making Long Macros

Menu items can get very long, especially macros that set up a large number of layers. You can continue a macro for many lines by ending each line with a "+" plus sign. Let's continue with the [BUBBLE] macro. As it gets too long for one line, continue the macro with a + sign.

Making A [BUBBLE] Macro Using the SELECT Command

Explore some more Osnap features, making the [BUBBLE] macro require just two picks. Use SELECT, an AutoCAD command seldom used outside macros, to enhance the [BUBBLE]'s features.

Command: **MNU**   Edit the [BUBBLE] item in the TEST.MNU.

Edit the menu to get the following two lines. Edit the TEST.MNU lines even if you have the CA DISK.

Making Long Macros 4—13

```
[BUBBLE ]^C^C^CCIRCLE \24 SELECT L ;TEXT S STD3-16 M @ 0 \+
LINE @ \ ID MID,QUI @ TRIM P ;ENDP,QUI @ ;
```

Save, exit, reload the menu.

```
Select [BUBBLE]
Command: CIRCLE 3P/2P/TTR/<Center point>: "\" Pick a point.
Diameter/<Radius>: 24
Command: SELECT
Select objects: L 1 found.
Select objects:                             ";" <RETURN>s to end selection.
Command: TEXT                               The text command scrolls by with "S STD3-16 M @ 0 ", then:
Text: A3                                    "\" Enter some text.
Command: LINE From point: @                 Starts at circle's center.
To point:                                   Pick a point.
To point:                                   ";" ends the line.
Command: ID Point: MID,QUI @                "Saves" line's midpoint as lastpoint.
X = 130'-0"  Y = 128'-0"  Z = 0'-0"
Command: TRIM
Select cutting edge(s)...
Select objects: P 1 found.                  Selects the circle SELECT "saved" as Previous.
Select objects:
Select object to trim: ENDP,QUI @           Trims end at the "saved" lastpoint
Select object to trim:                      And it's done.
```
Command: SELECT
Select objects: L 1 found.   This "saves" the Last circle as Previous.

Look at the "+" plus sign end of the first line. When AutoCAD sees a + at the end of the line, it treats the next line as part of the same item. The most important thing to remember about "+" is that you cannot put anything after the "+" character.

The Osnap trick "ID MID,QUI @ " uses the harmless ID command to reset lastpoint to the MIDpoint of the Line, which is picked by @. QUIck mode makes the macro find the most recent nearby MIDpoint, instead of the closest MIDpoint. Since you just drew the line, you know it will be most recent. This avoids MID finding the wrong line in heavy traffic. QUIck also speeds up Osnap. Use it when Osnapping an entity that you are sure is more recent than its neighbors. This is a frequent case in macros.

To trim off the extra bit of the line, "TRIM P ;" selects the Circle as a cutting edge, then "ENDP,QUI @ " picks the Line to trim. The macro uses ENDpoint to select the correct end, and "@" to pick the lastpoint that ID saved. The final ";" exits Trim and ends the macro. The macro uses ENDP (not END) to avoid accidentally ENDing the drawing.

Although both ends of any Line are equidistant from its midpoint (your lastpoint in the macro), AutoCAD is never random. AutoCAD will always find the first endpoint of the line if you Osnap ENDP from the midpoint of the line. You may not be used to worrying about the difference between the ends of lines, but you will have to consider it when using AutoLISP.

You can use the same techniques to form three handy stand alone macros to find the opposite end of a line, 1st end of a line, or the 1st end of a Pline. Add these new macros to your menu.

```
Command: MNU                              Edit TEST.MNU and add the following three macros.

[OppENDpt]^C^C^CLINE \MID @ ;ROTATE L ;@ 180;+
ID MID,QUI @ ID ENDP,QUI @ ERASE L ;ID ENDP @
[1stENDpt]^C^C^CID MID \ID ENDP @
[1st-PLpt]^C^C^CPEDIT \E M @ X X
```

Save, exit, reload the menu. Draw some Lines and Plines. Play with the macros.

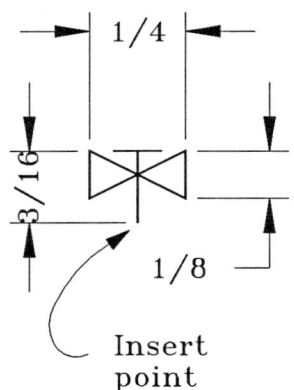

Pipe Flow Valve

Notice how the [OppENDpt] Rotates, flipping a line 180 degrees, then Osnaps to "walk" down it.

Layers and Macros

The next [PIPE-VLV] macro shows you how to control Layer names, Color and Linetype settings with a menu macro. Rather than assume that a layer was created in a prototype drawing, or by a setup routine, assume that it doesn't exist. Create the layer in macros that depend on it. This type of macro control avoids the problem of a user Purging unused layers from a drawing.

The macro [PIPE-VLV] contains several AutoCAD commands. It makes a layer, draws a Line, then Inserts a block named VALVE. If you have the CA DISK, you have VALVE.DWG. Otherwise you must make and Wblock a VALVE. If you are unfamiliar with the syntax of AutoCAD's LAYER command, test and record the macro sequence in AutoCAD.

[PIPE-VLV] Piping Macro with Layer Setup

 Use VALVE.DWG

 Draw a valve as shown. Wblock it to VALVE.

Command: **MNU**   Edit the TEST.MNU

Add the following:

```
[PIPE-VLV]^C^C^CLAYER M OBJ02 C RED ;;+
LINE \\;INSERT VALVE @ 96 ;DRAG
```

Save, exit, reload the menu.
Select **[PIPE-VLV]**   Pick 2 points. The Layer, Line and Insert commands scroll by. Drag rotation and pick a point to set it.

Command: **ZOOM**   Zoom in to see it better.

Here is how the [PIPE-VLV] macro works. First, "LAYER M OBJ02 C RED ;;" sets the Layer and its Color. Instead of Set, the macro uses M (Make) to set the Layer current. With "RED ;;" the <SPACE> enters RED, the first ";" defaults the current Layer to RED, and the second ";" exits the Layer command. The "+" continues the macro to the next line.

Next, "\\;" draws a single pipe Line when you pick two endpoints, and terminates the Line command with the ";." To "INSERT VALVE," the "@ 96 ;" places the insert at the @ lastpoint with a hardcoded X-scale of 96 and with ";" defaulting the Y scale equal to X. Finally, "DRAG" ensures that Drag is on while you rotate the Insert. Although your default is Drag Auto, some users may have it turned off. If a macro depends on DRAG to show the user orientation, you should explicitly DRAG to be sure.

Repeating Macros and Multiple Commands

A little automation always calls for more. Why should you have to repeatedly reselect a macro from the screen if you want to use it several times? Let's improve the [PIPE-VLV] macro, making the macro command repeat 5 times. Add " \+" to the last line. Then, add four more lines.

Command: **MNU**

Edit the TEST.MNU item [PIPE-VLV] to:

```
[PIPE-VLV]^C^C^CLAYER M OBJ02 C RED ;;+
LINE \\;INSERT VALVE @ 96 ;DRAG \+
LINE ;\;INSERT ;@ 96 ;DRAG \+
LINE ;\;INSERT ;@ 96 ;DRAG \+
LINE ;\;INSERT ;@ 96 ;DRAG \+
LINE ;\;INSERT ;@ 96 ;DRAG
```

Save, exit, reload the menu.
Select **[PIPE-VLV]**   It works as before, but draws 5 segments.

You could fit this repeating macro on fewer lines, but this format makes it easier to see what it does. The original macro didn't need a "\" after its ending "DRAG" at the Rotation prompt of the Insert command. However, since each subsequent line starts with another Line command, you have to fill the rotation prompt with a pause. The last line of an item never gets a "+" continuation character, nor does the last line need a " \."

Note the differences between the second line and the successive lines. The repetitive lines start with "LINE ;\" instead of "LINE \\," using a ";" to default the new start point to the end point of the previous Line. The macro is more efficient after the initial Line/Valve set is placed because you do not need to repeat the first line and the block name. The Layer is already set and the Insert command has a default block from the first Insert. The repetitive Inserts replace "VALVE" with a ";." Of course, if you want to draw less than five sections, you always can cut the macro short with a ^C.

4—16 Customizing AutoCAD

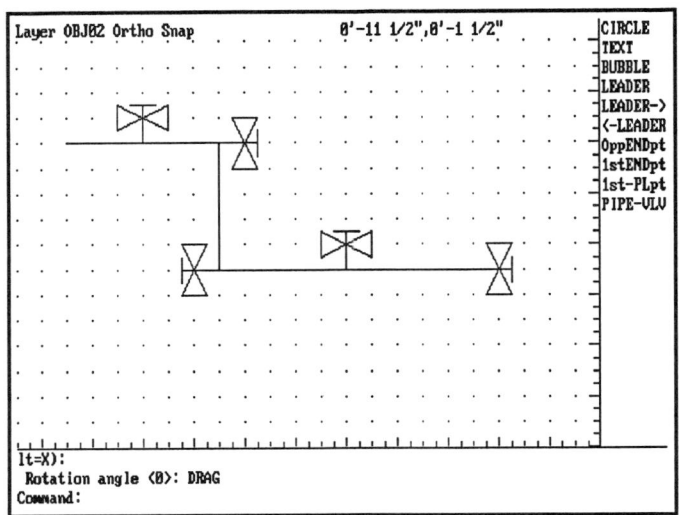

Pipe Lines and Valves

Indefinitely Repeating Macros and Commands

Using AutoCAD Release 9, you can repeat macros indefinitely by placing an asterisk "*" as the first character of the macro. If you use a screen [LABEL], put the "*" immediately after the closing right bracket.

Make [PIPE*VLV] a repeating macro, using Release 9's asterisk "*" feature for macros. Put an asterisk in the [PIPE*VLV] label as a reminder that it is a repeating macro.

Making Commands and Macros Repeat

Command: **MNU**   Copy the previous [PIPE-VLV] and edit it.

[PIPE*VLV]*^C^C^CLAYER M OBJ02 C RED ;;+
LINE \\;INSERT VALVE @ 96 ;DRAG \

Save, exit, reload the menu.
Select [PIPE*VLV]   It works like the first one but repeats until <^C> canceled.

The "*" triggers the menu interpreter to repeat the macro in its entirety. The "*" must be followed by at least one <^C> or <^X>. Otherwise, it is interpreted as a character and runs into the following command or input. AutoCAD would see *LAYER, not recognize it and cause an error. Since the "*" precedes the typical "^C^C^C" macro beginning, earlier versions of AutoCAD that don't recognize "*," simply ignore "*" and treat the macro as a standard single macro.

A "\" is needed at the end of the line, because the macro returns to the beginning and repeats. This macro also has the feature that, at the "LINE From point:" prompt of each repeat, you can hit a <RETURN> to continue from the previous line, or pick a point to start a new string of lines.

With Release 9, you can repeat single commands without an asterisk by using the MULTIPLE command. The MULTIPLE command modifier causes AutoCAD to repeat one command until you hit a ^C to cancel it. Multiple only repeats the command, ignoring any options or parameters used in the first execution. You can make your original CIRCLE and TEXT commands repeat automatically using MULTIPLE.

Command: **MNU**   Change the CIRCLE and TEXT entries to:

```
[CIRCLE   ]^C^C^CMULTIPLE CIRCLE
[TEXT     ]^C^C^CMULTIPLE TEXT
```

Save the file, reload your menu.
Select **[CIRCLE]**   It keeps drawing circles, repeating like the DONUT command.
Select **[TEXT]**   It repeats, but you have to reenter the parameters each time.

Multiple works for simple menu items like [CIRCLE] and [TEXT]. However, if you tried to make TEXT automatically repeat a mode like "M" for Middle, it would use the mode the first time, then ignore the mode on repeats. "^C^C^CMULTIPLE CIRCLE 3P" would draw a single 3-point circle, then repeat with normal prompts. Anything like "^C^C^CMULTIPLE CIRCLE \\LINE" causes an error as the "LINE" tries to become input to the repeat. Multiple is intended primarily for on-the-fly keyboard use or single commands. We find the "*" method works better for macros.

Macros that Edit Using Selection Sets

Selection sets give you a hand as a macro-writer. Selection sets let you control how your user selects objects. AutoCAD editing commands always ask for a set of entities before any edit tasks begin. This limits your macro writing by allowing only a fixed number of picks. Whether you use Window, Last, or another selection mode, you still need to predefine one "\" per pick. AutoCAD's SELECT command is indispensable in macros because it has the unique feature of pausing the macro indefinitely until the entire selection process is complete. It pauses even when other commands follow. Let's try an example.

Building Selection Sets in Advance

Make a Duplicate-to-another-Layer macro, called [DUP-LAYR].

Command: **MNU**   Edit TEST.MNU and add the macro:

```
[DUP LAYR]^C^C^CSELECT \COPY P ;0,0 ;CHANGE P ;PROP LA \;
```

Reload your menu. Try it on existing entities, or draw a few new ones to use.

Select **[DUP LAYR]**   It pauses for the new layer name. Use ANN02.
                                                    Switch the current and ANN02 Layers On/Off to see.

As you see in [DUP-LAYR], Select doesn't **do** anything except create a selection set of entities allowing an indefinite number of picks. This is invaluable because you can use a subsequent command, like COPY, to select that selection set with a "P" (Previous). "CHANGE" also uses a "P" in [DUP-LAYR], but it is technically reselecting Copy's set.

Notice the use of "0,0 ;" to do the Copy with 0,0 displacement. Many users are unaware that a <RETURN> in response to Copy's "Second point of displacement:" prompt causes it to take the first point as an actual relative displacement, 0,0 in this case. The same <RETURN> response applies to Move.

Selection Set Modes

AutoCAD Release 9.0 adds several selection modes to the old "Selecting by a point, Window, Last, Crossing, Add, Remove, Multiple, Previous, and Undo" selections set options. These are:

SIngle. Stays in selection mode until an object or set is successfully picked.

BOX. Acts like either Crossing or Window, depending on the order of the two points picked.

AUto. If the 1st point finds an object, AUto picks it. Otherwise AUto acts like BOX.

You can use SIngle combined with other modes, like Crossing or AUto. You also can use it with SELECT, if you want to force the creation of a "Previous" set with only a single selection. But, before you assume SIngle solves the problem of a missed pick ruining or aborting a menu macro, read on. Although SIngle ends object selection when an object or set is picked, it doesn't suspend the rest of a macro. SIngle will work OK for simple macros like:

[STRETCH1]*^C^C^CSTRETCH SI C

SIngle keeps you from having to hit a <RETURN> to end the selection process. However, if you miss your first selection and any additional commands or parameters follow, SI will try to use them as input. For example, this macro changes an entity to YELLOW:

[C-YELLOW]^C^C^CCHANGE SI \PROP C 2 ;

This macro works fine if you don't miss the entity, but if you do miss, SI tries to use PROP as object selection input.

We recommend using BOX for macros. Except for the STRETCH command, you should probably forget about using Window and Crossing in macros. Just use BOX. Play with it and get used to the order of picks. We think you will agree. BOX is clean and simple. Pick left-to-right for a "window" and right-to-left for a "crossing."

AUto also is useful. It gives the most flexibility, combining picking by point with the BOX mode. If your first point misses, you go into BOX mode, otherwise AUto selects the entity at the first pick point. This causes one quirk in menus if the macro has other parameters following it:

`[C-YELLOW]^C^C^CCHANGE AU \\;PROP C 2 ;`

This macro needs two backslashes for "crossing/window." It works fine if you use it as "window" or "crossing." However, if your first pick finds something, AUto is satisfied, but the macro is still suspended by the second backslash. Hit the button again at the same point. The macro will find the same object by pointing again. AUto is a good candidate for everyday editing and selection commands on the digitizer buttons.

Cleanup

It is time for a little cleanup to put your menu items in an easier to read format.

Improving the Menu Format

Command: **MNU**   Edit TEST.MNU to make it look like this:

```
[CIRCLE   ]^C^C^CMULTIPLE CIRCLE
[TEXT     ]^C^C^CMULTIPLE TEXT
[BUBBLE   ]^C^C^CCIRCLE \24 SELECT L ;TEXT S STD3-16 M @ 0 \+
LINE @ \ ID MID,QUI @ TRIM P ;END,QUI @ ;
[]
[LEADER   ]^C^C^CORTHO ON PLINE \\^O\ INSERT /CA-ACAD/ARROW @ 96 ;MID @ ^O
[LEADER-> ]^C^C^CORTHO ON PLINE \W 0 0 \^O\W 0 6 L -18 W 0 0 ;
[<-LEADER ]^C^C^CPLINE \\;PEDIT L E M @ X X ERASE L ;PLINE @ W 0 6 L 18 W 0 0 L 1
[]
[OppENDpt ]^C^C^CLINE \MID @ ;ROTATE L ;@ 180 +
ID MID,QUI @ ID END,QUI @ ERASE L ;ID END @
+
[1stENDpt ]^C^C^CID MID \ID END @
[1st-PLpt ]^C^C^CPEDIT \E M @ X X
[]
[PIPE-VLV ]^C^C^CLAYER M OBJ02 C RED ;;+
LINE \\;INSERT VALVE @ 96 ;DRAG \+
LINE ;\;INSERT ;@ 96 ;DRAG \+
LINE ;\;INSERT ;@ 96 ;DRAG \+
LINE ;\;INSERT ;@ 96 ;DRAG \+
LINE ;\;INSERT ;@ 96 ;DRAG
+
[PIPE*VLV ]*^C^C^CLAYER M OBJ02 C RED ;;+
LINE \\;INSERT VALVE @ 96 ;DRAG \
[DUP LAYR ]^C^C^CSELECT \COPY P ;0,0 ;CHANGE P ;PROP LA \;
```

Save, exit, reload the menu.

```
Select  [         ]                    Select and test each item to be sure they are error free.

Command: QUIT
```

 Save TEST.MNU for the next chapter.

 Copy TEST.MNU to MY04.MNU. Save TEST.MNU for the next chapter.

Make sure all menu items begin with ^C^C^C. We recommend spacing the closing label brackets] over to column 11. It is easier to read menu items if they are aligned, and if there is at least one space at the end of the label. The extra spaces have no other effect since they are within the label. Adding the three empty [] lines makes them appear on the screen menu as blank spaces separating items.

Adding blank lines containing only a solitary + plus sign, like the + before [1stEndPt] tells AutoCAD to continue to the next line. This + has no effect on the menu. While this is not the intended use of the +, it does make the file easier to read by providing a visual break. AutoCAD ignores the solitary +.

Practice with Macros

These aren't perfect macros. You will make more efficient, flexible and dependable macros with help from AutoLISP, but these macros show that you need to know AutoCAD's command and macro syntax to write good macros.

It also helps to pay attention to the graphic structure of entities, their on-screen behavior, their start points and endpoints that you can use as Osnappable points. Your macro and AutoLISP dependability will benefit. Play a little. It loosens up your imagination. The idea for the [OppENDpt] rotation trick came from play, making a worm walk across the screen!

Tips And Techniques

Here are some tips and techniques for making menu macros:

❑ Use Undo. Any error in a menu item crashes the rest of the item, but Undo cleans it up.

❑ Give all menu items [LABELS], including blank [] lines.

❑ A <SPACE> acts like a <RETURN>, except in text entry. Every one counts, so count them carefully.

❑ Never use two <SPACE>s in a row. You can't see them to count them. Use one <SPACE> and then ";" semicolon <RETURN>s for subsequent ones.

- ❏ The "\" always pauses to take input from a user, so you can't use it in a path name. Use "/" instead.
- ❏ Each "\" pauses for exactly one item of input. Count them carefully.
- ❏ The menu item resumes as soon as the "\" gets any form of input. It doesn't care whether it's correct or not.
- ❏ Look for tricks that you have picked up using AutoCAD, like the sequence "LINE;^C" that resets the last point "@" to the end of the last line drawn. It is great for Inserting Blocks at the end of a line.
- ❏ Use the MULTIPLE command modifier for simple single command macros.
- ❏ Use an asterisk "*^C^C^C" at the beginning of macros for automatic repeating. You also can use "*^X."
- ❏ Start new macros with three ^C^C^Cs to cancel the previous command.
- ❏ The Select command automatically pauses for a selection set. Use SELECT to make the set and then pass the set to other editing commands like COPY via the Previous option.
- ❏ Look for ways to use Osnaps as tools in macros, like using rotating Block Inserts to get correct angles.
- ❏ Make leaders using Plines. You can select them with one pick.
- ❏ Use AutoCAD's BOX selection mode to replace Window and Crossing options in your macros.
- ❏ The INSert key makes selections from the screen menu. Begin your macro labels with unique characters to make <INS> work effectively.

When you write complex menu items, your command syntax must be exact. Use the drawing editor interactively as a testbed so that you know what options and input parameters are expected in your macros. Write your sequences down. Make your mistakes in the drawing editor, not in your text editor.

Now that you have a few menu macros under your belt, let's look at how to put menus together.

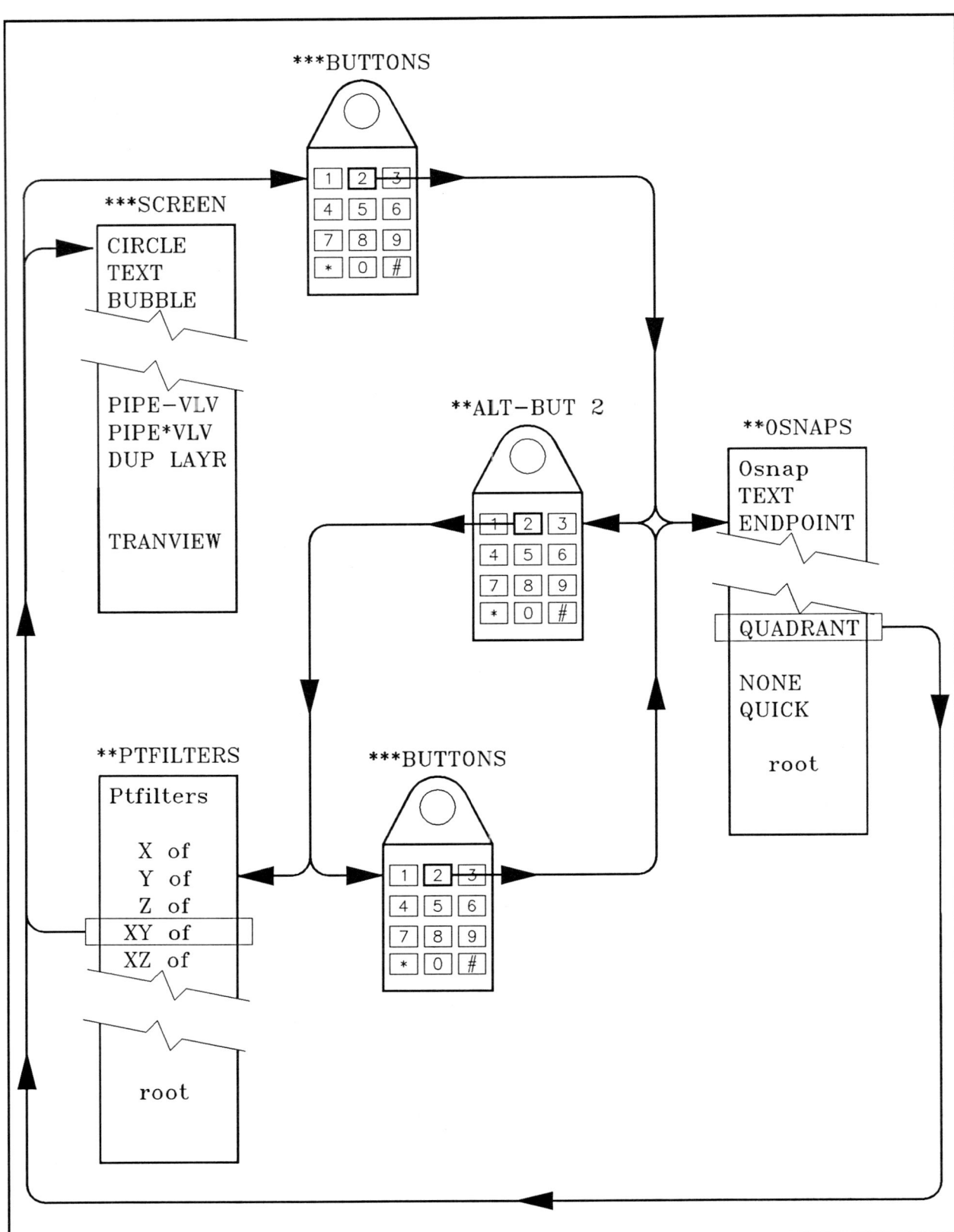

Buttons and Screen Toggling

CHAPTER 5

The Anatomy Of A Menu

So far, you have been using a single screen menu for your macros. As you add more commands to your menu you need a way to organize your menu system. This chapter shows you how to structure a menu system.

Pages of Macros

AutoCAD's menu works with "pages" of macros and commands. A MENU PAGE is simply a group of menu items that follow a special label. AutoCAD uses this label to find and activate the menu items as a set.

The Benefits of a Well Organized Menu

You gain several benefits from a well organized system of menu pages. Four important benefits are:

- Using menu pages helps you integrate your commands in a logical, intuitive way, informing your users of program settings as well as letting the pages control your program's operation.

- Using pages of symbols makes inserting symbols into the drawing faster and more accessible for the user.

- Designing your menu pages to group macros by application task instead of AutoCAD functions makes drawing straightforward and efficient.

- Designing your menu system pages to use screen, tablet, button, popup and icon menus takes advantage of the different features offered by each device.

How To Skills Checklist

Here are menu skills that you will acquire. You will learn how to:

- ❏ Put your symbols on screen menus.
- ❏ Use menu pages.
- ❏ Direct pages to different devices.
- ❏ Establish default pages for devices, using page and device labels.
- ❏ Make menu label toggles.
- ❏ Overlay partial pages of macros so that one macro replaces another.
- ❏ Use Popup menus.
- ❏ Use Icon menus.

5—2 Customizing AutoCAD

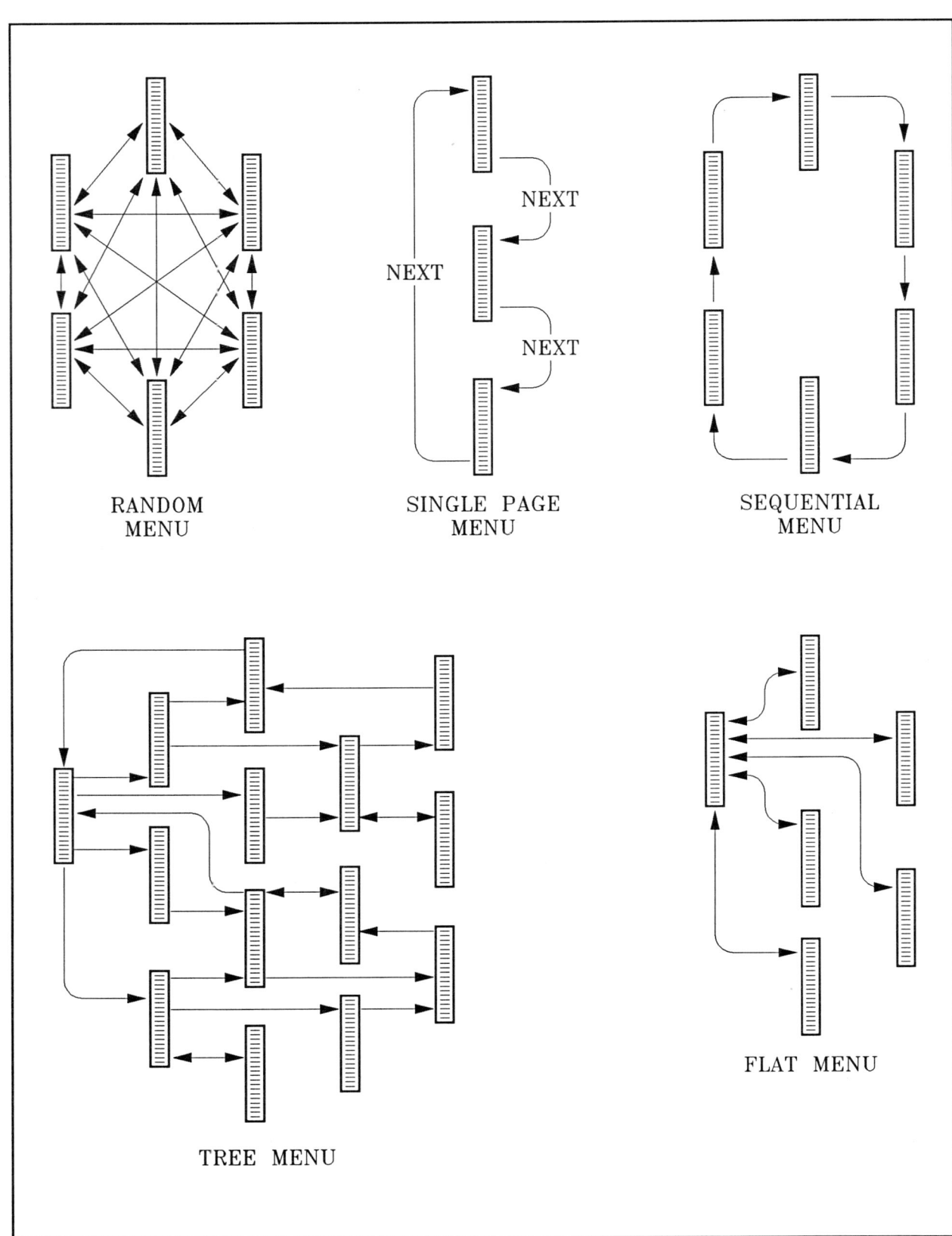

Types of Menus

Macros, AutoLISP Tools and Programs

MACROS
[SM VALVE] gives a variety of ways to insert, move and rotate nested symbols.
**FLOWSYMB is a full page of symbols for your basic library.
***BUTTONS is a device name, but it also serves as the default page name for a set of command entries, Osnaps, toggles and returns.
**OSNAPS is a menu page containing the Object Snap modes.
**PTFILTERS is a menu page with point filters.
**TRANVIEW is a page illustrating the use of transparent commands like 'ZOOM, 'PAN and 'VIEW.
**P1-SETVARS is a popup menu page. It contains system variable settings, like POINT MODE/SIZE and MIRROR of text.
**POINTS is an icon screen menu that presents AutoCAD's point options. The macros set the point mode and size.

Menu Layouts

AutoCAD provides flexibility in organizing your menu system. There are five basic ways to lay out menus: SINGLE PAGE, TREE, FLAT, SEQUENTIAL and RANDOM. A menu structure is created by dividing the menu up into "pages" identified by special labels. The five structures are created by controlling the page changes. You change pages by having menu items reference the special labels.

The SINGLE PAGE structure is the default, it uses simple NEXT paging.

The TREE structure is the most common layout. The TREE structure is useful when you need access to a library of interrelated macros. Each branch can apply to a single task or application. Branches for flow diagrams, schedules and site plans are good examples. Specific branches often incorporate FLAT and SEQUENTIAL menus.

The FLAT structure is rarely used for an entire menu system. It frequently is used in parts of a large menu. A FLAT structure works well to present specific command options, or subcommand parameters. A typical item in the page of a FLAT menu executes a command, pops up a list of choices, waits for the user to make a selection, then returns automatically to the calling page.

A SEQUENTIAL structure is used for applications that follow a strict pre-determined order, like an assembly process. You program the menu to enforce this order. You might use a SEQUENTIAL menu to force the order for drawing gears for a clock, or cogs for an engine. A page of a SEQUENTIAL menu can pause for any number of steps, but can only change to the next page containing instructions for the next step of the process.

It is easy to create a RANDOM structure menu system. Yes, random menus may be useful. They permit the change from one screen menu to another screen menu anywhere in the system. They are handy for notational menus, like special dimensioning routines, leader macros, text input macros, or any small frequently used set of items.

Which one should you use? Use them all! The menu tools you use to create them are the same. How you organize your menu pages makes the difference. To start working with menu pages, you need to put some more macros in your menu. Let's add a group of symbols.

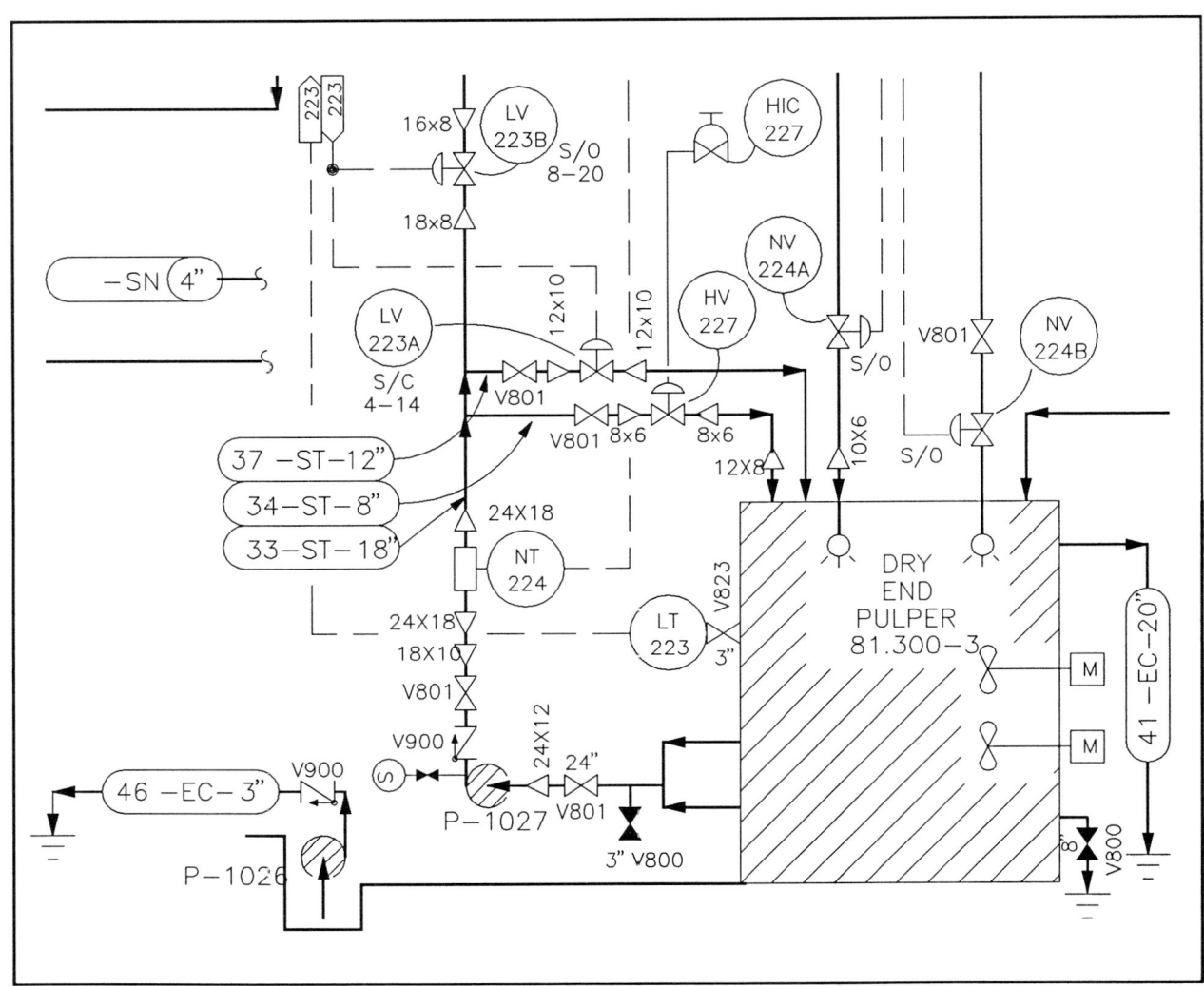

A Process Flow Diagram

Adding Some Symbols

In the following exercise you will create enough menu macros to fill two AutoCAD screens. In the process you will develop a small symbol library, making macros to insert the symbols. These symbols are process flow symbols, we got them from the Flow Diagram shown in the illustration. Don't put a narrow interpretation on the menu because it is made up of process flow symbols. The menu procedures that you will use are the same for any application that you choose to develop.

Adding Some Symbols 5—5

SYMBOL	FILE NAMES	SCREEN LABELS
	FLOWSMVL	SM VALVE
	FLOWARRO	ARROW
	FLOWPUMP	PUMP
	FLOWGATO	GATEOPEN
	FLOWXMTR	METER
	FLOWDRAN	DRAIN
	FLOWAGIT	AGITATOR
	FLOWMOTR	MOTOR
	FLOWCHKV	CHECK V
	FLOWSHOW	SHOWER
	FLOWDOT	DOT
	FLOWGATC	GATECLOS
	FLOWRED	REDUCER
	FLOWINC	INCREASE
	FLOWSAMP	SAMPLER

Symbols, Files and Labels

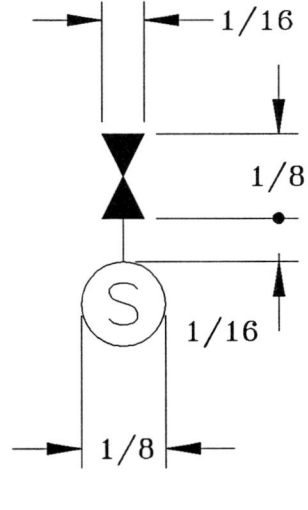

FLOWSMVL

A Small Valve

The symbols that you need are shown in the Symbols, Files and Labels drawing. If you have the CA DISK, you already have a set of symbol drawings. Otherwise, you need to create a set of symbols in a single "scratch" drawing. Draw and Wblock each symbol to a filename. Scale the dimensions. The file names and screen labels for the symbols are shown in the illustration drawing. The first symbol is shown in the drawing called Small Valve.

Let's look at the steps involved in doing the first symbol.

Adding Symbols to a Menu

Enter selection:	Begin a NEW drawing named CA05.
Command: **MENU**	Load TEST, the menu from the previous chapter.

 Skip to the menu writing part of the exercise where you add [SM VALVE].

 Draw and Wblock the valve as shown:

Command: **LIMITS**	0,0 to 36,24. A 1:1 schematic sheet size.
Command: **ZOOM**	Zoom Window 0,0 to 2,2. Get close for these small symbols.
Command: **GRID**	Set to 1/2.
Command: **SNAP**	Set to 1/32.
Command: **AXIS**	Reset to 1/4, if you use Axis.
Command: **LAYER**	You should be on Layer 0. If not, check and fix CA-ACAD.DWG.
Command: **STYLE**	Redefine STD1-16 to set Height to 1/16 for 1:1 scale.
Command: **SOLID**	Deliberately draw a "bow tie" by picking points clockwise.
Command: **LINE**	Then the vertical Line.
Command: **CIRCLE**	It's easiest with a two point circle, starting at the line.
Command: **TEXT**	Use Style STD1-16, Middle justified, and enter an "S".
Command: **WBLOCK**	To file FLOWSMVL, insertion basepoint top of the "bow tie".
Command: **INSERT**	Insert FLOWSMVL at 1:1 to test.
Command: **ERASE**	Then Erase it.
Command: **MNU**	Edit TEST.MNU from Chapter 4, even if you have CA DISK. If you skipped chapter 4, Copy CA04.MNU to TEST.MNU to continue this sequence).

Add the following [SM VALVE] to the end of your menu. Add the [FlowSymb] line as an identifying label. Use a [--------] line below [FlowSymb] to set the label off from [SM VALVE].

```
[DUP LAYR ]^C^C^CSELECT \COPY P ;0,0 ;CHANGE P ;P LA \;
[]
[FlowSymb ]
[-------- ]
[SM VALVE ]^C^C^CINSERT FLOWSMVL \1 ;\COPY L ;M @
```

Save, exit, and reload the TEST menu.

Select **[SM VALVE]**	Test it.
Command: INSERT FLOWSMVL	The Insert command scrolls by as you pick points.
Command: COPY	Copy scrolls by, selecting Last and leaves you in:
Second point of displacement:	Pick any number of points. It is Copy Multiple.

Here is how the [SM-VALVE] macro works. The first "\" pauses for the insertion point, then the macro sets X,Y scale to 1 with "1 ;." The second "\" pauses for rotation, then "COPY L ;M @" automatically copies it multiple times.

Your screen should show inserted valves.

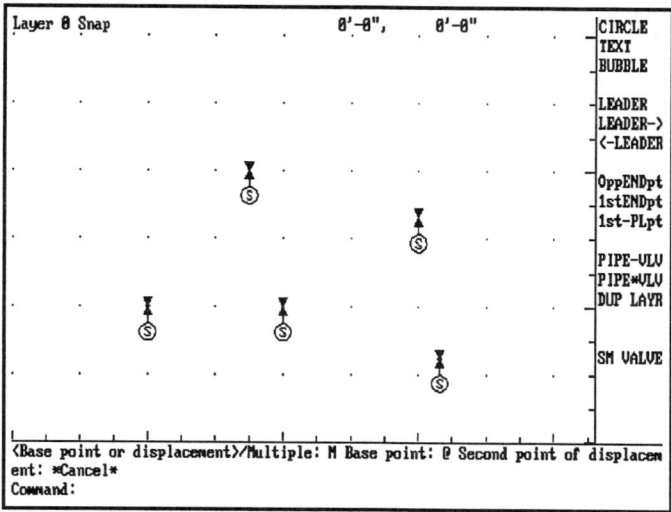

Inserted Valves

SM VALVE is a typical Symbol, using an Insert macro. You need to create a full page of similar symbol macros for the following menu.

Add the lines shown below to your TEST.MNU by duplicating the first line 14 more times. Then, go back and change the [LABEL]s and symbol Block names. Enter them in alphabetical order. This order places the original [SM VALVE] last.

Command: **MNU** Even if you have the CA DISK, edit the TEST.MNU to read:

```
[]
[FlowSymb ]
[-------- ]
[AGITATOR ]^C^C^CINSERT FLOWAGIT \1 ;\COPY L ;M @
[ARROW    ]^C^C^CINSERT FLOWARRO \1 ;\COPY L ;M @
[CHECK V  ]^C^C^CINSERT FLOWCHKV \1 ;\COPY L ;M @
[DOT      ]^C^C^CINSERT FLOWDOT  \1 ;\COPY L ;M @
[DRAIN    ]^C^C^CINSERT FLOWDRAN \1 ;\COPY L ;M @
[GATECLOS ]^C^C^CINSERT FLOWGATC \1 ;\COPY L ;M @
[GATEOPEN ]^C^C^CINSERT FLOWGATO \1 ;\COPY L ;M @
[INCREASE ]^C^C^CINSERT FLOWINC  \1 ;\COPY L ;M @
[METER    ]^C^C^CINSERT FLOWXMTR \1 ;\COPY L ;M @
[MOTOR    ]^C^C^CINSERT FLOWMOTR \1 ;\COPY L ;M @
[PUMP     ]^C^C^CINSERT FLOWPUMP \1 ;\COPY L ;M @
[REDUCE   ]^C^C^CINSERT FLOWRED  \1 ;\COPY L ;M @
[SAMPLER  ]^C^C^CINSERT FLOWSAMP \1 ;\COPY L ;M @
[SHOWER   ]^C^C^CINSERT FLOWSHOW \1 ;\COPY L ;M @
[SM VALVE ]^C^C^CINSERT FLOWSMVL \1 ;\COPY L ;M @
[]
```

Save, exit, and reload the menu.

Select **[NEXT]** Select [NEXT] at the bottom of the screen menus a few times.

5—8 Customizing AutoCAD

Your screen should look similar to:

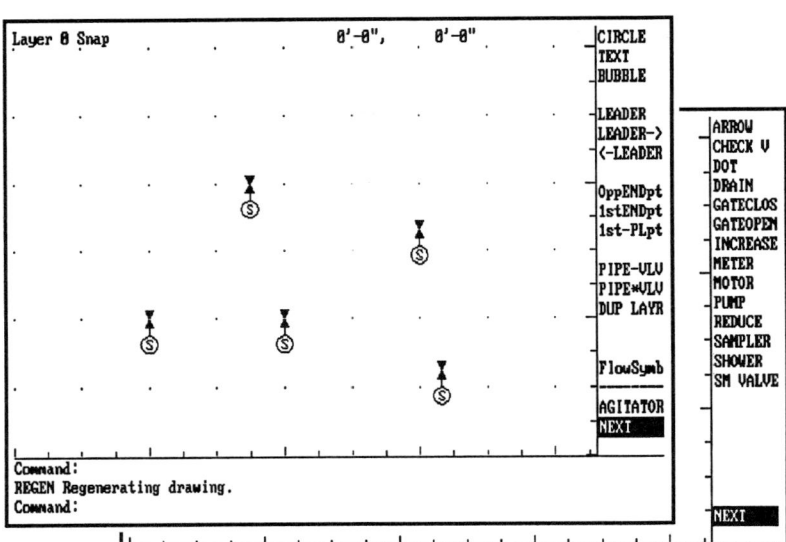

Screen Menu with NEXT Paging

You won't see all of the entries immediately if your menu exceeds one page. Instead you will see a "NEXT" at the bottom of the screen. AutoCAD automatically pages the screen menu for you, creating a NEXT item.

When you select the NEXT option another page of the macros appears. NEXT is a continuous loop with any number of pages. When you reach the last page, NEXT brings you to the first page. Notice that AutoCAD is smart enough to make the NEXTs come up in the same place on each page.

NOTE. There are 35 items in the menu. Most video displays support 20 to 30 items per page, but a few video displays support as many as 80 items per screen. If your display shows the entire menu, the next exercise will force the menu over into a second page.

You can control where a new menu page begins by using a leading asterisk "*" before a menu [LABEL].

Command: **MNU**   Edit the menu to force a new page. Add an * to [FlowSymb]:

[ ]
*[FlowSymb]
[--------]

Save, exit, and reload. All Flow items are forced to the 2nd page.

Select **[NEXT]**   Test NEXT.

Your screen should show a prettier screen menu.

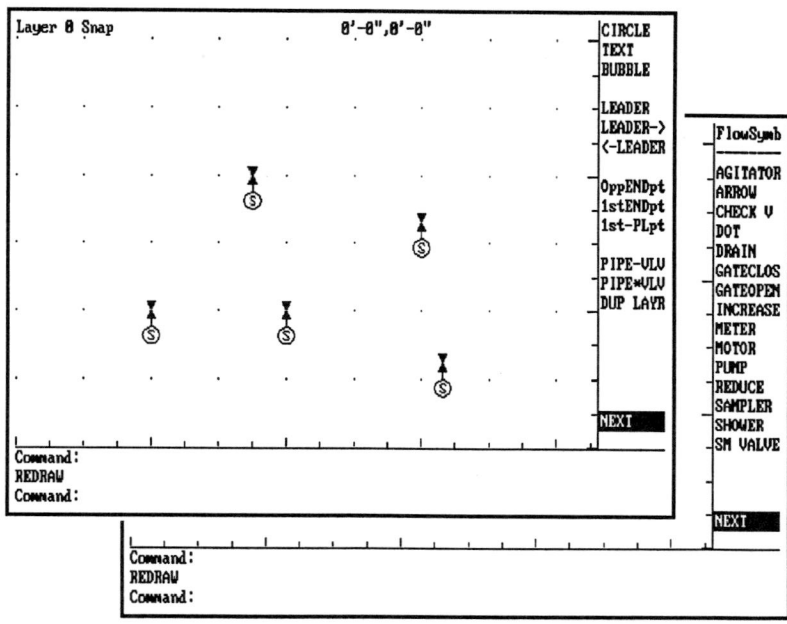

Prettier Screen Menu with NEXT Paging

Single page menus serve simple applications. You can use them for "on-the-fly" quick, temporary menus. Most applications need a more extensive, structured menu.

Naming Menu Pages

You can break menus into named pages of commands and macros by labeling each page with a unique name. You distinguish this page name or label from other lines in a menu by using two leading asterisks "**." The format is **name, where "name" is any name you like.

Let's break out the flow symbols as a menu page.

Labeling and Switching Pages

Command: **MNU**   Edit TEST.MNU

Insert the **FLOWSYMB page name above the [FlowSymb] label.

**FLOWSYMB
```
*[FlowSymb ]
```

Save, exit, and reload the menu.

Now you can't get to the symbol page, and the automatic [NEXT] is gone. To get to the symbol page, you need to tell AutoCAD which page to load on the device. You do this via the "$" menu special character code. Since you are working with a screen menu, you use a $S. The format is $S=name, where "name" is the page you are going to load. To make the menu switch to the FLOWSYMB page, add a line to the bottom of your first page.

When using **page names, single "*" asterisks, like *[FlowSymb] are ignored, but harmless. You can remove it from the file if you like.

Command: **MNU**   Edit TEST.MNU.

Add [FLOWSYMB] and [] blank lines to the first page. Count the [labels] to make sure [FLOWSYMB] is the 20th labeled item:

```
[]
[FLOWSYMB ]$S=FLOWSYMB
**FLOWSYMB
```

Save, exit, and reload the menu.

Select **[FLOWSYMB]**   This should flip to the symbol page.

Now, you can't get back to the first page menu, unless you reload the menu. You could add a **NAME to the top of the first page and a $S=NAME at the bottom of the second page. However, AutoCAD offers a better way to get to any previous menu page. You use a $S= without a name. AutoCAD saves the last used eight menu page names. You can step back through each one. At the bottom of your **FLOWSYMB page, add a [] as a "blank," and a [Last] label. You can call it anything, but we recommend using the label [Last].

Command: **MNU**   Edit TEST.MNU.
Count [labels] and make sure [Last] is the 19th labeled item.

```
[SM VALVE ]^C^C^CLAYER S OBJ02 ;INSERT FLOWSMVL \1 ;\COPY L ;M @
[]
[  Last   ]$S=
```

Save, exit, and reload the menu.
Select **[FLOWSYMB] then [LAST]**

Now you can flip back and forth, but the [FLOWSYMB] label remains at the bottom of the screen when you flip to the second page. This is not harmful, but it points out that menu items, once activated, remain active until they are displaced. With some thought, you can use this feature to design sophisticated menus, replacing only portions of your menu pages at a time. For now, just add another [] to the second page to force the [LAST] label to overwrite the [FLOWSYMB] label.

NOTE. Do not use a semicolon ";" as a <RETURN> following a page name call. In some cases, like "$S=NAME;+", AutoCAD will not recognize it as a <RETURN>. Use a <SPACE>, like "$S=NAME +".

```
Command: MNU                    Edit TEST.MNU, add []:

[]
[]
[. Last     ]$S=
```

Save, exit, and reload the menu. Test them.

Now you can select the Last menu page and get cleanly back to the original menu. At this point, your TEST.MNU should look about like the TEST.MNU Item drawing. We only show the [LABELS] and paging items.

Now, having made page changes, let's turn to look at a complete menu structure using different devices.

Structuring the Menu

Your menu structure is determined by your page structure and what devices you assign those pages to. So far, you have used the SCREEN device. Different devices have strengths and weaknesses. Your menu should take advantage of the strengths offered by each device. Let's assign menu items to the different devices so that you can see how the devices work.

Menu Devices

The devices available to you are: SCREEN, TABLET, BUTTON, POPUP, ICON, and AUXILIARY. The SCREEN device is the default device. If you do not assign a device name in a menu file, AutoCAD assigns the macros to the screen. The BUTTONS menu assigns macros to mouse or digitizer puck buttons. The TABLET menus assigns commands to your digitizer tablet. An AUXILIARY device is another electronic box that plugs into your computer. There are only one or two on the market, and they require an ADI driver. You can define POPUP and ICON menus with Release 9 and later versions of AutoCAD.

Assigning Pages to Devices

AutoCAD treats devices much the same as page names. Each device has a unique name identified by three consecutive asterisks *** followed by the name of the device, like ***SCREEN. AutoCAD supports 18 such devices in Release 9 and later versions. The TABLET is four devices, and POPUPS are ten devices. AutoCAD supports seven devices in earlier versions.

You can send any menu page to any device. You can select a page from any device. You have seen how this works with $S= for the screen menu. The next exercises show you how to direct menu pages to other devices, starting with the TABLET.

```
[CIRCLE   ]
[TEXT     ]
[BUBBLE   ]
[]
[LEADER   ]
[LEADER-> ]
[<-LEADER ]
[]
[OppENDpt ]
 +
[1stENDpt ]
[1st-PLpt ]
[]
[PIPE-VLV ]
 +
[PIPE*VLV ]
 +
[DUP LAYR ]
 +
[]
[]
[]
[FLOWSYMB ]$S=FLOWSYMB
**FLOWSYMB
[FlowSymb ]
[-------- ]
[AGITATOR ]
[ARROW    ]
[CHECK V  ]
[DOT      ]
[DRAIN    ]
[GATECLOS ]
[GATEOPEN ]
[INCREASE ]
[METER    ]
[MOTOR    ]
[PUMP     ]
[REDUCER  ]
[SAMPLER  ]
[SHOWER   ]
[SM VALVE ]
[]
[ LAST    ]$S=
```

TEST.MNU Item and
Page Labels

5—12 Customizing AutoCAD

Tablet Menu

We assume that you have a digitizer tablet. If you do not have a tablet, you won't have access to the tablet items. Just read this section.

Photocopy the Tablet Menu illustration and tape it on your tablet anywhere outside of the screen pointing area. Reconfigure for one tablet menu area.

The reconfiguration will affect only the ACADDG.OVL and ACAD.CFG files you copied into the \CA-ACAD directory. It won't change your normal AutoCAD setup.

Menu Device Names

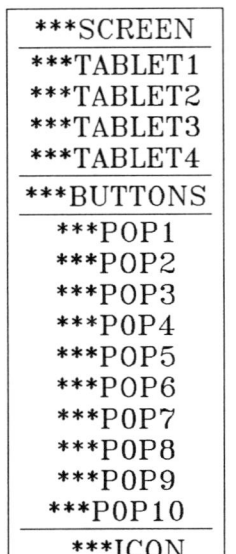

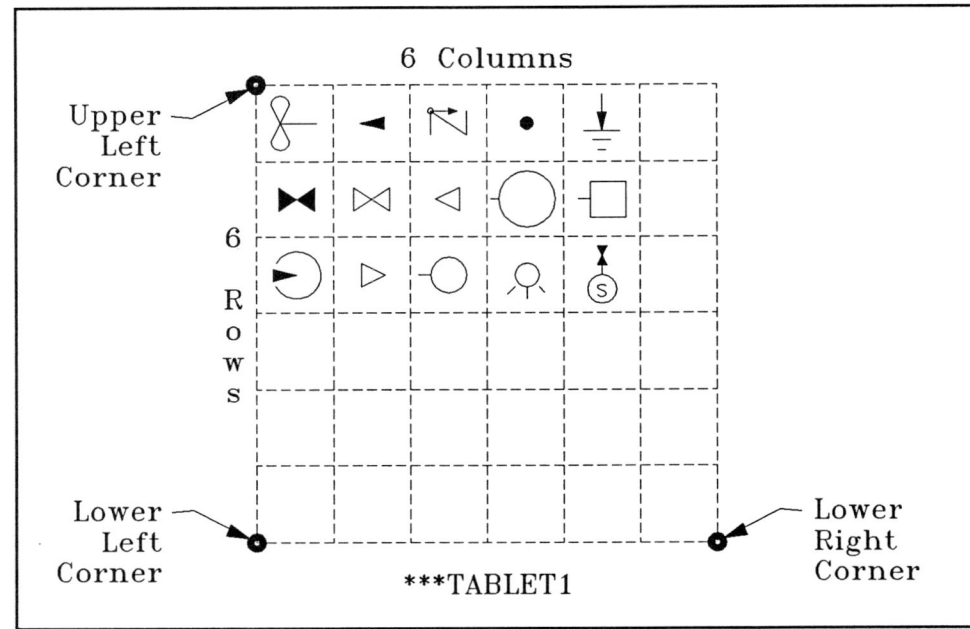

Tablet Menu

Assigning Menu Pages to a Tablet

```
Command: TABLET                                 Tape the TABLET MENU on. Reconfigure:
Option (ON/OFF/CAL/CFG): CFG
Enter number of tablet menus desired (0-4) <4>: 1
Digitize upper left corner of menu area 1:      Pick.
Digitize lower left corner of menu area 1:      Pick.
Digitize lower right corner of menu area 1:     Pick.
Enter the number of columns for menu area 1: 6
Enter the number of rows for menu area 1: 6
Do you want to respecify the screen pointing area? <N> Y   NO if you don't need to respecify.
Digitize lower left corner of screen pointing area:   Pick.
Digitize upper right corner of screen pointing area:  Pick.
```

If you have the CA DISK, you can stop typing. Examine TEST.MNU as we edit.

 COPY CA05.MNU to TEST.MNU

 COPY TEST.MNU to MY05A.MNU for backup, and continue editing TEST.MNU.

Command: **MNU**   Edit TEST.MNU.

Add a ***SCREEN device label as the first line of the file:

```
***SCREEN
[CIRCLE    ]^C^C^CMULTIPLE CIRCLE
```

Replace the [FLOWSYMB] line at the bottom of the first page with [].
Make sure you have four "[]" lines at the bottom for a total of 19 items.

Replace the old **FLOWSYMB page label with the ***TABLET1 device label.
Divide it into 3 groups of 5 items each and number them.
Delete the lines following [18] at the end of the TEST.MNU file:

```
***TABLET1
[1  AGITATOR ]^C^C^CINSERT FLOWAGIT \1 ;\COPY L ;M @
[2  ARROW    ]^C^C^CINSERT FLOWARRO \1 ;\COPY L ;M @
[3  CHECK V  ]^C^C^CINSERT FLOWCHKV \1 ;\COPY L ;M @
[4  DOT      ]^C^C^CINSERT FLOWDOT  \1 ;\COPY L ;M @
[5  DRAIN    ]^C^C^CINSERT FLOWDRAN \1 ;\COPY L ;M @
[6  ]
[7  GATECLOS ]^C^C^CINSERT FLOWGATC \1 ;\COPY L ;M @
[8  GATEOPEN ]^C^C^CINSERT FLOWGATO \1 ;\COPY L ;M @
[9  INCREASE ]^C^C^CINSERT FLOWINC  \1 ;\COPY L ;M @
[10 METER    ]^C^C^CINSERT FLOWXMTR \1 ;\COPY L ;M @
[11 MOTOR    ]^C^C^CINSERT FLOWMOTR \1 ;\COPY L ;M @
[12 ]
[13 PUMP     ]^C^C^CINSERT FLOWPUMP \1 ;\COPY L ;M @
[14 REDUCE   ]^C^C^CINSERT FLOWRED  \1 ;\COPY L ;M @
[15 SAMPLER  ]^C^C^CINSERT FLOWSAMP \1 ;\COPY L ;M @
[16 SHOWER   ]^C^C^CINSERT FLOWSHOW \1 ;\COPY L ;M @
[17 SM VALVE ]^C^C^CINSERT FLOWSMVL \1 ;\COPY L ;M @
[18 ]
```

Save, exit, and reload the menu.

Select **[17 SM VALVE]**   The 3rd tablet row, 5th column, box 17.
Command: INSERT Block name (or ?): FLOWSMVL   Executes the macro.

5—14   Customizing AutoCAD

Ten Button Cursor

The flow symbols are installed on the tablet. The first five boxes across each of the top three rows contain the symbols. Box 17 has the Block, FLOWSMVL. Since tablet locations depend on the file position of the menu items, we recommend that you always number and label items, even blank ones.

A ***BUTTONS Menu

Next, create and load some BUTTONS macros. The next exercise assumes that you have a ten button cursor that has a pick button and nine assignable buttons. If you have fewer than ten, or if you have a mouse, you won't be able to access all the macros. Choose which ones you want to include, changing the macro's menu order to suit your preference.

Making a ***BUTTONS Menu

Command: **MNU**   Edit TEST.MNU

   If you have the CA DISK, you need to temporarily change your [2 OSNAPS] button so the it reads like the [2 OSNAPS] line below.

If you don't have Release 9, modify the buttons shown as follows:

Change [3   ]AUTO \\   to   [3   ]CROSSING \
Change each double "\\" to a single "\"
Omit the button 9 item.

   Add a ***BUTTON label, and the 9 menu macros shown below, to the top of TEST.MNU:

```
***BUTTONS
[1 RETURN  ];
[2 OSNAPS  ]$S=OSNAPS
[3         ]AUTO \\
[4 SUSPENDS]\\
[5         ]NONE \\
[6         ]INT,ENDP \\
[7         ]INSERT \\
[8 LINE    ]^C^C^CLINE \
[9 MULTIPIK]\M NON @ NON @ NON @ NON @ NON @ NON @ NON @ ;\
```

Save, exit, and reload the menu.

Select **[Button 8]**   Draw some lines. The button enters LINE and first point.
Select [   ]   Draw other entities to test each button, except button 2.
                                         Undo and repeat for each.
Command: **ERASE**   Clean up.

Here is how the BUTTON macros work:

- Button 1 is a <RETURN>.

- Button 2 brings up an Osnap menu.

- Button 3 invokes the AUTO object selection mode.

- Button 4 picks a point, but adds another backslash. If you use it while another pending macro is waiting for input, it picks but lets the macro continue to wait.
- Buttons 5 through 7 are common Osnaps. Choose the ones you use most often. If you use Button 5, NONE, instead of the normal pick button, it overrides any running Osnap mode.
- Button 7 doubles as INS Osnap and an Insert command.
- Button 8 shows how to issue a command and simultaneously pick a point.
- Button 9 picks multiple items at one point with a single pick. It does not work with AutoCAD version 2.6.

Buttons 4 through 7, and 9, use a double backslash, \\, to keep any pending macros waiting.

Buttons are the most intimate and personal of the menu devices. Experiment to find what you like. You can't fit all the Osnaps on the buttons, so here is a comprehensive screenful for your menu.

Adding OSNAPs to the Screen Menu

Command: **MNU**

Add a page of OSNAPs just above the ***TABLET1 label. The last item returns to the default "root" screen:

```
**OSNAPS
[Osnaps   ]
[]
[ENDPOINT ]ENDP $S= $B=BUTTONS
[MIDPOINT ]MIDP $S= $B=BUTTONS
[INTERSCT ]INT $S= $B=BUTTONS
[PERPEND  ]PERP $S= $B=BUTTONS
[]
[NEAR     ]NEA $S= $B=BUTTONS
[INSERT'N ]INS $S= $B=BUTTONS
[NODE PT. ]NODE $S= $B=BUTTONS
[]
[CENTER   ]CEN $S= $B=BUTTONS
[TANGENT  ]TAN $S= $B=BUTTONS
[QUADRANT ]QUA $S= $B=BUTTONS
[]
[NONE     ]NON $S= $B=BUTTONS
[QUICK    ]QUICK,\
[]
[  root   ]$S=SCREEN $B=BUTTONS
```

Save, Exit and reload your menu.

Select **[Button 2]**   It calls a screen menu of the OSNAPS.
Select **[Button 8]**   Start a line.
Select **[PERPEND]**   Pick a point.
                               It osnaps and jumps back to the 1st screen page with "$S=".

The $B=BUTTONS is not needed yet, but we put it in to save adding it later. Experiment with the button and screen menu, trying other buttons and Osnaps with AutoCAD commands.

Menu Page Order

The order of your ***devices and **pages is not important. You can place the pages in any order you want, but you need to be careful not to duplicate any **page or ***device names in a menu file. AutoCAD will only recognize the first name. Menu **pages are completely independent of the device section under which they are listed. You can send any **page to any device, like sending the **OSNAPS screen to the ***BUTTONS.

Then, what are ***device labels for? Device labels are really special reserved names that AutoCAD uses to default load the first page following the device label. When AutoCAD loads a menu, it looks for each device label. AutoCAD loads each device with those items that follow its label, up to the next ***device or **page.

Switching Between Menu Pages

The key to menu structure is not where you put the **pages, the key is which $ codes call what pages. Organization will help you keep them straight. AutoCAD doesn't care!

Each device has a code. The code is used to send menu pages to that device:

CODE	ACTION
$S=	Screen menus
$P1= thru $P10=	POPUP screen menus 1 thru 10 (Release 9)
$B=	Button menu
$T1= thru $T4=	Tablet areas 1 thru 4
$I=	Icon menus (Release 9)
$A1=	Aux Box 1

A name following the code, like $B=ANYNAME, will send the named page to that device. You've used $S=SCREEN to load ***SCREEN to the screen, but any menu item can load or restore any page to any device.

In the button menu exercise example, Button 2 loads the **OSNAPS menu to the screen. The Osnaps items use $S= (without a name) to restore the previous menu page to the screen. Previous page loading is device specific, reloading the previous page that was loaded to its device without affecting any other device.

Let's do a test called SHOW.MNU to prove menus are really device independent:

Testing Menu Independence

Command: **ED**   Use your PGP command ED.

Start a new file called SHOW.MNU. Even if you have the CA DISK, enter:

```
**PAGE1
LINE
ARC
CIRCLE
[  Last  ]$S=
***SCREEN
[TO SCR  ]$S=PAGE1
[TO BUT  ]$B=PAGE1
[TO TAB1 ]$T1=PAGE1
[SCR2BUT ]$B=SCREEN
```

Save and exit.

Command: **MENU**   Load the SHOW menu and experiment with it.

Experiment and test each entry. See where the commands from **PAGE1 become active. Prove to yourself that you can make a **page available in two places at once. Here is how the test works:

[TO SCR] sends pages to the screen. Use [Last] to restore.
[TO BUT] activates pages on the BUTTONS.
[TO TAB1] makes pages selectable from TABLET1, tablet area 1.
[SCR2BUT] sends the ***SCREEN page to the buttons! To prove it, pick button 1, now containing [TO SCR], and the screen changes. Button 2 again, now [TO BUT], returns **PAGE1 to the buttons.

Command: **MENU**   When you're done, reload TEST.MNU

If you duplicate menu **LABELS in sections intended for different devices, you will get the wrong page on one device or the other. AutoCAD will find the first occurring label regardless of device.

Suspended Commands and Menu Macros

The AutoCAD menu observes the same command suspension rules as the AutoCAD command processor. If AutoCAD receives filtering modes, like Osnaps and Point Filters, it applies the filter to the current command. For example, when AutoCAD is paused for point input by a backslash in the menu, it suspends filling a macro backslash until it receives more input. In other words, these filters suspend and do not "use up" backslashes present in the macro. Let's look at suspended commands by making another menu page, this one for the point filters.

Making a Point Filter Menu Page

Command: **MNU**

Add this page above the **OSNAPS line:

```
**PTFILTERS
[Ptfilter ]
[]
[   X of    ].X $S= $S= $B=BUTTONS
[   Y of    ].Y $S= $S= $B=BUTTONS
[   Z of    ].Z $S= $S= $B=BUTTONS
[  XY of    ].XY $S= $S= $B=BUTTONS
[  XZ of    ].XZ $S= $S= $B=BUTTONS
[  YZ of    ].YZ $S= $S= $B=BUTTONS
[]
[   0,0     ]0,0 $S= $S= $B=BUTTONS
[    @      ]@ $S= $S= $B=BUTTONS
[]
[]
[]
[]
[]
[]
[]
[  root     ]$S=SCREEN $B=BUTTONS
```

You'll have to wait to try these until you add an item to access them.

Use your 2 Button to make a menu page toggle. Currently, your 2 Button calls up the OSNAPS screen. Add a "$B=ALT-BUT" to modify the 2 Button itself, then add an **ALT-BUT page menu.

Even if you have the CA DISK, change [2 OSNAPS]... to read:

```
[2 OSNAPS ]$S=OSNAPS $B=ALT-BUT
```

If you don't have the CA DISK, add the one line **ALT-BUT page just above ***SCREEN:

```
**ALT-BUT 2
[PTFILTER alt. def. of 2 button]$S=PTFILTERS $B=BUTTONS
```

Save, exit, and reload the menu.

Notice that the **ALT-BUT label is followed by the number "2." The number 2 is not part of the label name. The number tells AutoCAD to begin loading that page at that numbered position of the device. In this case, the position is the 2 button. This method of offsetting the loading position of a menu page applies to all devices. Negative numbers start at the bottom of the ***device. Use this basic technique to toggle individual menu items, or to toggle partial menus. The button/screen toggle menu system is shown in the Toggle Menu drawing at the beginning of the chapter.

Your only access to the **PTFILTERS page is by picking the 2 button twice, first flipping the screen to **OSNAPS, then to **PTFILTERS. Since this flips two screen menus, you have to use two $S= items per line in your **PTFILTERS menu to return to the previous screen menu. When a point filter item is selected, the screen restores. The $B=BUTTONS insures that the buttons menu is restored.

```
Select  [8 LINE ]                Start a line.
Select  [Button 2]               $S=OSNAPS $B=ALT-BUT invisibly toggles the screen to the
                                 OSNAPS and toggles button 2 to $S=PTFILTERS $B=BUTTONS.
Select  [Button 2]               $S=PTFILTERS $B=BUTTONS toggles the screen to
                                 **PTFILTERS and restores button 2 to $S=OSNAPS $B=ALT-BUT.
Select  [X of]                   It restores the screen and original button 2, then:
From point: .X of                Pick X component of point.
(need Y):                        Pick Y component.
Select  [Button 2]               Do it several times to see it flip-flop.
```

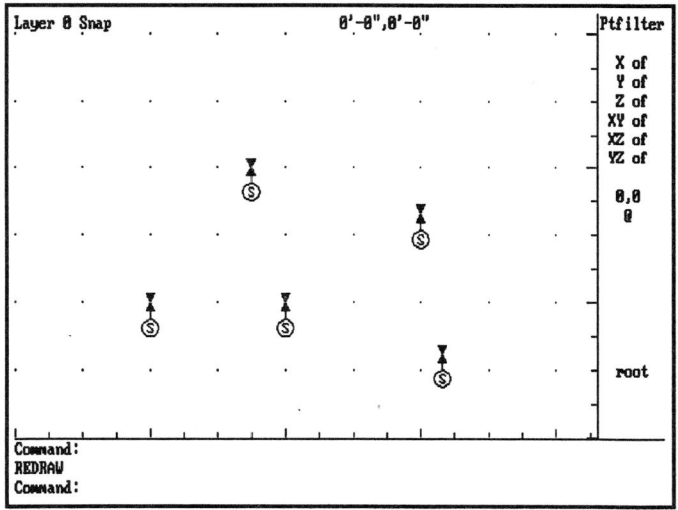

Point Filters Screen

Each time you hit the cursor button, the pages flip-flop between each other. The reason this works is that you can specify which screen line, box, or menu button each page will start on.

Macros and AutoCAD's Transparent Commands

As AutoCAD matures, the ability of commands to coexist improves. Page changes are always transparent. You may need to call transparent commands in the middle of other AutoCAD commands in your menu. Some common transparent commands are: 'GRAPHSCR, 'HELP, 'RESUME, 'SETVAR, 'TEXTSCR, 'ZOOM, 'PAN and 'VIEW. You makes these commands transparent by preceding the command with an ' apostrophe.

When a transparent command is called, AutoCAD interrupts the current command, processes the transparent command completely, then resumes the original command. If a menu item is currently suspended by a backslash \ and a transparent command is issued, the \ is not filled. The menu item remains suspended until after the transparent command is completed. To see how this works, make a screen menu that gives a selection of views in a drawing.

Making Transparent Commands Macros

Command: **MNU**   Add this page. Count to be sure you have 19 [labeled] items:

```
**TRANVIEW
[TranView ]
[]
[   All     ]'VIEW R A  $S=
[   A1      ]'VIEW R A1 $S=
[   A2      ]'VIEW R A2 $S=
[   A3      ]'VIEW R A3 $S=
[   A4      ]'VIEW R A4 $S=
[   B1      ]'VIEW R B1 $S=
[   B2      ]'VIEW R B2 $S=
[   B3      ]'VIEW R B3 $S=
[   B4      ]'VIEW R B4 $S=
[]
[]
[]
[]
[]
[]
[]
[   root    ]$S=SCREEN
```

Replace the 19th item at the bottom of your ***SCREEN page with:

`[TRANVIEW ]$S=TRANVIEW`

Save, exit, and reload. [TRANVIEW] should appear at the bottom.

Command: **VIEW**   Define several views, corresponding to the menu view names. View "A" for All was in your prototype drawing.

Command: **CIRCLE**   Start any command.

Select **[TRANVIEW]**   The TranView page should appear.
Select **[   ]**   Select a view name that you defined. It "zooms".
Select **[TRANVIEW] [All]**   It "zooms all".

NOTE. Do not begin transparent command macros with <^C>s. The <^C> cancels any pending command. If you issue a transparent command, like 'ZOOM, when no other command is pending, 'ZOOM functions as a normal ZOOM. The leading apostrophe causes no problem. AutoCAD does not permit transparent commands that cause a Regen.

PopUp Menus AutoCAD's Special Interface

PopUp Menus are specially implemented AutoCAD menus that "popup" to overlay a portion of your display. They are dynamic in nature. They only appear when they are opened. They disappear as soon as another action is taken. Unlike the screen menu, popups "rent" temporary space from the drawing.

Popup menus are devices in the menu file. There are 10 positions on the display for popups, so there are 10 device names: ***POP1 through ***POP10. Like other devices, the ***labels identify the defaults. You can access any number of **pages. Like the screen page code $S=, popups are loaded using $P1= through $P10= page codes. The ten positions are shown in the screen shot of PopUp Menu Locations.

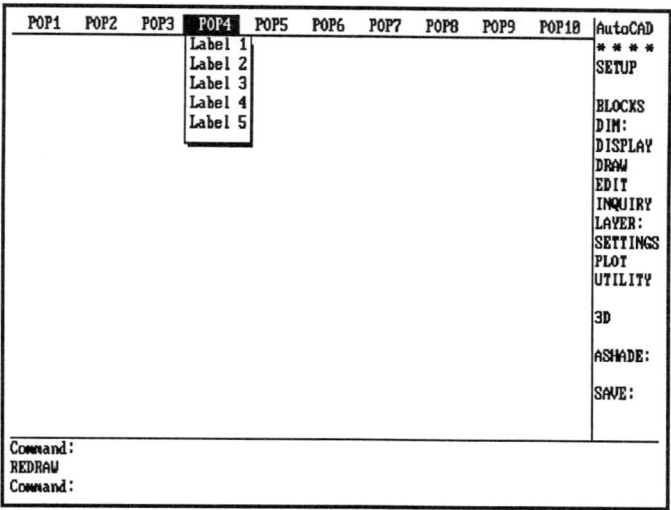

PopUp Menu Locations

Popups work differently than other devices. They need to be "opened" to be used. Using $P2=NAME loads **NAME to the P2 position, but it does not display it. You must use an asterisk "*" in combination with a $P code to tell AutoCAD to open a popup. $P2=* causes AutoCAD to display whatever is currently loaded to the POP2 device. Menu item toggling also works with popups, but you need to remember to explicitly open popups if you want them displayed.

You also can "open" a popup by moving the pointer up to the status area of the display. Then you pick it to "pull down" the popup currently loaded to that portion of the status line.

The following popup exercise transparently puts SETVARS on POP1. We are sure that you will think of many other useful applications for popups. Since you are using the leftmost pop device (POP1), you can use nearly the whole display screen width for your item labels. You can display more than 8 characters in the popup label on the status line.

Putting SETVARs on a PopUp Menu

```
Command: SETVAR                              If not sure, see if you can use popups.
Variable name or ?: POPUPS
POPUPS = 0 (read only)                       A "0" means no.
POPUPS = 1 (read only)                       A "1" means OK, then:

Command: MNU                                 Add this page at the bottom of TEST.MNU:

***POP1
**P1-SETVARS
[System Vars]
[APERTURE BOX SIZE - Target box height (entered in pixels)]'SETVAR APERTURE
[BLIPMODE - Toggles drawing blip marks off/on. 0=off, 1=on]'SETVAR BLIPMODE
[HIGHLIGHT - Toggles highlighting of entities. 0=off, 1=on]'SETVAR HIGHLIGHT
[POINT MODE/SIZE - Alters point display and size. ]$I=POINTS $I=*
[MIRROR TEXT - Toggles mirroring of text off/on. 0=off, 1=on]'SETVAR MIRRTEXT
[SKETCH WITH POLYLINE - Sketch creates lines or polylines. 0=lines, 1=pline]'SETVAR SKPOLY
[SNAP ROTATION ANGLE - Changes the snap rotation angle]'SETVAR SNAPANG
[SNAP INCREMENT - Changes snap increment. Give X,Y spacing.]'SETVAR SNAPUNIT
[PICK BOX SIZE - Object pick box size (entered in pixels)]'SETVAR PICKBOX
```

Save, Exit, and reload your menu. Play with the menu.

Move the cursor to highlight [System Vars] at the top left of the screen and:

```
Select [System Vars]                                           To pull it down.
Select [APERTURE BOX SIZE - Target box height (entered in pixels)]
'SETVAR                                                        The popup disappears.
Variable name or ? <POPUPS>: APERTURE
New value for APERTURE <10>: 30
```

Select something with an Osnap to see it, then UNDO or set Aperture back to 10.

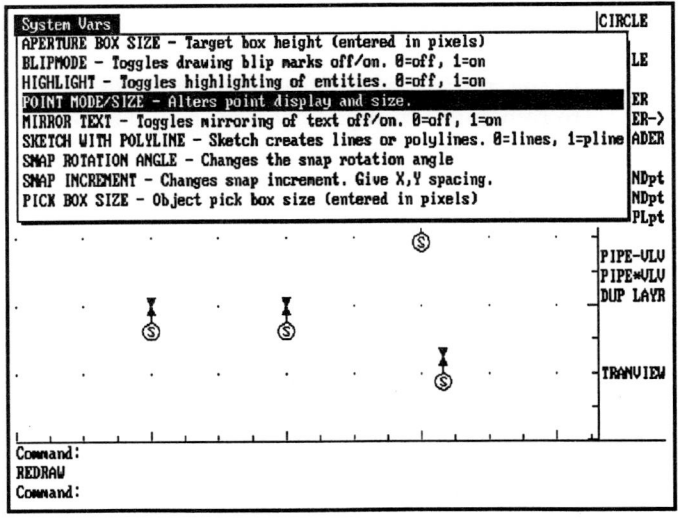

A Popup Menu Selection

ICON Menus   5—23

The PopUp menus selection called POINT MODE/SIZE item isn't ready yet. The $I= that it contains calls the ***ICONS device. Obviously, you need to create an ICONS menu.

ICON Menus

ICONS are menus using groups of AutoCAD slides as graphic labels. They show pictures that you can use to assist your users in choosing items.

Icons work like popups. You need to load the device with a $I=pagename and then pop it with $I=*. However, icons use the label [] differently. Icon labels supply AutoCAD with the name of the slide to show in the box. Each label corresponds to one box on the screen. These boxes are automatically arranged in groups of 4, 9 or 16. The label format is:

[library_file_name(individual_slide_name)]

AutoCAD evaluates these labels when the page is popped to the screen. If you put a space as the first character in the label, AutoCAD treats it like an ordinary label. AutoCAD displays any label text in that item's screen box instead of a slide. You can use this to show a "MORE" to the user, guiding the user to access additional pages of icons.

Icon menus use slide libraries that are created by AutoCAD's SLIDELIB.EXE program. Keep your images simple when making libraries. Small SLIDES display much faster than complex ones. The following exercise example uses a library of slides showing all the PDMODEs (Point Display MODEs). The slide library file is named POINT.SLB. Each slide is numerically named by its PDMODE. If you have the CA DISK, you have POINT.SLB.

Making Slide Libraries

   Read, but skip this sequence.

   Make Slides and a library file, POINT.SLB:

```
Command: ERASE                          Clean up your screen.
Command: POINT                          Draw a point in the center of your screen.
Command: 'SETVAR
Variable name or ?: PDSIZE              Point size.
New value for PDSIZE: -90               Makes it 90% of screen size.
```

Repeat the following to make the point Slides.

```
Command: SETVAR
Variable name or ? <PDSIZE>: PDMODE
New value for PDMODE : 1                Enter the number of the first Point Display Mode.
Command: REGEN                          Regen to force it to display.
Command: MSLIDE
Slide file <test>: 1                    Name the Slide for the Pdmode number used above.
```

Repeat the above for each of the twenty PDMODES listed below. Then:

Command: **ED**   Create a text file of the slide names. Call it TEMP.TXT.
Use no <SPACE>s. Be SURE to <RETURN> after the last line.

```
0
1
2
3
4
32
33
34
35
36
64
65
66
67
68
96
97
98
99
100
```

Save, exit, and reenter AutoCAD.

Command: **SHELL**   We assume SLIDELIB.EXE is in the \ACAD directory.

DOS Command: **\ACAD\SLIDELIB POINT <TEMP.TXT**
SLIDELIB 1.0   (7/29/87)
(C) Copyright 1987 Autodesk, Inc.
All Rights Reserved   If no errors, it created POINT.SLB.

Now that you have the slide library, you're ready to make an ICON menu.

Making an Icon Screen of PDMODEs

This menu is called by the [System Vars] popup menu. It graphically displays the PDMODE (Point Display MODEs) slides. When you select one, it SETVARs PDMODE, and prompts for PDSIZE (Point Display SIZE).

Command: **MNU**   Edit TEST.MNU and add to the bottom:

```
***ICON
**POINTS
[POINT DISPLAY OPTIONS]
[point(0)]'SETVAR PDMODE 0 'SETVAR PDSIZE
[point(1)]'SETVAR PDMODE 1 'SETVAR PDSIZE
[point(2)]'SETVAR PDMODE 2 'SETVAR PDSIZE
[point(3)]'SETVAR PDMODE 3 'SETVAR PDSIZE
[point(4)]'SETVAR PDMODE 4 'SETVAR PDSIZE
```

```
[point(32)]'SETVAR PDMODE 32 'SETVAR PDSIZE
[point(33)]'SETVAR PDMODE 33 'SETVAR PDSIZE
[point(34)]'SETVAR PDMODE 34 'SETVAR PDSIZE
[point(35)]'SETVAR PDMODE 35 'SETVAR PDSIZE
[point(36)]'SETVAR PDMODE 36 'SETVAR PDSIZE
[point(64)]'SETVAR PDMODE 64 'SETVAR PDSIZE
[point(65)]'SETVAR PDMODE 65 'SETVAR PDSIZE
[point(66)]'SETVAR PDMODE 66 'SETVAR PDSIZE
[point(67)]'SETVAR PDMODE 67 'SETVAR PDSIZE
[point(68)]'SETVAR PDMODE 68 'SETVAR PDSIZE
[ MORE]$I=MOREPTS $I=*
+
**MOREPTS
[ADDITIONAL POINT DISPLAY OPTIONS]
[point(96)]'SETVAR PDMODE 96 'SETVAR PDSIZE
[point(97)]'SETVAR PDMODE 97 'SETVAR PDSIZE
[point(98)]'SETVAR PDMODE 98 'SETVAR PDSIZE
[point(99)]'SETVAR PDMODE 99 'SETVAR PDSIZE
[point(100)]'SETVAR PDMODE 100 'SETVAR PDSIZE
[ PREVIOUS]$I=POINTS $I=*
[ ]
[ ]
[ ]
[ ]
```

Save, exit, and reload the menu.

```
Pulldown [System Vars]
Select   [POINT MODE/SIZE - Alters point display and size. ]
```

                                              The POINT DISPLAY OPTIONS icon menu displays.

```
Select  [point(0)]                          Highlight the top left little box and pick.
Command: 'SETVAR                            The menu disappears and sets the mode:
Variable name or ? <APERTURE>: PDMODE
New value for PDMODE <100>: 0
Command: 'SETVAR Variable name or ? <PDMODE>: PDSIZE
New value for PDSIZE <-90.0000>: -12   Enter size. Negative means % screen height.
```

Delete TEST.MNU.

Rename TEST.MNU to MY05.MNU.

Command: **QUIT**

Macros following ICON menu labels work just as you expect them to. The [MORE] line is an example of using a text label in an icon screen. The [MORE] macro loads the second page of points to the ICON device, then pops it. The extra [ ] blank labels in the second page force the page past nine items. Without the blanks, the page would display as 3 x 3 items and not match the 4 x 4 items in the main page.

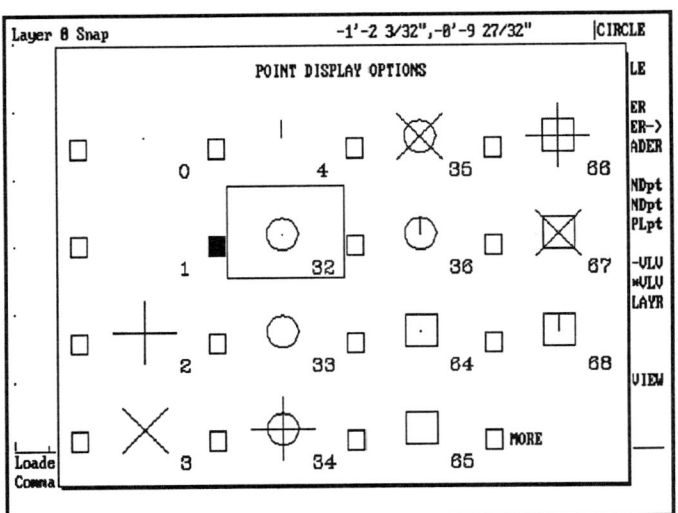

Icons Screen Called from a Macro

To develop a good menu, you need working macros and working symbols. Use your text editor interactively to develop and test your working macros and symbols. Your menu layout depends on what you want to achieve in your application. If you are uncertain about layout, use a simple tree structure. Use branches for different operations. How you set up and use your pages and devices are the keys to a successful menu.

Summary Tips and Techniques for Menu Devices

Selecting and using different menu devices depends on the dynamics that you want to achieve in your application. Here are some tips for using different devices:

- ❑ The SCREEN device is the default. The screen menu is convenient. It is right in your line of vision. The screen menu offers dynamic flipping of visible "pages" with definable [LABELS]. Screen menus are unlimited in size. You can easily modify them "on-the-fly." Use screen menus for logical organization and to order your work flow.

- ❑ BUTTONS menus assign macros and menu items to mouse or digitizer puck buttons. Having user defined macros at the fingertips is one of AutoCAD's best features. Buttons are too often overlooked in customization. You can make a button simultaneously pick a macro and issue a coordinate point. Make your button menus personal, standard and static to get a "touch typing" efficiency. Put the commands that you use the most often on your buttons.

- ❑ The TABLET menu assigns commands to your digitizer tablet. You can divide it into four rectangular areas, each containing a different set of commands. Tablet menus offer the advantage of using template "icons" to visually indicate macros and commands. Use the Tablet menu for complex, application specific macros, or for general drafting routines that need visual cues. Tablet menus are static. If you change overlays very often you will make mistakes.

Users and developers often favor tablet menus because they add a tangible element to electronic menus. The AutoCAD "factory standard" Template has a small screen pointing area, and the entire menu uses only a 10x10 area regardless of your tablet's active area. If you plan to market your developed system, you probably need to follow the standard. If not, layout your own tablet menu to your best advantage.

❑ POPUP and ICON menus come with AutoCAD Release 9 and later releases. Popups are good for single-pick type menu actions. They are not suited for multiple picks from the same menu, since they disappear. Popups also are good for giving users a long description of a menu item, and to toggle status items. Use Icons for actions that require graphic clues. They are excellent as graphic reminders of your symbols.

❑ An AUXILIARY device is another electronic box plugged into your computer. It requires an ADI driver. Generally, these devices have about the same advantages as an "extra" digitizing TABLET without the ability to act as a pointing device.

❑ The KEYBOARD is not an AutoCAD menu device. But you should not neglect it when you design your menus. AutoCAD nearly always looks for input from the keyboard. You can assign commands to two or three key abbreviations using macro utilities like PRD+ and JOT!. The standard AutoCAD commands are static and so is the keyboard. You can access any command with a quick two-step. Once you learn the abbreviations, it is quick one-handed two-finger touch typing to access any command.

You have menu pages and devices to work with. The next step is to build a menu system. This book is built around a menu system. The next chapter helps you develop the system.

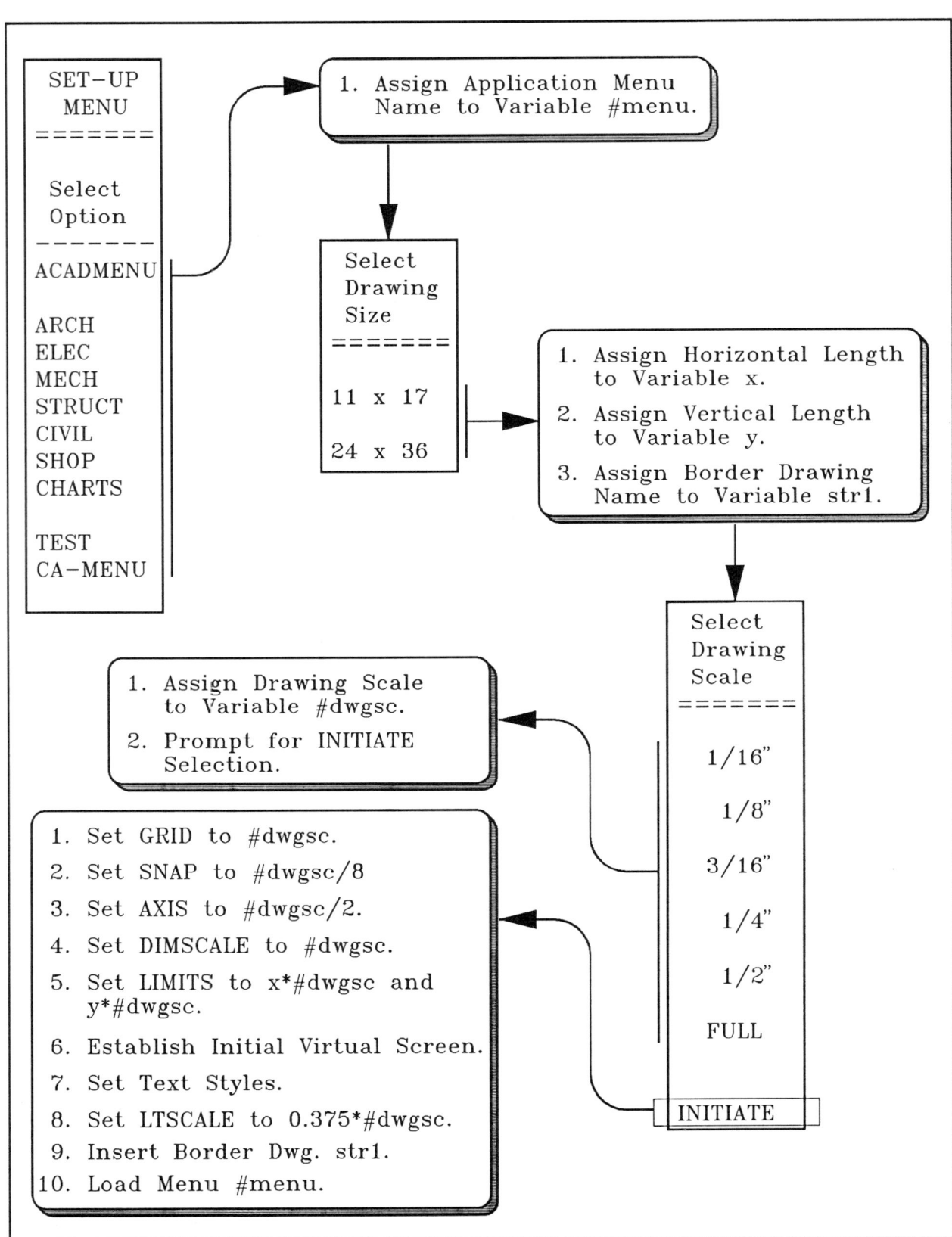

The SET-UP Menu System

CHAPTER 6

A Menu System

Writing menu macros is a simple task. Structuring your menu macros into a working system takes a little more time and effort. This chapter shows you how to make a setup menu, called CA-SETUP.MNU, and an applications menu, called CA-MENU.MNU. Structuring these two menus will provide you with a menu model for your own custom applications. You will save a subset of the CA-MENU, called MAIN.MNU, to use as a development menu template in the rest of the book.

Surveying Your Menu Needs

There is a common set of tasks that any custom menu performs. A custom menu system helps you:

- Standardize.
- Setup, format and fill in data for title sheets.
- Place text on the drawing.
- Locate, draw and insert components, assemblies and materials.
- Dimension components.
- Annotate components.

All these tasks are common to most drawing applications. If you examine these tasks, you will find that they divide into two groups: "general drafting tools" and "application specific tools." This chapter follows this basic two-part breakdown in organizing a menu.

The Benefits of Developing a Menu System

Tools to set up a drawing for border sheet size and drawing scale are basic to any menu system. If you work on different paper sizes at a variety of scales, you can automate your setup to deal efficiently with the different paper sizes and scales. You also can automate your layering scheme with menu macros. A setup menu lets you avoid the same old routine. You will find that you make fewer setup mistakes. Besides these common benefits, other benefits that you gain from developing a menu system are:

- Presenting screen menus of border, scale and applications helps tell your users what is available in your system.
- Making your menus standardize text keeps your drawings uniform.
- Controlling the access to screen menus ensures that you get your drawing setup in the way your programs expect to find it.

How To Skills Checklist

The exercises in this chapter show you how to:

- ❑ Make a setup menu for border sheets, scales, grids and snaps.
- ❑ Use Setvar to establish drawing settings.
- ❑ Automate loading of applications menus.
- ❑ Organize an application screen menu.
- ❑ Use screen menu page footers.
- ❑ Make a standard text screen menu.
- ❑ Use a subset of your applications menu as a development TEST.MNU.
- ❑ Use menu page access entry points to control your drawing environment.
- ❑ Use AutoLISP to help set up menus.

Macros, AutoLISP Tools and Programs

MACROS
CA-SETUP.MNU is a menu file containing routines to initialize a new drawing. The root page selects the application menu to load.
**DWG is a screen page to select drawing size.
**SCALE is a screen page of supported plot scales.
[INITIATE] is the main CA-SETUP.MNU macro. It sets up the drawing with the correct limits, snaps, grids, inserts the border sheet and loads the application menu.

CA-MENU.MNU is the applications menu.
**ROOT is the main applications page. It gives access to application and task specific branches.
**TEXT is a typical menu page with standard text sizes and styles.

AutoLISP TOOLS
The [INITIATE] macro contains AutoLISP commands that perform setup functions. We defer an AutoLISP discussion until the next chapter, but test the waters a bit now.

Automating a Setup Menu

Setup is common to all drawings. What does a setup menu do? It does as much as possible. A setup menu shifts control from the prototype drawing to the menu. It provides a selectable drawing environment. A good setup menu, like CA-SETUP, does the following:

- ■ Determines drawing scale.
- ■ Determines drawing limits.

The Setup Menu CA-SETUP.MNU

- Determines linetype scale.
- Establishes text styles.
- Sets GRID, SNAP, and AXIS.
- Sets DIMSCALE equal to #DWGSC, an AutoLISP variable to save it.
- Uses AutoLISP variables to save settings.
- Sets an initial screen menu.
- Inserts a border drawing.
- Does an expanded ZOOM to establish virtual screen size.
- Resets VIEW A (All).
- Loads application menus.

Once the drawing is setup, these settings usually do not change. We recommend that you keep your setup menu separate from your application menus. This menu separation lets you lead your user through setup to get to your application menu.

The Setup Menu CA-SETUP.MNU

The book's Setup Menu is a basic three page sequential screen menu. It starts the user with a menu showing the application choices available. Later, you can use the book's menu model to substitute your own labels and applications.

If you have the CA-DISK, you have the CA-SETUP.MNU. Examine CA-SETUP as you edit, and ignore those items that you haven't "created" yet.

A Blank Template Page

If you do not have the CA-SETUP.MNU, you need to create it. To create the menu, make a 20 line template page. The first line should contain **. Follow this line with 19 lines containing []s. This template is shown in the Template drawing. Copy and edit the template to write each screen page. The template insures that each menu page has exactly 19 items, so items align on the screen.

```
**
[ ]
[ ]
[ ]
[ ]
[ ]
[ ]
[ ]
[ ]
[ ]
[ ]
[ ]
[ ]
[ ]
[ ]
[ ]
[ ]
[ ]
[ ]
[ ]
```

Making CA-SETUP.MNU

Enter selection: Begin a NEW drawing named CA06.

   Copy CA-SETUP.MNU to TEST.MNU.

   Start a new file named TEST.MNU.

Command: **MNU**
Edit TEST.MNU, make your 20 line template, copy it twice. Make the first page read:

6—4 Customizing AutoCAD

```
***SCREEN
[ SET-UP  ]
[  MENU   ]
[========]
[]
[ Select ]
[ Option ]
[--------]
[ACADMENU]^C^C^C(setq #menu "ACAD");$S=DWG
[]
[ARCH    ]^C^C^C(prompt "Architectural Menu is not available yet!^M");
[ELEC    ]^C^C^C(prompt "Electrical Menu is not available yet!^M");
[MECH    ]^C^C^C(prompt "Mechanical Menu is not available yet!^M");
[STRUCT  ]^C^C^C(prompt "Structural Menu is not available yet!^M");
[CIVIL   ]^C^C^C(prompt "Civil Menu is not available yet!^M");
[SHOP    ]^C^C^C(prompt "Shop Drawing Menu is not available yet!^M");
[CHARTS  ]^C^C^C(prompt "Charts Menu is not available yet!^M");
[]
[TEST    ]^C^C^C(setq #menu "TEST");$S=DWG
[CA-MENU ]^C^C^C(setq #menu "CA-MENU");$S=DWG
```

Save and exit to AutoCAD.

Command: **MENU**   Reload the new TEST menu. Test all items.

Select **[CIVIL]**   Inactive items issue an AutoLISP prompt, like:
Command: (prompt "Civil Menu is not available yet!
1> ")
Civil Menu is not available yet!
Lisp returns: nil

Select **[TEST]**
Command: (setq #menu "TEST")   If you have the CA DISK, it also flips pages.
Lisp returns: "TEST"

The first page of the menu lets you select an application. Notice, there is a ^M inside the prompt strings. The menu interpreter reads the ^M as a special menu character and issues a <RETURN>. This <RETURN> cleans up the prompt, forcing the "nil" to a separate line.

When you complete the menu pages in this chapter, you will have three active selections: [ACADMENU], [CA-MENU], and [TEST]. Each menu item saves the name of the application menu selected, TEST, CA-ACAD, or the standard ACAD menu. These menu items use the AutoLISP function SETQ to save the menu name in an AutoLISP variable, #MENU. After saving the selected menu item name, the menu changes to a menu page of border sheet sizes.

The Border Sheet menu establishes the drawing size. It stores the drawing size along with the appropriate border Block name. The size and block are stored as the AutoLISP variables X, Y and STR1. Create the border sheet menu.

Making a Border Sheet Page

Command: **MNU**
Edit TEST.MNU and add the following below the first screen page.

```
**DWG-DOC
[This selection establishes the drawing size and saves it as variables.]
[The final limits will be determined by a scale factor in the next selection.]
**DWG
[ Select ]
[ Drawing]
[  Size  ]
[=======]
[]
[11 x 17 ](setq x 17.0) (setq y 11.0) (setq str1 "SHEET-B");$S=SCALE
[]
[24 x 36 ](setq x 36.0) (setq y 24.0) (setq str1 "SHEET-D");$S=SCALE
[]
[]
[]
[]
[]
[]
[]
[]
[]
[]
```

Save, exit, and reload the menu. Test both sizes. First:

```
Select [11 x 17]
Command: (setq x 17.0)  17.000000              All this scrolls by ...
Command: (setq y 11.0)  11.000000
Command: (setq str1 "SHEET-B")
Lisp returns: "SHEET-B"                        It flips pages if you have the CA DISK.
```

Although this exercise only provides two sheet sizes, B and D, you can add additional sheet sizes. If you have the CA DISK, you will find two border drawings: SHEET-B.DWG and SHEET-D.DWG in your CA-ACAD directory. Each border .DWG contains a complex border on Layer BORDER, and a simple outline reference border on Layer REF00. You can freeze the complex BORDER layer, and use the reference outline on Layer REF00. Please note that the book's drawing borders have very wide margins. These margins are meant to accommodate a variety of plotters. You can adjust the margins to fit your plotter.

Notice the **DWG-DOC page on the border sheet menu page. **DWG-DOC is a dummy page. It shows how to include non-functional items as additional documentation in your menu files.

6—6   Customizing AutoCAD

```
                                        36,24                                           17,11
         ┌───────────────────────┐              ┌───────────────────────┐
         │                       │              │                       │
         │                       │              │                       │
         │                       │              │                       │
         │                       │              │                       │
         │                       │              │                       │
         0,0                                    C,0
              Layer: BORDER                          Layer: BORDER
                                        36,24                                           17,11
                              34.75,22.25                                  16.375,10.375
         │                       │              │                       │
         │                       │              │                       │
         │                       │              │                       │
         │                       │              │                       │
         1.75,1.5                              .875,.625
         0,0                                    0,0
              Layer: REF00                         Layer: REF00
                SHEET-D                              SHEET-B
          Limits 0,0 & 36,24                   Limits 0,0 & 17,11
```

Border Drawings

Making Border Drawings

If you are not using the CA-DISK, you will have to create the border drawings. Create the simple outline reference border for both sheets.

   If you have the CA DISK, skip this sequence.

Command: **QUIT**

Enter selection: Begin a NEW drawing named SHEET-B=

Command: **LIMITS**   Set 17x11 Limits.
Command: **LAYER**   Make and Set layer REF00 current.
Command: **PLINE**   Draw the simple border illustrated.
Command: **END**

```
Enter selection:                        Begin a NEW drawing named SHEET-D=.
Repeat the same process for 36x24.

Enter selection:                        Begin a NEW drawing named CA06 again.
```

Selecting the drawing size from the border sheet menu switches the screen menu to the Scale Menu page. The scale menu page is the third page in the setup menu system. The user selects the drawing scale from this menu.

Making a Drawing Scale Menu

The Drawing Scale Menu contains a set of pre-calculated plot scale factors, Limits, LTscales, Text Style standards, and Dimension Variables settings.

Here is the table of the drawing scale factors. Later, you may wish to add or eliminate scale factors in your own menu.

DRAWING FACTORS									
DRAWING SCALE	SCALE FACTOR	HORZ LIMITS	VERT LIMITS	LTSCALE	TEXT HEIGHT				
					1/16"	3/32"	1/8"	3/16"	1/4"
FULL	1	3'	2'	.375	.0625	.09375	.125	.1875	.25
3"	4	12'	8'	1.5	.25	.375	.5	.75	1
1 1/2"	8	24'	16'	3	.5	.5625	1	1.5	2
1"	12	36'	24'	4.5	.75	1.125	1.5	2.25	3
3/4"	16	48'	32'	6	1	1.5	2	3	4
1/2"	24	72'	48'	9	1.5	2.25	3	4.5	6
3/8"	32	96'	64'	12	2	3	4	6	8
1/4"	48	144'	96'	18	3	4.5	6	9	12
3/16"	64	192'	128'	24	4	6	8	12	16
1/8"	96	288'	192'	36	6	9	12	18	24
1/16"	192	576'	384'	72	12	18	24	36	48

Drawing Factors

The setup menu closes with an [INITIATE] macro. [INITIATE] establishes the drawing settings by running several AutoCAD commands. It Zooms, Inserts the border, and then exits to the application menu that the user selected on the first menu page.

```
Command: MNU
```
Edit the TEST.MNU and add the following page:

```
**SCALE
[ Select ]
[ Drawing]
[ Scale  ]
[=======]
[]
[  1/16" ](setq #dwgsc 192.0) (prompt "Select [INITIATE]...^M");
[]
[   1/8" ](setq #dwgsc 96.0) (prompt "Select [INITIATE]...^M");
[]
[  3/16" ](setq #dwgsc 64.0) (prompt "Select [INITIATE]...^M");
[]
[   1/4" ](setq #dwgsc 48.0) (prompt "Select [INITIATE]...^M");
[]
[   1/2" ](setq #dwgsc 24.0) (prompt "Select [INITIATE]...^M");
[]
[  FULL  ](setq #dwgsc 1.0) (prompt "Select [INITIATE]...^M");
[]
[]
[INITIATE]^C^C^CGRID !#dwgsc ^GSNAP (/ #dwgsc 8) AXIS (/ #dwgsc 2) AXIS OFF;+
SETVAR DIMSCALE !#dwgsc LIMITS 0,0 (list (* x #dwgsc) (* y #dwgsc));+
REGENAUTO ON ZOOM W 0,0 (getvar "LIMMAX") ZOOM .75X REGENAUTO OFF;+
STYLE STANDARD SIMPLEX (* 0.125 #dwgsc) ;;;;;+
STYLE STD1-16 SIMPLEX (* 0.0625 #dwgsc) ;;;;;+
STYLE STD3-32 SIMPLEX (* 0.09375 #dwgsc) ;;;;;+
STYLE STD3-16 SIMPLEX (* 0.1875 #dwgsc) ;;;;;+
STYLE STD1-4 SIMPLEX (* 0.25 #dwgsc) ;;;;;+
STYLE STD1-8 SIMPLEX (* 0.125 #dwgsc) ;;;;;+
VIEW S A LTSCALE (* 0.375 #dwgsc) INSERT !str1 0,0 !#dwgsc ;;MENU !#menu ^GAXIS ON
```

Save, exit, and reload the menu.

Here is how the [INITIATE] macro works. First, [INITIATE] sets Grid to the drawing scale #DWGSC. One grid space on screen will equal one inch on the plot. It sets Snap to 1/8 #DWGSC, Axis to 1/2 of #DWGSC, and DIMSCALE to #DWGSC. Then, [INITIATE] resets the Limits, and Zooms out to allow a healthy margin around the drawing edges. Regenauto is toggled On to avoid a possible "About to regen -- proceed? Y" prompt. Next, [INITIATE] uses AutoLISP to recalculate the Limits, and feeds the value to Zoom with the LIMMIN and LIMMAX system variables. [INITIATE] turns Grid and Axis off to avoid multiple redraws during the initiation process. Grid and Axis are turned back on at the end of the macro.

After recalculating the Limits, [INITIATE] resets the standard text Styles for the text scale. An AutoLISP calculation provides the height of the text. Next, the [INITIATE] macro inserts the border sheet, saved as STR1, and scales it to the current scale. Last, the [INITIATE] macro loads a new menu file using the menu name that was selected by the user on the first menu page. This selected menu name is stored in the #MENU variable.

It is time to test the Setup Menu.

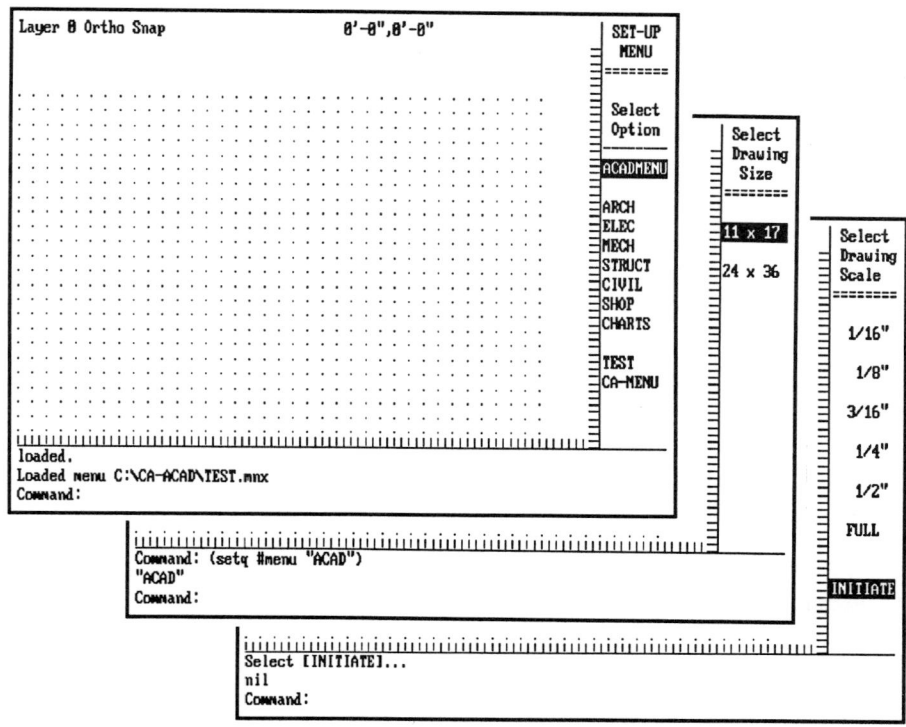

TEST.MNU with the CA-SETUP Menu

Testing the Setup Menu and Updating CA-PROTO.DWG

To make sure that there are no errors in the setup menu, test the menu by selecting each item on each menu page. You should be in drawing CA06 with the setup TEST.MNU loaded.

```
Select [ACADMENU]                                You see:
Command: (setq #menu "ACAD")                     Saves menu name.
Lisp returns: "ACAD"                             If OK, it flips to the Select Drawing Size screen menu.

Select [11 x 17]
Command: (setq x 17.C) 17.000000                 Saves X limit.
Command: (setq y 11.C) 11.000000                 Saves Y.
Command: (setq str1 "SHEET-B")                   Saves border name.
Lisp returns: "SHEET-B"                          If OK, you get a page of scales and [INITIATE].

Select [1/16"]
Command: (setq #dwgsc 192.0) 192.000000                     Saves plot scale.
Command: (prompt "Select [INITIATE]...
1> ")
Select [INITIATE]...                             Prompts you.
Lisp returns: nil

Select [INITIATE]                                GRID, SNAP, AXIS, SETVAR, LIMITS, REGENAUTO, STYLE,
                                                 VIEW, and LTSCALE scroll by. It INSERTs the 11x17 border.
                                                 MENU loads ACAD.MNU.
```

6—10   Customizing AutoCAD

```
Command: U                                  Type U to Undo and return to 1st page of setup TEST.MNU.
Select [TEST] [24 x 36] [FULL]
Select [INITIATE]                           It scrolls and reloads the TEST Menu.

Command: U                                  Undo it. Test the rest of the items.
Command: QUIT
```

If you have the CA-DISK, [INITIATE] will find the CA-MENU file when it goes to load the CA-MENU.MNU file.

If you don't have the CA DISK, you need to create the CA-MENU file.

Enter selection: Edit the EXISTING drawing CA-PROTO.

  Delete TEST.MNU.

  Rename TEST.MNU to CA-SETUP.MNU.

  When all the menus work, add CA-SETUP to the prototype drawing. Even if you have the CA DISK, add CA-SETUP to the prototype drawing.

```
Command: MENU                               Load the CA-SETUP menu.
Command: END
```

Controlling Access to Your Menu Pages

Knowing when to take and release control of your menu system to your users is part of the customization process. Your initial drawing setup relied purely on the prototype drawing to establish your drawing environment. By creating a Setup Menu and the start of an Applications Menu, you have shifted control of the drawing environment to these menus.

It helps to think of menus in terms of controlling the drawing environment. Understanding where to establish your control points and how to manage the drawing data, gives you control of the environment. Good menu controls still give users flexibility in operating the system. The intent is not to limit the user, but to make their drawing work easier and to make their drawings consistent. Your users will make fewer mistakes with layers, colors, linetypes and sizing inserted blocks if you manage these operations at control points in your application menus.

The control point that you have started with is switching menu pages. Using page changing is a natural custom control point. You can make your menu page display the current environment, or change the environment in preparation for the next set of commands. You can run a series of setup and checking commands in the stage between a user selecting a menu item and the presentation of a supporting screen page. Some control steps to consider are:

- Make or set a layer, and store the current one.
- Change the text style.
- Set the grid and snap.
- Load AutoLISP commands.
- Set AutoLISP variables.
- Save the drawing.
- Perform a Zoom.
- Restore a View.
- Display status information in screen menu boxes.
- Update a time log.

These are common tasks performed at control points. In the next menu, you will use page access to control a layer setting. You also will make sure the #DWGSC variable is reset if the drawing is ended or restarted. #DWGSC, like many of these controls, requires a little use of AutoLISP.

Application Menus

Your application specific menu pages will form the core of your menu system. They may not be the largest portion, but they are the most important. The application menu in this book **is the book**. The book's application menu pages are designed to show a range of applications rather than a single pure application, but the menu structure is typical of any application menu.

Making the Main Application Screen

The book's menu has a primary tree structure. The default **ROOT screen page provides access to the individual menu pages that you make as you work through the book.

Enter selection:

Begin a NEW drawing, again named CA06.DWG.
The CA-SETUP menu should automatically load. It was set in the previous exercise.

 Copy CA-MENU.MNU to TEST.MNU.

 Begin a new menu file TEST.MNU.

Command: **MNU**   Edit the new TEST.MNU.
Make a "**" and 19 line "[]" template as before, copy it twice, then edit the first page to:

```
***SCREEN
**ROOT
[CA-MENU ]
[]
[FLOWLINE]^C^C^CLAYER S OBJ01;;$S=FOOTER $S=FLOWLINE +
(if #dwgsc nil (setq #dwgsc (getvar "DIMSCALE")))
[PIPEFITS]
[--------]
[TEXT    ]
[SPECIALS]
[ANNOTATE]$S=FOOTER $S=ANNOT LAYER S ANN02;;+
(if #dwgsc nil (setq #dwgsc (getvar "DIMSCALE")))
[ISO DIMS]
[HATCH   ]
[--------]
[SETTINGS]$P1=P1-SETVARS $P1=*
[TITLEBLK]
[  3D   ]
[EXAMPLES]$S=FOOTER $S=EXAMPLES +
(if #dwgsc nil (setq #dwgsc (getvar "DIMSCALE")))
[]
[EDIT-LSP]LSP (load "TEST")
[EDIT-MNU]MNU MENU TEST
[]
```

Add one more short page:

```
**FOOTER -1
[  root  ]$S=SCREEN LAYER S 0 ;SNAP ON GRID ON ORTHO ON
```

Change the ** at the top of the template page into the **EXAMPLE** page, like:

```
**EXAMPLES
[Examples]
[]
```

Save, exit, and reload the TEST menu.

Select [EDIT-MNU]	Puts you in your text editor in TEST.MNU. Exit to AutoCAD.
Select [EDIT-LSP]	Puts text editor in a new TEST.LSP file. Exit back to AutoCAD.
Select [EXAMPLES]	It flips to the Examples page.
Command: (if #dwgsc nil (setq #dwgsc (getvar "DIMSCALE")))	
Lisp returns: nil	And sets #DWGSC, if needed.
Select [root]	It returns to the CA-MENU page.

The LAYER and "(if #dwgsc nil..." expressions in the [ANNOTATE] and [FLOWLINE] items, and the [EXAMPLES] settings are examples of page access control.

The **FOOTER is the menu's "footer." Each time a selection from the main menu is made, it places an entry at the **bottom** of the screen, letting you get back to the main menu. The -1 loads the page from the bottom of the device.

The pages have 19 items to allow room for the footer. We recommend holding the total screen menu length to 20 lines to make your menu compatible with all displays.

Making a Development Menu Template MAIN.MNU

You will use the current menu as your development template menu, called the MAIN.MNU, for the rest of the book. At the start of each chapter, you will copy MAIN.MNU to start a new TEST.MNU. If you have the CA DISK, you have MAIN.MNU so you can skip the following sequence. If you do not have the CA DISK, you need to name a copy of your current menu to MAIN.MNU.

 Skip this sequence.

 Copy TEST.MNU to MAIN.MNU.

Command: **ED**   Edit MAIN.MNU.

Change the **ROOT page [CA Menu] label to [CA00TEST], and strip the code following the [FLOWLINE], [ANNOTATE] and [SETTINGS] labels. Their pages aren't in the test menus.

```
***SCREEN
**ROOT
[CA00TEST]
[]
[FLOWLINE]
[PIPEFITS]
[--------]
[TEXT    ]
[SPECIALS]
[ANNOTATE]
[ISO DIMS]
[--------]
[SETTINGS]
```

The rest is unchanged.

Save MAIN.MNU and exit to AutoCAD.

Command: **MENU**   Load MAIN and test it.
Select [   ]   Test each item.
                                   Only the [EXAMPLES], [EDIT-MNU], [EDIT-LSP], and [root] items should function the same as they did before.

Despite the best of intentions, you are likely to build your menus in fits and starts, merging and reorganizing your menus as you go. We are going to do some menu reorganizing, merging the macros and menus that you created in the last chapter. This merging will create the book's CA-MENU.MNU.

6—14 Customizing AutoCAD

If you have the CA DISK, you already have CA-MENU.MNU. Read along but skip the following exercise.

If you don't have the CA DISK, you will get some practice appending and merging files with your text editor.

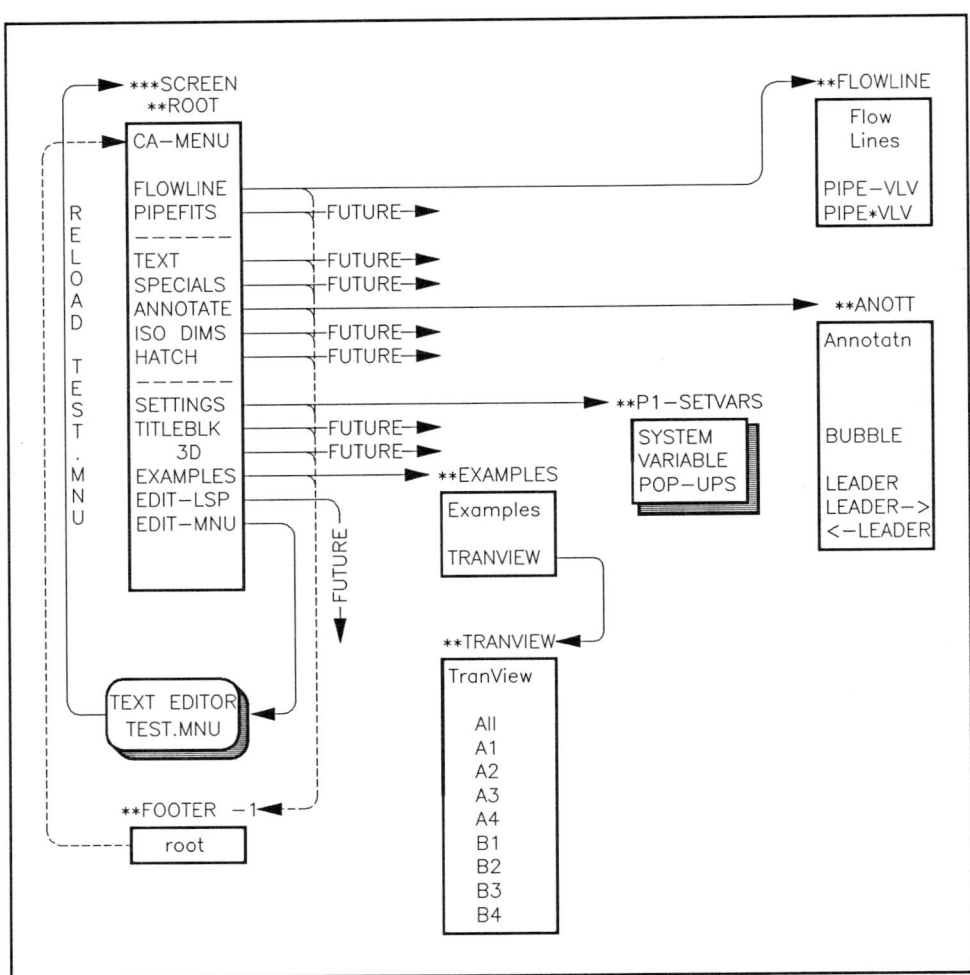

The CA-MENU Menu

Merging Menu Files

You don't have to type MNU to edit the TEST.MNU. The [EDIT-MNU] item you wrote in the last exercise automatically edits TEST.MNU and reloads it when you reenter AutoCAD.

Select **[EDIT-MNU]**

Edit TEST.MNU.

💾 Read, but skip the following.

Merge the following files.

Merge the entire MY05.MNU file onto the bottom of TEST.MNU, just above the ** [] template.

Only the names of the **pages and ***devices to be merged are shown.

```
***BUTTONS
**ALT-BUT 2
***SCREEN
**PTFILTERS
**OSNAPS
**TRANVIEW
***TABLET1
***POP1
**P1-SETVARS
***ICON
**POINTS
**MOREPTS
```

Save it.

You are almost done merging the menus, but you need to reorganize some of your macros. You also have two ***SCREEN labels. You will change most of the ***SCREEN menu from MY05 into a page of annotation and leader macros. Reorganize the old ***SCREEN into three pages, **ANNOT, **FLOWLINE, and **EXAMPLES, by adding and moving the lines shown below.

Change the "***SCREEN" label from the MY05 menu to read "**ANNOT".
Add a [Annotatn] label and a [] blank line to the top and add "blank" []s to make 19 items in page.
Replace the [CIRCLE], [TEXT], [OppENDpt], [1stENDpt], [1stPLpt] and [DUP LAYR] items with [] blanks.
The additions and changes are shown in **bold**:

```
**ANNOT
[Annotatn ]
[]
[]
[]
[BUBBLE   ]^C^C^CCIRCLE \24 SELECT L ;TEXT S STD3-16 M @ 0 \+
LINE @ \ ID MID,QUI @ TRIM P ;END,QUI @ ;
[]
[LEADER   ]^C^C^CORTHO ON PLINE \\^O\ INSERT C:/CA-DWGS/ARROW @ 18 ;MID @ ^O
[LEADER-> ]^C^C^CORTHO ON PLINE \W 0 0 \^O\W 0 6 L -18 W 0 0 ;
[<-LEADER ]^C^C^CPLINE \\;PEDIT L E M @ X X ERASE L ;PLINE @ W 0 6 L 18 W 0 0 L 1
[]
[]
```

The rest of the 19 items are [] blanks.

Move the [PIPE-VLV] and [PIPE*VLV] items (9 lines) from above to a new **FLOWLINE page. Add the new **FLOWLINE, [Flow] and [Lines] labels and [] blanks to make 19 items (not all [] blanks are shown).

```
**FLOWLINE
[   Flow   ]
[   Lines  ]
[]
```

```
[PIPE-VLV ]^C^C^CLAYER M OBJ02 C RED ;;+
LINE \\;INSERT /CA-ACAD/VALVE @ 16 ;DRAG \+
LINE ;\;INSERT ;@ 16 ;DRAG \+
LINE ;\;INSERT ;@ 16 ;DRAG \+
LINE ;\;INSERT ;@ 16 ;DRAG \+
LINE ;\;INSERT ;@ 16 ;DRAG
+
[PIPE*VLV ]*^C^C^CLAYER M OBJ02 C RED ;;+
LINE \\;INSERT /CA-ACAD/VALVE @ 16 ;DRAG \
[]
[]
[]
```

Move [TRANVIEW] to the **EXAMPLES page. Make sure the [labeled] items and [] blanks total 19 items:

```
**EXAMPLES
[Examples]
[]
[TRANVIEW]$S=TRANVIEW
[]
```

Find the **TRANVIEW page. Now it is accessed through the **EXAMPLES page. Add a second $S= to the end of **each** item to return it back to the previous menu before EXAMPLES. A typical line will be:

```
[ All     ]'VIEW R A $S= $S=
```

Save, Exit and return to AutoCAD. TEST.MNU should reload automatically.

Note that the [PIPEFITS], [TEXT], [SPECIALS], [ISO DIMS], [HATCH], [SETTINGS], [TITLEBLK], and [3D] items are inactive. The TEST.MNU pages and devices are not in any particular order. If you want to place them in a more logical order, you can block move the pages.

Testing the CA-MENU Test Menu

Make sure the TEST.MNU is reloaded. Print out the TEST.MNU file so that you can study it as you test it. Test all the page changes and see that the screen positions match those shown in the book's illustration.

```
Select [ANNOTATE]        It brings up Annotatn. Try some of the macros again.
Select [root]            To get back to CA-MENU.
Select [FLOWLINE].       It brings up Flow Lines. Make sure the screen changes correctly.
Select [root]            To get back to CA-MENU.
Select [Button 2]        It calls the OSNAPS and PTFILTERS toggling screens.
Select [      ]          Continue, testing each of the items.
Select [EDIT-MNU]        Edit TEST.MNU if you get any errors.
```

  Delete TEST.MNU.

  Rename TEST.MNU to CA-MENU.MNU.

How To Work in TEST.MNU Menus

You will be happy to know the most tedious task of putting together menus is finished. You have roughed out a menu structure with a number of functional sections filled in. It is organized for the rest of the book. From now on, you will create and work with menu pages in small TEST.MNU sections.

How To Work in TEST.MNU Menus

The next exercise is an example of how to edit a typical chapter menu in the book. Each new menu page in the rest of the book is designed so that it "plugs into" the CA-MENU structure. The book gives instructions at the end of each chapter for merging the chapter menu with the CA-MENU. Some chapters contain example macros and AutoLISP material that you do not need to merge into the CA-MENU.

Using individual menus for each chapter's development makes menu editing easier, reloading menus faster, and protects the CA-MENU from editing errors. If you have the CA DISK, you already have the menus for each chapter. These are copies of MAIN.MNU with new menu pages added for the chapter. These menus are named for the chapter, like CA06.MNU, CA07.MNU, CA08.MNU, etc.

If you don't have the CA DISK, we will ask you to copy the MAIN.MNU to a new TEST.MNU, edit it and add new pages. Always remember to rename the previous TEST.MNU to save it. Using the MAIN.MNU template keeps your test menus short and easy to use.

Making a Standard Text Screen Menu

Your version of the prototype drawing has several standard text styles. The **TEXT menu page automates the changing of Styles when you enter text. If you have the CA DISK, you already have this menu, called CA06.MNU. You do not need to edit it. Examine the menu with your editor.

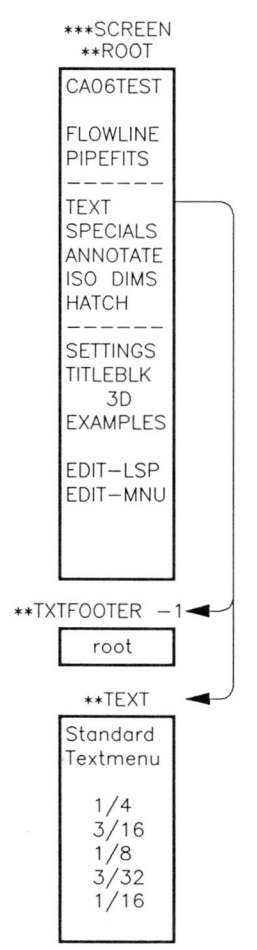

A Standard Text Screen

Creating the **TEXT Screen

 Copy CA06.MNU to TEST.MNU.

 Copy MAIN.MNU to TEST.MNU.

Select **[EDIT-MNU]**   Edit the TEST.MNU.
Replace [CA00TEST] with [CA06TEST] and add the statements shown to the [TEXT] label.

```
***SCREEN
**ROOT
[CA06TEST]
[ ]
[FLOWLINE]
[PIPEFITS]
[--------]
[TEXT     ]$S=TEXT $S=TXTFOOTER
```

Edit **FOOTER to **TXTFOOTER and add the following text.

```
**TXTFOOTER -1
[ root   ]$S=SCREEN LAYER S 0 ;SNAP ON GRID ON ORTHO ON STYLE STD1-8 ;;;;;;;
```

Edit the example page by adding the following material and changing **EXAMPLES to **TEXT.
Make sure [labels] and [] blanks total 19 items.

```
**TEXT
[Standard]
[Textmenu]
[]
[  1/4   ]^C^C^CLAYER S TXT02;;DTEXT S STD1-4
[  3/16  ]^C^C^CLAYER S TXT02;;DTEXT S STD3-16
[  1/8   ]^C^C^CLAYER S TXT01;;DTEXT S STD1-8
[  3/32  ]^C^C^CLAYER S TXT01;;DTEXT S STD3-32
[  1/16  ]^C^C^CLAYER S TXT01;;DTEXT S STD1-16
[]
```

Save, Exit and return to AutoCAD. The menu should reload.

Select [TEXT]   Test them. Each should set layer and start Dtext with its style.

  Delete TEST.MNU.

  Rename TEST.MNU to MY06.MNU.

Command: QUIT

Appending the Chapter Menus

The CAMASTER.MNU file on the CA DISK includes all the important menu pages which you will develop in the following chapters. If you don't have the CA DISK, you will find instructions at the end of each chapter to combine the chapter menu with your CA-MENU.MNU. If you combine your chapter menus, your CA-MENU will match the book's CAMASTER.MNU file at the end of the book. The following section shows how the book provides instructions on merging menus.

Integrating the Text Menu

If you don't have the CA DISK, add the text menu to CA-MENU by appending and changing the following lines:

- Append the **TXTFOOTER and **TEXT page of your MY06.MNU to the end of the CA-MENU.MNU file.

- Change the [TEXT] label on the **ROOT page of the CA-MENU to:

```
[TEXT    ]$S=TEXT $S=TXTFOOTER
```

Here is the chapter's menu.

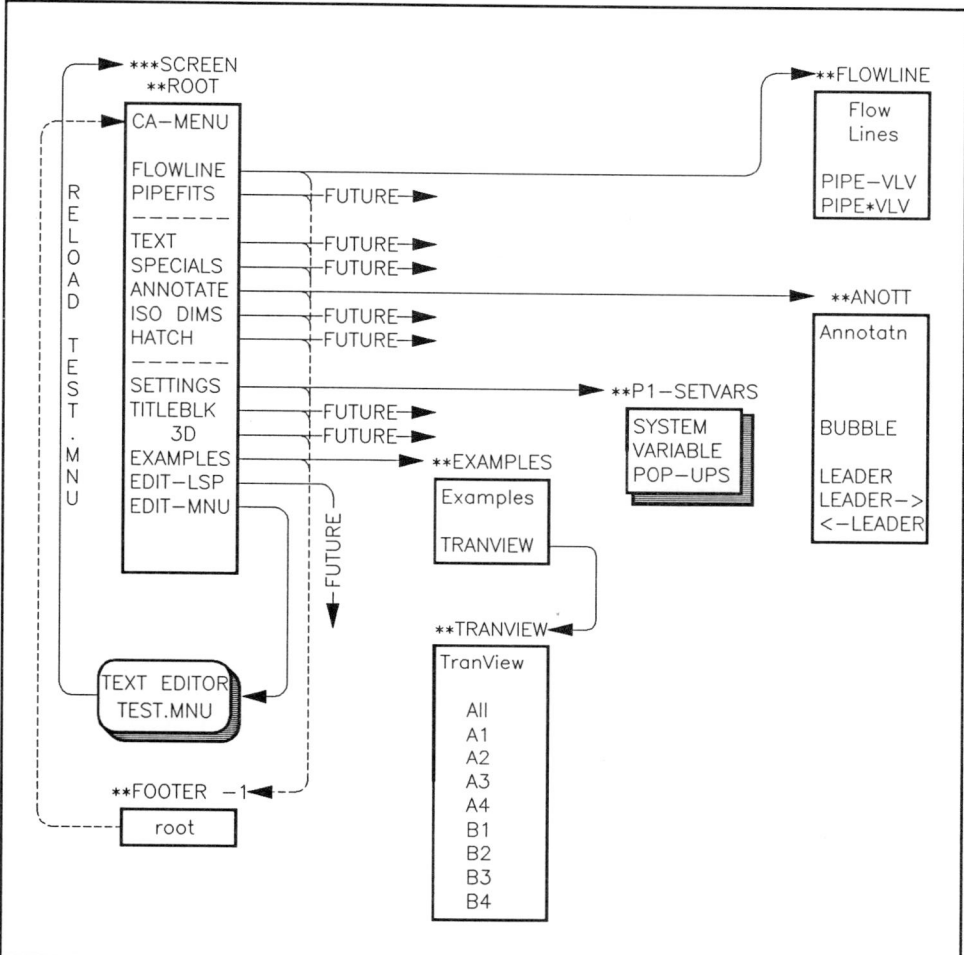

The CA06 Menu

Tips and Techniques

Here are a few last reminders and tips for developing your menus.

- ❑ Use an automatic setup routine to establish your environment. You also can use your setup menu to set units.
- ❑ Set global variables for drawing plot scale and related scaling information.
- ❑ Design your setup menu to lead to your applications menu.
- ❑ Design your root setup menu page to access all your application task menus.
- ❑ Design your footers to always get back to the main menu and restore settings.
- ❑ Use menu page switching as control points for your application program.

7—0 Customizing AutoCAD

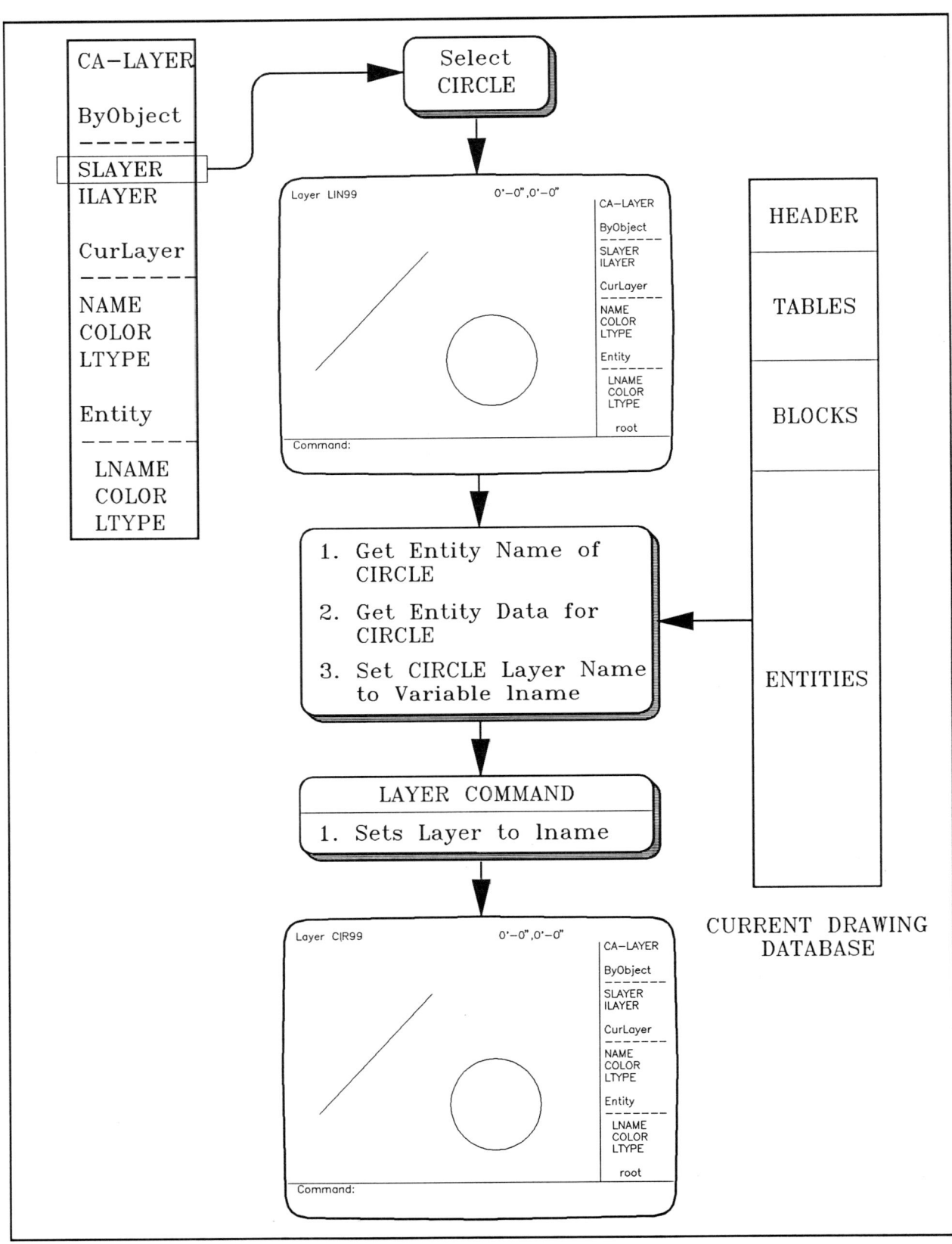

[SLAYER] Menu Macro Using AutoLISP

CHAPTER 7

A Little More LISP

The macro commands that you have made so far have been sequences of standard AutoCAD commands. This chapter shows you how to add AutoLISP to your menu macros.

What Is AutoLISP?

AutoLISP is a user friendly programming language that runs with AutoCAD. AutoLISP contains a group of predefined general purpose subroutines that provide access to the AutoCAD database, perform calculations and can control the AutoCAD drawing editor. These predefined routines are contained in AutoLISP's atomlist.

Introducing AutoLISP into your macro commands has a profound effect on your macros. The "fast typist" way of looking at macros is a good analogy, but by using AutoLISP you can transform your macros, making them perform calculations and logical decisions.

The Benefits of Using AutoLISP in Macros

You gain substantial benefits by using AutoLISP in your menu macros. Using AutoLISP lets you:

- Save data in variables, process that data, and send it back to AutoCAD.

- Use AutoCAD system variables to control the drawing editor. You can store AutoCAD system variables in AutoLISP variables, present drawing status to your users and change the settings.

- Change system settings transparently in AutoCAD commands, making your macros more responsive.

- Prompt for basic types of input to AutoCAD, store the input in variables, and pass the input along to your commands and macros.

- Access entity data, extract entity data and use the data in your programs.

- Look up reference data in AutoCAD's tables. For example, you can look up data in AutoCAD's layer table to help track layer information in your drawing.

How To Skills Checklist

In this chapter you will learn how to:

- ❏ Use AutoCAD system variables.

- ❏ Use AutoLISP variables.

- Extract information about the drawing and use it in macros.
- Write an AutoLISP expression, including the syntax, function names and arguments.
- Add AutoLISP routines to your menu.
- Pass the AutoLISP data back to AutoCAD.
- Get user input using AutoLISP.
- Use the rubberband feature of AutoCAD in AutoLISP.
- Write your own dimensioning command.
- Work with AutoLISP lists, build them, and retrieve any element from them.
- Access the entity database using selection sets, entity names and entity data.
- Use entity data in a macro.
- Look up AutoCAD table data.

Macros, AutoLISP Tools and Programs

MACROS using AutoLISP
[DWG NAME] prints the current drawing name to a string of text.
[CA DIMS] is a macro using AutoLISP to get two points and put in a dimension string of text.
**CA-LAYER is a screen page that performs layer operations based on entity selections.
[SLAYER] sets the current layer to the layer of a selected entity.
[ILAYER] isolates the layer of a selected object, turning all other layers off.
**LISPSAVE is a screen menu page containing AutoLISP expressions to save and retrieve user input data.

Variables

AutoLISP uses variables. A VARIABLE is simply a name representing something that is not constant. A variable changes its value from time to time. You use variables all the time. You multiply your tax RATE by your INCOME and, if it were only that simple, send the result to the IRS. RATE and INCOME are variables.

AutoLISP lets you substitute variable names in place of constant values. You attach a value to a variable name. You can attach the information to the variable name explicitly in a macro, or you can attach the information by having a user supply the information on the fly. You can use the variable names within AutoLISP expressions to perform calculations and make logical decisions.

AutoCAD's System Variables

When you change the value of Snap, set an Osnap mode, Toggle Ortho On or Off, AutoCAD saves the newly set condition or value as a variable. These variables are called System Variables. You can see all of their current settings within AutoCAD, using the Setvar command.

Looking at AutoCAD's System Variables with SETVAR

Enter selection: Begin a NEW drawing named CA07.

 Copy CA07.MNU to TEST.MNU.

 Copy MAIN.MNU to TEST.MNU.

Select **[TEST] [24 x 36] [FULL] [INITIATE]**   It should bring up TEST.MNU. Then:

Command: **SETVAR**
Variable name or ?: **?**   ? Gives a listing including:

```
ACADPREFIX    "C:\CA-ACAD\"              (read only)
ACADVER       "9.0"                      (read only)
AFLAGS        0
ANGBASE       0
ANGDIR        0
APERTURE      10
AREA          0.0000                     (read only)
ATTDIA        0
ATTMODE       1
ATTREQ        1
AUNITS        0
AUPREC        0
AXISMODE      1
AXISUNIT      0'-0",0'-0"
BLIPMODE      1
CDATE         19870921.124043350         (read only)
CECOLOR       "BYLAYER"                  (read only)
CELTYPE       "BYLAYER"                  (read only)
CHAMFERA      0'-0"
DWGNAME       "CA07"                     (read only)
```

Although you can guess what most of the variables mean, the list itself doesn't explain what each variable does. See the SYSTEM VARIABLE table in Appendix C for a complete listing, along with descriptive comments on the variables.

Variable Types

AutoCAD uses three types of system variables: Strings, Integers, and Reals.

A STRING variable has text as its value. In AutoLISP, you place the value of a string variable in "" quotes to identify the value as a string. AutoCAD has several System Variables that have text strings as values. For example, the

System Variable DWGNAME is currently "CA07." Some valid AutoLISP strings are:

```
"1.0", "JOHN", "-12134", "Albert Einstein once thought..."
```

An INTEGER is a positive or negative whole number, without fractions, decimal places or a decimal point. AutoCAD often uses values of 0 and 1 to indicate whether a system variable toggle, like SNAPMODE and ORTHOMODE, is turned off (0) or on (1). Integers **must** be between -32768 and +32767. Some valid integers are:

```
1, 3234, and -12134
```

REALs are positive or negative numbers with decimal points. In AutoLISP, you cannot begin or end a real with a decimal. If the value is less than 1.0, you need to put a 0 before the decimal point (0.123). Examples of Real system variables are FILLETRAD, LTSCALE and AREA. Unlike integers, you can make reals very large or very small. AutoLISP formats reals in scientific notation when the values are very large or very small. Valid Reals look like:

```
1.0, 3.75218437615, -71213.7358 and 1.234568E+17
```

Using system variables, you can control many AutoCAD drawing editor settings. You can update these system settings with the SETVAR command. If you are in the middle of another AutoCAD command, you can use the transparent 'SETVAR command to change settings, like the cursor pick box size. Try changing the pick box size.

Changing Settings in AutoCAD with 'SETVAR

```
Command: LINE                           Draw a Line.

Command: ERASE                          Issue an Erase.
Select objects: 'SETVAR
Variable name or ?: PICKBOX
New value for PICKBOX <3>: 10           The size of the pick box has changed.
Resuming ERASE command.
Select objects: ^C                      Cancel and Undo.
```

You can change some, but not all, of the system variable settings. Certain variables are "Read-Only," meaning you can only extract their values, not update their values. The Setvar command tells whether the variable is "Read-Only." The Table in Appendix C also tells whether a variable is "read only."

```
Command: SETVAR
Variable name or ?: CLAYER
CLAYER = "0" (read only)
```

You can retrieve and use AutoCAD's system variables using AutoLISP's GETVAR and SETVAR functions. GETVAR is a function, not a command. The AutoLISP SETVAR function has the same name as the standard AutoCAD command. Try using SETVAR and GETVAR to see how they work. Then, make a macro to get the current drawing name and place the name in the drawing as a text string.

Accessing System Variables with AutoLISP's GETVAR and SETVAR

```
Command: (getvar "DWGNAME")
Lisp returns: "CA07"

Command: (setvar "PICKBOX" 3)
Lisp returns: 3

Command: (setvar "CLAYER" "ANN02")
error: AutoCAD rejected function                It's read only!
(SETVAR "CLAYER" "ANN02")

Command: TEXT                                   First try this. Pick a point:
Start point or Align/Center/Fit/Middle/Right/Style:
Rotation angle : 0
Text: (getvar "DWGNAME")

Command: ZOOM                                   Zoom in so you can read it.
```

You wanted the drawing name, stored as the system variable value of DWGNAME. You didn't get it. You got a string of text that says: (getvar "DWGNAME").

AutoCAD can cause a problem within AutoLISP when AutoCAD wants a literal text string. There are two places where AutoCAD wants a literal text string: when prompting for text, or prompting for an attribute. A system variable, called TEXTEVAL, controls the situation. Setting TEXTEVAL to 1 (ON) forces AutoCAD to send the expression to AutoLISP for evaluation and prints the results (instead of the expression itself).

Let's set TEXTEVAL and try it again by making a menu item.

```
Select [EDIT-MNU]                               Edit TEST.MNU.
```

Change [CA00TEST] to [CA07TEST] and add the following to the **EXAMPLES page:

```
***SCREEN
**ROOT
[CA07TEST]
[]
```

```
**EXAMPLES
[Examples]
[]
[]
[DWG NAME ]^C^C^CSETVAR TEXTEVAL 1 TEXT \0 (getvar "DWGNAME");SETVAR ;0
[]
```

Save and Exit to AutoCAD. TEST.MNU reloads.

```
Select [EXAMPLES] [DWG NAME]                              You get:
Command: SETVAR Variable name or ?: TEXTEVAL
New value for TEXTEVAL <0>: 1
Command: TEXT Start point or Align/Center/Fit/Middle/Right/Style: Pick a point.
Rotation angle <0>: 0
Text: (getvar "DWGNAME")                                  The name CA07 prints.
Command: SETVAR Variable name or ? <TEXTEVAL>:
New value for TEXTEVAL <1>: 0
```

You got what you wanted, the drawing name CA07. Notice how we terminated the input to the Text command with a semicolon ";." Remember that the only way to tell AutoCAD you are done entering a text string is to use a ; or <^M> for a <RETURN>.

AutoLISP Variables

AutoCAD has variables. You can have your own variables using AutoLISP. Your variables have to follow the same rules. Like AutoCAD's system variables, each AutoLISP variable has a data "type" that is automatically assigned when you create the variable. AutoLISP variable names **may** duplicate AutoCAD System Variable names. AutoLISP variables are completely independent of AutoCAD System Variables.

An AutoLISP VARIABLE is a pointer to a value that has been stored in memory. Each time you use the variable name or refer to the variable name in a macro, program or expression, the program replaces the variable name with its associated value. If you use the same name twice, the last value assigned to the name will be used.

You can make variables any printable combination of letters and numbers except those reserved for other purposes. You cannot use reserved characters in variable names because of their special meanings in AutoLISP. There also are some ill-advised characters that may cause confusion or interfere with AutoLISP when you use them in menu macros. The reserved and ill-advised characters that you want to avoid using are:

RESERVED AND ILLEGAL CHARACTERS: . ' " ; ( )   or <SPACE>
and: ~ * = > < + - /   (which are already AutoLISP functions.)

ILL-ADVISED CHARACTERS: ? ` ! \ ^   or any control character.

In addition to these characters, there are other reserved words for AutoLISP's built-in functions that you cannot use as variable names. The ATOMLIST is an

AutoLISP variable that stores all defined functions and variable names. You can see these functions by looking at AutoLISP's Atomlist.

Looking at the Atomlist

AutoLISP will list the rest of its function names if you type:

```
Command: !ATOMLIST
Lisp returns: (#DWGSC STR1 Y X #MENU INTERS GRREAD GRTEXT GRDRAW GRCLEAR TBLSEARCH TBLNEXT
ENTUPD ENTMOD ENTSEL  ENTLAST ENTNEXT ENTDEL ENTGET SSMEMB SSDEL SSADD SSLENGTH SSNAME SSGET
ANGTOS RTOS COMMAND OSNAP REDRAW GRAPHSCR TEXTSCR POLAR DISTANCE ANGLE INITGET GETKWORD
GETCORNER GETINT GETSTRING GETORIENT GETANGLE GETREAL GETDIST GETPOINT MENUCMD PROMPT SETVAR
GETVAR  TERPRI PRINC PRIN1 PRINT WRITE-LINE READ-LINE WRITE-CHAR READ-CHAR CLOSE OPEN
STRCASE ITOA ATOF ATOI CHR ASCII SUBSTR STRCAT STRLEN PI MINUSP ZEROP NUMBERP FLOAT FIX SQRT
SIN LOG EXPT EXP COS ATAN 1- 1+ ABS MAX MIN NOT OR AND > >= /= = <= < ~ GCD BOOLE LSH LOGIOR
LOGAND REM * - + ASSOC MEMBER SUBST LENGTH REVERSE LAST APPEND CDDDDR CDDDAR CDDADR CDDAAR
CDADDR CDADAR CDAADR CDAAAR CADDDR CADDAR CADADR CADAAR CAADDR CAADAR CAAADR CAAAAR CDDDR
CDDAR CDADR CDAAR CADDR CADAR CAADR CAAAR CDDR CDAR CADR CAAR CDR CAR CONS COND LISTP TYPE
NULL EQUAL EQ BOUNDP ATOM NTH PAGETB PICKSET ENAME REAL FILE STR INT SYM LIST SUBR T MAPCAR
APPLY LAMBDA EVAL *ERROR* / QUIT EXIT _VER VER IF UNTRACE TRACE DEFUN FOREACH REPEAT WHILE
PROGN FUNCTION QUOTE READ LOAD SETQ SET MEM VMON ALLOC EXPAND GC ATOMLIST)
```

The first five **bold** variable names in the ATOMLIST are the variable names created by your CA-SETUP menu. The rest are the standard AutoLISP functions. You will use most of these functions in the course of the book, but some of these variables are used exclusively by the AutoLISP evaluator and have no meaning for your programs. The AutoLISP Programmer's Reference documents the built-in AutoLISP functions for your use. We recommend that you read the AutoLISP Programmer's Reference as you encounter each built-in function in the book.

Your use of character case (UPPER or lower) in a variable name has no effect on its performance. However, variable names over six characters in length require more memory. We recommend that you keep your names under six characters. You also are not supposed to begin a variable name with a number, although it currently seems to work.

INVALID VARIABLE NAMES:

123 (represents an integer number)
10.5 (represents a constant real value of 10.5)
LIST (redefines the built in function named LIST)
A(1) (contains invalid characters)
OLD SUM (contains space)

VALID VARIABLE NAMES:

PT1
TXT
Rang
#DWGSC ("#" is book's global variables)
OLD-SUM (use - or _ for spaces.)

AutoLISP binds every stored value to a variable name. Binding a value to a name is similar to a function in algebra. In algebra, you write **y=3**, but in AutoLISP you enter **(setq y 3)**. Both the "=" of the algebraic expression and the "setq" of the AutoLISP expression are called "functions." Each function binds (sets) the value 3 to the variable "y." The open (left) and closed (right) parenthesis, "(" and ")," are required by AutoLISP syntax to form an expression.

All of these expressions bind values to variable names:

`(setq #dwgsc 1.0)   (setq draftname "JOHN")   (setq var1 1000)`

You also have some expressions in your setup menu:

```
[CA-MENU ]^C^C^C(setq #menu "CA-MENU");$S=DWG
[11 x 17 ](setq x 17.0) (setq y 11.0) (setq str1 "SHEET-B");$S=SCALE
```

Flavoring Your Menu with AutoLISP

Once you bind a value to a variable name, you can supply that value to AutoCAD at any time. You use the exclamation point "!" as a special character to tell AutoCAD that the name following the exclamation point is an AutoLISP variable name. The ! must precede the variable name. Each time the AutoCAD command processor sees the ! character it passes the variable name to AutoLISP. AutoLISP looks up the name in memory and passes its value back to the AutoCAD command processor. You did this value passing at the end of the [INITIATE] macro in the CA-SETUP menu:

`INSERT !str1 0,0 !#dwgsc ;;MENU !#menu`

All AutoLISP "(" open parenthesis in expressions must have matching ")" closing parenthesis. Failure to close a parenthesis will give you a prompt like 1> or 2>. The AutoLISP error prompt form is n> where the number n indicates how many closing parentheses are missing. You saw this in CA-SETUP when you selected a scale, like [1/16"]:

```
Command: (setq #dwgsc 192.0) 192.000000
Command: (prompt "Select [INITIATE]...
1> ")
Select [INITIATE]...
```

Any AutoLISP " open quotes must have matching " closing quotes. If you get a n> error prompt, and adding additional)) parentheses doesn't help, then you left off a quote ". Type a single " and then type as many)) parentheses as you need.

Simple AutoLISP Expressions and Evaluation

AutoCAD needs to have a way to distinguish between AutoCAD commands and AutoLISP expressions. AutoCAD identifies AutoLISP expressions by an open parenthesis "(" as the first character passed to AutoCAD. Each time AutoCAD detects an open "(" parenthesis, it passes the entire expression to AutoLISP. AutoLISP evaluates the expression and returns the result to AutoCAD. AutoCAD uses the result and continues. AutoLISP expressions can contain other AutoLISP expressions. You must nest each expression in its own pair () of parenthesis. Here are the rules for the game:

- Every expression has an opening "(" and closing ")."
- Every expression has a function name. The name **must** immediately follow the opening "(" parenthesis.
- Every expression gets evaluated (executed) and returns a result. The result may be NIL.

A typical LISP expression has this syntax:

```
(function argument argument)
```

A FUNCTION is a subroutine that tells AutoLISP what task to perform. Tasks include: add, subtract, multiply or divide. A function may have no arguments, or any number of arguments.

An ARGUMENT provides data to the function. Arguments may be variables, constants, or other functions. Some arguments are "flags" that alter the action of the function. If you define a function to take an argument, you **must** provide the function with a value for the argument.

Math with AutoLISP

In algebra, you add two numbers as:

`1 + 2`

In LISP you write:

`(+ 1 2)`

These expressions have two constants as arguments. Some built-in AutoLISP math functions, like **+ - *** and **/**, can handle a varying number of arguments. User defined functions must have a predefined number of arguments.

Doing Math Functions in AutoLISP

Let's type a quick math function:

```
Command: (* 6.0 3.0 2.0)            Multiplies all three numbers.
Lisp returns: 36.000000
```

AutoLISP evaluates expressions left to right at the same nesting level. When a nested expression is encountered, the entire nested expression is evaluated before the next expression on the right. One good thing about having all these parentheses is that you never write a formula that evaluates in a different order than you expect.

Let's try nesting expressions:

```
Command: (* 6.0 (- 3.0 1.0))              The inner expression is nested.
Lisp returns: 12.000000

Command:   (setq a 1)                     Assigns a variable the value 1.
Lisp returns: 1

Command: (+ (setq a (* a 3)) (+ a 2))     Assigns variable before second expression.
Lisp returns: 8

Command: (+ (+ a 2) (setq a (* a 3)))     Uses variable, then reassigns it.
Lisp returns: 6
```

In algebra, you say **1 + 2** "equals" **3**, but in AutoLISP you say **(+ 1 2)** "evaluates" to **3**. Although this may seem a silly distinction, it is helpful to think in terms of "evaluation," particularly when you are working with complicated expressions.

Using GET Functions for User Input

All interactive computer programs have methods for requesting input. Most of the typical input data forms are present in AutoLISP.

You use the GET family of AutoLISP functions to request types of data input. You use the GET functions, like all AutoLISP functions, between () parentheses within an AutoLISP expression. GETs use an optional number of arguments. One of the arguments is a prompt. The prompt can be any string value. You usually use the prompt to ask a question or to give an instruction.

You can request Reals and Strings as input. Their AutoLISP GET functions are: GETREAL and GETSTRING. Like all GET functions, they pause for input.

Getting Input with GET

```
Command: (getreal "Multiply totals by: ")
Multiply totals by: ZXC                   Enter a string.
Requires numeric value.                   It rejects it.
Multiply totals by: 24                    Enter a number.
Lisp returns: 24.000000

Command: (getstring "Perform solar analysis (Y/N): ")
Perform solar analysis (Y/N): Y           Enter Y for Yes.
Lisp returns: "Y"
```

In these examples, each expression contains one argument, the prompt to display. The user's reply to these prompts will automatically become the data type requested. Invalid responses, answers that are not the requested data type, are rejected. GETSTRING will accept numbers as string data, and GETREAL will accept INTegers, but converts the integers to floating point REALs. The GETREAL and GETSTRING functions have no ability to interact with the graphics screen of AutoCAD.

Getting Distances, Angles and Points

AutoLISP provides special GET functions for requesting user input of points, distances and angles, and graphic input. These specialized GET functions are:

`GETDIST   GETANGLE   GETORIENT   GETPOINT   GETCORNER`

Using GETDIST is preferable to using GETREAL because you can enter the distance either by picking points or by typing values. You can pick a distance by picking two points on the screen, using a rubberband line and the COORDS display. GETDIST treats the distance as a REAL data type. You can input either decimal or current UNITS. For architectural or engineering units, a distance of 12'0" is automatically converted to 144.000000, and either 137'4" or 1648 returns 1648.000000.

Using GETDIST, GETANGLE and GETPOINT

```
Command: (getdist "Distance between supports: ")
Command: Distance between supports: 10'          Enter a number in current units.
Lisp returns: 120.000000
```

If you use an integer, like 120, within an AutoLISP expression where it expects a real (120.0), AutoLISP converts it to 120.000000. If you give an AutoLISP GETREAL or GETDIST function input of 120.0, 120. or 120, AutoLISP accepts and converts these values to 120.000000. However, if you give AutoLISP a real when it expects an integer, it will cause an error.

You use the function GETANGLE to get an angle. AutoLISP calculates the rotation angle between the zero angle origin and the angle of the rubberband line drawn. Angles are REAL data types, but based in radians. Radians are the internal angular units used by AutoLISP and AutoCAD. 180 degrees equals PI, or 3.14159 radians. An angle entered as 90 gets converted to 1/2 of PI, or 1.571 radians. Although the angle is in radians, it is still a REAL data value.

```
Command: (getangle "Rotational angle of pipe run: ")
Rotational angle of pipe run: 45        Type an angle.
Lisp returns: 0.785398                   The angle in radians.
```

AutoLISP has a GETPOINT function that accepts point input. Points are defined by an X and Y coordinate. A Z coordinate is an optional value. Coordinates are REAL values in decimal units. An X coordinate of 10' would be converted to 120.00000 units. The GETPOINT function automatically forms a list of the two real X and Y coordinates. A POINT is simply a list of two REALs.

7—12 Customizing AutoCAD

```
Command: (getpoint "Starting point of pipe run: ")
Command: Starting point of pipe run:    Pick one.
Lisp returns: (120.0000 240.0000)       Really your point.
```

Notice there is no comma between the coordinates. You **must** separate the items in a list with a space. You **must** always imbed the decimal point within digits: 111.0 or 0.123, not 111. or .123. If you fail to imbed the decimal point, you will get an "Invalid dotted pair" error.

Making a Menu to Save AutoLISP Data

Select **[EDIT-MNU]**
Add the following page by copying and editing the **EXAMPLES page:

```
**LISPSAVE
[LISPsave]
[]
[Save str](setq str (getstring "String: " t)) $S=
[Use str ]!str $S=
[]
[Savereal](setq rea (getreal "Number: ")) $S=
[Use real]!rea $S=
[]
[Save int](setq int (getint "Integer: ")) $S=
[Use  int]!int $S=
[]
[Save pt1](setq pt1 (getpoint "Point: ")) $S=
[Use pt1 ]!pt1 $S=
[]
[Save ang](setq ang (getangle "Angle: ")) $S=
[Use ang ]!ang $S=
[]
[Savedist](setq dist (getdist "Distance: ")) $S=
[Use dist]!dist $S=
[]
[]
```

Also edit the **EXAMPLES page to add [LISPSAVE]:

```
[]
[DWG NAME ]^C^C^CSETVAR TEXTEVAL 1 TEXT \0 (getvar "DWGNAME");SETVAR ;0
[]
[]
[]
[LISPSAVE]$S=LISPSAVE
[]
```

Save and Exit to AutoCAD. TEST.MNU reloads.

Select **[EXAMPLES] [LISPSAVE]**   Test each, from top to bottom.
                                        Notice how you are reprompted if you enter the wrong type of input.

GETSTRING has an optional flag argument, T, that lets you enter spaces in the string. This is the reason for the "t" in the getstring expression.

Base Point Arguments and the GET Functions

In addition to their prompt argument, the built-in functions GETDIST, GETANGLE, GETPOINT and GETCORNER can use an optional base point argument. AutoCAD uses rubberband lines when a user shows a distance or angle by selecting points. Rubberbanding lets you dynamically display a distance in the coordinates box of the screen. In the relative distance Coords mode, the distance<angle is displayed. The GETDIST, GETANGLE, GETORIENT, GETPOINT and GETCORNER functions all can rubberband input. Let's try GETPOINT and GETCORNER.

Select **[SAVE pt1]**   Pick point to set base point. Rubberband GETPOINT by typing:

Command: **(getpoint pt1 "Enter second point: ")**
Enter second point: Pick any point.
Lisp returns: (10.164329 120.135036)   You get the point you picked.

Try GETCORNER.

Command: **(getcorner pt1 "Other corner: ")**
Other corner: Pick a point.
Lisp returns: (13.000000 102.500000)   You get the point you picked.

Placing a GET function without a base point in the middle of an ordinary AutoCAD command interferes with the normal AutoCAD command's rubberbanding. Try this:

Command: **LINE**
From point: Pick a point.
To point: **(getpoint "Pick it: ")**   And your rubberbanding disappears.

When should you use a base point? GETDIST, GETANGLE and GETORIENT all rubberband automatically if you pick a "first" point. Use an optional base point if you need to tie the point down. If you want GETPOINT to rubberband to a previous point, use an optional base point argument. GETCORNER **must** have a base point.

You can get around the disappearing rubberband using a base point. The next macro, called [CA DIMS], will show you how.

GET Functions in Menu Commands

When you use the keyboard, the GET functions automatically pause for input. However, when you use GET functions in a menu macro command, you need to use the backslash "\" character to pause the execution of the macro for input.

This gets tricky when you use a base point. Functions like GETDIST need only one "\" with a base point, but two "\\"s without a base point.

The next simple dimensioning macro, called [CA DIMS], shows you how to use base points and rubberbanding in your menus. It asks for two points, calculates the distance between these points, draws a line, inserts an arrow at each end, places the text, and lets you move the text.

Creating a Dimensioning Macro [CA DIMS]

Select **[EDIT-MNU]**

Edit TEST.MNU and add these lines on the **EXAMPLES page:

```
[CA DIMS   ]^C^C^C(setq pt1 (getpoint "First point: "));\+
(setq pt2 (getpoint "Second point: " pt1));\+
(setq dis (rtos (distance pt1 pt2))) LINE !pt1 !pt2;;+
INSERT ARROW !pt1 !#dwgsc ;!pt2 INSERT ;!pt2 !#dwgsc ;!pt1;+
SETVAR TEXTEVAL 1 TEXT C !pt1 !#txtht !pt2 !dis;SETVAR ;0 MOVE L ;INS @
[]
[LISPSAVE]$S=LISPSAVE
```

Save and Exit to AutoCAD. TEST.MNU reloads.

Select **[EXAMPLES] [CA DIMS]**

Command: (setq pt1 (getpoint "First point: "))	
First point:	Pick point. AutoLISP returns it as a list.
Command: (setq pt2 (getpoint "Second point: " pt1))	
Second point:	Pick point. AutoLISP returns it as a list and calculates distance. The Line draws from PT1 to PT2 and both arrows Insert.
Command: SETVAR Variable name or ?: TEXTEVAL	Set to 1 to recognize AutoLISP.
Command: TEXT Start point or Align/Center/Fit/Middle/Right/Style: C	
Center point: !pt1	Temporary location
Rotation angle : !txtht nil	
Rotation angle : !pt2	
Text: !dis	The calculated distance.
Command: SETVAR Variable name or ? <TEXTEVAL>:	Restored to 0.
Command: MOVE	
Select objects: L 1 found.	
Base point or displacement: INS of @	
Second point of displacement:	Pick point to locate text.

Your screen should show the dimension text.

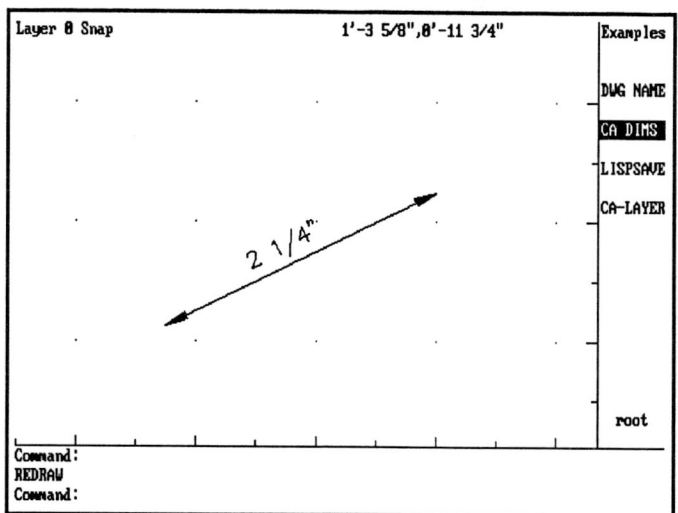

The [CA DIMS] Macro

The backslash "\" after each GETPOINT prompt string pauses the menu macro. The second GETPOINT has the optional basepoint to make it rubberband. The second getpoint got its value from the first getpoint. The third line in the macro uses DISTANCE to return the distance between the two points. The RTOS (Real TO String) function converts the distance into a string in the current drawing units. The LINE is drawn with the two points, using ! to tell AutoCAD to "see" the variables. Your old ARROW Block is Inserted at each PT1 and PT2 and rotated to the opposite point at scale #DWGSC. #DWGSC was set in your setup menu.

Last, the variable !PT places the text using Center justification. !PT2 rotates it, and !DIS enters the text. MOVE Last repositions the text.

But what was !#TXTHT? Your text Styles are fixed height, so AutoCAD doesn't even prompt for height. #TXTHT has no value. It is nil. You never setq'd it. When AutoCAD sees #TXTHT, it sees nil, which has no effect. If you were using variable height styles, you would have SETQed a value, and the macro would still work.

With very little AutoLISP, you have written a new dimensioning tool!

Pausing Problems with GET Functions in Macros

Two AutoLISP GET functions pose a problem when you use the backslash method of pause control. The GETDIST and GETANGLE functions offer the user the choice of explicitly typing the distance/angle. Typing needs only one \ backslash. If the user shows the value by picking two points, you need \\ two backslashes.

The following exercise shows four alternate ways of using a GETDIST statement in a menu line. The exercise shows you how to deal with different numbers of backslashes.

Pausing for Get Functions

Select **[EDIT-MNU]**
Add these to the ****EXAMPLE** page for comparison:

```
[TypeDist]^C^C^C(setq var1 (getdist "Type in distance: "));\CIRCLE \!var1
[BptDist ]^C^C^C(setq bpt (getpoint "Enter basepoint: "));\+
(setq var1 (getdist Bpt "Draw or type distance: "));\CIRCLE !bpt !var1
[DrawDist]^C^C^C(setq var1 (getdist "Draw distance: "));\\CIRCLE \!var1
[Distance]^C^C^C(setq var1 (getdist;+
"Draw distance, or Type distance <RETURN><RETURN>: "));\\^CCIRCLE \!var1
```

Save and Exit to AutoCAD. TEST.MNU reloads.

Select **[EXAMPLES]**   Then select and compare each of the above.

[TypeDist] requires one backslash, but if the user picks a point, it causes an error. [BptDist] has the optional base point argument so one backslash is sufficient to pause for the input of the second point. If the macro provides no basepoint, the GET function still accepts two-point input. Since you must decide whether to include one or two backslashes, you should inform the user what is expected. If two pauses are allowed, as in [DrawDist], the function will behave normally if the distance/angle is picked by points. If you type a distance for [DrawDist], the menu is still paused for input and further input will cause an error.

[Distance] provides a workaround. If your macro is at the command prompt, a <^C> is harmless. You can use a prompt and an imbedded <^C> to solve the number of backslashes problem. The second <RETURN> reinvokes the prior command, but the <^C> cancels it and the macro continues.

LISP Lists

A LIST is a group of elements of any data type, treated as one expression and stored as a single variable. An AutoLISP LIST may contain any number of REALS, INTEGERS, STRINGS, variables, or other LISTS. Anything between an opening parenthesis "(" and closing parenthesis ")" is a LIST. You use LISTS to organize and process groups of information. The LASTPOINT System Variable is a list:

Working with LISP Lists

Command: **(getvar "LASTPOINT")**
Lisp returns: (17.968750 13.125000)   The command SETVAR shows 1'-5 31/32",1'-1 1/8".

Sample AutoLISP LISTs are:

```
("A" "B")
("NAME" 10.0 10.0 "DESK" "WS291A")
(5.327 3.513)
("6IN-ELBOW" (6.0 0.375 0.25 90.0))
```

AutoLISP has many built-in functions to manipulate lists of information. You have seen some of these functions already in the SHADOW BOX function that you built in the first chapter. The SHBOX-TX function used the built-in functions CAR and CADR.

Since a list is a group of elements, you need a way to extract the element that you want. CAR is a shorthand for the first element of a list and CADR for the second element. In SHBOX-TX you used the CAR function to extract the X coordinate of the lower left LL and the upper right UR points picked by the user. Let's look at CAR and CADR ("cadder") again; and look at a similar function called CDR ("could-er").

Select **[EXAMPLES] [LISPSAVE] [Save pt1]**
Command: **(setq pt1 (getpoint "Point: "))**
Point: **1,2**
Lisp returns: (1.000000 2.000000)

Command: **(car pt1)**   Get the first element.
Lisp returns: 1.000000   The X coordinate.

Command: **(cadr pt1)**   Get the second element.
Lisp returns: 2.000000

Command: **(cdr pt1)**   Returns a LIST.
Lisp returns: (2.000000)

CDR returns a LIST of all elements, except the first element. The list is still in parentheses indicating a LIST, not a REAL. It is a LIST containing a single REAL.

AutoLISP has other built-in functions, called CAAR, CADDR, and CADAR, that will retrieve lists with more than two elements. You also can use another function, called the NTH function, to access any element of a list. NTH retrieves the element number that you want. Make a list using several data types and use NTH.

Command: **(setq a (list "A" "B" 10.0 25))**
Lisp returns: ("A" "B" 10.000000 25)

Command: **(nth 0 a)**   Using the NTH function.
Lisp returns: "A"

Command: **(nth 2 a)**   The third element, index of 2.
Lisp returns: 10.000000

Notice how we asked for the 0 element of the "a" list. NTH starts with a "zero base," meaning that the first element starts at a zero position. This is typical computer counting. Watch your 0's and 1's carefully with AutoLISP functions. All AutoLISP functions do not count the same way. The NTH function takes the index number as its first argument and the list name as the second argument.

AutoLISP provides other built-in functions to manipulate lists. These are: LAST, REVERSE, LENGTH and APPEND. REVERSE flips the order of the list. LAST will give you the last element of a list. LAST is more efficient than the equivalent (car (reverse a)). But how do you **remove** the last element of a list, leaving the rest of the elements?

Command: **(last a)**
Lisp returns: 25

Gets the last element.

Command: **(reverse a)**
Lisp returns: (25 10.000000 "B" "A")

Transposes the order.

Command: **(car (reverse a))**
Lisp returns: 25

Gets the last element.

Command: **(cdr (reverse a))**
Lisp returns: (10.000000 "B" "A")

The best way to remove last element of a list. But it leaves it reversed. Try:

Command: **(reverse (cdr (reverse a)))**
Lisp returns: ("A" "B" 10.000000)

Command: **(length a)**
Lisp returns: 4

LENGTH tells how many items are in the list.

Command: **(append a "C" "D")**
Lisp returns: error: bad argument type

Adds to a list. APPEND adds only other lists. "C" "D" isn't a list. Try:

Command: **(append a (list "C" "D"))**
("A" "B" 10.000000 25 "C" "D")

Keep your AutoLISP Programmer's Reference handy and refer to it to keep these list functions and their cousins straight!

A Little Entity Stuff

How about a little fun! We want to show you some of the power AutoLISP gives you over the drawing. First, let's see how AutoCAD reveals data about entities. Then, you can use this data in your macros.

Every AutoCAD entity, whether a Line, Arc, or Circle, has a name that is recognized by AutoCAD. There is nothing special about the name itself. In computer terms, the name is a "hexadecimal pointer location in a data segment of computer memory," but who cares. AutoCAD needs to remember these names, but you don't need to remember them. Since the names currently change every time a drawing is entered from the main menu of AutoCAD, you don't even want

A Little Entity Stuff 7—19

to try to remember them. Instead you use AutoLISP's ENTLAST function to ask for the name of the last entity in the database.

Editing with Entity Data

Command: **ERASE**	Clean up your drawing.
Command: **LINE**	You need a line. Draw one.
Command: **(setq a (entlast))**	And now, introducing... Entity Names!
Lisp returns: <Entity name: 600006F4>	The line's entity name.
Command: **ERASE**	
Select objects: **!A**	Gets the last entity.
Select objects:	<RETURN> to erase it.
Command: **OOPS**	Bring it back.

You can erase entities that you can't see by feeding AutoCAD entity names. AutoCAD's standard object selection depends on visibility, but you bypass visibility using entity names.

A name is nothing without a face. In AutoCAD's case, the face is Entity Data! AutoLISP has a function called ENTGET, to return the data of a selected entity. ENTGET looks up the data associated with the name you provide.

Command: **(setq b (entget a))**
Lisp returns: ((-1 . <Entity name: 600006F4>) (0 . "LINE") (8 . "0") (10 0.721101 -0.329621) (11 2.875779 1.362630))

Entity data is returned in a list format with special groups of DXF (Drawing eXchange Format) codes that flag which type of data is contained in the sublist. Each sublist has two parts.

The first part is the DXF code. This code (always an integer) gives you a way to access the inner parts of an entity list by association. In effect, you say "Give me the sublist that has a 10 DXF code." AutoLISP will look at the entity list, find the sublist that has the matching 10 code and return that sublist. AutoLISP's ASSOC function gives you the ability to associate lists.

Command: **(assoc 10 b)**
Lisp returns: (10 0.721101 -0.329621)   Your points will vary.

The DXF 10 code identifies the first point of the entity. In the case of a line, it would be the first endpoint. In the case of a circle, it would be the center of the circle. The book provides a complete table of the DXF codes in Appendix C for easy reference.

Since the association list still contains the DXF code, you need to strip this code before passing the list to AutoCAD. You need to get the CDR (the list less the

7—20   Customizing AutoCAD

first item) of the association list. Let's do this, but let's do it in the context of using the AutoCAD Insert command. Notice how the example uses the second point of the line (DXF 11) to rotate the inserted object.

```
Command: INSERT
Block name (or ?): ARROW
Insertion point: (cdr (assoc 10 b))        1st point of line.
X scale factor <1> / Corner / XYZ:
Y scale factor (default=X):
Rotation angle <0>: (cdr (assoc 11 b))     2nd point of line.
```

A Little Table Searching

AutoCAD keeps items like Block, Layer, Style, Linetype and View names in reference tables. AutoLISP has the ability to look up information from these tables. AutoLISP's TBLSEARCH function looks through a table and returns the information from it. You will use a little table access in the next few chapters, but the full discussion of table access is given in the AutoLISP chapter on Table Access.

The TBLSEARCH function takes two pieces of information: the table to search and the name of the item to look for. Let's try two searches.

A Quick Look at Table Access with TBLSEARCH

```
Command: (tblsearch "LAYER" "0")                          Looks up layer 0.

Lisp returns: ((0 . "LAYER") (2 . "0") (70 . 64) (62 . 7) (6 . "CONTINUOUS"))

Command: (tblsearch "STYLE" "STD1-8")                     Looks up style "STD1-8".

Lisp returns: ((0 . "STYLE") (2 . "STD1-8") (70 . 0) (40 . 12.000000) (41 . 1.000000)
(50 . 0.000000) (71 . 0) (42 . 0.125000) (3 . "SIMPLEX") (4 . ""))
```

To show you how simple entity access and table searching works, let's write a menu page with a few useful macros.

Integrating AutoLISP Entity Access with Screen Menus

```
Enter selection:                           Edit the EXISTING drawing BLOCKS from the
                                           Symbols chapter.
```

BLOCKS has a variety of Layers, Linetypes and Colors. If you don't have the drawing, begin a NEW drawing named BLOCKS. Create a variety of entities.

```
Command: MENU                              Load TEST.

Select [EDIT-MNU]
```

A Little Table Searching 7—21

Add the following new page:

```
**CA-LAYER
[CA-LAYER]
[]
[ByObject]
[--------]
[SLAYER  ]^C^C^C(setq lname (cdr (assoc 8 (entget (car (entsel))))));\LAYER S !lname ;
[ILAYER  ]^C^C^C(setq lname (cdr (assoc 8 (entget (car (entsel))))));\LAYER S !lname OFF * ;;
[]
[CurLayer]
[--------]
[NAME    ](getvar "CLAYER")
[COLOR   ](cdr (assoc 62 (tblsearch "LAYER" (getvar "CLAYER"))))
[LTYPE   ](cdr (assoc 6 (tblsearch "LAYER" (getvar "CLAYER"))))
[]
[Entity  ]
[--------]
[ LNAME  ](cdr (assoc 8 (entget (car (entsel)))))
[ COLOR  ](cdr (assoc 62 (entget (car (entsel)))))
[ LTYPE  ](cdr (assoc 6 (entget (car (entsel)))))
[]
```

Add the following **bold** page switch to the **EXAMPLES page:

```
[LISPSAVE]$S=LISPSAVE
[]
[CA-LAYER]$S=CA-LAYER
[]
```

Save, Exit and return to AutoCAD. The menu should reload.

Select **[EXAMPLES] [CA-LAYER]**   Then try each item, selecting various entities.

Remember, if you select a Block Insert, you get the information for the Insert, not the entities within it. You can EXPLODE blocks in the BLOCKS drawing to get more entities to play with.

Here is a brief description of the menu macros:

[SLAYER] sets the current Layer to the Layer of the selected entity.

[ILAYER] isolates the Layer of the selected entity, turning all other layers off.

[NAME] returns the current Layer's name.

[COLOR] and [LTYPE] return the color and linetype of the current Layer. Current layer color and linetype are available through the Status command, but these macros show you how you can access the same values through AutoLISP.

[Entity] [LNAME], [COLOR], and [LTYPE] return the Layer, Color or Linetype of the entity that you select.

7—22 Customizing AutoCAD

It is time to quit and do some file handling.

Command: **QUIT**

 Delete TEST.MNU.

 Rename TEST.MNU to MY07.MNU.

Integrating the Example Menu

You do not need to integrate this chapter's menu into the CA-MENU. If any of the macros in this chapter are useful to you, add them to your own menus.

Here is the chapter's menu.

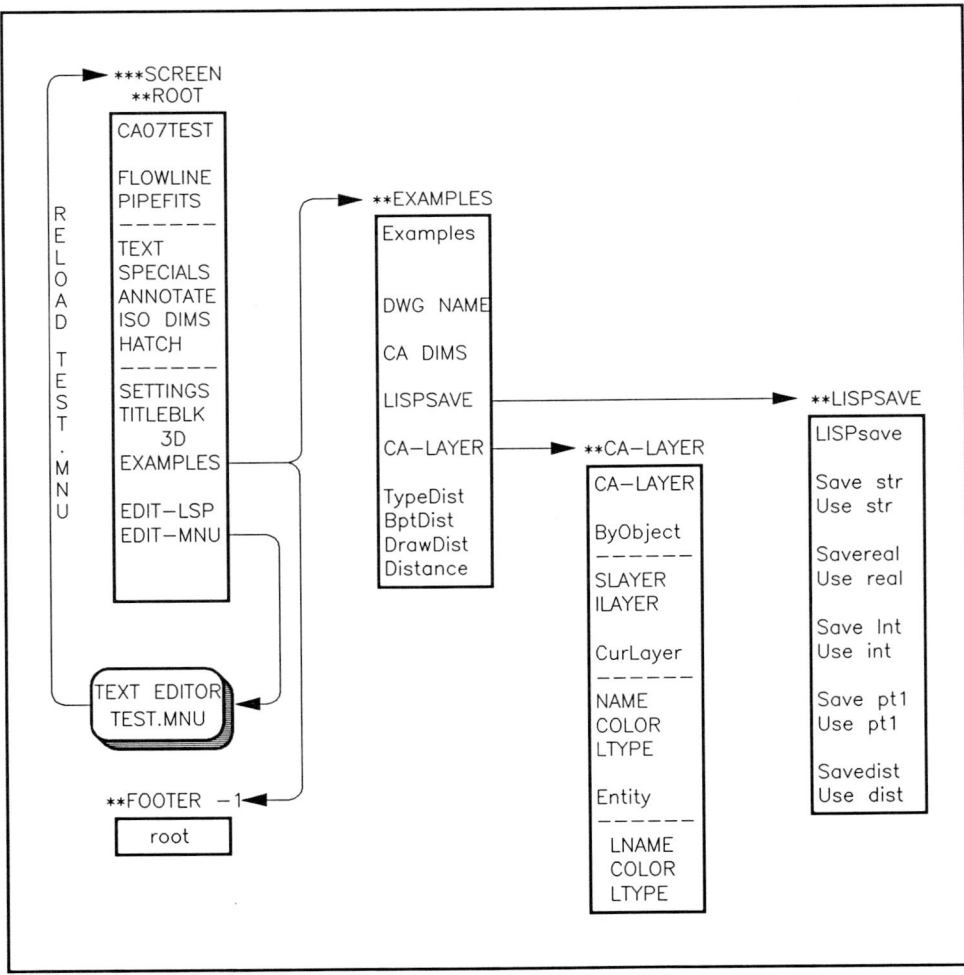

The CA07 Menu

Tips and Techniques

Here are some summary tips and techniques for using variables and expressions to integrate AutoLISP into your menu macros.

- ❑ Use AutoLISP variables to eliminate menu selections that depend on drawing scale.
- ❑ Use AutoLISP to create intelligent macros to replace repetitive menu macros.
- ❑ When possible, calculate values to feed to AutoCAD commands, rather than pause for user's input.
- ❑ Add prompts to menu macros to clarify use. Coordinate input with backslashes.
- ❑ Use GET functions to obtain user input.

This is the end of the book's basic tools section for menus and macros. It is time to move on to fonts, hatches and fills.

8—0 Customizing AutoCAD

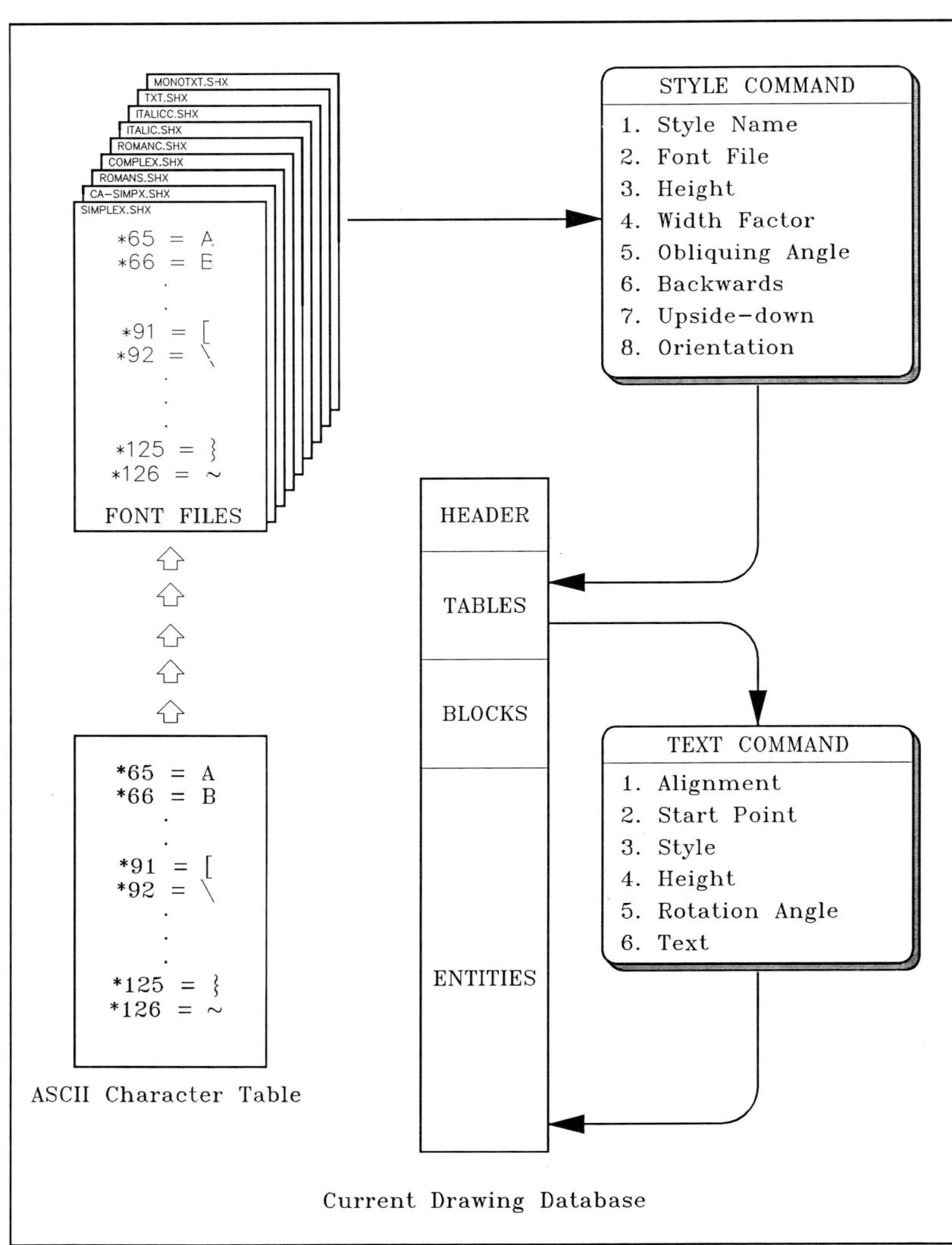

Text Style and Font Organization

PART III. SUPPORT LIBRARIES, FONTS AND HATCHES

CHAPTER 8

Text Styles and Fonts

Text Styles and Fonts offer a unique avenue to AutoCAD customization. Fonts and Styles give you control over annotation, the update of drawing text size, and the stylization of script characters. In this chapter, we will give you the basics you need to alter AutoCAD's standard font files, create your own characters and integrate them with standard AutoCAD fonts.

AutoCAD Text Fonts

Since AutoCAD can't display the regular DOS generated text on the graphics screen, it generates text through font simulation. Font characters are collections of vectors, which AutoCAD uses to display text characters on the graphics screen. AutoCAD uses a file, called a FONT file that contains a description for each printable character of the keyboard. Describing a character is simple. You tell AutoCAD to draw a vector for "some length" in "some direction." The book uses the term vector instead of line so we don't confuse a vector stroke of a Text entity with the Line entity.

Unlike using menus and AutoLISP, there are no easy ways to customize fonts. You need to use shape files with ASCII codes and vector coded definitions. Font customization involves making your own shapes, and controlling how the shapes are placed, sized and aligned. This chapter uses a simple form of font customization. It shows you how to create new "characters" by using references to existing characters. Then, it shows you how to use vectors to position the new "characters."

If you have no need to make special fonts, then you may want to skim this chapter and move on to AutoLISP.

The Benefits of Customizing AutoCAD Text Fonts

Text is one of the most widely used AutoCAD entities, yet it is seldom customized. What can you do with a font? You can make your own dimension fractions, switch your drawings from printed to handwritten script, or create your own company logo.

Why not just use a block and insert it each time you need a special character? One simple reason is you cannot place blocks inside a string of text. Every time you need a 0 with a slash to indicate a zero instead of the letter O, you have to go around and insert a block. If you write formulas and fractions in your drawing, isn't there a good way of subscripting and superscripting? There is with font customization.

Here are three main benefits that you gain with font customization:

- Fonts and Styles can give your drawing a distinctive professional look.
- Creating special characters overcomes any limitations of AutoCAD's standard text characters.
- Creating fractional characters lets you control how you show fractions on your drawings.

How To Skills Checklist

Here are the skills that you will acquire working through font customization. You will learn:

- ❑ How to use the Style command, make new styles, and change style definitions.
- ❑ How to make a center line CL symbol as a text font character.
- ❑ How to make fraction characters.
- ❑ How to make Bigfont characters and integrate Bigfonts into your standard styles.

Macros, AutoLISP Tools and Programs

MACROS
**SPECIALS is a menu page that types in special characters.

FILES
CA-SIMPX.SHP is a Bigfont shape file containing a CenterLine character, a subscript and a superscript, and a set of 1/16 increment fractions.

AutoCAD Styles

Text is a matter of style. AutoCAD gives you many choices. The Style command lets you alter the appearance of any set of font characters. Here is the Style command along with some comments to help refresh your AutoCAD memory.

```
Command: STYLE
Text style name (or ?) <STD>:        Name as referenced inside AutoCAD.
Font file <SIMPLEX>:                 The font character set to SIMPLEX.SHX.
Height <1'-0">:                      A Value presets the height, or 0 lets it vary.
Width factor <1.00>:                 Less than 1.0 squeezes the characters.
                                     More than 1.0 expands the character.
Obliquing angle <0>:                 Slants characters. - for left, + for right.
Backwards? <N>                       Draws them backwards, like a mirror.
Upside-down? <N>                     Puts characters upside down.
Vertical? <N>                        Defines a vertically oriented style.
```

You can use Style to preset text height so that you get the same height every time that style is used. Using Style not only saves answering the Text Height question, but reduces errors and lets you change font styles to resize text or stylize text.

Changing Text Size Using the STYLE Command

This exercise shows how AutoCAD can update restyled text on the screen with the Change command.

Enter selection: Begin a NEW drawing named CA08.

  Copy CA08.MNU to TEST.MNU.

  Copy MAIN.MNU to TEST.MNU.

Select **[TEST]**
Select **[24 x 36]**
Select **[FULL]**   Do this at 1:1 scale.
Select **[INITIATE]**   It should bring up TEST.MNU.

Command: **ZOOM**   Zoom Center, <RETURN> to default point, height 2 inches.

Command: **DTEXT**   Pick a point, rotation 0 degrees.
Text: **THIS IS 1/8" TEXT**   Hit return to continue text line.
Text: **PLACED IN AUTOCAD**
Text: **LINE BY LINE.**

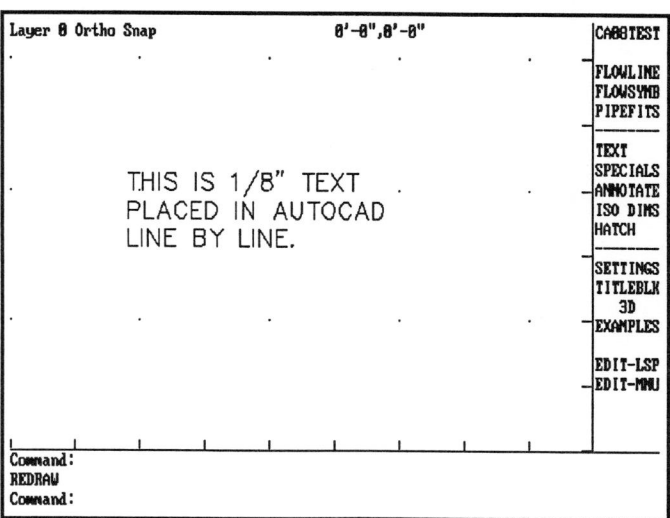

Some Standard Text

```
Command: CHANGE                                    Change the text height.
Select objects:                                    Pick the 2nd string, hit two <RETURN>s.
Enter text insertion point:
Text style: STD1-8
New style or RETURN for no change: STD1-16         Give a new style name.
New rotation angle <0>:
New text <PLACED IN AUTOCAD>:

Command: STYLE                                     Create a new Style.
Text style name (or ?) <STD1-8>: SLANT
Font file <txt>: TXT                               The TXT.SHX font character set.
Height <0'-0">: 1/8                                Height 1/8 inch and <RETURN> width for default 1.0.
Obliquing angle <0>: 10                            Slant 10 degrees to right and default the rest.

Command: CHANGE
Select objects:                                    Pick the 3rd string of text.
Enter text insertion point:                        <Return>
Text style: STD1-8
New style or RETURN for no change: SLANT           Give a new style name.
New rotation angle <0>:
New text <LINE BY LINE>:                           Enter some text or <Return>.
```

Your screen should look like a Slanted Style.

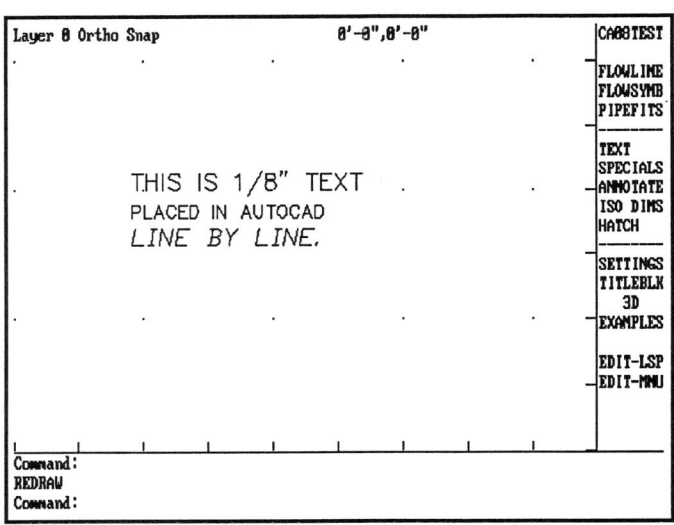

A Slanted Style

If you want to play around with the styles and fonts, type **STYLE ?** to see all the Styles.

Making a Centerline Character

The next exercise will show you how to develop your first shape character, a CenterLine. The book builds on this basic exercise to help you create special fractions for dimensioning. If you feel rusty trying to remember how AutoCAD handles fonts, look up "font definition" in the AutoCAD User Reference.

Making SIMPX.SHP

You will be working with one of the AutoCAD support files, SIMPLEX.SHP. Locate SIMPLEX.SHP in your \ACAD directory, or in the \SOURCE subdirectory from your original AutoCAD SUPPORT disks. Once you've found it, make a copy of the file, but call it SIMPX.SHP. Copy the file to the \CA-ACAD subdirectory.

```
Command: DUP                                Copy SIMPLEX.SHP, even if you have the CA DISK.
SOURCEfile(s) TARGETfile(s): \ACAD\SIMPLEX.SHP \CA-ACAD\SIMPX.SHP

Command: ED
File to edit: SIMPX.SHP                     Look at the SIMPX.SHP file.

*0,4,Standard Font
6,2,2,0
```

Looking at the first line, the *0 code alerts AutoCAD to the fact that this is a font file and gives information about the font strokes. It says, "Hey AutoCAD, the next line contains 4 pieces of information that you should look at." The second line says: "Each uppercase character extends 6 vectors above the baseline, a descender (for lower case letters) goes 2 vectors below the baseline and the font may be drawn either in a side by side manner, or it can be drawn vertically. When I give you a 0 code that means I'm finished talking."

AutoCAD needs to know how much space to put between successive lines of text. This is called a line feed and is treated just like any other character in the font file. Look in the file to find the line feed character.

```
*10,7,lf
2,0AC,14,8,(9,10),0
```

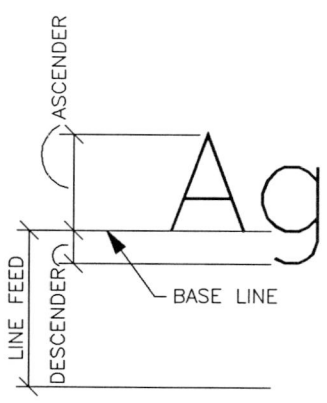

Character Ascender, Descender, and Baseline

The line feed character, ASCII 10, is 7 bytes long and instructs AutoCAD to move down 10 vector lengths.

Font characters are identified by a number from 1 thru 256, like the *10 above. Each number tells AutoCAD to draw a different text character. The number and the associated character it draws are standardized by the ASCII code system. AutoCAD does not define this system, it just adheres to it. You can find the standard ASCII character set in almost any book on DOS or BASIC. When you look them up, refer to the ASCII decimal value for the character, not the "hex" value.

Now, let's make a new font character, a capital C with a L superimposed over it to indicate a "Centerline."

Go to the end of your file and add the line:

`*150,11,Centerline`

Each asterisk marks the beginning of a new item in the file. The 150 is the ASCII number for the font character, 11 tells AutoCAD how many data entries to expect for the character definition.

You delimit all fields within a definition by commas. You can use parentheses to improve readability, but they are not required. Parentheses are ignored by the shape compiler.

The AutoCAD's font definition language has 15 key instruction codes. Look at the table to see what each one means. Here, you will only use codes 000 through 004, and the 007 code.

AutoCAD FONT DEFINITION INSTRUCTION CODES

```
0  Marks the end of a shape definition.
1  Places the "pen" in a down position.
2  Lifts the "pen" up.
3  Scales vector lengths by division of next number.
4  Scale vector by multiplication by next number.
5  Saves the current location in memory. (Pushes.)
6  Recalls the previously saved location. (Pops.)
7  Draws a subshape character as given by next number in file.
8  XY movement as given by next two numbers in file.
9  Continuous XY displacements. Terminate with (0,0).
A  Arc as defined by next two numbers in file.
B  Arc portion defined by the next 5 numbers.
C  Arc is defined with XY displacements and a bulge.
D  Continuous Arc defined by XY & bulge. Terminate with (0,0).
E  Vertical flag. Optional command.
```

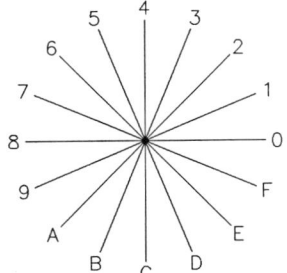

Vector Directions

The centerline font character does not create new vectors and shapes. It references the C character, 7,67, and the L character, 7,76 present in the file to create the character shape. To get vector directions, AutoCAD uses 16 predefined vector directions, 0 thru 9, and A through F. AutoCAD uses a hexadecimal-based numbering system to define vector direction to conserve memory. AutoCAD does not use numbers all the way up to 16 because numbers above 10 need two spaces in memory.

The Vector Direction diagram shows the vector directions. The standard vector lengths also are hexadecimal: 1 through 9 are normal, A=10, B=11, C=12, D=13, E=14, F=15, 10=16, 11=17, 12=18, 13=19.....20=32. IT'S BASE 16!

Add the following line to the end of your SIMPX.SHP file:

`2,064,7,67,008,-12,-13,7,76,064,0`

Hit <RETURN> at the end of line.

Making a Centerline Character 8—7

To follow the vector directions, look at the Centerline Diagram. Here is how the character is put together. First, you start a shape with a pen up 2 code and move the pen 6 vectors in the 4 direction. Then, call the subshape using the 7 instruction code and the number of the shape, ASCII 67 for uppercase C. Next, make a move to place the L character.

The 008 calls a pen movement by specifying an XY displacement. The code immediately following the 008 code moves the pen -12 vectors in the X direction, followed by -13 vectors in the Y direction. Call the 7 code again to place shape ASCII 76, uppercase L. Then, move the pen 6 vectors in the 4 direction to place the pen back at the base line. Finally, end the shape definition with the 0 code.

AutoCAD doesn't care where you break the lines of the shape file. You can write the same definition, using the lines shown below. The comments, "See #," refer the numbers in the CenterLine Diagram.

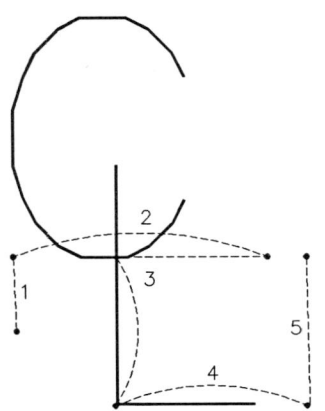

CenterLine Diagram

```
2,              Pen-up. Always put a pen up at the beginning.
064,            Move 6 units in 4 direction (90 degrees). See 1.
7,67,           67 "C". See 2.
008,-12,-13     Offset -12 in X direction, -13 in Y. See 3.
7,76,           76 "L". See 4.
064,            Move 6 units in 4 direction (90 degrees). See 5.
0               0 indicates the end of the shape definition.
```

Now, have AutoCAD compile the shapes. AutoCAD makes a .SHX file from the .SHP file. The file name will be SIMPX.SHX. While compiling, you are given a message and the size of the output file.

```
Command: END                                    End the drawing.
Enter Selection: 7                              Compile shape/font description at the Main Menu.
Enter NAME of shape file (default ca08): SIMPX  File named SIMPX.

Compiling shape/font description file
Compilation successful.  Output file SIMPX.SHX contains 2709 bytes.
Press RETURN to continue:
```

AutoCAD only sees characters 10, and 32 to 128, as standard keys on the keyboard. If you want to enter ASCII characters above 128, you need to type them in a special way. Characters above 128 are referenced by an ASCII number, preceded by a double %% percent sign. You may also be able to enter the characters as an "Alt -code," depending on your system configuration. Hold down either your <ALT> or <ALT>+<LEFT-SHIFT> keys while you type the number 150 **with the numeric keypad keys,** then release. The advantage to this method is that you can see the character in DTEXT, instead of seeing %%150.

```
Enter selection:                Go back into the drawing, CA08.

Command: STYLE                  Define new style SIMPX using font SIMPX and height 1/8".
Command: DTEXT                  Use either method. Enter %%150 or <ALT-150>.
```

The %% is not special for ASCII numbers above 128. You also can type a capital A by typing %%65 or <ALT-65>. AutoCAD treats all %%nnn (where nnn is the ASCII number) codes the same.

Developing Special Fraction Characters

Many dimensions require fractions. AutoCAD's standard character set lacks fractions. Rather than live with this limitation, you can make special fractions for your dimensions. Let's start by making the fraction 1/8. Again, you use three characters that are already defined in the font file, ASCII code 49 for **1**, code 47 for **/**, and code 56 for **8**.

Making Special Fractions

The exercise defines a 1/8 fraction as ASCII code 151.

Command: **ED**   Edit the SIMPX.SHP file.

At the bottom of the file add the following 2 lines.

*151,29,oneeigth
2,0F4,3,06,4,05,020,7,49,3,05,4,06,0CB,7,47,0CB,3,06,4,05,7,56,3,05,4,06,094,0

Save and Ext, be sure to <RETURN>.

The following table shows the definition so that it is easier to follow. The annotations, "See #," refer to the Special 1/8 Fraction vectors diagram to help you follow the vector directions.

*151,29,oneeigth	Defines it as ASCII 151 with 29 bytes.
2,	Lift pen with 2 instruction.
0F4,	Move F vectors (15) in the 4 direction. See 1.
3,06,4,05,	Divide by 6, Multiply by 5.
	A 5/6ths scale for the following strokes.
020,	Move 2 vectors in the 0 direction. See 2.
7,49,	Call the ASCII 49 subshape "1". See 3.
3,05,4,06,	Reverse the scale factor by 6/5 to normal.
0CB,	Move C vectors (12) in the B direction. See 4.
7,47,	Insert ASCII 47 "/". See 5.
0CB,	Move C vectors (12) in the B direction. See 6.
3,06,4,05,	Flop the scale factors back to 5/6.
7,56,	Call the subshape "8" (ASCII 56). See 7.
3,05,4,06,	Reset the scale factor to normal.
094,	Move 9 vectors in the 4 direction. See 8.
0	End the shape with the 0 code.

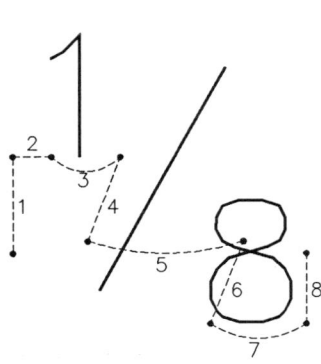

A Special 1/8 Fraction

Command: **END**   End the drawing.
Enter Selection: **7**   Compile shape/font description.
Enter NAME of shape file (default ca08): **SIMPX**   File named SIMPX.

```
Enter selection:

Command: DTEXT
```
Restart the drawing and test it.
The SIMPX Style should be current, defined with the SIMPX font.
Use either %%151 or <ALT-151>.
Try a dimension too, using one of these codes.

The fraction definition uses the 3 and 4 codes to scale the characters so they are drawn slightly smaller than normal characters. The 3 code indicates that the next number divides the current vector length. The 4 code indicates the next number multiplies the vector length. The 3,06,4,05, first divides the vector length by 6, then multiples that length by 5, creating a vector length that is 5/6 the size of the normal vector. Only integers are acceptable as input.

You always need to reset the scale factors at the end of the shape. AutoCAD remembers the factors between each shape. If you don't reset the scaling factors, AutoCAD will draw the next character with the last scale calculated.

BigFonts

A BIGFONT character is a character that has a number above the ASCII limit of 256. AutoCAD developed Bigfont characters to work with non-ASCII foreign language character sets, and sets with more than 256 characters in the language. The Japanese KANJI character set is an example. Bigfonts require two key strokes for each character. Instead of typing something like %%151 for the 1/8 character, you might call a bigfont character by ~1, a tilde followed by a regular character.

Bigfonts are stored in shape font files that are handled in a special way by AutoCAD. AutoCAD's special handling lets you append your font characters to any font file already in use with your system. You can append bigfonts without modifying existing font files.

Using bigfonts to make special characters lets you use the stroke style of any existing compiled font file. For example, if your current text style, say STANDARD, uses the SIMPLEX.SHX file, your special characters can use the 1, 8 and slash character from the SIMPLEX file. Your special character definitions are independent of the font strokes and font file you use.

Bigfont definitions also override duplicate standard font characters. For example, if you redefine your zero, ASCII 48, in a bigfont file to put a slash through the zero, the zero will override the standard zero character of any font that is loaded along with it.

BIGFONTS Files

A bigfont file starts with a header that alerts AutoCAD to the special file format. The line:

*BIGFONT 20,1,126,126

identifies the file as a bigfont file. The header has four pieces of information that tells AutoCAD how to treat the contents of the file.

The first number, 20 is the number of character definitions in the file. It doesn't have to be exact, but it does have to be within 10% of the actual number. The second number, 1 in our example, tells AutoCAD how many successive keyboard pairs, issued as input, it takes to "sense" a complete character. You only need a single pair of characters to call a bigfont character. However, fonts like the KANJI set need two pairs (4 keys) to call a single character. You might key in a KANJI character as ~M~C (mother, child) meaning "Peace on Earth."

The tilde ~ flags the beginning of a paired key set. The tilde is an ESCAPE CHARACTER. It instructs AutoCAD to treat the character that follows it with special meaning. The escape character can be any character you wish. It should be a character that you use infrequently, but should be accessible on the keyboard. There are not many good keys to choose from, some choices are:

```
`        Reverse apostrophy.
|        Vertical bar.
~        Tilde.
```

The book uses the tilde. The tilde has an ASCII code of 126. AutoCAD needs to know which character you have defined as the escape. This information is contained in the header label for the file. The code 126 must appear as the fourth and fifth item of the header line to define the tilde as a single pair and double pair character set.

You will need more codes, so try to plan ahead. Try to devise a mnemonic system. The book uses an easy one. Tilde will be your escape character, and the numerators of the fractions will be the numbers along the top of the keyboard. This works until you run out of keys and spill over onto the QWERTY keys! Here is the book's BIGFONT MAP.

BIGFONT CHARACTER MAP

Key	Fraction	ASCII code	Bigfont number
~1	1/16	49	32305
~2	1/8	50	32306
~3	3/16	51	32307
~4	1/4	52	32308
~5	5/16	53	32309
~6	3/8	54	32310
~7	7/16	55	32311
~8	1/2	56	32312
~9	9/16	57	32313
~0	5/8	58	32314
~q	11/16	113	32369
~w	3/4	119	32375
~e	13/16	101	32357
~r	7/8	114	32370
~t	15/16	116	32372
~C	CL	67	32323
~S	Superscript	83	32339
~s	subscript	115	32371
~~	the tilde	126	32382

NOTE. All keys are lower case, except the Superscript and Centerline.

You still need your font file definition lines in the file, like a normal font. The 6,2,2,0 (above,below,modes,0) is the same as the SIMPLEX font. The book uses:

```
*0,4,SPECIALS   Customizing AutoCAD
21,7,2,0
```

We will make use of the existing definition of the 1/8 special fraction, giving it a new code number. To compute the number you assign to a special character in a bigfont file, take the ASCII code of your escape code character, 126 in this case, and multiply it by 256. This yields 32256. This number is called the base offset.

There are 256 characters in the ASCII set. Each of the 256 ASCII characters can be an escape character for a character set. Each character set can have up to 256 characters. This yields 256 times 256 potential characters, a total of 65,536.

The book's base offset is 32256. Its set of 256 potential characters starts at the 32256th position in the 65,536 total. For each character in your set, you add the base offset to the value of the ASCII character that is used for the second key of the pair. For a 1/8 character we use a ~2 combination. Since 2 is ASCII 50, the code is 32256+50 or 32306.

Since you use the regular tilde as the escape character, you have to recreate a tilde in your upper level of characters. This ensures that you can type one, if you ever need it. To do get a tilde, type ~~.

Integrating Special Characters as Bigfonts

  Examine the CA DISK file CA-SIMPX.SHP.

  Copy SIMPX.SHP to CA-SIMPX.SHP.

Command: **ED**   Edit CA-SIMPX.SHP.

You don't need the standard Simplex characters, except for the tilde (ASCII 126).
Delete all lines above the *150,11,CenterLine line except for:

```
*126,49,ktilde
2,14,8,(-9,-12),064,1,024,8,(1,3),021,020,02F,8,(4,-3),02F,020,
021,023,2,8,(-18,-2),1,023,021,020,02F,8,(4,-3),02F,020,021,
8,(1,3),024,2,8,(8,-12),14,8,(-17,-4),0
```

Enter the header lines.
Renumber *150 to *32323, for ~C, and add a -C to indicate that it is called by ~C.
Renumber the 1/8 font character from 151 to 32306, for ~2, and add a -2. It is called by ~2.
Renumber the tilde *126,49,ktilde to **32382,49,ktilde** (32382 is 32256+126). You should have:

```
*BIGFONT 20,1,126,126
*0,4,SPECIALS   Customizing AutoCAD
21,7,2,0
*32323,11,CenterLine-C
2,064,7,67,008,-12,-13,7,76,064,0
*32306,29,oneeigth-2
2,0F4,3,06,4,05,020,7,49,3,05,4,06,0CB,7,47,0CB,3,06,4,05,7,56,3,05,4,06,094,0
*32382,49,ktilde
2,14,8,(-9,-12),064,1,024,8,(1,3),021,020,02F,8,(4,-3),02F,020,
021,023,2,8,(-18,-2),1,023,021,020,02F,8,(4,-3),02F,020,021,
8,(1,3),024,2,8,(8,-12),14,8,(-17,-4),0
```
Don't forget to <RETURN> after the last line.

Save. Exit and test it. Remember to recompile the font. Then, come back in the file to add some more.

To make the rest of the fractions, use the following procedure:

1. Copy both lines of the 1/8 definition 15 more times.

2. Change the *32306,29,one eighth-2 to the appropriate code and description.

3. Substitute the appropriate 7,nn, ASCII codes for the 7,49, "1" and/or 7,56, "8".

4. For characters without a "1" for the numerator, delete the **020,** code before the **7,49,** numerator code. It adjusts for the narrower character. If you delete the 020, you must change the **29,** the number of bytes in the definition line, to **28**.

5. For the characters with 16ths denominator, put both byte pairs **7,49,7,54,** in place of the 7,56, "8" code. You must add **2 (bytes)** to the 28 or 29 (number of bytes in the definition line), making it **30** or **31**.

6. For characters with 11, 13, or 15 numerators, put both byte pairs **7,49,7,nn,** where nn is 49, 51 or 53, in place of the 7,49, "1" code. You must add **2 more bytes**, making it **32** or **34**.

To make the 1/4 fraction, for example, use the subshape ASCII 52, for "4", in place of subshape ASCII 56.

After you add the rest of the special fractions to the CA-SIMPX.SHP file, you will have:

Command: **ED**   Edit your CA-SIMPX.SHP file to:

```
*BIGFONT 20,1,126,126
*0,4,SPECIALS   Customizing AutoCAD
21,7,2,0
*32323,11,CenterLine-C
2,064,7,67,008,-12,-13,7,76,064,0
*32305,32,onesixteenth-1
2,0F4,020,3,06,4,05,020,7,49,3,05,4,06,0CB,7,47,0CB,3,06,4,05,7,49,7,54,
3,05,4,06,094,0
*32306,29,oneeigth-2
2,0F4,3,06,4,05,020,7,49,3,05,4,06,0CB,7,47,0CB,3,06,4,05,7,56,3,05,4,06,094,0
*32307,32,threesixteenth-3
2,0F4,020,3,06,4,05,020,7,51,3,05,4,06,
```

```
0CB,7,47,0CB,3,06,4,05,7,49,7,54,3,05,4,06,094,0
*32308,29,onequarter-4
2,0F4,3,06,4,05,020,7,49,3,05,4,06,0CB,7,47,0CB,3,06,4,05,7,52,3,05,4,06,094,0
*32309,32,fivesixteenth-5
2,0F4,020,3,06,4,05,020,7,53,3,05,4,06,
0CB,7,47,0CB,3,06,4,05,7,49,7,54,3,05,4,06,094,0
*32310,28,threeeigths-6
2,0F4,3,06,4,05,7,51,3,05,4,06,0CB,7,47,0CB,3,06,4,05,7,56,3,05,4,06,094,0
*32311,32,sevensixteenth-7
2,0F4,020,3,06,4,05,020,7,55,3,05,4,06,
0CB,7,47,0CB,3,06,4,05,7,49,7,54,3,05,4,06,094,0
*32312,29,onehalf-8
2,0F4,3,06,4,05,020,7,49,3,05,4,06,0CB,7,47,0CB,3,06,4,05,7,50,3,05,4,06,094,0
*32313,32,ninesixteenth-9
2,0F4,020,3,06,4,05,020,7,57,3,05,4,06,
0CB,7,47,0CB,3,06,4,05,7,49,7,54,3,05,4,06,094,0
*32304,28,fiveeigths-0
2,0F4,3,06,4,05,7,53,3,05,4,06,0CB,7,47,0CB,3,06,4,05,7,56,3,05,4,06,094,0
*32369,34,elevsixteenth-q
2,0F4,020,3,06,4,05,020,7,49,7,49,3,05,4,06,
0CB,7,47,0CB,3,06,4,05,7,49,7,54,3,05,4,06,094,0
*32375,28,threequarters-w
2,0F4,3,06,4,05,7,51,3,05,4,06,0CB,7,47,0CB,3,06,4,05,7,52,3,05,4,06,094,0
*32357,34,thirteensixteenth-e
2,0F4,020,3,06,4,05,020,7,49,7,51,3,05,4,06,
0CB,7,47,0CB,3,06,4,05,7,49,7,54,3,05,4,06,094,0
*32370,29,seveneigths-r
2,0F4,3,06,4,05,030,7,55,3,05,4,06,0CB,7,47,0CB,3,06,4,05,7,56,3,05,4,06,094,0
*32372,34,fifteensixteenth-t
2,0F4,020,3,06,4,05,020,7,49,7,53,3,05,4,06,
0CB,7,47,0CB,3,06,4,05,7,49,7,54,3,05,4,06,094,0
*32382,49,ktilde
2,14,8,(-9,-12),064,1,024,8,(1,3),021,020,02F,8,(4,-3),02F,020,
021,023,2,8,(-18,-2),1,023,021,020,02F,8,(4,-3),02F,020,021,
8,(1,3),024,2,8,(8,-12),14,8,(-17,-4),0
```

For good measure, add some subscript and superscript codes. They are pen up moves:

```
*32339,3,superscript-S
2,044,0
*32371,3,subscript-s
2,04C,0
```

Don't forget to <RETURN> after the last line.

Save, exit to the AutoCAD Main Menu and recompile the CA-SIMPX file.

```
Enter selection:           Re-enter your drawing and test it.
```

Using Bigfonts with AutoCAD

Let's integrate the Bigfonts. Then, make a menu page for the Big fonts.

It is easy to integrate a bigfont file into your system, using the AutoCAD Style command. At the font file prompt, answer with both the standard font file you use, like SIMPLEX, and the new CA-SIMPX file. Separate the font names by a comma.

Integrating Bigfonts

```
Command: STYLE
Text style name (or ?) <SIMPX>: CA-SIMPX
New style.
Font file <txt>: SIMPLEX,CA-SIMPX      Give both names. Set height 1/8". Default the rest.
CA-SIMPX is now the current text style.
```

Add the following menu, then test everything.

```
Select [EDIT-MNU]                      Edit the TEST.MNU.
```

Make the following changes to the root menu.

```
***SCREEN
**ROOT
[CA08TEST]

[SPECIALS]$S=FOOTER $S=SPECIALS
```

Relabel the **EXAMPLES page to **SPECIALS and add the following:

```
**SPECIALS
[Specials]
[TEXT     ]^C^C^CTEXT
[ 1/16 ~1]~1 $S=
[ 1/8  ~2]~2 $S=
[ 3/16 ~3]~3 $S=
[ 1/4  ~4]~4 $S=
[ 5/16 ~5]~5 $S=
[ 3/8  ~6]~6 $S=
[ 7/16 ~7]~7 $S=
[ 1/2  ~8]~8 $S=
[ 9/16 ~9]~9 $S=
[ 5/8  ~0]~0 $S=
[11/16 ~q]~q $S=
[ 3/4  ~w]~w $S=
[13/16 ~e]~e $S=
[ 7/8  ~r]~r $S=
[15/16 ~t]~t $S=
[ CL   ~C]~C $S=
[]
```

Save, exit, reload the menu and test it.

```
Select [SPECIALS]
Select [TEXT]                          Select one of the fraction characters. Test them all.
```

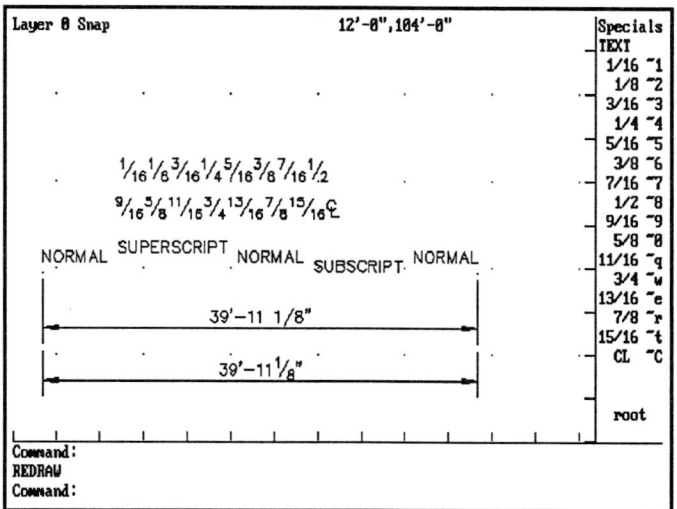

Fractions, Subscripts and Superscripts

Use Dtext to try the Subscripts and Superscripts. Remember, the DTEXT command only accepts keyboard input and will not accept input from the menu. But, if you type these ~S and ~s codes, you see the text in the drawing as you type, like the <ALT-nnn> method.

```
Command: DTEXT                          Try the sub-, superscripts.
Start point or Align/Center/Fit/Middle/Right/Style:
Height <1'-0">:
Rotation angle <0>:
Text: NORMAL ~SSUPERSCRIPT ~sNORMAL ~sSUBSCRIPT ~SNORMAL

Command: QUIT
```

 Delete TEST.MNU.

 Copy TEST.MNU to MY08.MNU.

Integrating the Specials Menu

If you don't have the CA DISK and want to add the text menu to CA-MENU:

- Append the **SPECIALS page of your MY08.MNU to the end of the CA-MENU.MNU file.
- Change the [SPECIALS] label on the **ROOT page of the CA-MENU to:

[SPECIALS]^C^C^C$S=TEXT $S=FOOTER LAYER S TXT01;;TEXT S STD1-8

Here is the chapter's menu.

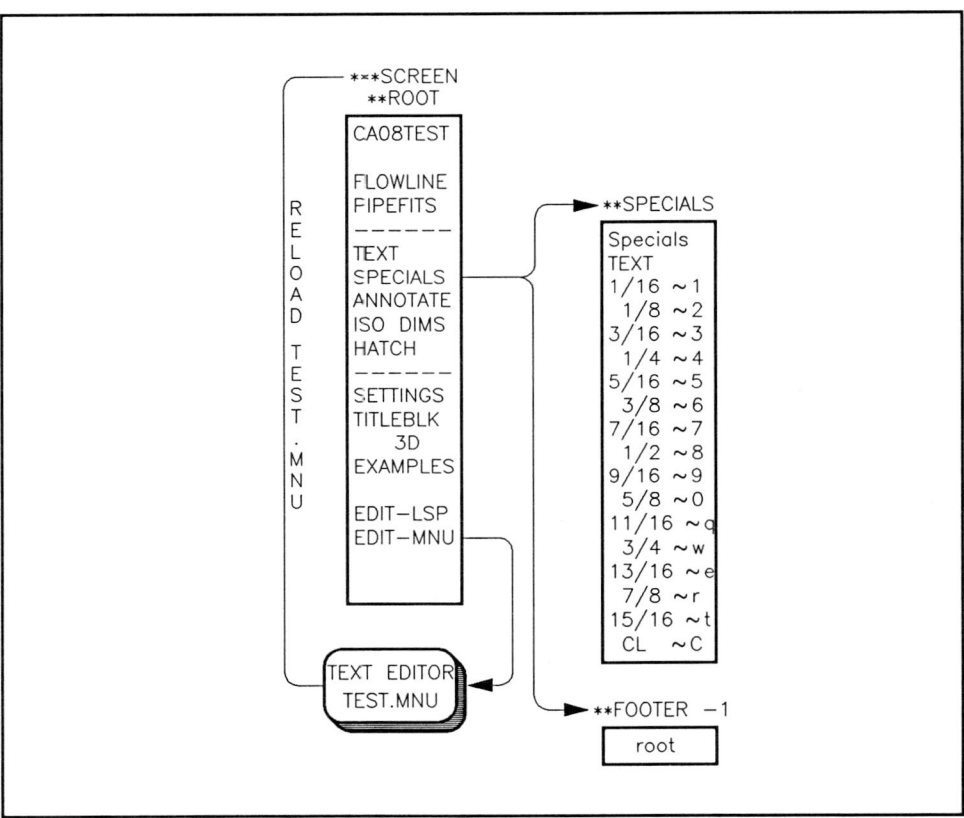

The Chapter 8 Menu Added to CA-MENU

Summary Tips and Techniques

Here are a few tips.

- Add a *48 slash-zero character to your bigfont file to redefine the character.

- Use a bigfont to define standard boilerplates and notes. String each phrase together as a single bigfont character.

- If you do use a custom font, remember that it needs to accompany the drawing when it is displayed or plotted.

There are no easy ways to text customization. If you need special text characters, you have to get "down and dirty" and create them.

The same holds true if you need special linetypes, hatches and fills. Let's move on to look at their customization.

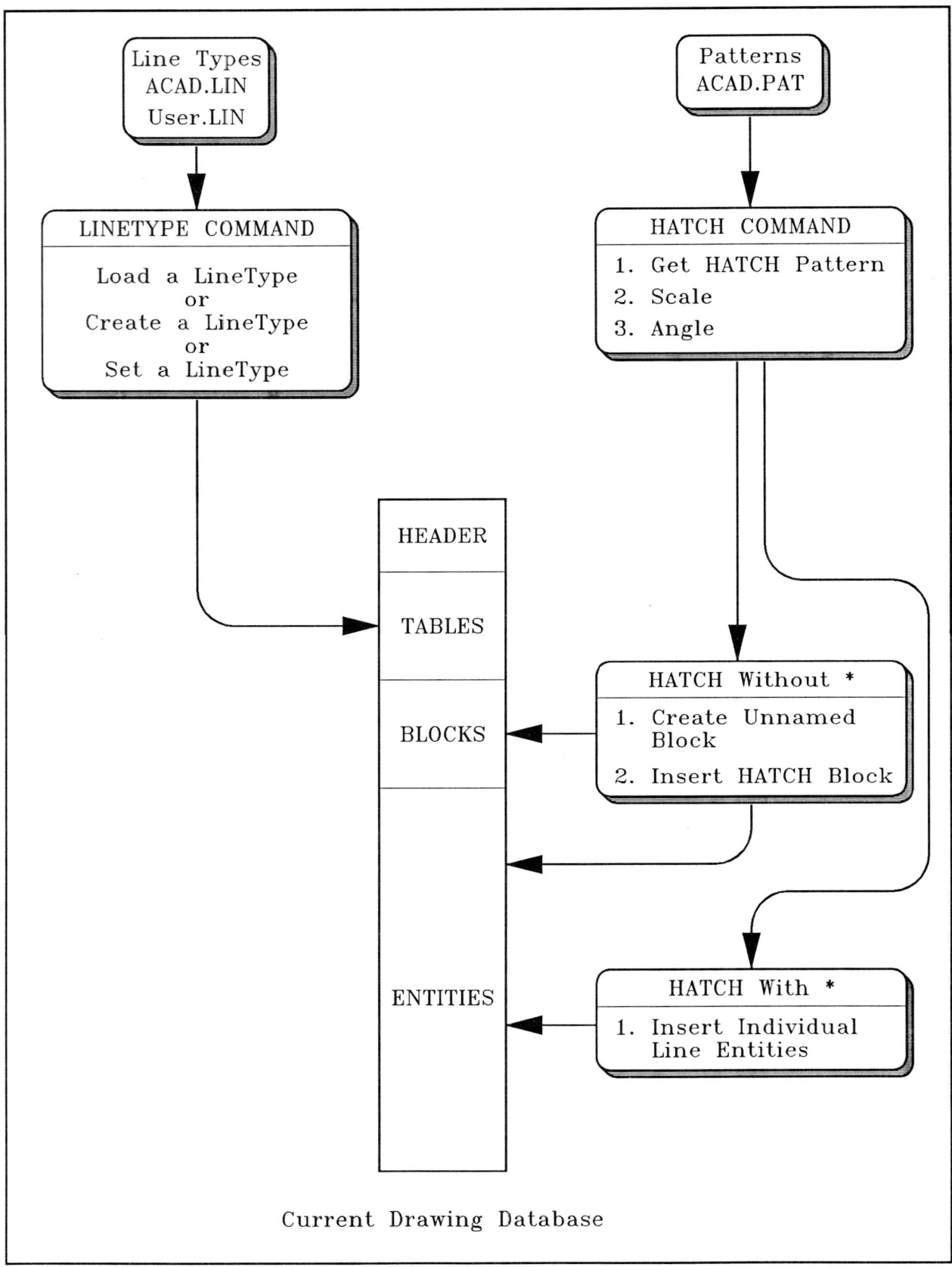

Linetype and Hatch Organization

CHAPTER 9

Linetypes, Hatches and Fills

You can add visual excitement, clarity and textured meaning to your drawings by using patterned lines, areas and solid fills. This chapter shows you how to customize linetypes, hatch patterns and irregular fills.

The Benefits of Customized Hatches and Fills

There are two basic benefits to customizing your hatches and fills:

- You can extend your drawing tools to indicate materials or components by using custom linetypes and hatches.

- You can make your materials labeling faster by combining macros with hatches.

How To Skills Checklist

You will learn how to:

- Create Linetypes and control their scales.
- Write Hatch patterns, using a development box and using AutoCAD to do the hatch calculations.
- Put together a screen menu that uses basic blocks to automatically fill in irregular areas with hatch patterns.

Macros, AutoLISP Tools and Programs

MACROS
**HATCH is a screen page with irregular hatch fill blocks.
[CONCRE] is a macro which calls the concrete hatch pattern definition.
[CHECKP] is a macro to draw checkered plates.

CONCRE is a hatch pattern definition which draws concrete.
CHECKP is a hatch pattern that draws checkered plate hatches.

Patterns of Dots and Dashes

In AutoCAD's world, a Linetype PATTERN is a set of instructions used to draw a broken line. A Hatch PATTERN adds angle and spacing to draw a set of lines. You write a pattern by describing a dot-dash sequence. You numerically define the dot-dash pattern to tell AutoCAD how long each dash is and how far to skip between them.

9—2 Customizing AutoCAD

There are two basic controls present with patterns. They are a "pen up" and "pen down" motion of the graphics "pen." It doesn't matter whether you are dealing with screen vectors or plotting vectors, the terms, pen up and pen down, are used generically to mean the pen is actively drawing a segment or is skipping a segment. A negative number means "lift" the pen up and move in the "up" position for some distance.

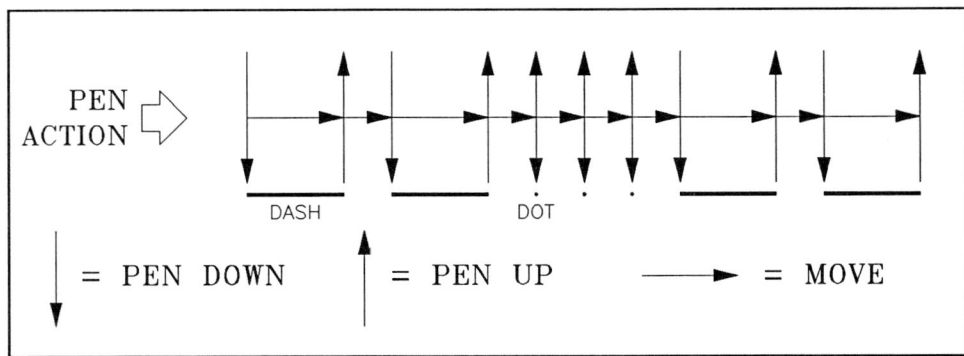

Dash and Dot Patterns

Linetypes

A LINETYPE is a pattern that AutoCAD applies "on-the-fly" to drawing entities. You can pattern all entities except TEXT, TRACEs and SOLIDs. AutoCAD applies a dot-dash pattern along the length or circumference of an entity. AutoCAD automatically adjusts the beginning and ending portion of the line pattern to make it fit equally between the end points of segments.

Linetypes are stored in a default library support file called ACAD.LIN. AutoCAD provides a Create option to the Linetype command to make linetype definitions. You can edit the ACAD.LIN file with any ASCII text editor and create other name.LIN files.

To help you get a feel for working with linetypes, let's create a simple linetype called DASH3DOT. DASH3DOT has one unit long dashes and three dots, each dot is separated by a quarter unit space.

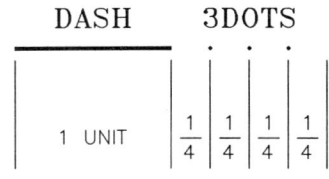

DASH3DOT Linetype

Creating a DASH3DOT Linetype

Enter selection: Begin a NEW drawing named CA09.

  Copy CA09.MNU to TEST.MNU.

  Copy MAIN.MNU to TEST.MNU.

Command: **DUP**   Copy your ACAD.LIN file from your AutoCAD directory
                                                    to \CA-ACAD\CA-ACAD.LIN.

Select **[TEST]** **[24 x 36]** **[FULL]** **[INITIATE]**   It should bring up TEST.MNU.

```
Command: ERASE                                          Erase the border.

Command: LINETYPE                                       Use the Linetype command to create a new pattern.
?/Create/Load/Set: C
Name of linetype to create: DASH3DOT                    AutoCAD requests a name for the linetype definition.
File for storage of linetype <ACAD>: CA-ACAD            Use your copy of the standard file.
Wait, checking if linetype already defined...
```

After AutoCAD verifies that the linetype definition does not exist, you get a prompt for a description of the line pattern. It helps to use a visual reminder with underscores and periods, as well as a narrative description in parentheses to describe the pattern.

The numerical values defining the pen up and down motions are preceded by an "alignment" code. Currently, the alignment code is limited to only one: "A." AutoCAD automatically puts the code in for you.

```
Descriptive text: ___...___...___...___  (Dash with 3 dots)    Enter the description

Enter pattern (on next line):                                   Enter the line:

A,1,-0.25,0,-0.25,0,-0.25,0,-0.25

New definition written to file.
?/Create/Load/Set: L
Name of linetype to load: DASH3DOT                      Linetypes must be loaded before use.
File to search <CA-ACAD>:
Linetype DASH3DOT loaded.
?/Create/Load/Set:
```

Here is how the pattern is put together. The definition starts with a 1 unit dash. Then, it lifts the pen one-quarter unit and puts a dot (-0.25,0). Dots are drawn by giving a 0 unit length. The other two dots are drawn. The sequence ends by moving over another quarter unit. The four pen up codes relate to the four blank segments of each pattern segment.

Scaling Linetype Patterns

Linetype patterns are influenced by a global AutoCAD scale factor. The LTSCALE command sets and applies this factor to the unit values of the linetype pattern definition. We normally suggest you use a linetype scale factor of 0.375 times the plotting scale factor. For a 1/8-inch drawing, the plotting scale (DIMSCALE variable) is 96.0, making LTSCALE equal to 36.0. However, to make it easier to see unit relationships in Linetypes, set LTSCALE to 1 for the following exercise.

Adjusting Line Scale

To see how AutoCAD matches the pattern definition to entities, draw a few lines and circles and put them on a layer using the new linetype.

```
Command: LTSCALE
New scale factor <0.3750>: 1
```
Change for a 1:1 relationship to the linetype definition.

```
Command: UNITS
Command: SNAP
Command: GRID
```
Set to decimal units, 4 places.
Set to 0.25.
Set to 1.

```
Command: LAYER
?/Make/Set/New/ON/OFF/Colcr/Ltype/Freeze/Thaw: LT
Linetype (or ?) <CONTINUOUS>: DASH3DOT
Layer name(s) for linetype DASH3DOT <TEST>:           <RETURN><RETURN>
```
Make layer TEST with Color Green for the linetype. Then:

```
Command: ZOOM
```
Zoom Left corner 0,0 and Height 6.

```
Command: LINE
```
Draw 4 lines, 1, 2, 3.5 and 4 units long. (You're in decimal).

```
Command: CIRCLE
```
Now draw 3 circles with 0.25, 0.5, 2 unit radius.

Your screen should look like Linetype Pattern Adjusting.

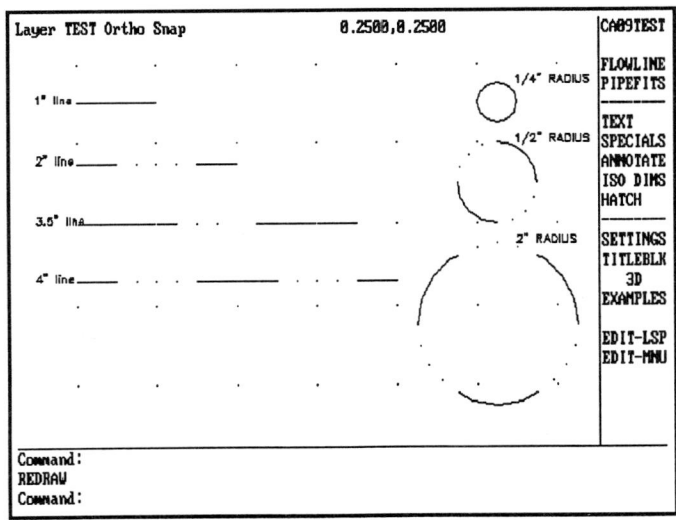

Linetype Pattern Adjusting

Look at the set of lines to see how AutoCAD fits the linetypes. AutoCAD can't fit the pattern to the 1 unit long line. It starts to fit the pattern to the 2-unit line, applying the dash equally between the line's endpoints. The 3.5-unit line shows how AutoCAD makes the adjusted ends longer until there is sufficient space to fit in at least one dash segment. The entire line pattern appears when the line is 4-units long.

Again, look at how AutoCAD applies the pattern to circles. The 0.25-radius circle (circumference 1.57 units) shows no linetype breaks. The circle with a circumference of 3.1416 shows two sets of dots. The 2-unit circle shows three sets of dots at 120 degree intervals.

Hatch Patterns

Both linetype and hatch patterns follow the same concept of a pen up/down motion. Hatch patterns have several differences. A linetype has an alignment code, but hatches are aligned by the snap base point of the drawing. Unlike a linetype, hatch patterns are families of patterned lines forced to specific angles. Hatches patterns are not applied to entities but instead create new entities to fill in a bounded area.

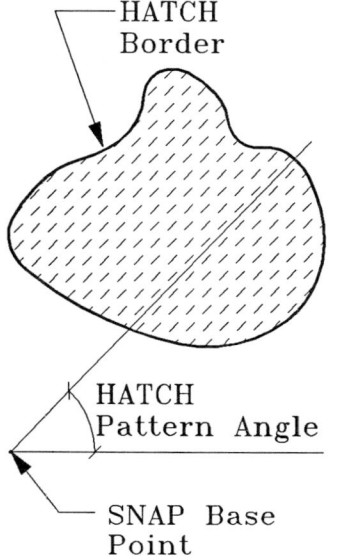

Hatch Pattern Components

Hatches can create Blocks or individual Line entities. To create individual Lines, precede the Hatch pattern name with an "*," like "Insert *" for blocks. If you don't use the *, an unnamed block is created for each Hatch insertion. These blocks actually have hidden names like *X1, *X2, etc. Leaving hatches as Blocks saves no file space, since each has a unique Block Def which includes each individual line within it. Exploding a Hatch temporarily doubles the entity count, but then the Block Def is automatically purged upon the next load of the drawing file. There are no commands to help create Hatch patterns, but you will develop a customized one later.

Hatch patterns are stored in the ACAD.PAT file. Unlike linetypes, pattern files cannot have other names. **Be careful to not overwrite your original ACAD.PAT file in your ACAD directory.**

A typical entry in a hatch pattern file may look like:

`45, 0.125,0.125, 0.25,0.25, 0.25,-0.25`

where each line has items in the format:

`Angle, X-origin, Y-origin, X-offset, Y-offset, dash-1, dash-2`

The angle, tells AutoCAD which direction to draw the lines, forty-five degrees in the example. The next two fields, X-origin and Y-origin, control the starting point of the line segment. This is not an AutoCAD coordinate, but rather a relative distance from the current Snap base point of the drawing. The X-offset and Y-offset control the incrementing of the origin point to repeat the line. The last part of a hatch definition is the dot-dash pattern.

When AutoCAD draws a Hatch, each linetype in the pattern continues at the specified angle in both directions until it hits a boundary. It dashes and dots according to the pattern. Each line is repeated parallel to itself. The origin of each repeat is offset from the previous line by the X-offset and Y-offset values, and the dash-dots are applied to that line. The X,Y offset is measured perpendicular to the angle of the line, like a rotated axis. Each line is repeated parallel until it fills the boundary being hatched. Groups of parallel lines created by each line of the Hatch pattern are referred to as families of lines.

The next exercise shows how to create a simple checkered plate hatch pattern.

Making a Checkered Plate Pattern

Command: **ERASE**	Get a fresh screen. Erase the previous entities.
Command: **SNAP**	Set to 0.1
Command: **ZOOM**	Left, corner 0,0 and Height of 3.5.
Command: **PLINE**	Draw a 1x1 box with a Polyline. Close the last segment.
From point: **0,0**	

 You have the custom ACAD.PAT file in your \CA-ACAD directory. Examine it.

 Do the following:

Command: **ED**
Start and edit a new ACAD.PAT file in the \CA-ACAD directory. Enter the lines:

*CHECKP, Checkered Plate
0, 0,.09375, .25,.25, .25,-.25
90, .125,.21875, .25,.25, .25,-.25   <RETURN>

Save, Exit and return to AutoCAD.

Be sure to <RETURN> at the last line or you'll get: "Bad pattern definition file:" when you try to use it. If you don't <RETURN>, you may also crash AutoCAD.

Command: **HATCH**	
Pattern (? or name/U,style): **?**	List the patterns in the file.
CHECKP - Checkered Plate	There should only be one, (unless you have the CA DISK).
Command: **HATCH**	Try it.
Pattern (? or name/U,style): **CHECKP**	Give the name.
Scale for pattern <1.0000>: **.5**	Trust us!
Angle for pattern <0>: **45**	Rotate the pattern.
Select objects: **L**	The Pline is Last.

Your screen should look like the Checkered Plate Pattern.

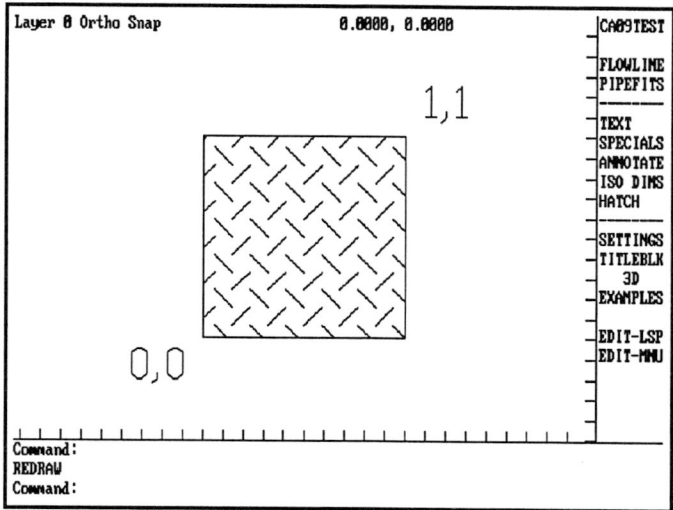

The Checkered Plate Pattern

Even though you defined the pattern with 0 and 90 degree lines, you can rotate the pattern to get a desired effect. We made the pattern at 0 and 90 degrees to avoid calculating angles.

This exercise didn't show you how to figure out the origins and offsets. The next exercise developing a concrete hatch pattern shows you how to figure the origins and offsets.

Making a Hatch Pattern

The concrete hatch pattern example will give you a handy technique for creating seemingly complex irregular patterns with little effort. The challenge presented by non-regular hatch patterns is to wade through a set of calculations to make the pattern appear random.

You can make hatches appear random by developing a tiny pattern and applying it at a larger scale. This exercise helps you develop a simple concrete hatch pattern. Later on, we will show you how to automate the whole hatch generation process using AutoLISP. Think of it. You'll have custom hatches without writing hatch patterns.

9—8   **Customizing AutoCAD**

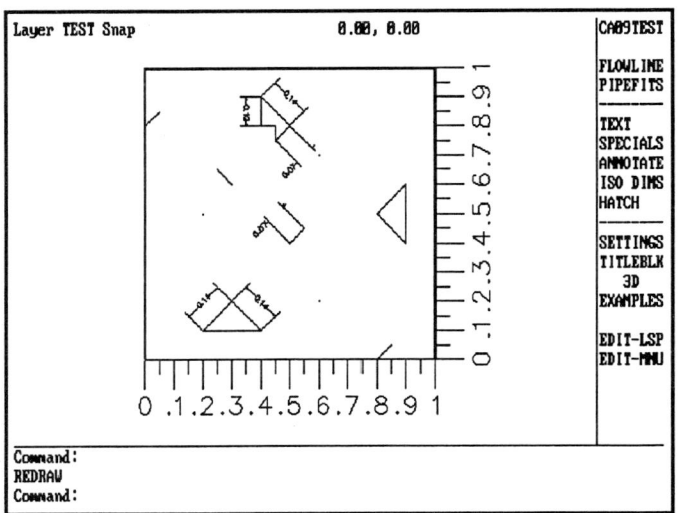

The Proposed Hatch Pattern

The hatch pattern is shown in the screen shot of the Proposed Hatch Pattern. Once you've decided on your pattern, you need to put it together. Make three other copies of the pattern and box to make a four box window pane image. Use all four boxes and the objects within them to calculate the pattern.

Developing a Concrete Hatch Pattern

   You have the custom ACAD.PAT file in your \CA-ACAD directory. Examine it as we edit.

   Create the pattern.

Command: **ERASE**	Clean out the previous hatch pattern, but leave the 1x1 box.
Command: **LINE**	Draw and dimension the pattern as illustrated.
Command: **ARRAY**	Select the 1x1 box and pattern. Make 2 Rows x 2 Cols.
Command: **COPY**	Copy the Pline box again, to the bottom right corner of the array. This test box is used to develop the pattern.
Command: **PLINE**	Draw a 2x2 test box, LowerLeft corner at the LowerLeft of the previous 1x1 test box. Close last segment.

Your screen should look like the Development Boxes.

Making a Hatch Pattern

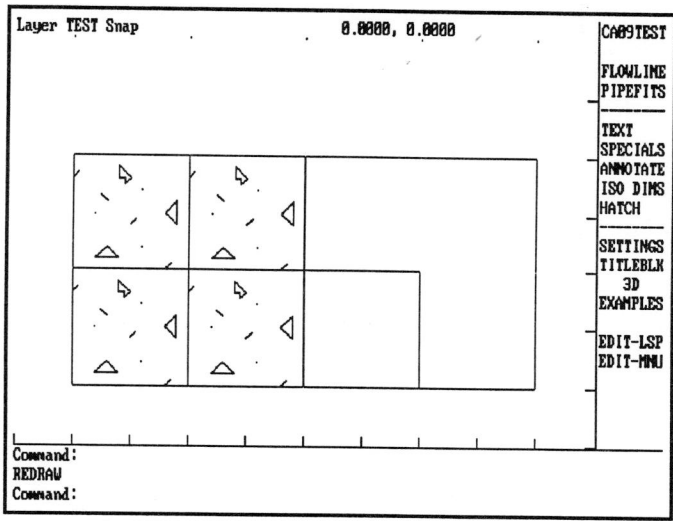

The Development Boxes

NOTE. The rest of this exercise flips in and out the text editor to create the pattern. In actual practice, we recommend that you write down the lines to enter. Go through all the measurements first, then edit the ACAD.PAT file. To speed up your pattern development, set up a drawing area and let AutoCAD calculate the distances and angles.

First, you need to create a header for the hatch pattern so that AutoCAD knows it by name and can give us a description about it if you ask. Then, create the pattern definitions. You need to write one definition line for each member of the hatch pattern.

Start with the dots in the pattern. Since there are 3 dots, you will have three lines to write. They are much like linetypes. Measure the distances and offsets from 0,0 to each of the points using DIST. The first dot is located 6/10ths of a unit over and 2/10ths of a unit up from the origin of the pattern. The origin at coordinate 0,0 is for convenience. The X-origin and Y-origin are .6 and .2 respectively. The line starts at .6,.2. Here is the start of the definition sequence:

Command: **ED**   Edit the ACAD.PAT file.

Add this header and first definition line to the end:

```
*CONCRE, Customizing AutoCAD Concrete
0, .6,.2, 0,1, 0,-1
```

The origins and offsets are shown in the First Dot Line Family drawing.

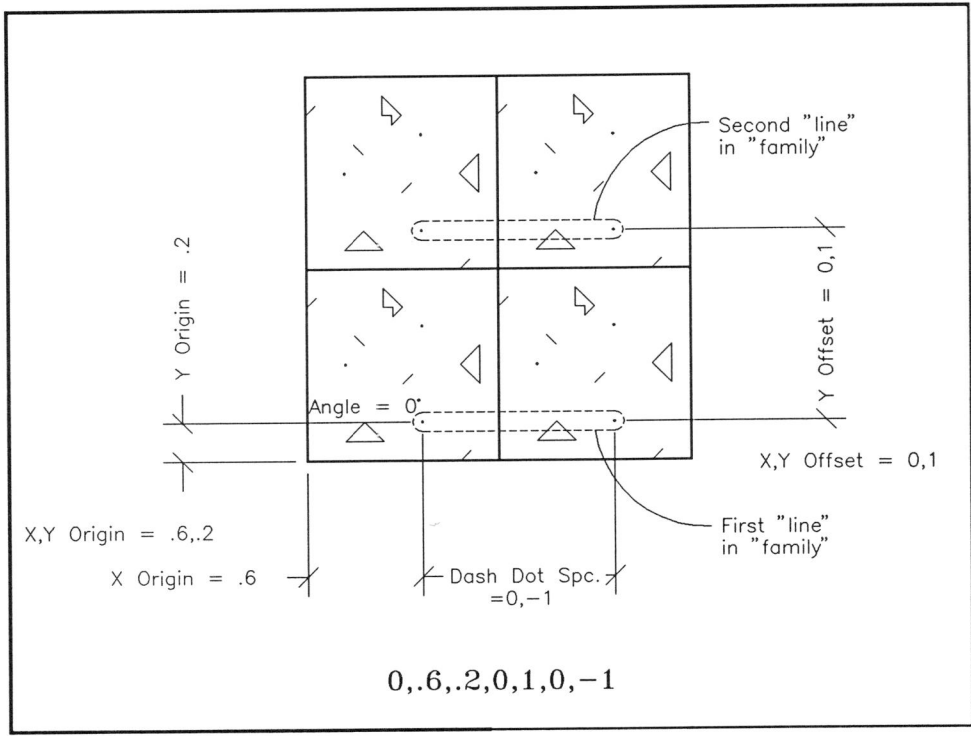

The First Dot Line Family

In this first line, the first entry, **0**, is the angle, which we keep at 0 when defining dots. The **.6,.2** is the X,Y origin of the first dot, from 0,0. The **0,1** is the X,Y offset to the parallel brothers and sisters in this family.

An X of 0 makes the next line's endpoint start perpendicular to the first. Any any other value would stagger the starting origin for dashed and dotted lines. The dotted line repeats parallel at every 1-unit so the Y offset is set to 1. The last two items, **0,-1** are the dot-dash measurements. 0 means put the pen down to draw a dot. The -1 means lift the pen and move over one unit before putting a dot.

Repeat the same steps for the next two dot patterns. The only difference is the X,Y origin of the first dot of the line. Make sure you <RETURN> after the last line, each time you add to the ACAD.PAT file. Do not add any blank lines.

Add these two lines:

0, .2,.5, 0,1, 0,-1
0, .6,.7, 0,1, 0,-1

Save, Exit and get back to AutoCAD.

Command: **HATCH**   Type ? to see the listing.

Command: **HATCH**
Pattern (? or name/U,style) <CHECKP>: **CONCRE**
          Use 1:1 scale, 0 angle, select the 2x2 test box.

The hatch pattern should be identical to the dots in your sample dimensioned pattern. Next, make the isolated dash segments. First measure the distances. The origins and offsets are shown in the First Dash Line Family drawing.

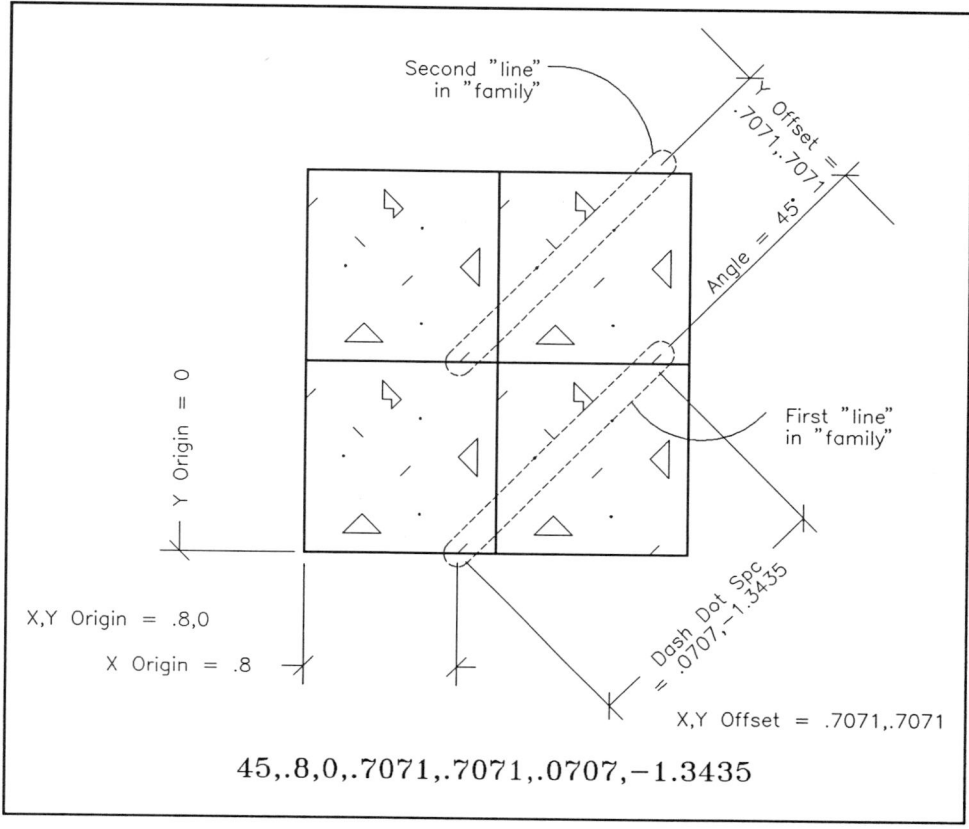

The First Dash Line Family

Measuring the distance perpendicular to the line between the line in the sample box and its sister line in the box to the right, you get a Y offset of 0.7071. The X-offset of the lines is 0.7071 because the line is on a 45 degree angle.

If you measure the distance from the endpoint of the first line, .85, to the start of the next line in the upper right box, 1.8,1, AutoCAD will return 1.3435, the pen up part of the line.

Command: **ED**   Edit the ACAD.PAT file.

Add the line below:

45, .8,0, .7071,.7071, .0707,-1.3435

The **45** means it is at a 45 degree angle. It is 0.0707 long (2/(sqrt of 2)). The **.8,0,** means the first one starts at X,Y origin of .8,0 in the sample box. The **.7071,.7071,** are the family X,Y offset. The **.0707,-1.3435** defines a pen down dash of .0707.

9—12 Customizing AutoCAD

The other three isolated dash line definitions are determined in the same manner.

Add these three lines:

```
45, .5,.4, 0.7071,.7071, .0707,-1.3435
135, .3,.6, 0.7071,.7071, .0707,-1.3435
45, 0,.8, 0.7071,.7071, .0707,-1.3435
```

Save, Exit and get back to AutoCAD.

Command: **ERASE**	Erase the previous test hatch.
Command: **HATCH**	Test the updated pattern, selecting the 2x2 test box.

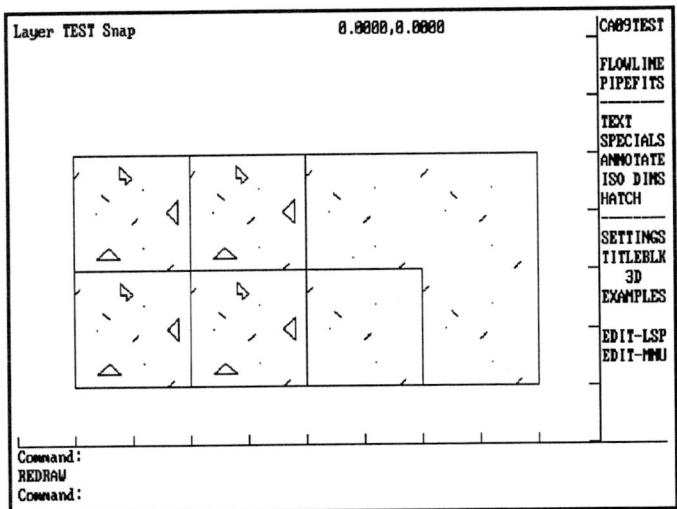

Concrete Dots and Lines

Your pattern should match part of the sample pattern, Concrete Dots and Lines. You've done the hard part. While a hatch pattern requires a little patience, it is straight forward using the methods we've shown. Finish up by writing the code for a 5 sided aggregate stone and then two triangles.

Look at the figure of the 5-Sided Aggregate. Measure the short 45 degree line first. It starts at .45,.75. It has the same delta X and Ys you used with the earlier dashed lines.

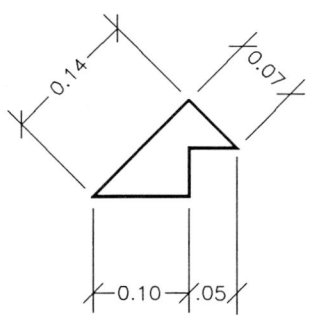

Five-sided Aggregate

Command **ED** Edit the ACAD.PAT file.

Add the short 45 degree line, then these other four lines:

`45, .45,.75, 0.7071,.7071, .0707,-1.3435`

The short vertical line starts at the same point,.45,.75. The dash is .05 long and the pen up length is .95 (1 unit -.05). So the line is defined as:

`90, .45,.75, 0,1, .05,-.95`

The horizontal line starts at .45,.8, is .05 long and runs at an angle of 180 degrees. You could choose the other endpoint and have it run at zero degrees, but you can't give a negative length. It means pen up! The line is defined as:

```
180, .45,.8, 0,1, .05,-.95
```

The long vertical line is defined as:

```
90, .4,.8, 0,1, .1,-.9
```

The closing side runs at 135 degrees, starts at .5,.8, has a dash length of .1414 and a pen up length of 1.2728. Use DIST to measure it. The perpendicular distance is 0.7071, making the definition:

```
135, .5,.8, 0.7071,.7071, .1414,-1.2728
```

Save, Exit and get back to AutoCAD.

Command: **ERASE** Erase the previous hatch.
Command: **HATCH** Test the updated pattern, selecting the 2x2 test box.

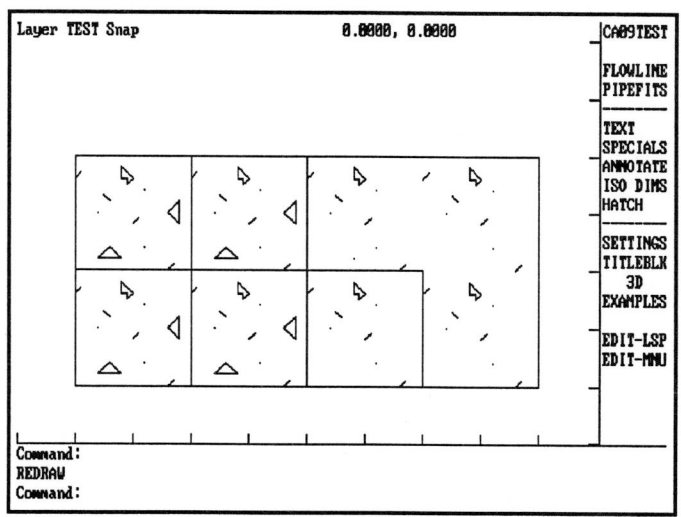

Concrete with Triangle Added

Command: **ED**

Edit the ACAD.PAT file. Measure the triangles and add the lower left triangle.

```
0, .2,.1, 0,1, .2,-.8
45, .2,.1, 0.7071,.7071, .1414,-1.2728
135, .4,.1, 0.7071,.7071, .1414,-1.2728
```

and the middle right triangle as:

```
90, .9,.4, 0,1, .2,-.8
45, .8,.5, 0.7071,.7071, .1414,-1.2728
135, .9,.4, 0.7071,.7071, .1414,-1.2728
```

Remember that last <RETURN>!

Save, Exit and get back to AutoCAD.

Command: **ERASE** Erase the previous hatch.
Command: **HATCH** Test the updated pattern, selecting the 2x2 test box.

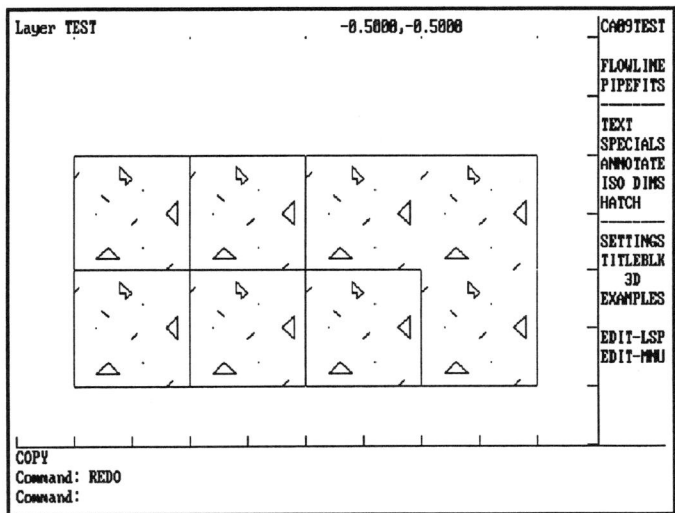

Final Pattern

Putting all this together, you have a concrete hatch pattern. Your screen should look like Final Pattern.

The final definition is shown below.

```
*CONCRE, Customizing AutoCAD Concrete
0, .6,.2, 0,1, 0,-1
0, .2,.5, 0,1, 0,-1
0, .6,.7, 0,1, 0,-1
45, .8,0, 0.7071,.7071, .0707,-1.3435
45, .5,.4, 0.7071,.7071, .0707,-1.3435
135, .3,.6, 0.7071,.7071, .0707,-1.3435
45, 0,.8, 0.7071,.7071, .0707,-1.3435
45, .45,.75, 0.7071,.7071, .0707,-1.3435
90, .45,.75, 0,1, .05,-.95
180, .45,.8, 0,1, .05,-.95
90, .4,.8, 0,1, .1,-.9
135, .5,.8, 0.7071,.7071, .1414,-1.2728
0, .2,.1, 0,1, .2,-.8
45, .2,.1, 0.7071,.7071, .1414,-1.2728
135, .4,.1, 0.7071,.7071, .1414,-1.2728
90, .9,.4, 0,1, .2,-.8
45, .8,.5, 0.7071,.7071, .1414,-1.2728
135, .9,.4, 0.7071,.7071, .1414,-1.2728
```

Achieving a more irregular appearance comes at great expense of Hatch size, speed and your time. Deviating from 45 or 60 degree angular increments gets very complex, so does mixing 45 and 90 degree angles in a single pattern. When you write your AutoLISP hatch generator, you can make this same hatch pattern in about 5% of the time!

Partial and Irregular Fills

Hatches do not have Block-like efficiency in their drawing data storage unless they are copied or arrayed. Large areas do not require a complete area fill. You can get around Hatch inefficiencies by using standard "irregular" shaped Blocks as hatch boundaries, stretching scale to fit, and only partially filling large areas. This saves data storage.

You can automate this process by making a custom screen page to Insert and Hatch "irregular" boundary Blocks. The following exercise shows how to make a Hatch Menu. If you are using Release 9, put *s at the beginning of the macros to auto-repeat.

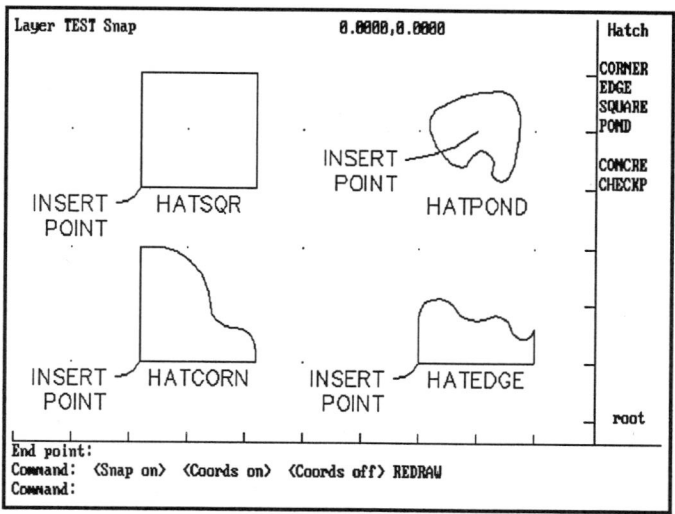

Irregular Boundaries

If you don't have the CA DISK, you need to make four 1x1 hatch boundaries. These are shown in the screen shot Irregular Boundaries. You need to Wblock the hatch boundaries with the names shown in the diagram.

Making a Hatch Menu

 You have the hatch boundaries HATCORN, HATEDGE, HATSQR, and HATPOND as drawing files.

 Draw and WBLOCK the four boundaries before proceeding.

Command: **ERASE**   Erase to clean up the screen.

Select **[EDIT-MNU]**

Edit the TEST.MNU. Change the [CA00TEST] label at the top of the screen and change the [HATCH] label on the root menu.

[CA09TEST]
[HATCH]^C^C^CLAYER M HAT01 C 3 ;;$S=HATCH $S=FOOTER.

Change the **EXAMPLES page as shown below.

9—16 Customizing AutoCAD

```
**HATCH
[ Hatch  ]
[]
[CORNER  ]*^C^C^CINSERT HATCORN \C \\HATCH \\\L ;ERASE P ;
[EDGE    ]*^C^C^CINSERT HATEDGE \C \\HATCH \\\L ;ERASE P ;
[SQUARE  ]*^C^C^CINSERT HATSQR \C \\HATCH \\\L ;ERASE P ;
[POND    ]*^C^C^CINSERT HATPOND \C \\HATCH \\\L ;ERASE P ;
[]
[CONCRE  ]*^C^C^CHATCH CONCRE
[CHECKP  ]*^C^C^CHATCH CHECKP
[]
[]
```
Count []s for 19 items total.

Save, Exit and return to AutoCAD. The menu should reload.
Test each item and Block, like:

```
Select [CORNER]
Command: INSERT Block name (or ?): HATCORN
Insertion point:                                    Pick it.
X scale factor <1> / Corner / XYZ: C Other corner:  Drag it to fit.
Rotation angle <0>:

Select [CONCRE]
Command: HATCH Pattern (? or name/U,style): CONCRE
Scale for pattern <1.0000>:
Angle for pattern <0>:
Select objects: L 1 found.        It selects Last
Command: ERASE
Select objects: P 1 found.        Then erases the boundary.

Command: QUIT
```

 Delete TEST.MNU.

 Rename TEST.MNU to MY09.MNU.

Copy **\CA-ACAD\ACAD.PAT** to **CA-ACAD.PAT** for backup.
If you want it for general use, append it to the end of your standard \ACAD\ACAD.PAT file.

You need to rename or append the file because AutoCAD only supports one pattern file, ACAD.PAT. You won't be able to use the standard patterns from the ACAD directory if the custom file gets in the way of the standard file. If you want to have both available, they must be merged.

Integrating the Hatch Menu

If you don't have the CA DISK and want to add the hatch menu to CA-MENU:

- Append the **HATCH page of your MY09.MNU to the end of the CA-MENU.MNU file.

- Change the [HATCH] label on the **ROOT page of the CA-MENU to:

```
[HATCH]^C^C^CLAYER M HAT01 C 3 ;;$S=HATCH $S=FOOTER
```

The chapter's menu is shown below.

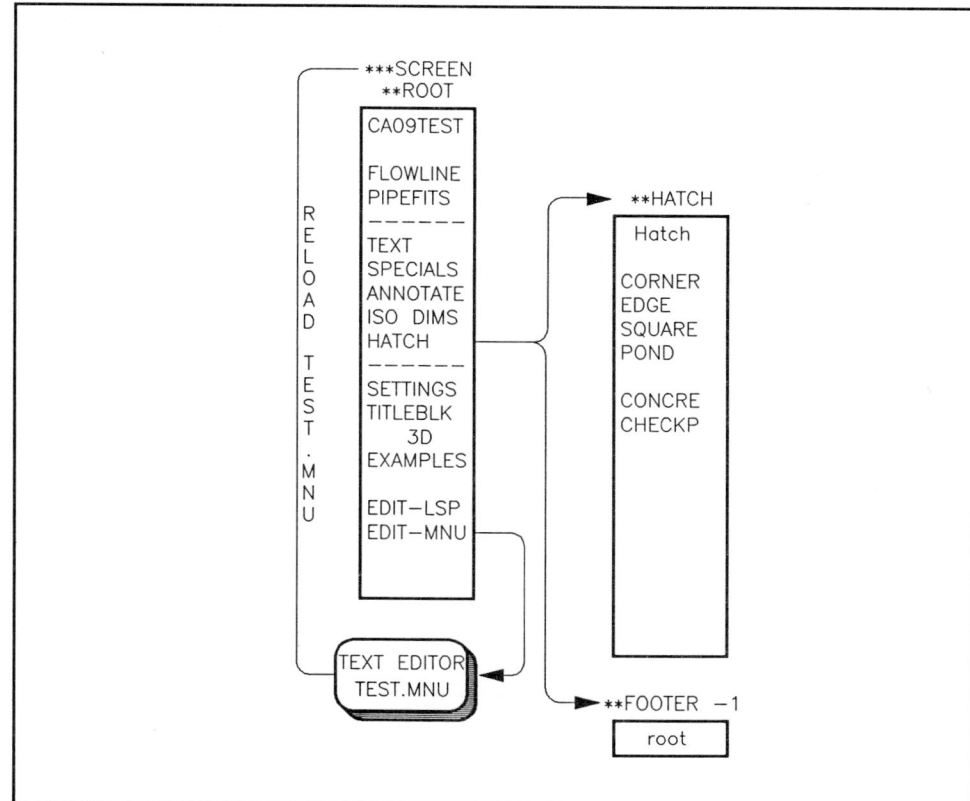

The Chapter 9 Menu Added to CA-MENU

Tips and Techniques

You don't have to rely on AutoCAD's standard linetypes and hatch patterns. You can supplement both by creating your own. Here are some summary tips for creating your own hatch patterns.

- ❑ When you create your pattern, set a Snap and Grid so all the lines and dots align to some reasonable increment. The smaller the increment, the larger and slower the Hatch.
- ❑ Draw your lines and dots, trying to make as many lines as possible align with other lines. This creates more efficient patterns.
- ❑ Write down your measurements as you go.
- ❑ Write the pattern in the ACAD.PAT file. Test it, then use it.

You can get a solid filled area by using closely spaced Hatch lines, but it takes a lot of drawing data and file space. An alternative, even for irregular areas, is to use wide Polylines, Donuts and Solids in any combination that works.

This marks the end of customizing linetypes and hatches. It is time to move on to using AutoLISP!

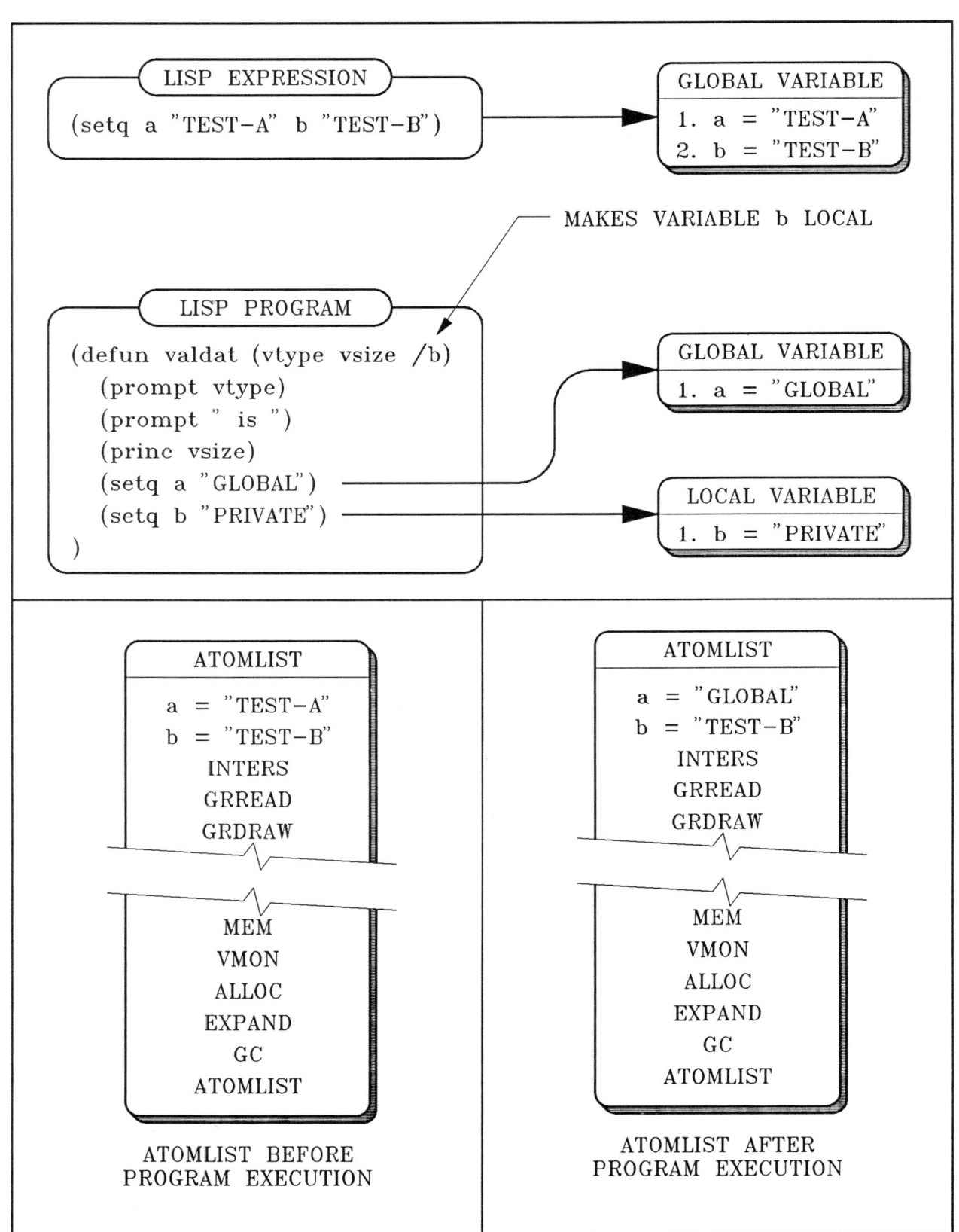

AutoLISP LOCALS, GLOBALS and the ATOMLIST

PART IV. AutoLISP: AutoCAD's PROGRAMMING LANGUAGE

CHAPTER 10

AutoLISP: Theory and Program Structure

PART IV. of CUSTOMIZING AutoCAD provides a complete description of AutoLISP. This chapter will teach you the basics of the AutoLISP language, show you its logical structures, and demonstrate how to create functions in AutoLISP by giving you hands-on examples.

LISP and AutoLISP

Although you will hear the common joke that says "LISP is an acronym for Lost-In-Stupid-Parentheses," it really stands for LISt Processing. This is the important key to understanding LISP, and AutoLISP. The LIST is all important, even if the list contains only one element.

LISP is built on the idea that a program is "data." Programs are not groups of instructions telling a computer how to operate; they are groups of data that a computer "reacts" to. Programs in AutoLISP are actually LISTs of data. By making a LIST containing functions, strings, integers and real numbers, you can make a program.

AutoLISP is a dialect of the LISP language that was derived from XLISP. AutoLISP coexists with the AutoCAD program. Every programming language, including AutoLISP, provides general tools. These include: ways to structure the flow of a program; tools to manipulate program data; and means to input and output data to computer devices.

In addition to these general tools, AutoLISP has AutoCAD-specific tools. These tools let it:

- Access and update AutoCAD's drawing entity data.
- Access AutoCAD's tables for Blocks, Layer, Views, Styles and Linetypes.
- Control AutoCAD's graphics screen and device input.

The Benefit of Using AutoLISP

What is the benefit of using AutoLISP?

The benefit to using AutoLISP for customization is quite simple. It is the only tool that you have to access AutoCAD drawing entities, reference tables, and to pass files in and out of AutoCAD. Think of AutoLISP as the only direct pipeline that you have to get inside AutoCAD!

How To Skills Checklist

How you structure your AutoLISP program's flow determines how your programs will behave. The flow is set by how you organize your program statements using AutoLISP's logical functions and relational operators. One key skill to defining AutoLISP functions is to distinguish between global and local environment variables. Besides gaining an understanding of AutoLISP's global and local environments, you will learn how to:

- ❑ Create an executable AutoLISP program, using NIL and NON-NIL to control your program's flow.
- ❑ Define functions with DEFUN.
- ❑ Store functions in external .LSP files and load the functions into AutoCAD's memory when you need them.
- ❑ Use AutoLISP's COMMAND function to pipeline instructions to AutoCAD.
- ❑ Make C: AutoLISP commands that act like other AutoCAD commands.
- ❑ Use multiple test conditions, including the use of AND, OR and comparative functions like LESS THAN and EQUAL TO.
- ❑ Make looping programs to construct data lists, using AutoLISP's REPEAT function.
- ❑ Batch process data lists using AutoLISP's FOREACH function.

Macros, AutoLISP Tools and Programs

MACROS
[ASK ME ?] is macro to illustrate the way IF conditions work.
[FITTING?] shows how the COND conditional structure works.

AutoLISP TOOLS
[MAKE PTS] builds a set of points. It is a general utility-type command.

AutoLISP PROGRAMS
SLOT.LSP draws a solid filled slot with a pline. The AutoLISP command lets the user pick the center of the slot, and supply the length and diameter before drawing the slot.

AutoLISP Program Vs. Data

Earlier when you worked with AutoLISP, you created simple LISTs. You put together a series of data and made it a LIST by putting the data within a set of parentheses. Let's write a a simple AutoLISP program to look at the idea of program vs. data in AutoLISP. First, you need to set up a drawing and set some AutoLISP variables.

VALDAT Is It Program or Is It Data?

Enter selection: Begin a NEW drawing named CA10

 Copy CA10.MNU to TEST.MNU.

 Copy MAIN.MNU to TEST.MNU.

```
Select [TEST] [24 x 36] [FULL] [INITIATE]
Command: ERASE                              Erase the border. You need the Setup menu for settings.

Command: ZOOM                               Left, corner 0,0 and height 12.

Command: (setq vt "VALVE" vs 1.5)           Sets variables
Lisp returns: 1.500000                      The last evaluated atom.

Command: (setq valdat '(nil (prompt vt) (prompt " is ") (princ vs) ))
Lisp returns: (nil (PROMPT VT) (PROMPT " is ") (PRINC VS))   The program.
```

Notice what AutoLISP returns. When you set the variables, the value AutoLISP returned was what you set VS to. When AutoLISP "sees" an expression, whether it is a program that you make or a built-in AutoLISP function, it executes the function. AutoLISP executes the function unless you tell it not to.

When AutoLISP evaluates an expression, it always "returns" the results of the last, and only the last, atom evaluated. When you made the program, the little quote ' told AutoLISP **not** to evaluate the list contained in the following parentheses. Remember, every item in a list is an "atom," even if the atom is itself a nested list. In this VALDAT example, the last atom evaluated is the nested list itself, not its contents.

VALDAT is a variable. The AutoLISP TYPE function returns the data type of an atom. Use AutoLISP's TYPE function to prove that VALDAT's value is the list.

VALDAT also is a program, a user defined function. Prove it by running the program VALDAT. To run VALDAT, put it inside parentheses just like you would to evaluate any other AutoLISP expression.

```
Command: (type valdat)
Lisp returns: LIST

Command: (valdat)
Lisp prints and returns: VALVE is 1.5000001.500000
```

PRINC is AutoLISP's print function. Besides printing the value 1.500000, it also returns the value because vs was the last atom evaluated by PRINC.

The Structure of Functions in AutoLISP

You can define your own functions in AutoLISP. Your AutoLISP functions create a local self-contained environment. You pass data into the function's local environment, your program statements use and manipulate the data, then you pass the data back to the general AutoLISP-AutoCAD environment.

You create AutoLISP functions with the DEFUN (DEFine-FUNction) function. DEFUN defines a function by constructing a structured list of the program statements. Here is the general format and an example of a DEFUN statement.

GENERAL FORMAT	EXAMPLE
`(defun NAME (ARGUMENTS / LOCALS)`	`(defun valdat (vtype vsize / b)`
`PROGRAM STATEMENTS...`	`  (prompt vtype)`
	`  (prompt " is ")`
	`  (princ vsize)`
	`  (setq a "GLOBAL" b "PRIVATE" )`
`)`	`)`

You can make the NAME of a function any name that you wish, using upper and/or lower case characters. Like variable names, you must avoid using reserved names from the ATOMLIST. If you use a reserved name, you will redefine the original AutoLISP function. If you do redefine an AutoLISP function, the original meaning will be unavailable until you start a new drawing. Certain characters are illegal. They should not be used as or within function names. These illegal and ill-advised characters are:

RESERVED AND ILLEGAL CHARACTERS: . ' " ; ( ) or <SPACE>
and: ~ * = > < + - /   These characters are already functions.

ILL ADVISED CHARACTERS: ? ` ! \ ^ or <any control character>.

You can put the reserved characters ' " ; ( ) immediately after a function or variable name without an intervening space. In general, we recommend that you not use any reserved characters as part of a function name.

ARGUMENTS are variable names used to refer to the data passed into the function's environment. The number of arguments must match the number of pieces of data passed to the function. In the VALDAT example in the table, two arguments VTYPE and VSIZE are passed to the function. When you execute VALDAT by typing **(valdat "VALVE" 1.5)**, the function sets VTYPE to "VALVE" and VSIZE to 1.5. The arguments are assigned in order in a 1:1 relationship to the values supplied.

LOCALS are variable names that you need and use only within the function. AutoLISP makes a small localized environment in which you store values of the locally defined variables. You can think of it as a sort of private ATOMLIST that works much the same way as the general ATOMLIST. When variable names are included on the local list, they are listed but do not put their value on the global ATOMLIST. You usually don't want your function's variables to be global or to change the values of global variables. In most cases, you'll want to put all your internal variables on the LOCAL list. ARGUMENTS, like VTYPE, are always LOCAL whether you like it or not. If you want an argument's value available outside the function, you have to use a different name.

GLOBALS are variables that have a value outside of the function in which they were created, as well as within the creating function. Every variable or function that you create, global or local, goes on the ATOMLIST during the current drawing session. Except for arguments, variables are global unless you declare them local. Local variables have no value outside of their parent function, unless they were also set outside of the function. Their value is LOCAL to each execution of the function. If they also have an external GLOBAL value, it is unaffected by the LOCAL value in the function.

PROGRAM STATEMENTS are the core of your function. Program statements follow the general rules of AutoLISP evaluation. The results of the last evaluated statement are returned to the global AutoLISP environment. In the VALDAT table example, the global variable **A** is reset within the function. The local variable **B** is also reset, but the global value does not change. Since **B** has been localized, it has no effect on the outside values. Even though its value is returned by the VALDAT function, its outside value is unchanged.

Defining a Subroutine with DEFUN

Let's look at this idea of a local environment by defining VALDAT with DEFUN. First, check VTYPE and VSIZE's values, assign values to A and B. Then, redefine the VALDAT function with DEFUN, and run it. Remember if you encounter any 1>s, it is AutoLISP prompting you how many closing parentheses you need.

DEFUN of VALDAT

```
Command: !VT                                      A leading ! identifies the variable to AutoCAD.
Lisp returns: "VALVE"                             It still has the previous global value.
Command: !VS
Lisp returns: 1.500000
Command: (setq a "TEST-A" b "TEST-B")             More globals.
Lisp returns: "TEST-B"
Command: !A                                       Check them.
Lisp returns: "TEST-A"
Command: !B
Lisp returns: "TEST-B"

Command: (defun valdat (vtype vsize / b)          Define the function
1> (prompt vtype) (prompt " is ")
1> (princ vsize)
1> (setq a "GLOBAL" b "PRIVATE")
1> )                                              Close the parentheses.
Lisp returns: VALDAT                              The function name.
                                                  Compared to SETQ which returned a list of program statements.

Command: (valdat)                                 Run the function.
error: incorrect number of arguments to a function    Oops!
Lisp returns: (VALDAT)

Command: (valdat vt vs)
```

```
VALVE is 1.500000
Lisp returns: "PRIVATE"                    B's value was returned.

Command: !B
Lisp returns: "TEST-B"                     But its global value is unchanged.
Command: !A
Lisp returns: "GLOBAL"                     It changed. It was not listed in the LOCAL list.

Command: !VTYPE                            Check your variables.
Lisp returns: nil                          VTYPE was kept local.
Command: !VSIZE
Lisp returns: nil                          Also local.

Command: !atomlist                         Look again at the top of the ATOMLIST:
Lisp returns: (VSIZE VTYPE B A VALDAT VS VT #DWGSC STR1 Y X #MENU INTERS GRREAD GRTEXT
```

The expression following the function name, (vtype vsize /b), declares arguments and local variables. The arguments precede the /slash and local variables follow it. Arguments are automatically local. When declaring locals, put a <SPACE> before and after the slash. Although only required if there are no arguments, it is always a good practice.

Notice that the ATOMLIST now includes all your new variables and functions. Functions are simply special variables!

You are limited in defining your own functions unless you can establish a line of communications between AutoLISP and AutoCAD. It is nice to see the data, but it would be more useful to make the data change a drawing. One method of transferring data to AutoCAD is to let the value returned by a function act as drawing input. Try transferring some data with VALDAT.

Transferring Data with VALDAT

```
Command: SETVAR                            Set TEXTEVAL to 1, so TEXT will recognize the AutoLISP.
Command: TEXT                              Pick a point, default 0 rotation.
Text: (valdat "ROO" "TWO")
Command: ROO is TWO                        PROMPT and PRINC print this, but "PRIVATE" returns as text.
```

Notice that AutoLISP doesn't care whether the arguments are numbers or strings. It is up to you to call your functions with the right type of data. AutoLISP will kick out an error if an internal AutoLISP function expects a different data type. PRINC prints anything, but PROMPT needs a string value.

The VALDAT example put some AutoLISP results into the drawing. Say, however, that you wanted to use "ROO is TWO" as text. You could have a function make it a single string with **(setq str (strcat vtype " is " vsize))**. STRCAT is a simple built-in AutoLISP function that merges two or more strings. STRCAT cares what type of data you feed it. It will crash if it doesn't get pure stringfood. Try feeding your drawing some text.

```
Command: (defun vtext (vtype vsize)
1> (setq str (strcat vtype " is " vsize)))

Command: TEXT                          Enter (vtext "ROO" "TWO") at the text: prompt.
```

The expression (vtext "ROO" "TWO") as text sent "ROO is TWO" to the text command.

If you wanted to feed a LINE command, you could make the last evaluated atom a POINT variable. Again, it is constraining to use a new function for each bit of data that you want to feed to AutoCAD. AutoLISP's COMMAND function provides a much better way.

The COMMAND Pipeline

General LISP languages do not have the ability to draw graphic images. AutoLISP does. AutoLISP can instruct AutoCAD to draw. In fact, AutoLISP can tell AutoCAD to do anything inherent in AutoCAD. The key is the COMMAND function —"The Command Pipeline." It lets you run AutoCAD commands within an AutoLISP statement.

The rules for using the COMMAND function are simple.

- Put your commands, options and text in quotes.
- Do not use GET input functions within a COMMAND pipeline.
- Do not precede your variable names with an ! exclamation point.
- Use only true AutoCAD commands, not C: commands defined with DEFUN.

AutoLISP takes each item of the COMMAND's argument list and sends it to AutoCAD. If the argument is an AutoLISP variable or expression, it is evaluated and the results are passed to AutoCAD. Variables can be used anywhere in the statement to supply AutoCAD with data. Quoted words are taken as "literal strings" by AutoCAD. Unquoted words are taken as AutoLISP variables. You can use other built-in AutoLISP functions within the pipeline.

Of course, AutoCAD will reject the command input if there are any errors in syntax or data type. AutoCAD doesn't care where the command data comes from as long as it is in the right form.

Each item of the COMMAND list must be one complete instruction to AutoCAD. You cannot enter half an AutoCAD command word, like "ERA," and later add the SE. You also must treat each instruction separately. AutoCAD treats "CIRCLE 5,5 0.25" as one erroneous instruction, not three instructions. Here is the correct format with examples for the COMMAND function.

GENERAL FORMAT   **EXAMPLE**

(command item1 item2...)   (command "CIRCLE" "5,5" 0.25)
                                        (command "CIRCLE" '(5 5) "0.25")

You can interrupt a command sequence. AutoCAD is independent of AutoLISP. You can let AutoCAD do half of a job. Then, let AutoLISP intervene and finish the job, or pass control back to AutoCAD. Try an example with the Circle command.

Using the COMMAND Function

```
Command: (command "CIRCLE")
CIRCLE 3P/2P/TTR/<Center point>: nil
3P/2P/TTR/<Center point>: 5,5                 Pick a point.
Diameter/<Radius>: (command 0.25)             Supply the radius.
Lisp returns: 0.250000000000005
```

Note that we didn't "quote" the argument data. You can quote numbers, but you don't have to.

The AutoLISP COMMAND sends the Circle command to the AutoCAD command interpreter. AutoCAD issues the prompt for the command. Since the COMMAND function always returns a NIL after evaluation, AutoCAD sees the NIL, does not understand it and reissues the "3P/2P/TTR/<Center point>:" prompt. Although a NIL **issued** by the COMMAND cancels, <^C>, a **returned** NIL is generally ignored.

Using the COMMAND pipeline, you can send AutoCAD the <RETURN> and the <^C> cancel codes. When AutoLISP sees the "" (null string) it sends a <RETURN>, and a NIL sends a <^C> cancel. Try using a "" <RETURN>, then try a NIL.

```
Command: (defun vtext (vtype vsize)
1> (command "TEXT" "4,4" "" (strcat vtype " is " vsize) ) a )
                                              A is the last evaluated atom, for comparison.
Lisp returns: VTEXT                           The function name.

Command: (vtext "AutoLISP" "powerful")        AutoLISP issues:
Command: TEXT Start point or Align/Center/Fit/Middle/Right/Style: 4,4
Rotation angle <0>:                           The "" <RETURN> defaulted the angle.
Text: AutoLISP is powerful                    Is written on screen as text.
Lisp returns: "GLOBAL"                        The value of A.

Command: (command "CIRCLE" nil)
CIRCLE 3P/2P/TTR/<Center point>:              The command is cleanly canceled by the NIL.
Lisp returns: nil                             COMMAND always returns NIL.
```

Adding Command Names to AutoCAD

AutoCAD maintains a List of command names. All the primitive AutoCAD commands, like LINE, ARC, ERASE, COPY, are on this list. AutoCAD can receive commands from any of the supported devices. You can enter a command from the keyboard, from a menu macro, a script file, or the COMMAND pipeline. You can make AutoLISP function names into AutoCAD commands by adding the function names to AutoCAD's Command List. Once you add the function, you can use any method to call the function.

If you precede the function name with a "C:," AutoCAD will place the function name on AutoCAD's Command List. However, these AutoLISP defined "commands" cannot be used as AutoCAD commands by the COMMAND pipeline. You cannot enter a DEFUNed AutoLISP function when another function is active. Let's make the VALDAT function a "command."

Making VALDAT a Command

```
Command: (defun c:valve() (valdat vt vs))         Define the function.
Lisp returns: C:VALVE                              The function name.

Command: VALVE                                     Run the "command".
Lisp returns: VALVE is 1.500000"PRIVATE"           Same as before.
```

If you had originally defined the VALDAT function as a "command," the first line would have been:

(defun C:VALDAT (/ vtype vsize b)

Note that the defun has no arguments. VTYPE and VSIZE are declared as local variables. C: functions never have arguments. They use GET functions or global values to obtain input.

External Function Storage and AutoLISP's File Format

You can store AutoLISP expressions in external files. Since a function is an AutoLISP expression, you also can store a function in an external file. In most cases, you will store your functions in external files. You assign a file extension of ".LSP" to these files. The .LSP file may contain any number of function definitions, and other expressions.

There are significant differences between AutoCAD's menu file format and AutoLISP's file format. AutoLISP's format differences are:

- Function files are not divided into device or page sections.
- Spaces, semicolons and extra lines do not act like a <RETURN>.
- The plus + continuation of line character is not used.
- Function files do not need the backslash \ to make GET functions pause.
- Prompt strings can include escape codes, like \n for <RETURN>.
- Comments and documentation may be freely placed in the file, if the comment is preceded by a semicolon.

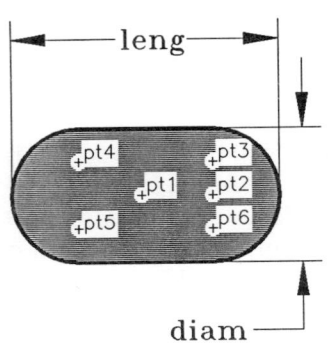

Solid Slot Hole

Let's make a function in a .LSP file. This function, called C:SLOT.LSP, draws solid slotted holes. It asks for the center of the slot, the slot diameter and slot length, and uses a Pline entity to fill in the slotted area. The calculations are done by using AutoLISP's basic divide, subtract, and multiply functions. SLOT uses PI and POLAR. PI is a predefined real that saves you the trouble of defining it. POLAR calculates a point at a specified angle and distance from a base point.

Type the following function, called C:SLOT, shown with **bold** lines, in your text editor. Use two spaces or more to indent each level of nested AutoLISP expressions. Don't use tabs in your text editor for indenting AutoLISP expressions. The points for the function are shown in the drawing Solid Slot Hole.

C:SLOT.LSP Making a Function in a .LSP File

 Copy SLOT.LSP to TEST.LSP.

 Create a new LISP function file called TEST.LSP.

[EDIT-LSP]

Edit TEST.LSP, and enter:

```
;* C:SLOT.LSP draws solid slotted Pline holes from prompted input.

(prompt "Loading C:SLOT program...")
(defun C:SLOT( / diam leng pt1 pt2 pt3 pt4 pt5 pt6)
                                                          ;Input
  (setq pt1 (getpoint "Insertion point: ")                ;Center of slot
        diam (getdist "Slot diameter: ")
        leng (getdist "Slot length: ")
  );setq inputs
                                                          ;Calculations
  (setq pt2 (polar pt1 0.0 (/ (- leng diam) 2.0))         ;Center of end radius
        pt3 (polar pt2 (/ pi 2.0) (/ diam 4.0))           ;Endpt of straight segment
        pt4 (polar pt3 pi (- leng diam))                  ;Endpt of straight segment
        pt5 (polar pt4 (* pi 1.5) (/ diam 2.0))           ;Endpt of straight segment
        pt6 (polar pt5 0.0 (- leng diam))                 ;Endpt of straight segment
  );setq calculations
                                                          ;Output commands to AutoCAD
  (command "PLINE" pt3 "W" (/ diam 2.0) ""
    pt4 "ARC" pt5 "LINE" pt6 "ARC" "CLOSE"
  );command
);defun C:SLOT
;*
```

Save, Exit and Reenter AutoCAD. TEST.LSP should load.

Loading a Function

AutoCAD loads function files much like it loads menu files. Upon loading a function file, AutoLISP reads the function definitions and stores them in memory. Expressions in the file, like the prompt "(prompt "Loading C:SLOT program...")," are executed while loading. AutoLISP does not execute the user defined functions within the file until it is explicitly instructed to execute them.

AutoLISP's LOAD function loads the function file. Loading is not done via AutoCAD's Load command. (AutoCAD's Load command loads shape definitions.) AutoLISP's LOAD automatically assumes the file extension .LSP unless you give it another extension.

To LOAD the TEST.LSP file you could type (load "test"), but we put the load in the [EDIT-LSP] menu macro so that it automatically reloads when you reenter AutoCAD from your text editor.

Test the SLOT function. Then, clean up your files by deleting or renaming TEST.LSP.

Testing the Slot Function

```
Command: (load "TEST") Loading C:SLOT program...      The prompt in the file tells us this.
Lisp returns: C:SLOT

Command: SLOT                          Run the command.
Insertion point:                       Pick a point.
Slot diameter: 0.5                     Enter diameter.
Slot length: 2                         Then the length.
```

 Delete TEST.LSP.

Rename TEST.LSP to SLOT.LSP.

NIL and NON-NIL Conditional Program Branching

A BRANCH is the method you use to direct the flow of your AutoLISP program. Every program has a flow, direction or logic which it follows. You structure an AutoLISP program to execute in a predetermined order. Conditional statements are your branching tools for controlling the actions of your AutoLISP programs.

AutoLISP has two branching functions, IF and COND. All branching conditions need a conditional test to perform a branch. These conditional test expressions usually use Logical and Relational Operators, but a conditional test may use any LISP expression.

AutoLISP's conditional functions work on a "NIL" or a "NON-NIL" basis. NON-NIL means that as long as there is some value, the condition "passes" the test. It helps to think of everything in AutoLISP as either True or NIL. If a condition is NON-NIL, then it has to be True, T. Since everything is either NIL or T, any expression can act as a conditional test.

Logical and Relational Operators

A LOGICIAL operator is a function that determines how two or more items are compared. The basis for comparison is whether something is NIL or non-NIL. Logical operators return either a T (NON-NIL) or a NIL (False) condition. The basic functions available for logical operations are:

AND   OR   NOT

The table below gives the general format and examples for the logical operations. The table extends the VALDAT example. As you look at the table, remember that you have three variables already set: A is "GLOBAL," B is "TEST-B" and VSIZE is nil. The getpoint is explained in the exercise following the table.

GENERAL FORMAT	EXAMPLES	RETURN
(and arg1 arg2 ...)	(and a b vsize)	NIL
	(and a b)	T
	(and b (getpoint "Pick: "))	
(or arg1 arg2 ....)	(or vsize a b)	T
	(or vsize)	NIL
(not arg)	(not (or a b))	NIL
	(not vsize)	T

Understanding Logical and Relational Operators

Both the AND and the OR functions can take any number of arguments. The AND function returns NIL if **any** of its arguments are NIL, otherwise, it returns T. The OR function returns True if **any** of its arguments are non-NIL, otherwise it returns NIL. Reading this carefully explains why, with no arguments, (and) is T, but (or) is NIL!

NOT is simple. NOT takes a single argument and returns the opposite. NOT returns T if its argument is NIL; and returns NIL if its argument is non-NIL.

Logical Operations with GETPOINT

	First, check the values of the variables, & set if needed:
Command: **!A**	Returns "GLOBAL".
Command: **!B**	Returns "TEST-B".
Command: **!VSIZE**	Returns nil.
Command: **(and b (getpoint "Pick: "))**	All must be non-NIL to return T.
Pick:	Hit a <RETURN>.
Lisp returns: nil	GETPOINT returned NIL, so must AND.
Command: **(and b (getpoint "Pick: "))**	
Pick:	Pick a point.
Lisp returns: T	GETPOINT returned T and so must AND.
Command: **(or 1 a stuff)**	Only one needs to be NON-NIL.
Lisp returns: T	
Command: **(and b (getpoint "Pick: "))**	Only asks for a point if B is NON-NIL.
Pick:	Pick a point.

Lisp returns: T

```
Command: (or (getpoint "Pick or RETURN for distance: ")     Only prompts "Distance:" if <RETURN> is entered.
1> (getdist "Distance: "))                                  Hit <RETURN>.
Pick or RETURN for distance:                                Enter a distance.
Distance: 3
```
Lisp returns: T

In the last example, OR stops evaluating and returns T as soon as it sees the first NON-NIL atom. In the same way, AND quits evaluating and returns NIL as soon as it encounters a NIL argument. You need to exercise care when you put other functions inside AND or OR. Whether an argument is evaluated depends on the values of preceding arguments.

Relational Operators

A RELATIONAL operator is a function that evaluates the relationship between two or more items. Relational operators include: LESS THAN, GREATER THAN, EQUAL TO, and NOT EQUAL TO. The basis for comparison is whether something is NIL or NON-NIL. Relational operators return either a T, if the expression is true (NON-NIL), or return NIL if the expression is false.

The following table gives the general format and examples for the relational operations. In the examples, VS is 1.500000, Z is '(1 a 2) and X is '(1 a 2).

GENERAL FORMAT	EXAMPLE	READ AS	RETURNS
(< arg1 arg2 ...)	(< 2 vs)	2 is less than VS -- false	NIL
(> arg1 arg2 ...)	(> 2 vs 3)	2 is greater than VS or 3 -- false	NIL
(<= arg1 arg2 ...)	(<= 1.5 vs)	1.5 is less than or equal to VS	T
(>= arg1 arg2 ...)	(>= 2 vs)	2 is greater than or equal to VS	T
(= arg1 arg)	(= 1.5 vs)	1.5 is equal to VS	T
(equal arg1 arg2)	(equal 1.5 vs)	1.5 evaluates to same as VS	T
(eq arg1 arg2)	(eq z x)	Z is identical to X -- false	NIL
(equal arga arg2)	(equal z x)	Z evaluates to same as X	T
(/= arg1 arg2)	(/= 2 vs)	2 is not equal to vs	T

Except for EQ, EQUAL, = and /=, these operations may have multiple arguments, comparing the first argument to all other arguments. Use the EQ function to test lists to see if they are SETQed (bound) to the same object. EQ generally is equivalent to = and EQUAL for numerical and string comparisons.

The IF Structure

The simplest and most frequently used program branch is the IF structure, sometimes called IF-THEN-ELSE. In plain english, AutoLISP thinks, "IF the condition is T, THEN execute the first expression, ELSE (it is nil) execute the second expression." The structure of an IF statement looks like the IF Branch diagram.

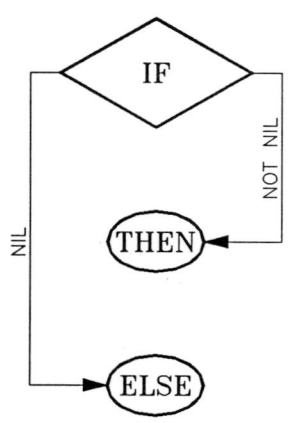

The IF Branch

Here is the general format and an example:

GENERAL FORMAT	EXAMPLE
	`(setq ans (getstring "Are you ready to go yet Y/N? "))`

`(if test-condition`
`   then...`
`   else...`
`)`
	`(if (= ans "Y")`
	`   (prompt "Good - THEN come on.")`
	`   (prompt "ELSE stay home!")`
	`)`

IF has two possible paths to travel in the example. If the person answers "Y" for yes, then the condition is true. The THEN step is executed so they get a warm "Good - then come on." However, if the user responds with anything but "Y," the ELSE step is executed and they get the "stay home."

To help you look at the IF function, the book provides [ASK ME ?], a menu macro that prompts for Yes or No and executes the correct response.

Trying IF with [ASK ME ?]

Select **[EDIT-MNU]**

Edit TEST.MNU. Change [CA00TEST] to [CA10TEST]. Then add the following to the EXAMPLES page.

```
***SCREEN
**ROOT
[CA10TEST]
[]

**EXAMPLES
[Examples]
[]
[ASK ME ?]^C^C^C(setq ans (getstring "Are you ready to go yet? Y/N? "));\+
(if (or (= ans "y")(= ans "Y")) (prompt "Good - THEN come on.");+
(prompt  "ELSE stay home!"));
[]
```

Save, Exit and return to AutoCAD. The menu should reload.

Select **[EXAMPLES]**	Test it.
Select **[ASK ME ?]**	Enter a Y. Notice how OR deals with UPPER/lower case.
Good - THEN come on.nil	
Select **[ASK ME ?]**	Enter an N.
ELSE stay home!nil	

The PROGN Structure

Limiting IF statements to only one THEN and one ELSE statement is confining. If you want to execute several statements, AutoLISP provides PROGN. PROGN groups multiple AutoLISP expressions into one expression. You use PROGN to tell AutoLISP to treat the next series of statements as one statement. It always returns the last atom evaluated by the last expression within it. PROGN's structure is shown in the PROGN Structure drawing.

Here is PROGN's format along with some examples.

```
GENERAL FORMAT              EXAMPLE

                            (if (and vt vs)
                              (valdat vt vs)
(progn                        (progn
                                (setq vt (getstring "Valve type: "))
   statements....             (setq vs (getdist "Size: "))
                              (valdat vt vs)
)                             )
                            )
```

If VT and/or VS are not already set, the else PROGN executes all three arguments in the table example.

The COND Structure: A Multiple IF

COND works much like IF, except COND can evaluate any number of test conditions. You can think of COND as a kind of a multiple IF routine. Once COND finds the first condition that is NON-NIL, it processes the statements associated with that condition. COND only processes the first NON-NIL condition. The structure of a COND is shown in the COND Structure.

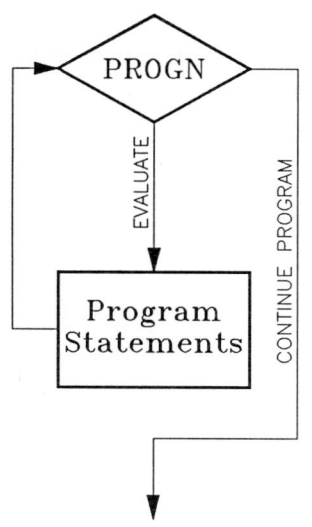

The PROGN Structure

The general format and an example is shown below.

```
GENERAL FORMAT                    EXAMPLE

                                  (setq ans (getstring "Fitting type:"))

(cond                             (cond

   (first test condition...          ( (= ans "E90")
        first statements...             (prompt "Ninety deg elbow.")
   )                                 )

   (second test condition...         ( (= ans "E45")
        2nd statements...               (prompt "Forty-five deg elbow.")
   )                                 )

   (more tests statements..)
   (T                                ( T
        last statements...              (prompt "Invalid type!")
   )                                 )
)                                 )
```

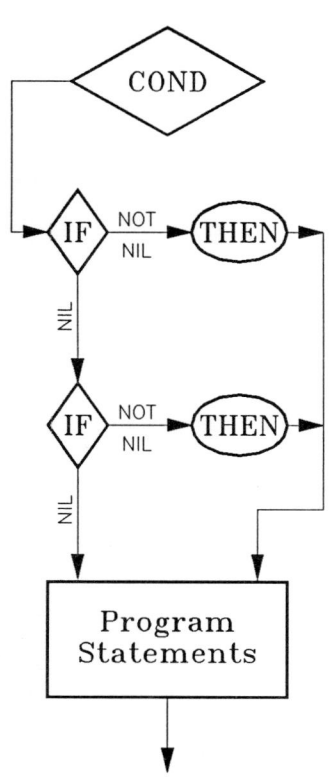

The COND Structure

COND takes any number of LISTs as its arguments. Each argument must be a list containing a test followed by any number of expressions to be evaluated. COND interprets the first item of each LIST as that list's test condition. It evaluates all of the expressions within the list of the first NON-NIL test.

Since COND looks for the first NON-NIL condition, you want to make sure you test the most likely conditions before the least likely. Putting your most likely

NON-NIL conditions first, increases your program's speed. The COND function is a good way of making programs branch based on a series of conditions. The last test can be a test that is always NON-NIL, like the symbol T. Its expression will be evaluated if none of the others were NON-NIL, making it good for error prompts.

Using COND with a [FITTING?] Function

[EDIT-MNU]

Add the following to the **EXAMPLES page:

```
[]
[FITTING?]^C^C^C+
(setq ans (strcase (getstring "Fitting type (E90/E45/RED/TEE): ")));\+
(cond ((= ans "E90") (prompt "Ninety deg elbow.^M"));+
((= ans "E45") (prompt "Forty-five deg elbow.^M"));+
((= ans "RED") (prompt "Reducer.^M"));+
((= ans "TEE") (prompt "Tee, non-reducing.^M"));+
(T (prompt "Error: Unknown fitting type.^M"));+
);
[]
```

Save, Exit and return to AutoCAD. The menu should reload.

Select **[EXAMPLES] [FITTING]**
Fitting type (E90/E45/RED/TEE):   Enter RED. It all scrolls by, then:
Reducer.
Lisp returns: nil

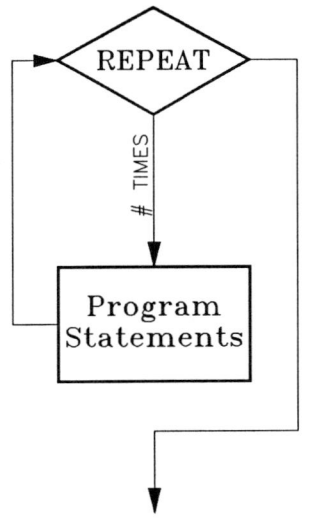

The REPEAT Structure

The [FITTING?] macro uses the STRCASE function to force its test argument to UPPER case.

Program Looping Structures

Like many other programming languages, AutoLISP has several methods to cause a series of program steps to loop, executing over and over again. These looping structures are used to:

- Reduce the number of statements in the program.
- Continue a routine until a user action terminates it.
- Converge on a mathematical solution.
- Batch process a list of data.

The REPEAT Structure

AutoLISP's REPEAT is a simple looping structure. Many application programs repeat some task. The REPEAT function executes any number of statements a specific number of times. Like PROGN, all of its expressions get evaluated, but they get evaluated once each loop. REPEAT returns the value of the last expression on the last loop. The structure of REPEAT is shown in the REPEAT Structure drawing.

Here is the general format and an example.

```
GENERAL FORMAT                       EXAMPLE
(repeat number                       (repeat 20
    statements to repeat...             (prompt "Step around.")
)                                       (command "ROTATE" "L" "" "5,5" "3")
                                        (prompt "..again..")
                                     )
```

You can type a simple repeating statement at the AutoCAD command line. Draw a small circle and animate it.

REPEAT Animation with AutoCAD

Command: **CIRCLE**   Draw a circle near right side of screen.

Command: **(setq pt (getpoint "Center of rotation: "))**
Center of rotation: Pick point in center of screen.

Command: **(repeat (getint "Enter steps: ")**
1> **(command "ROTATE" "L" "" pt "3"))**
Enter steps: **20**   And it moves on around.

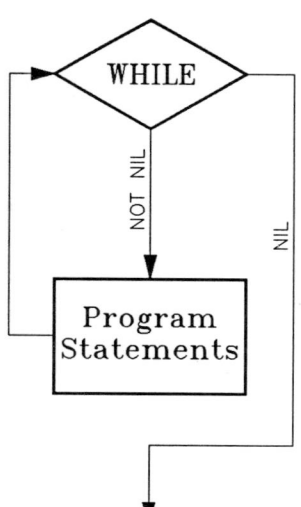

The WHILE Program Loop

The WHILE Program Structure

The function WHILE loops like REPEAT, except WHILE has a conditional test. WHILE continues to loop through its series of statements until the condition is NIL. The drawing WHILE Program Loop shows its structure.

WHILE statements are structured like those in the following table:

```
GENERAL FORMAT                       EXAMPLE

(while condition                     (while (setq pt (getpoint))
    statements to execute...            (command "point" pt)
)                                    )
```

Unlike the IF function, WHILE does not have an alternate "else" set of statements to execute if the condition fails the test. However, WHILE lets you include an unlimited number of statements in the loop. WHILE allows an indefinite, but controllable number of loops. Each loop of a WHILE function tests the condition and, if NON-NIL, evaluates each of the statements included within the closing parenthesis. WHILE returns the last evaluation of the last completed loop. If no loops are completed, it returns NIL. Try this next example to see how the WHILE function behaves.

Using WHILE with [MAKE PTS]

[EDIT-MNU]

Add the following to the [EXAMPLES] page of TEST.MNU:

```
[MAKE PTS]^C^C^C(setq ptlist nil)
(while (setq pt (getpoint "^MEnter point or RETURN when done: "));+
(setq ptlist (append ptlist (list pt))));+
);
```

Save and Exit to AutoCAD. The menu should reload.

Select **[EXAMPLES] [MAKE PTS]**   Test it:

Enter point, or RETURN when done: Enter as many as you like.
Enter point, or RETURN when done: Then <RETURN>.
Lisp returns: ((4.375000 5.625000) (4.375000 6.250000) (5.000000 6.250000) (5.375000 5.625000) (6.125000 6.000000) (6.500000 5.625000) (6.500000 5.000000) (7.250000 5.000000))

The [MAKE PTS] example keeps asking for a point via GETPOINT and adding the point to the list of points, called PTLIST. It uses the APPEND function to merge the new point list to PTLIST. As soon you hit a <RETURN> the loop stops and the list, saved as PTLIST, is returned.

Use WHILE to validate input, looping until the input meets the test.
You also can use the WHILE function for program iteration. ITERATION means that a loop is continued until the results of one or more expressions, calculated within the loop, determine whether the loop is terminated. The conditional test for an iteration usually contains some variable whose value gets changed during the course of the loop. A common form of iteration is to increment a counter.

Try the following counter at the keyboard.

Program Iteration Using WHILE

Command: **(setq count 0)**
Lisp returns: 0

Command: **(while (< count 10) (princ count)**
1> **(setq count (1+ count)))**
Lisp returns: 012345678910

Command: **(while (>= 8 (setq dist (getdist "Enter a distance greater than 8: "))))**
Enter a distance greater than 8:   Enter 7.
Enter a distance greater than 8:   Enter 8.
Enter a distance greater than 8:   Enter 9.
Lisp returns: nil

Command: **!dist**
Lisp returns: 9.000000

The 1+ is a function that increments a little faster than a (+1 count) expression.

Processing Lists with the FOREACH Function

The GETDIST WHILE returned NIL because there were no statements in the loop following the test. WHILE is often used like this to see if input is a member of a list.

Processing Lists with the FOREACH Function

Another type of AutoLISP looping statement is the FOREACH function. This function pops out each item of a list and uses it in a temporary variable in following statements. FOREACH is used in many applications to perform a function on each member of a list, looping once for each member.

FOREACH is shown in the drawing diagram. Here is the format of a FOREACH structure.

```
GENERAL FORMAT                      EXAMPLE

(foreach item list                  (foreach pt ptlist

    statements...                       (command pt)

)                                   )
```

The ITEM is any alias name that you want to use as the temporary variable name for the loop's current item from the list. The statements inside the loop must refer to the current item of the list by its alias. The value of ITEM is local to the FOREACH.

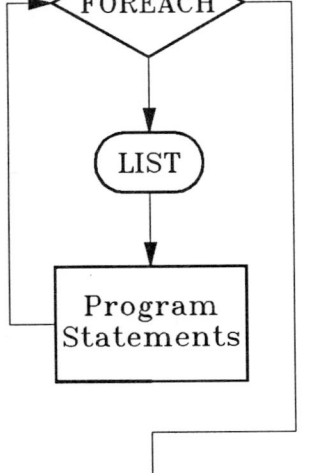

The FOREACH Structure

Use the PTLIST from the previous WHILE exercise to see how the FOREACH function works. Rerun the [MAKE PTS] macro from the [EXAMPLES] menu page to ensure that PTLIST is defined. Then, start the line command and try FOREACH.

Using the FOREACH Function with [MAKE PTS]

Select **[MAKE PTS]**   Pick several points.

Command: **LINE**
From point: **(foreach pt ptlist (command pt))**
To point: Prompts scroll as the Lines are drawn.
To point: nil
To point:

Command: **QUIT**

 Delete TEST.MNU.

 Rename TEST.MNU to MY10.MNU.

Integrating the Example Menu

You do not need to integrate this chapter's menu into the CA-MENU. If any of these macros are useful to you, add them to your own menus. The following figure shows the chapter's menu.

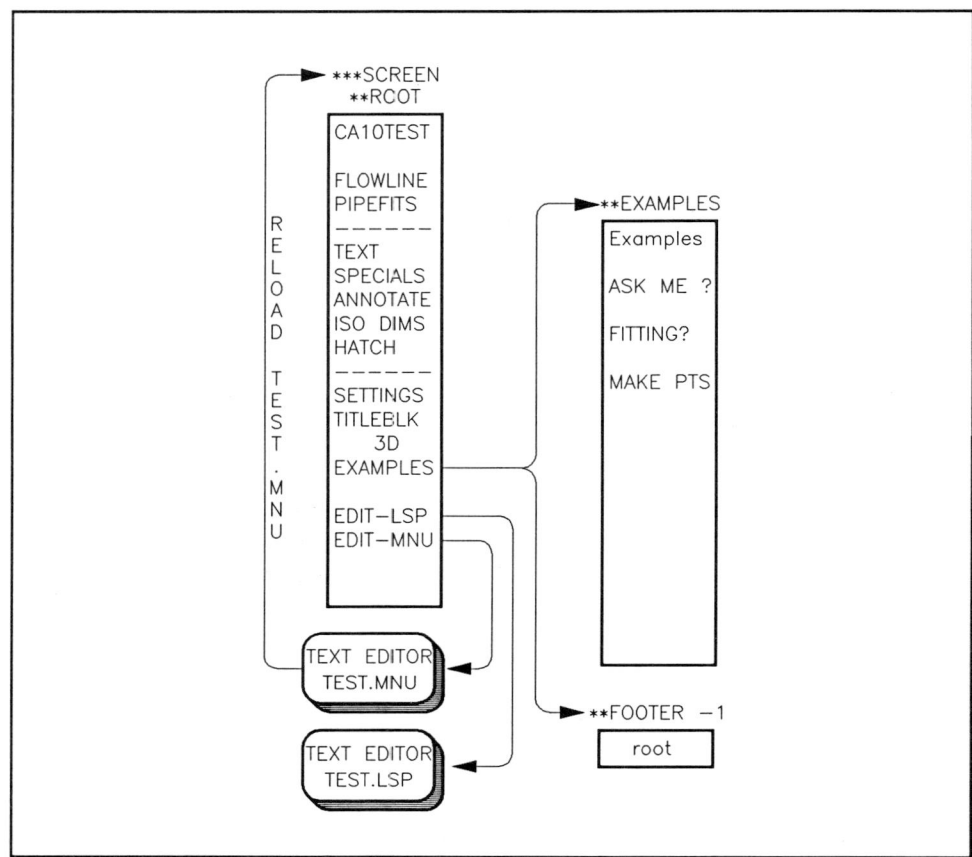

The Chapter 10 Menu

Tips and Techniques

The functions and techniques presented in this chapter apply to any application. All the functions are used throughout the rest of the book. Here are some summary tips:

- ❑ Use a standard prefix, like #, to indicate global variables.
- ❑ Declare variables as local, unless you have a specific reason to make them global.
- ❑ Remember to use AutoLISP's COMMAND function to pipeline instructions to AutoCAD.
- ❑ Remember that you can add AutoLISP C: commands to AutoCAD's command list.
- ❑ Use AND/OR functions for condition prompts.
- ❑ Use WHILE to filter and validate input.

- Use OR or STRCASE to make your input test independent of the case entered by the user.
- Make looping programs to construct data lists, using the REPEAT function.
- If you write loop intensive programs, make your statements within the loop as efficient as possible.
- Batch process data lists using the FOREACH function.

You've seen the basic logic built into AutoLISP. Let's turn to look at how you process data with AutoLISP.

11—0 Customizing AutoCAD

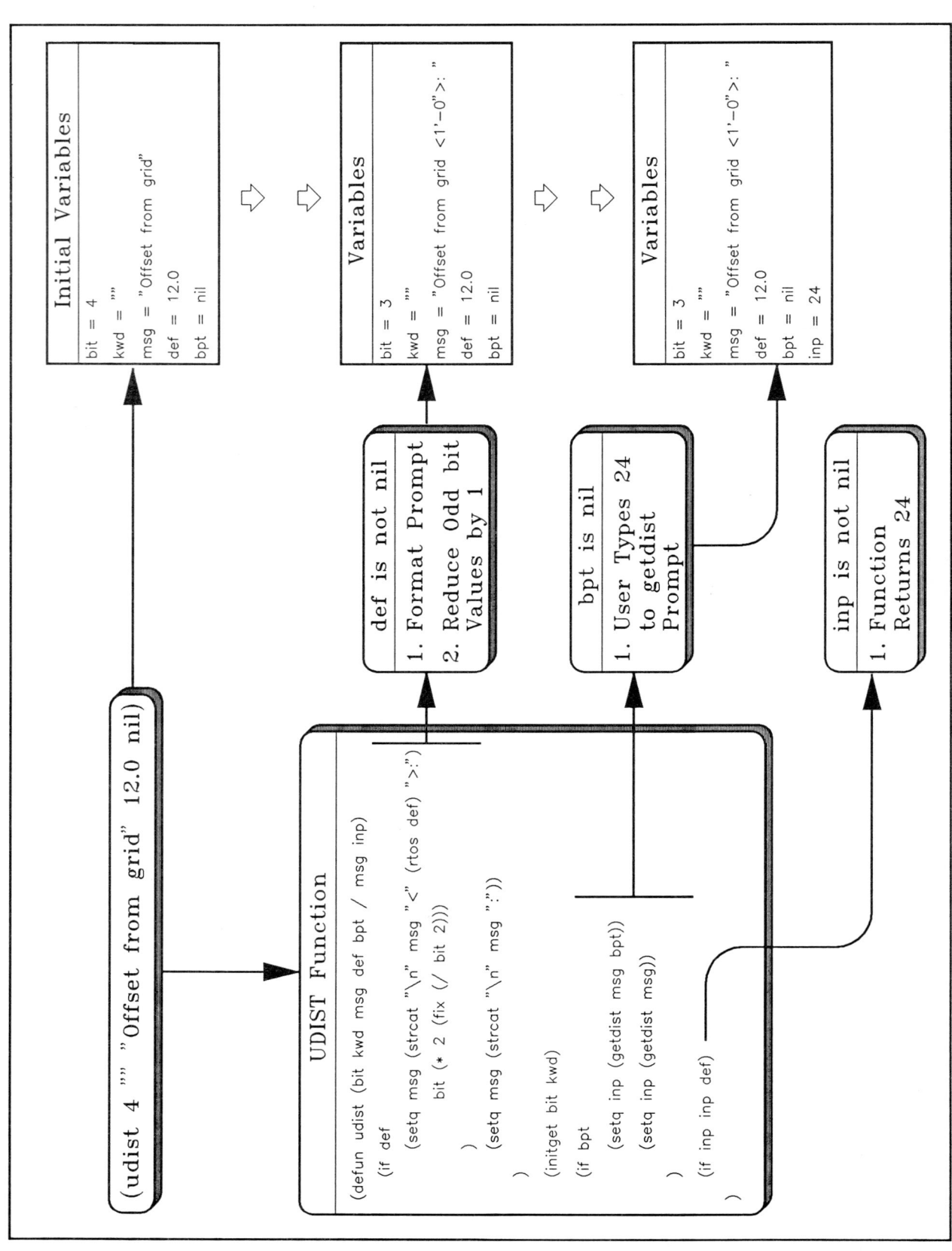

A Custom UDIST Function

CHAPTER 11

AutoLISP Data Processing

To customize your applications with AutoLISP you need to master AutoLISP's tools for working with different data types. This chapter shows you how to test for data types, process strings, format, display and print strings, convert numbers to strings, and strings to numbers. Finally, the chapter shows you to manage and combine data types to create new tools for your personal applications.

The Benefits of AutoLISP Data Processing

There are three main benefits to gaining a solid understanding of working with AutoLISP's data-types:

- You will write more dependable, efficient programs. For example, formatting your program prompts using AutoLISP's built-in functions will speed your program's execution.

- You can broaden the utility of your program by knowing how to format certain data-types. For example, linear and angular distances require special handling with AutoLISP to pass data to AutoCAD.

- You can create easier-to-use programs by displaying numbers to your users in familiar formats, and controlling your users' input by showing default values so that they understand the type of input your program is requesting.

How To Skills Checklist

In this chapter, you learn how to:

- ❑ Test data to determine its type, using BOUNDP, LISP, MINUSP and ZEROP functions.
- ❑ Process strings using STRCASE, STRLEN, STRCAT and SUBSTR functions.
- ❑ Format prompts and strings, then display and print them using PROMPT, PRINT, PRIN1 and PRINC.
- ❑ Make AutoCAD write text along a curve.
- ❑ Convert numbers to strings, format them as linear or angular dimensions using ATOF, ATOI and ANGTOS.
- ❑ Write your own AutoLISP dimension command.
- ❑ Make your own set of the GET functions that automatically present the user with default values, check for proper input and accept key response type words.

Macros, AutoLISP Tools and Programs

MACROS
[ARC-TEXT] implements the ATEXT function that writes text on an arc, and handles setting and resetting layers.
[GODIM] calls the GODIM function, a simple dimension routine.

AutoLISP TOOLS
ANGTOC is a function used to format AutoLISP angles in preparation for passing them to AutoCAD's commands.

UFUNS.LSP is a collection of functions that filter and format user input.
UDIST is a function to format, show defaults and filter user input. It buffers the GETDIST AutoLISP function.
UANGLE is a function similar to UDIST, but for the GETANGLE function.
UPOINT is a function that buffers the GETPOINT function.
UINT formats prompts and filters input for the GETINT function.
UREAL uses the GETREAL function, but filters input and formats prompts.
USTR formats a string prompt and uses the GETSTRING function.
UKWORD implements the GETKWORD function of AutoLISP, but formats prompts and filters input.

AutoLISP PROGRAMS
ATEXT.LSP is a program to draw text along a curve or arc segment.
GODIM.LSP is a sample dimension line text program. It places the text midway between the dimension points, and shows text layer management.

Determining Data Types: What are You?

Sometimes AutoLISP programs fail, stumping you with the cause of the failure. User defined functions and the AutoLISP built-in functions expect to receive and process specific types of data. Understanding the types of data will help you to eliminate data-type errors in your programs. Here is a recap of the simple data types supported by AutoLISP:

DATA TYPE:	TYPE RETURNS:
STRings	"STR"
REALs	"REAL"
INTegers	"INT"
LISTs	"LIST"
SYMbols (variable names)	"SYM"

If you give an AutoLISP built-in function the wrong type of data, it generates an error message. You can use the AutoLISP TYPE function to determine the data type of any variable or expression. TYPE returns the data type as shown above, or it returns NIL if the type is undefined. AutoLISP also provides a set of predicates to test data types. A PREDICATE is a test function that returns T or NIL. The data type predicates are:

BOUNDP	Tests a SYMbol variable to see if it is bound to a value.
LISTP	Tests if an item is a LIST.
MINUSP	Determines if a REAL or INTeger number is negative.
NUMBERP	Checks to see if an item is a REAL or INTeger number.
ZEROP	Tests to see if a REAL or INTeger number is zero

Using a predicate is a more efficient way to determine data types than using a comparison statement. Compare the following AutoLISP expressions:

```
(= (type a) "LIST"))         or          (listp a)
```

By comparing the returned string from the TYPE function to the word "LIST," you arrive at the same conclusion as the LISTP predicate. You get a T or NIL. The predicates are more efficient in program speed and size, since the predicate test requires only one AutoLISP evaluation.

In some cases, you may need to determine if a variable has been assigned a value, other than a nil value. Use the BOUNDP predicate to perform the test.

When testing with BOUNDP to determine if a variable is bound to a value, you need to quote the variable name. AutoLISP typically evaluates each item it sees. The QUOTE function tells AutoLISP not to evaluate the contents of the item, but to take it "verbatim."

If you do not quote the variable A, then the contents of A get tested, not A. Using 'A is AutoLISP shorthand for quote A. Since BOUNDP is the only predicate that directly tests a SYMbol, it is the only predicate that needs a quoted argument. TYPE also needs a quoted argument.

Let's see how TYPE and BOUNDP work.

Processing Data with BOUNDP

Enter selection:	Begin a NEW drawing named CA11 and try these tests.
Command: **(type a)**	Returns nil. A is not defined.
Command: **(type 'a)**	Returns SYM. It's an undefined symbol.
Command: **(boundp 'a)**	Returns nil, still undefined.
Command: **(setq a "TEST")**	Returns "TEST", the value A is set to.
Command: **(setq aa a)**	Returns "TEST", the value of A passes through to AA.
Command: **(setq aa 'a)**	Returns A. Reset AA to symbol A, not its value.
Command: **(boundp aa)**	Returns T, the value of AA is A, which is bound to "TEST".
Command: **(type aa)**	Returns SYM. Tests the contents of AA, the symbol A.
Command: **!AA**	Returns A. The contents of AA is a symbol.
Command: **(boundp a)**	Still returns nil. You have to quote it.
Command: **(boundp 'a)**	Returns T. Now it is bound to "TEST".
Command: **(if (boundp 'a) "TRUE" "NIL")**	Returns "TRUE".
Command: **(if a "TRUE" "NIL")**	Returns "TRUE". Everything's T or NIL. A is "TEST" so it's T.

As the last example shows, you do not need to use BOUNDP in a conditional test. The variable alone acts as a test.

Let's look at the other predicates to test the value of a symbol. LISTP works on any list, but you need to be careful if the variable is undefined. Strangely, (listp nil) returns T. Don't let this test trip you up. NUMBERP and MINUSP work on either REAL or INTeger data types, but they do not distinguish between the two types. Use a TYPE test to see if a number is a REAL or an INTeger. MINUSP and ZEROP will return an error if the item tested is not a number. The predicate tests shown below provide safer alternatives for LISTP, MINUSP and ZEROP. Try the predicate tests.

More Predicate Tests

Command: **(setq a '(0.0 0.0))**   Set "a" to a list.

Command: **(listp 'a)**   Returns nil. 'A is not a list but a symbol.

Command: **(listp a)**   Returns T. Without the ' you test A's contents.

Command: **(listp b)**   Returns T. Although B is undefined!

Command: **(and b (listp b))**   Returns nil. A safer list test.

Command: **!b**   Returns nil. It is undefined.

Command: **(numberp 1.0)**   Returns T. It is a number.

Command: **(minusp -1)**   Returns T. It is negative.

Command: **(zerop 0)**   Returns T. It is a zero value.

Command: **(zerop b)**   A dangerous test.
error: bad argument type
Lisp returns: (ZEROP B)

Command: **(= 0 b)**   Returns nil. A safer zero test.

Command: **(minusp b)**
error: bad argument type
Lisp returns: (MINUSP B)

Command: **(< b 0)**   Returns T. Not safe. B isn't negative. It's NIL!

Command: **(and (numberp b) (minusp b))**   Returns nil. A safer minus test.

Command: **(= (type 1.0) "REAL")**   Returns T. Identifies what type of number it is.

The last example determines if the number is a REAL, or an INTeger, by using the TYPE function and comparing the returned string. You can use all of these predicates in IF, WHILE and COND tests.

Processing STRINGs

Working with a STRING is simple and straight forward. In most applications you make the string UPPERCASE, find out how many characters it contains, determine what those characters are, then add strings together or format them.

This next section looks at AutoLISP string manipulations. The string functions provided by AutoLISP are:

STRCASE	Makes UPPERCASE/lowercase.
STRLEN	Returns the number of characters.
STRCAT	Combines strings.
SUBSTR	Returns a portion of a string.

Because you can make STRINGs any length in a program, you cannot anticipate the exact amount of memory required to store the string. For this reason, AutoLISP establishes an upper limit on the input/output string length. This limit is 132 characters. You cannot exceed this limit with input from the GET series of functions, nor can you exceed this limit with any single string in a prompt message. However, you can make strings longer than 132 characters if you join them together with the STRCAT function. Try a few string functions.

String Functions

Command: **(setq ca "Customizing AutoCAD")**
Lisp returns: "Customizing AutoCAD"

Command: **(strcase ca)**   Force to UPPERCASE, the STRCASE default.
Lisp returns: "CUSTOMIZING AUTOCAD"

Command: **(strcase ca T)**   Make it lower case by using a T flag.
Lisp returns: "customizing autocad"

Command: **(strlen ca)**   Test the length.
Lisp returns: 19

Command: **(setq ca (strcat ca " inspires me!"))**   Add the two strings together.
Lisp returns: "Customizing AutoCAD inspires me!"

Note that CA was still a mixed case in the last example. It was not SETQed when you used STRCASE.

In many applications, you have to be able to break a string apart. Use the SUBSTR function to return a portion of a string. SUBSTR takes three arguments: the string to dissect; the starting position of the first character you want; and the length of string to return. The length to return is an optional argument. If you do not specify a length, or if the length exceeds the string length, the remainder of the string is returned. Try getting the word AutoCAD.

Command: **(substr ca 13 7)**　　　　　Start at 13, 7 characters long.
Lisp returns: "AutoCAD"

Character Functions

There are several AutoLISP functions that operate on individual characters. These functions are useful for working with ASCII code characters. The character functions are:

CHR　　Takes an INTeger and returns its ASCII character.
ASCII　Takes an ASCII character and returns its INTeger value.

These characters functions are case sensitive because UPPER and lower case have different ASCII codes. Give CHR and ASCII a try:

Using the CHR and ASCII Functions

Command: **(ascii "A")**　　　　　Returns A, the integer for capital A.

Command: **(ascii "a")**　　　　　Returns 97, the ASCII value for lowercase a.

Command: **(chr 65)**　　　　　　Returns "A", a string.

Command: **(chr 97)**　　　　　　Returns "a".

The READ function is another function that can convert a string to another data type. READ returns the first list or atom in the string. The string cannot contain blanks or it will return NIL. You can use READ to convert a string to a symbol. This conversion is useful if you want to create variables from STRCAT created names.

Command: **(type (read "b"))**　　　　Returns SYM. B is a symbol.

Command: **(setq inc "2")**　　　　　Returns "2", setting up for:

Command: **(set (read (strcat "b" inc)) 123)**　　Returns 123.

Command: **!b2**　　　　　　　　　Returns 123, the value set to b2.

The STRCAT of "b" and "2" creates "b2," which READ returns as B2, a symbol. There is a difference in the use of SET in the example above and the usual SETQ. SET evaluates its first argument, the READ expression in the example, and sets its value, B2, to the value of the second argument, 123. SETQ treats its first argument as if the argument were quoted. The expressions (set 'b 321) and (setq b 321) are equivalent.

Formatting STRINGs

Programming languages give you a way to control the format of your strings. Formatting includes positioning prompts, setting the columns in which strings start, as well as setting their color and underscoring. In the next few examples, we will look at simple cases of string formatting in AutoLISP, but leave the more stylized color and underscoring formatting features to the chapter on AutoLISP Input/Output.

Formatting strings involves using ASCII control characters to cause the screen to format a string. These control codes are generally invisible to the user. AutoLISP provides two ways to include these characters within a string. You have already seen the CHR function. You can use CHR along with STRCAT to put control codes in strings.

A more straightforward way is to use "expanded" ASCII codes. You "expand" the AutoLISP string code by representing it as a backslash and an alpha or numeric code, like "\n." This code expands a single ASCII character into a 2 character AutoLISP code. When the string is sent to a device, like the screen, the AutoLISP codes are converted back to ASCII control characters and give the desired formatting effect.

Formatting codes all begin with the "\" backslash character and are followed by one or more letters and codes. The AutoLISP defined control codes for expanded code formatting are:

EXPANDED CODE	ASCII CODE	MENU CODE	FUNCTIONAL MEANING
\n	010	^J	Causes a new line (linefeed). The cursor moves to the far left and down one line.
\t	009		Moves cursor one tab position to the right.
\10	008	^H	The cursor moves back one space. (10 is the octal value for ASCII 8.)
\r	013	^M	A true return. Moves the cursor to the far left, but doesn't start a new line.
\e	027		This is the <ESC> escape character. See the AutoLISP: Input/Output chapter.
\\	092		Use this when you need an actual "\" backslash to appear in a string.
\nnn			nnn is octal code for an ASCII character. Convert ASCII decimal to base 8 for nnn.

The keyboard <RETURN> actually issues the ASCII 13 (return) plus ASCII 10 (linefeed). However, when AutoLISP receives keyboard input, it sees the <TAB>, <RETURN> and <SPACE> only as ASCII 10 codes.

A frequent formatting task that you will encounter is to make a prompt start on a new line. You do this by including a "\n" in the string. The format code characters, like \n, must be lowercase. Typically, you place the format string "\n" first in the string.

The following example shows you how to set a string with the "\n" new line expanded code, then display it with PROMPT. The example also shows a tab formatting code that gives you a convenient way to align text in columns. The

"\t" character will advance to invisible stops on the screen. A typical screen has 10 such stops, column one and 9 tabs. Try making a string of 9 tabs with numbers following each tab.

Using Expanded ASCII Codes

```
Command: (setq ca "\nCustomizing AutoCAD\nis fun!\n")          Set a string with expanded codes.
Lisp returns: "\nCustomizing AutoCAD\nis fun!\n"

Command: (prompt ca)                                            Sends string to the screen:
Customizing AutoCAD                                             The \n gives you a new line.
is fun!                                                         The \n returns again.
Lisp returns: nil

Command: (setq b "1 \t2 \t3 \t4 \t5 \t6 \t7 \t8 \t9 \t10\n")   Set the tabs with expanded codes.
Lisp returns: "1 \t2 \t3 \t4 \t5 \t6 \t7 \t8 \t9 \t10\n"

Command: (prompt b)                                             Sends string to the screen.
1     2     3     4     5     6     7     8     9     10
Lisp returns: nil
```

If you have a single screen system, you may have to flip screens for some of these codes. The TEXTSCR and GRAPHSCR functions accomplish these simple chores. Another useful screen function is TERPRI. TERPRI's sole purpose in life is to start a new line. You may use it occasionally with REPEAT to clear the textscreen. These functions take no arguments. Try them.

Using TEXTSCR, GRAPHSCR and TERPRI

```
Command: (textscr)              Returns nil and flips to the text screen.

Command: (repeat 25 (terpri))   Returns nil, but does 25 newlines first.

Command: (graphscr)             Returns nil and flips back to graphics.
```

You cannot use these expanded formatting codes in menu macros because the "\" causes AutoCAD to pause the menu. Even in an AutoLISP string within a menu macro, the AutoCAD menu interpreter reads and interprets the control characters as real control characters first, and then feeds them to the AutoLISP string.

Recall the menu control codes (^M and ^H) from the menu macro chapter? When you need to force a return in a menu macro string, use a ^M. Type a carat, then M. Use ^H for a backspace. The ^I for tab does not space correctly, so avoid it. The ^J acts the same as a ^M. Edit your menu and try a few of these codes.

Using Menu Control Codes

 Copy CA11.MNU to TEST.MNU.

 Copy MAIN.MNU to TEST.MNU.

```
Select [TEST] [24 x 36] [FULL] [INITIATE]          It brings up TEST.MNU.

Command: ZOOM                                      Center to height 6".
Select [EDIT-MNU]
```

Change the [CA00TEST] label to [CA11TEST]. Add the following to the EXAMPLES page:

```
***SCREEN
**ROOT
[CA11TEST]
[]

**EXAMPLES
[Examples]
[]
[^M TEST ]^C^C^C(textscr) (prompt "^MCustomizing AutoCAD^Mis fun!^M");
[]
```

Save and Exit to AutoCAD. TEST.MNU reloads.

```
Select [EXAMPLES] [^M TEST]         It displays:

Customizing AutoCAD
is fun!
Lisp returns: nil
```

Displaying and Printing Strings

AutoLISP has four built-in functions that you can use to output a string to the screen. These are: PROMPT, PRINT, PRIN1, and PRINC.

For simple screen printing, we recommend using the PROMPT function. PROMPT is the only function that displays its message on both screens of dual screen systems. The difference between PROMPT and the functions, PRINT, PRIN1 and PRINC, is that PROMPT can only accept a STRING-type argument. The other three functions can print any expression. You can only direct PROMPT to the screen. You can use the other three functions to print data to external files, or to any device.

PRINT and PRIN1 print any control codes in their data arguments as expanded codes, like "\n." PRINT also adds a new line before the data and adds a space after the data. PRIN1 does not add extra lines or spaces. PRINC uses control codes instead of expanded codes in its output, so "\n" or (chr 10) will cause an actual new line.

If you call any of these three functions with a file argument, they print the data to the device. If the data is a string, PRINC prints unquoted. PRINT and PRIN1 print quoted strings. They all also return the original data from the function.

If called without arguments, PRINT, PRIN1 and PRINC return no visible characters. PRINT, PRIN1, and PRINC print a new line to the screen; yet, unlike everything else in AutoLISP, they return nothing. These functions finish very cleanly. You can take advantage of this feature to enhance your user interface. The book often puts PRINC at the end of AutoLISP functions and function files for a clean finish. Try a few print functions.

Using PROMPT, PRINT, PRIN1 and PRINC

```
Command: (prompt "\tCustomizing AutoCAD")
         Customizing AutoCADnil
```
Prompts with tab. Returns NIL.

```
Command: (print "\tCustomizing AutoCAD")
"\tCustomizing AutoCAD" "\tCustomizing AutoCAD"
```
An extra line before and space after are printed and returned.

```
Command: (prin1 "\tCustomizing AutoCAD")
"\tCustomizing AutoCAD""\tCustomizing AutoCAD"
```
What goes in is printed and returned without change.

```
Command: (print (strcat (chr 9) "Customizing AutoCAD"))
"\tCustomizing AutoCAD" "\tCustomizing AutoCAD"
```
Expands the tab as ^t.

```
Command: (princ)
```
Finished cleanly. Prints and returns null, ASCII 00.

```
Command: (princ "\tCustomizing AutoCAD")
Lisp returns:      Customizing AutoCAD"\tCustomizing AutoCAD"
```

Tabs over and prints, still returns it expanded.

Converting Numbers to STRINGs and STRINGs to Numbers

Converting a number to a STRING is easy. AutoLISP provides several functions to do conversions. The ITOA function, INTeger to ASCII/string, takes only one argument, the number. ITOA only deals with INTeger types.

To convert STRINGs into REALs or INTegers, AutoLISP gives you the ATOF, Ascii TO Floating point REAL, and ATOI, Ascii TO INTeger, functions. Each function takes a string-type argument and returns the desired data type. Try these.

String Conversions and Units Handling

Command: **(itoa 24)**   Returns "24". It's a string

Command: **(atof "24.5")**   Returns 24.500000. A floating point REAL, 6 decimal places.

Command: **(atoi "24.5")**   Returns 24, an INTeger with fractional values discarded.

These are simple formats for numbers. AutoCAD handles linear distances and angular rotations in a special manner.

Formatting for Linear Distances

In many application programs, you will need to take a number and either show it to a user, or place it as text in the drawing. AutoLISP has several functions that handle these situations. The RTOS function extends AutoCAD's numeric formatting into AutoLISP.

RTOS performs a data-type conversion from REAL to STRING. RTOS represents the string as an AutoCAD linear distance. It accepts an optional format code argument to control the string format based on the AutoCAD units and precision settings. The linear unit settings and the associated AutoLISP format settings are:

CODE:	SYSTEMS OF UNITS:	EXAMPLES:
1	1. Scientific	1.55E+01
2	2. Decimal	15.50
3	3. Engineering	1'-3.50"
4	4. Architectural	1'-3 1/2"
5	5. Fractional	15 1/2

RTOS called with no optional argument uses the current linear and precision settings of AutoCAD. The AutoCAD dimension variables also can have an effect on the formatted string. When the units are set to Architectural, the AutoCAD Dimension variable DIMZIN controls the Zero INches and feet displayed. Try some examples.

Using RTOS

Command: **(rtos 8.5)**   Called with no arguments,
Lisp returns: "8 1/2""   it uses default settings.

Command: **(rtos 12.5 2)**   Mode argument converts to
Lisp returns: "12.500000"   different settings

Command: **(rtos 12.5 2 2)**   Now a precision of 2 places
Lisp returns: "12.50"   gives this result.

```
Command: SETVAR                 DIMZIN to 1
Command: (rtos 8.5)             With DIMZIN set to display both zero inches and feet.
Lisp returns: "0'-8 1/2""       Notice the zero.

Command: SETVAR                 Reset DIMZIN to 0.
```

There are several other modes for DIMZIN. These are listed in the AutoCAD User Reference.

AutoCAD Angles

Internally, AutoCAD keeps track of angles in radian units. All angles are stored in radians but displayed in the users current angular units setting.

AutoLISP angular calculations use only radians. One radian is PI degrees. One degree is 1/PI radians, or 1/3.141593.

```
Command: (getangle "Angle? ")
Angle? 180
Lisp returns: 3.141593
```

The AutoCAD UNITS angular formats and their AutoLISP format codes are:

CODE:	SYSTEM OF ANGLE MEASURE:	EXAMPLES:
0	1. Decimal degrees	45.0000
1	2. Degrees/minutes/seconds	45d0'0"
2	3. Grads	50.0000g
3	4. Radians	0.7854r
4	5. Surveyor's units	N 45d0'0" E

Please note that the angular format code numbers are not the same as the AutoCAD UNITS command angle option number. Unlike linear numbers, this code begins with 0, so they are one less.

The angle to string function, ANGTOS, can pose problems if you are unfamiliar with its intent. Both RTOS and ANGTOS format a string for visual display, and not for input to AutoCAD commands. RTOS and ANGTOS are formatted the same way as the AutoCAD status line. For example, the decimal form works fine when passed along to any AutoCAD command requesting "Rotation angle." However, AutoCAD balks at the radian, grads and surveyor's unit formats returned by ANGTOS.

Using ANGTOS

Command: **(angtos pi 0 3)**
Lisp returns: "180.000"

Command: **(angtos pi 3 3)**   Radian format.
Lisp returns: "3.142r"   The "r" is a problem.

Command: **(angtos pi 2 3)**   Grads format.
Lisp returns: "200.000g"   Problem is with the "g".

Command: **(angtos (* pi 0.25) 4 3)**
Lisp returns: "N 45d0'0" E"   Oh no - Spaces!

You generally do not use the RTOS function to format strings sent to AutoCAD because AutoCAD treats all linear distances the same, regardless of format. AutoCAD does not treat angles equally. Because a number in radians is not the same number in grads or degrees, AutoCAD angles are dependant on the users current settings. So 3.14 radians is not equal to 3.14 grads or 3.14 degrees.

However, AutoCAD provides a way around this. If you precede the angle passed to AutoCAD with a double angle sign <<, AutoCAD considers it as decimal degree input, regardless of the current angular units.

When you want to convert angles for display only, use ANGTOS. If you intend to pass the string to AutoCAD, say for the text rotation angle, use your own function. We will call this function ANGTOC, for ANGgle TO Command. Try the expression at the AutoCAD command prompt, then write the function.

Creating ANGTOC

Command: **(defun angtoc (ang) (strcat "<<" (angtos ang 0 8)))**

Command: **(angtoc 3.14159265)**   Try it with PI.
Lisp returns: "<<179.99999979"

Command: **(rtos (atof "179.99999979") 2 6)**   This is a way to roundoff.
Lisp returns: "180.000000"

Roundoff error is a common problem with AutoCAD and AutoLISP calculations. You can use RTOS and ATOF to round off first, then use STRCAT. Make an AutoLISP file called ANGTOC.LSP and type in the ANGTOC function. ANGTOC is a good tool that you will use later on. Give it a try.

 Copy ANGTOC.LSP to TEST.LSP

 Create a short new LISP function file called TEST.LSP.

11—14 Customizing AutoCAD

Select **[EDIT-LSP]**
Edit TEST.LSP, and enter:

```
;* ANGTOC is an angle formatting function that takes an angle
;* argument in radians and returns it with 6 decimal places
;* in a form universally acceptable to AutoCAD command input.

(defun angtoc (ang)
  (setq ang
    (rtos (atof (angtos ang 0 8)) 2 6)
  )
  (strcat "<<" ang)
)
;*
```

Save, Exit and Reenter AutoCAD. TEST.LSP should load.

Command: **(angtoc 3.14159265)**   Give it PI.
Lisp returns: "<180.000000"

 Delete TEST.LSP.

 Rename TEST.LSP to ANGTOC.LSP.

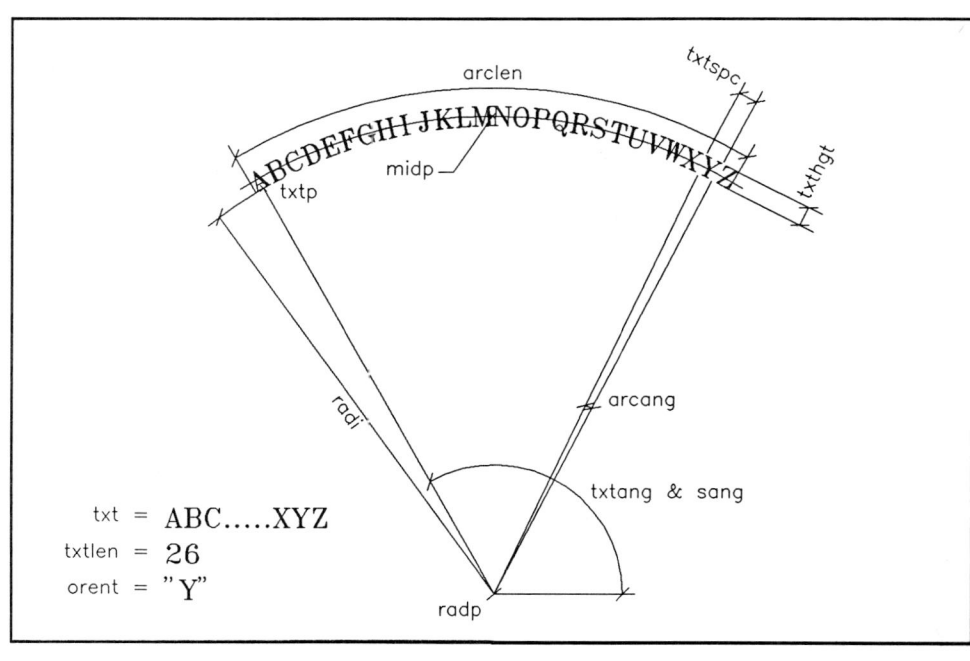

ATEXT.LSP Diagram

It's time for an example putting these formatting controls together. The ATEXT command you are about to create places text in an arc form. Each character is placed as a separate TEXT entity. ATEXT gets the user text string, midpoint, radius and orientation, and checks the Style defaults. It then calculates each character's placement and executes a TEXT command in a REPEAT loop.

The ATEXT.LSP function is shown in the previous ATEXT.LSP diagram.

ATEXT an Arc Text Command

 Copy ATEXT.LSP to TEST.LSP.

 Create a new LISP function file called TEST.LSP.

Select **[EDIT-LSP]**

Edit TEST.LSP, and enter the header information and initialization:

```
;* ATEXT types text in an arc. The function requires the midpoint of the
;* txt, the radius point, and the orientation. It determines if the text
;* height is fixed and prompts for height if not. It use the ANGTOC function.

(defun C:ATEXT ( / midp radp txt radi txtlen txtspc txthgt cmd arclen
  arcang sang orent txtang txtp char)

  (setvar "HIGHLIGHT" 0)           ;Turn highlite off
  (setvar "CMDECHO" 0)             ;Turn command echo off
```

Use the GET functions to get input from the user. You need the center of the arc text, a point on the radius, and the string of text to use. Add the following input and variable assignments:

```
(setq                            ;Assign variables
  radp (getpoint "\nPick radius point: " )        ;Get radius point of text
  midp (getpoint "\nPick middle point of text: " radp) ;Get midpoint of text
  txt (getstring "\nText: " T)                    ;Get text string
  radi (distance radp midp)                       ;Determine radius
  txtlen (strlen txt)                             ;Determine string length
  txtspc (cdr (assoc 41 (tblsearch "STYLE" (getvar "TEXTSTYLE"))))   ;Determine char width
);end variable assignment
```

Next, the program looks up the current text style to see if the text being used has a fixed height. If the text does not have a fixed height, the program asks the user for a height. Notice the CMD variable. CMD saves a list containing a COMMAND function that you will execute later in a REPEAT loop.

Save what you've done and continue adding to TEST.LSP.

```
(if (= (setq txthgt (cdr (assoc 40 (tblsearch "STYLE" (getvar "TEXTSTYLE"))))) 0)
    ;Test for fixed text height
  (setq                        ;If height not fixed...
    txthgt (getdist "\nText height: ")   ;Get text height
    cmd '(command "TEXT" "C" txtp txthgt txtang char) ;assign command list
  )
  (setq                        ;If height fixed...
    cmd '(command "TEXT" "C" txtp txtang char)      ;assign command list
  )
);end if
```

You need to find out if the text faces towards the center of the arc, or away from it, and compute the arc length required for the text. Calculate the space between characters and the start angle for the first character.
Save what you've done and continue adding to TEST.LSP.

```
(setq orent            ;Get text orientation
  (strcase (getstring "\nIs base of text towards radius point <Y>: ")) ;upper
)

(if (or (= orent "") (= orent "Y"))    ;calculate new radius length based on orientation
  (setq radi (- radi (/ txthgt 2)))
  (setq radi (+ radi (/ txthgt 2)))
);end if

(setq                ;Calculate variables
  arclen (* txtlen txtspc txthgt)          ;calculate arc length
  txtspc (/ arclen txtlen)                 ;calculate arc length of one character
  arcang (/ arclen radi)                   ;calculate arc angle of one character
  sang (- (+ (angle radp midp) (/ arcang 2)) (/ txtspc radi 2))  ;calc start angle of text
  count 1            ;Initialize counter
)
```

Use a REPEAT structure to loop the program. The loop calculates the placement of each character along the arc, and executes the text command. You don't see the COMMAND function below. Instead of using the COMMAND function directly, you evaluate the variable CMD. You set CMD to a quoted list containing the COMMAND function when you did the TBLSEARCH for Style. Evaluating CMD executes its contents, which are a COMMAND function list with arguments.

```
(repeat txtlen        ;Insert character loop
  (if (or (= orent "") (= orent "Y"))   ;Test text angle
    (setq txtang (angtoc  ;Preface angle w/ < for universal angular units in dec. string
                  (- sang (/ pi 2))   ;Convert start angle minus 90deg
                )
          txtpos count
    )
    ;Calc angle for character towards radius and character position in string
    (setq txtang (angtos (- sang (* pi 1.5)) 0)  ;angle to a command function
          txtpos (- (1+ txtlen) count)
```

```
    )
      ;Calc angle for character away from radius and character position in string
  )
  (setq txtp (polar radp sang radi))      ;Calculate character point
  (setq char (substr txt txtpos 1))       ;Get text character
  (eval cmd)                              ;Execute command list
  (setq count (1+ count))                 ;Increment counter
  (setq sang (- sang (/ txtspc radi)))    ;Calculate new start angle
);End repeat loop
```

Reset the system variables and end the function. Save what you've done and continue adding to TEST.LSP:

```
  (setvar "HIGHLIGHT" 1)        ;Turn highlight on
  (setvar "CMDECHO" 1)          ;Turn command echo on
  (princ)                       ;ends program cleanly
);End defun
;*end of ATEXT.LSP
;*
```

Save, Exit and Reenter AutoCAD. TEST.LSP should load.

Lisp returns: ATEXT

Select **[EDIT-MNU]**

Edit TEST.MNU. Add the following to the EXAMPLES page:

[ARC-TEXT]^C^C^C(setq lay (getvar "CLAYER")) LAYER M ANN02 C 4 ;;ATEXT \\\\LAYER S !LAY;;

Save and Exit to AutoCAD. TEST.MNU reloads.

(load "ANGTOC")	It is used by ATEXT.
Select **[EXAMPLES] [ARC-TEXT]**	Gets ATEXT

Command: (setq lay (getvar "CLAYER")) "0"   Saves layer name, then sets to ANN02.

Command: ATEXT   Then runs ATEXT.
Pick radius point: Pick the radius point.
Pick center point of text: Pick the center point.
Text: **THIS IS CURVED TOWARD THE CENTER**   Enter the text.
Is base of text towards radius point <Y>:   <RETURN>for the default.

Command: LAYER   Then it resets original layer.

Try other options as well. It works in any orientation.

  Delete TEST.LSP.

  Rename TEST.LSP to ATEXT.LSP.

Your screen should look like the Arc Text Routine.

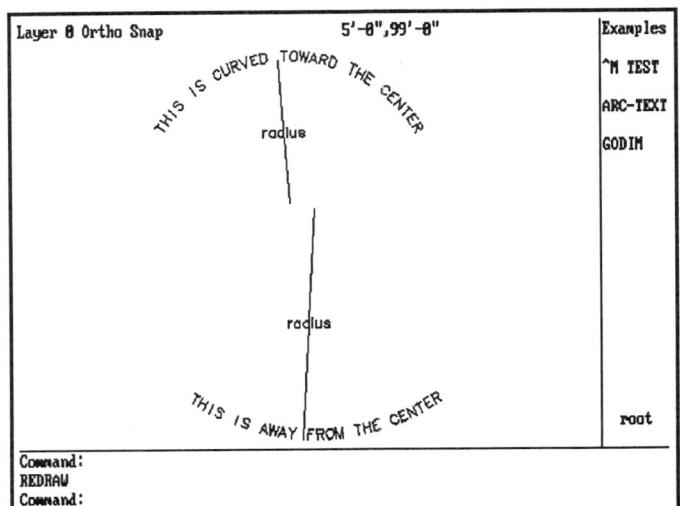

The Arc Text Routine

Making A Dimensioning Command

The GODIM function shows how easily you can write your own dimensioning commands. GODIM takes two points and an existing layer name as input. When you test it, you will have to supply these. GODIM automatically puts in text at the correct angle. The screen shot shows GODIM.

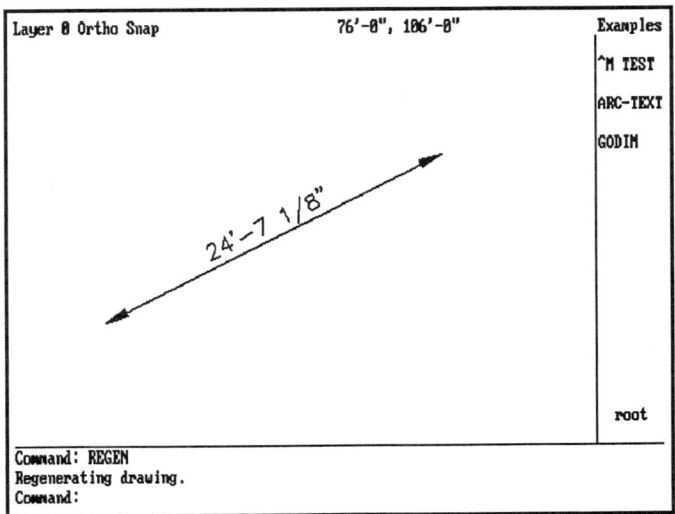

GODIM Dimensioning Command

Start with the first part of the function that calculates the angle and distance between the points.

GODIM A Custom Dimensioning Command

  Copy GODIM.LSP to TEST.LSP.

  Create a new LISP function file called TEST.LSP.

Select **[EDIT-LSP]**

Edit TEST.LSP, and enter:

```
;* GODIM is a LISP tool that accepts two points and an existing layer name.
;* It constructs a dimension line, verifies the dimension with the user and
;* places the formatted string as a dimension.
;* It assumes a fixed height text with the desired style as current.

(setq #pi90 (* 0.5 pi))

(defun godim(pt1 pt2 dtxtla / cang len tang mpt)
   (setq cang (angle pt1 pt2)                      ;angle between points
         len  (distance pt1 pt2)                   ;leng of dim
   );setq
```

Next, you need to determine the angle of the text. You want your text to appear at an angle between 270 and 90 degrees (quadrants 4 and 1) on your drawing. After determining the angle, the function finds the middle of the dimension line and calculates a point that is 1/16-inch off the dimension line. Then, it tests to see if the text needs rotation to the correct orientation. Save what you've done and continue adding to TEST.LSP.

```
(if (and (> cang #pi90) (<= cang (* pi 1.5)))     ;test rotation
    (setq tang (+ cang pi))                        ;then statement
    (setq tang cang)                               ;else statement
);if

(setq mpt (polar pt1 cang (* len 0.5))             ;midpoint
      mpt (polar mpt (+ tang #pi90) (* #dwgsc 0.0625))   ;1/16" clearance
);setq
```

The next part of GODIM draws the dimension line, inserts the dimension arrow and formats the distance using the user's current units. It passes the distance as a string to the text command. Then, it changes the layer of the text. Save what you've done and continue adding to TEST.LSP.

```
    (command "line" pt1 pt2 "" "insert" "ARROW" pt1 #dwgsc "" pt2
             "insert" "" pt2 #dwgsc "" pt1
             "text" "C" mpt (angtoc tang) (rtos len)
    );command

    (command "change" "L" "" "P" "LA" dtxtla "") ;change text to proper layer

);defun
;*end of GODIM.LSP
;*
```

Save, Exit and Reenter AutoCAD. TEST.LSP should load.

Lisp returns: GODIM

[EDIT-MNU]   Edit TEST.MNU.
                                         Add the following to the EXAMPLES page. It's a quick menu item to test GODIM:

[GODIM   ]^C^C^C(godim (getpoint "^MFrom point: ") (getpoint "^MTo point: ") "ANN02");

Save and Exit to AutoCAD. The menu should reload.

Select [EXAMPLES] [GODIM]   Test it, it prompts:

From point: Pick point.
To point: Pick point, and a lot scrolls by.
Lisp returns: nil   It drew lines and arrow and put text on layer ANN02.

Delete TEST.LSP and TEST.MNU.

 Rename TEST.LSP to GODIM.LSP. Rename TEST.MNU to MY11.MNU.

Formatting Prompt Strings

Many program problems arise from poor user interfaces. Two common problems are: confusing prompts, and lack of error checking. In this last section, there are several examples that show you how to format user prompt strings and do error checking. These exercises help you develop a customized version of the GET family of functions.

It would be useful, if the typical GETxxx function would format prompts, show the user default values, and impose restrictions on the user's input. Let's make a function called UDIST. UDIST concatenates several strings and issues the AutoLISP GETDIST built-in function.

Writing a User Interface Function UDIST

 Copy UDIST.LSP to TEST.LSP.

 Create a new LISP function file called TEST.LSP.

Select **[EDIT-LSP]**
Edit TEST.LSP, and enter:

```
;* UDIST Userinterface distance function
(defun udist (msg def / msg inp)
  (if def                              ;test for a default
    (setq msg (strcat "\n" msg " <" (rtos def) ">: ")) ;string'em with default
    );setq
    (setq msg (strcat "\n" msg ": "))  ;without default
  );if
  (setq inp (getdist msg))             ;and use it in the GET commands.
  (if inp inp def)                     ;Compare the results, return appropriate value
);defun
;*
```

The UDIST function takes two arguments: the prompt string, and the value to show in the prompt's default angle brackets. First, it tests to ensure that the default has a value. If the default is NON-NIL, it resets the message string to include the converted default. The function uses RTOS to format the raw value and display it in the user's current units. The closing parenthesis of the first SETQ is placed down one line to make adding more code to the function easier.

GETDIST is issued with the prompt. The "inp" variable is tested to see if the user actually entered a value, or if they hit a <RETURN> to accept the default. If a value is entered, the function pops out of the IF statement. Otherwise, the default value is passed out. Since the IF statement is the last statement in the function, its results are passed out of the function.

Save, Exit and Reenter AutoCAD. TEST.LSP should load. Test it:

```
Command: (udist "Distance between columns" 240.0)
Distance between columns <20'-0">:      Hit a <RETURN> to accept the default.
Lisp returns: 240.000000

Command: (udist "Distance between columns" nil)    Give a NIL default.
Distance between columns: 10'4                     No default is shown.
Lisp returns: 124.000000                           Returns the entered value.

Command: (udist "Distance between columns" nil)    Try it again with NIL.
Distance between columns:                          Just hit a return this time.
Lisp returns: nil                                  You get nothing!
```

The book uses UDIST frequently within other functions to prompt the user for a distance. It is worthwhile to extend UDIST by adding another argument to pass along a base point for the GETDIST function. The new part of the function is shown in **bold** in the example below. The new parts are "bpt" on the second line, and four lines near the end.

Even if you have the CA DISK, edit the TEST.LSP as shown.

Select **[EDIT-LSP]**

Edit TEST.LSP:

```
;* UDIST Userinterface distance function
(defun udist (msg def bpt / msg inp)
  (if def                                      ;test for a default.
    (setq msg (strcat "\r" msg " <" (rtos def) ">: ")) ;string'em with default
                            );setq
    (setq msg (strcat "\n" msg ": "))          ;without default
  );if
  (if bpt
     (setq inp (getdist msg bpt))    ;and use it in the GET commands.
     (setq inp (getdist msg))
  );if a base point exist
  (if inp inp def)                    ;Compare the results, return appropriate value
);defun
;*
```

Save, Exit and Reenter AutoCAD. TEST.LSP should load.

Command: **(setq bpt (getpoint "Grid point: "))**	First pick a base point.
Grid point:	
Lisp returns: (1248.219000 1220.373000)	Yours is different.
Command: **(udist "Offset from grid" nil bpt)**	Rerun without a default distance.
Offset from grid: **479.089000**	Draw in a distance.

You will encounter cases where a negative value from the user does not make sense to the application of your program. Or you may not want a NIL answer, yet you have not supplied a default value. The UDIST function does not handle these cases in its current form. Let's look at additional means that you can use to control user input.

Controlling User Input with INITGET

AutoLISP gives you a means to control input through the INITGET function. INITGET is a built-in AutoLISP function that initializes the types of input allowed or disallowed for a GET function call. For instance, you can disallow a NIL response, making the GET function automatically repeat until a suitable answer is given. This INITGET-type of control is established by passing integer bits (numbers) to the function. The bits and their resulting controls are shown in the following table:

INITGET CONTROL BIT	RESULTING CONTROL OVER INPUT
1	A value must be given. Null values are unacceptable input.
2	Requires input other than zero.
4	Input must be greater than zero. Disallows negative input.
8	No limit checking of points, regardless of LIMCHECK System Variable.
16	Returns full 3D (x, y, z) points instead of 2D (x, y) points.
32	Shows a dashed rubberband line only if popups are available.

You can use any combination of integer bits to gain the level of input control that you need. Adding all the Control Bits together (+ 1 2 4 8 16 32), you get 63. This is your full 3D and dashed arsenal of input control.

You must use INITGET before each GET function that you wish to control. INITGET's settings are not saved after the GET function is evaluated. INITGET takes two arguments, an optional string and an integer value, where the integer value is a combination between 1 and 63. The integer value establishes the type of input or point controls for the GET call.

When given unacceptable input, INITGET causes AutoLISP to make AutoCAD reissue the original prompt. AutoLISP will accept input only when the proper type of input is given. The following example shows how to use the INITGET call.

Command: **(initget 1)** Filters for a NIL input.
Lisp returns: nil

Command: **(getdist "Offset: ")** Try a distance.
Offset: Just hit a <RETURN>.

Requires numeric distance or two points.
Offset: **12** Give a numeric value.
Lisp returns: 12.000000

There are cases where you may want a function like GETPOINT to accept either a point, or a key word that you want to use in an application. You can use the optional string argument of INITGET to pass a string of key words to the GET functions. This lets the GET function accept key words from the user instead of the natural data type of the particular GET function.

The usual INITGET argument for key words is a space delimited string, consisting of upper and/or lower case characters. The book's call to INITGET is shown in the following sequence.

```
Command: (initget "Working R Origin")                    Try it.
Lisp returns: nil

Command: (getpoint "Working, Reference, Origin or pick point: ")
                                                         Use the GETPOINT function.
Working, Reference, Origin or pick point: w              Enter either a lower or upper case W.
Lisp returns: "Working"
```

Notice how AutoLISP returns the key word, not what you entered. You could have entered "wor," or "WORK," or just "W." Your program was not concerned with what you actually entered. Your input case has no effect. As long as the characters you enter match all or part of the key word. AutoLISP returns the entire key word, "Working." Experiment a little and verify that AutoLISP does return a consistent value.

You only need to enter the capitalized portions of the key words for the match to succeed. In cases containing key words with similar spelling, like ON and OFF, you have to tell INITGET to distinguish between the two by capitalizing up to the second character. Some users prefer to use the full key word spelling because it aids in the readability of their programs, others prefer to keep it short. Follow your own preference.

You can extend the UDIST function to control the user's input by adding two more arguments to the function line, and call the INITGET before the GETDIST function.

Even if you have the CA DISK, edit TEST.LSP by adding the header, adding two words to the second line, and adding two lines in the middle of the function:

Select **[EDIT-LSP]**

Edit TEST.LSP and add the new **bold** code as shown:

```
;* UDIST Userinterface distance function
(defun udist (bit kwd msg def bpt / msg bit inp)
  (if def                                              ;test for a default
    (setq msg (strcat "\n" msg " <" (rtos def) ">: ")  ;string'em with default
          bit (* 2 (fix (/ bit 2)))) ;a default and no null bit code conflict so
  );setq                             ;this reduces bit by 1 if odd, to allow null
    (setq msg (strcat "\n" msg ": ")) ;without default
  );if
  (initget bit kwd)
  (if bpt                                              ;check for a base point
    (setq inp (getdist msg bpt))     ;and use it in the GET commands.
    (setq inp (getdist msg))
  );if a base point exist
  (if inp inp def)                   ;Compare the results, return appropriate value
);defun
;*
```

You may encounter a case where the function tries to disallow null input, yet has a default value to show. In this case, there is no way to let the user accept the default via the <RETURN> key to return the null. An INITGET bit code of 1 is a "no null." Using a bit code of 1 will always make BIT an odd value. To control this, if a default exists and the bit is odd, the (* 2 (fix (/ bit 2))) expression reduces BIT by one. Both the default and control bit then work together. This effect is shown in the following example.

Save, Exit and Reenter AutoCAD. TEST.LSP should load. Try a few:

```
Command: (udist 0 "" "Offset from grid" 12.0 nil)
Offset from grid <1'-0">:
Lisp returns: 12.000000
```
First use 0 and "" as the control codes:
Accept the default.

```
Command: (udist 4 "" "Offset from grid" 12.0 nil)
Offset from grid <1'-0">: -24
Value must be positive.
Offset from grid <1'-0">: 24
24.000000
```
A 4 control Bit (no negatives):
Enter a negative value.

```
Command: (udist 4 "Working R Origin" "Offset from grid" 12.0 nil)
Offset from grid <1'-0">: w
"Working"
```
Try the key word argument:
Same as before.
Show default, but disallow null inputs:

```
Command: (udist 1 "" "Offset from grid" 12.0 nil)
Offset from grid <1'-0">:
Lisp returns: 12.000000
```
Accept the default.
The default

More UFUNS Functions

The book uses several other user interface functions, called UFUNS, in later AutoLISP programs. Unless you have the CA DISK, you need to make them now. These functions are: UANGLE, UPOINT, UINT, UREAL, USTR and UKWORD.

Making More UFUNS

  Copy UFUNS.LSP to TEST.LSP, overwriting the previous TEST file.

  If you don't have the CA DISK, continue to edit TEST.LSP.

Select **[EDIT-LSP]**   Edit TEST.LSP.

Add this ;* header at the top of the file:

```
;* UFUNS.LSP is a file of custom user GET function to format the
;* prompts, show default values and restrict user input.
;*
```

Beneath the existing UDIST function, add these new functions:

```
;* UINT User interface integer function
(defun uint (bit kwd msg def / msg bit inp)
  (if def                                            ;test for a default
    (setq msg (strcat "\n" msg " <" (itoa def) ">: ")   ;string'em with default
          bit (* 2 (fix (/ bit 2)))  ;a default and no null bit code conflict so
    )                                ;this reduces bit by 1 if odd, to allow null
    (setq msg (strcat "\n" msg ": "))                ;without default
  );if
  (initget bit kwd)
  (setq inp (getint msg))                            ;the GETINT function
  (if inp inp def)              ;Compare the results, return appropriate value
);defun
;*

;* UREAL User interface real function
(defun ureal (bit kwd msg def / msg bit inp)
  (if def                                            ;test for a default
    (setq msg (strcat "\n" msg " <" (rtos def 2) ">: ") ;string'em with default
          bit (* 2 (fix (/ bit 2)))  ;a default & no null bit code conflict so
    )                                ;this reduces bit by 1 if odd, to allow null
    (setq msg (strcat "\n" msg ": "))                ;without default
  );if
  (initget bit kwd)
  (setq inp (getreal msg))                           ;the GETREAL function
  (if inp inp def)              ;Compare the results, return appropriate value
);defun
;*

;* UPOINT User interface point function
(defun upoint (bit kwd msg def bpt / msg bit inp)
  (if def                                            ;Check for a default
    (setq pts (strcat (rtos (car def)) "," (rtos (cadr def))) ;formats for 2D point
          msg (strcat "\n" msg " <" pts ">: ")       ;string them with default
          bit (* 2 (fix (/ bit 2)))  ;a default and no null bit code conflict so
    )                                ;this reduces bit by 1 if odd, to allow null
    (setq msg (strcat "\n" msg ": "))                ;or without
  );if a default was specified
  (initget bit kwd)
  (if bpt                                            ;Check for base point
    (setq inp (getpoint msg bpt))                    ;and use it
    (setq inp (getpoint msg))                        ;but not if nil
  );if a base point exist
  (if inp inp def)              ;Evaluate results and return proper value
);defun
;*

;* UANGLE User interface angle function
(defun uangle (bit kwd msg def bpt / msg bit inp)
  (if def
    (setq msg (strcat "\n" msg " <" (angtos def) ">: ")
          bit (* 2 (fix (/ bit 2)))
    )
    (setq msg (strcat "\n" msg ": "))
  )
  (initget bit kwd)
  (if bpt
    (setq inp (getangle msg bpt))
    (setq inp (getangle msg))
```

```
    );if a base point exist
    (if inp inp def)
);defun
;*

;* USTR User interface string
(defun ustr (msg def spflag / msg inp)
  (if (and def (/= def ""))                         ;Test for both nil and null string
    (setq msg (strcat "\n" msg " <" def ">: "))     ;Include the default string
    (setq msg (strcat "\n" msg ": "))               ;Make string without the default
  );if
  (if spflag                                        ;check to see if space allowed flag is set
    (setq inp (getstring msg spflag))               ;get input - allow spaces
    (setq inp (getstring msg))                      ;disallow spaces
  );if
  (if (/= inp "") inp def)                          ;Test and return value
);defun
;*

;* UKWORD User key word. DEF, if any, must match one of the KWD strings.
(defun ukword (bit kwd msg def / msg bit inp)
  (if def
    (setq msg (strcat "\n" msg " <" def ">: ")
          bit (* 2 (fix (/ bit 2)))
    )
    (setq msg (strcat "\n" msg ": "))
  )
  (if (and def (> bit 0)) (setq bit (1- bit)))
  (initget bit kwd)
  (setq inp (getkword msg))
  (if inp inp def)
);defun
;*
(princ)
;*end of UFUNS.LSP
;*
```

Save, Exit and Reenter AutoCAD. TEST.LSP should load.

Test all of the functions, like you tested UDIST.

Command: **QUIT**   Quit the CA11 drawing.

  Delete TEST.LSP.

  Copy TEST.LSP to UFUNS.LSP.

Test these functions well if you don't have the CA DISK. The functions are used by other functions throughout the rest of the book.

Integrating the Text Menu

If you don't have the CA DISK and want to add the [ARC-TEXT] selection to the CA-MENU, you need to:

- Append the [ARC-TEXT] selection of your MY11.MNU to the end of the **TEXT menu page of the CA-MENU.MNU file.

The chapter's menu is shown below.

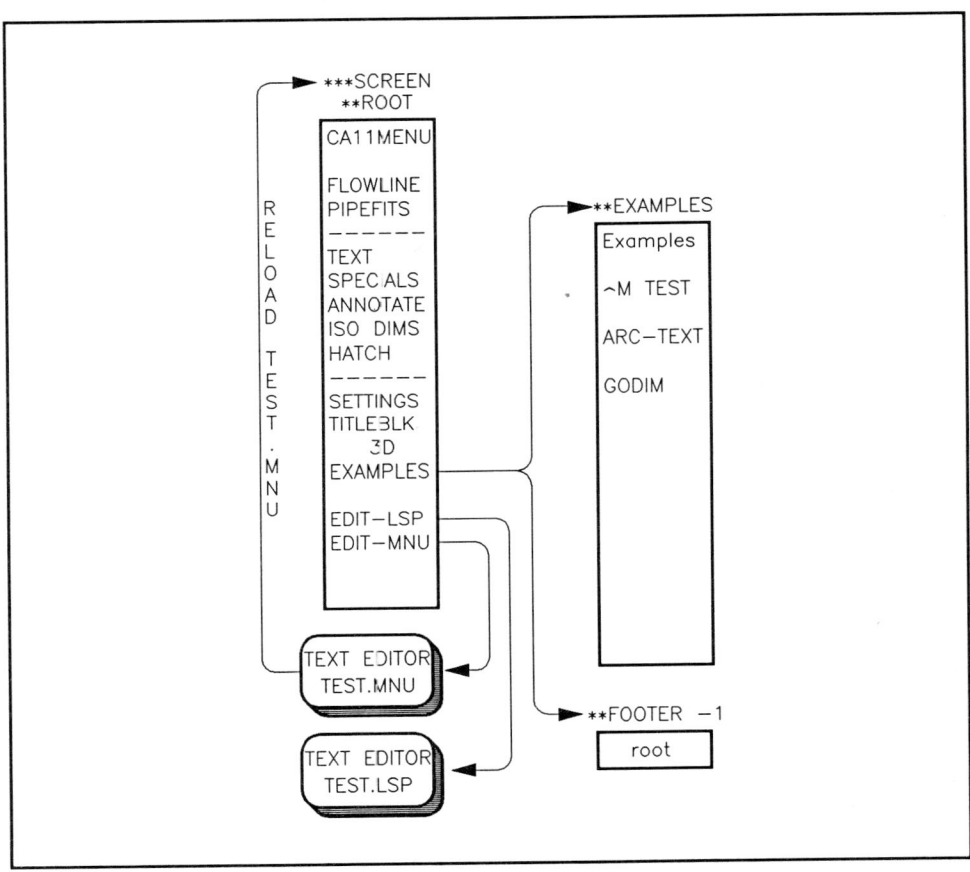

The Chapter 11 Menu

Summary Tips and Techniques

Think about your own application requirements. You can easily extend the special user functions provided in this chapter to other applications. Think about making GET functions to accept certain types of input, like numbers in a given range, or strings with less than nn characters.

Here are some tips.

❑ Use STRCAT to create long strings.

- Use ANGTOS for angle prompts and status. Use the book's ANGTOC for command input.
- Use the custom GODIM.LSP as a core function to build your own dimensioning program.
- Use the customized UFUNS GET functions in your programs. They simplify programming, formatting prompts and getting user input.
- When setting variables to long lists or long strings, clear the current value first (set to NIL) to save memory. Otherwise, AutoLISP allocates memory for both the old and new values.

12—0 Customizing AutoCAD

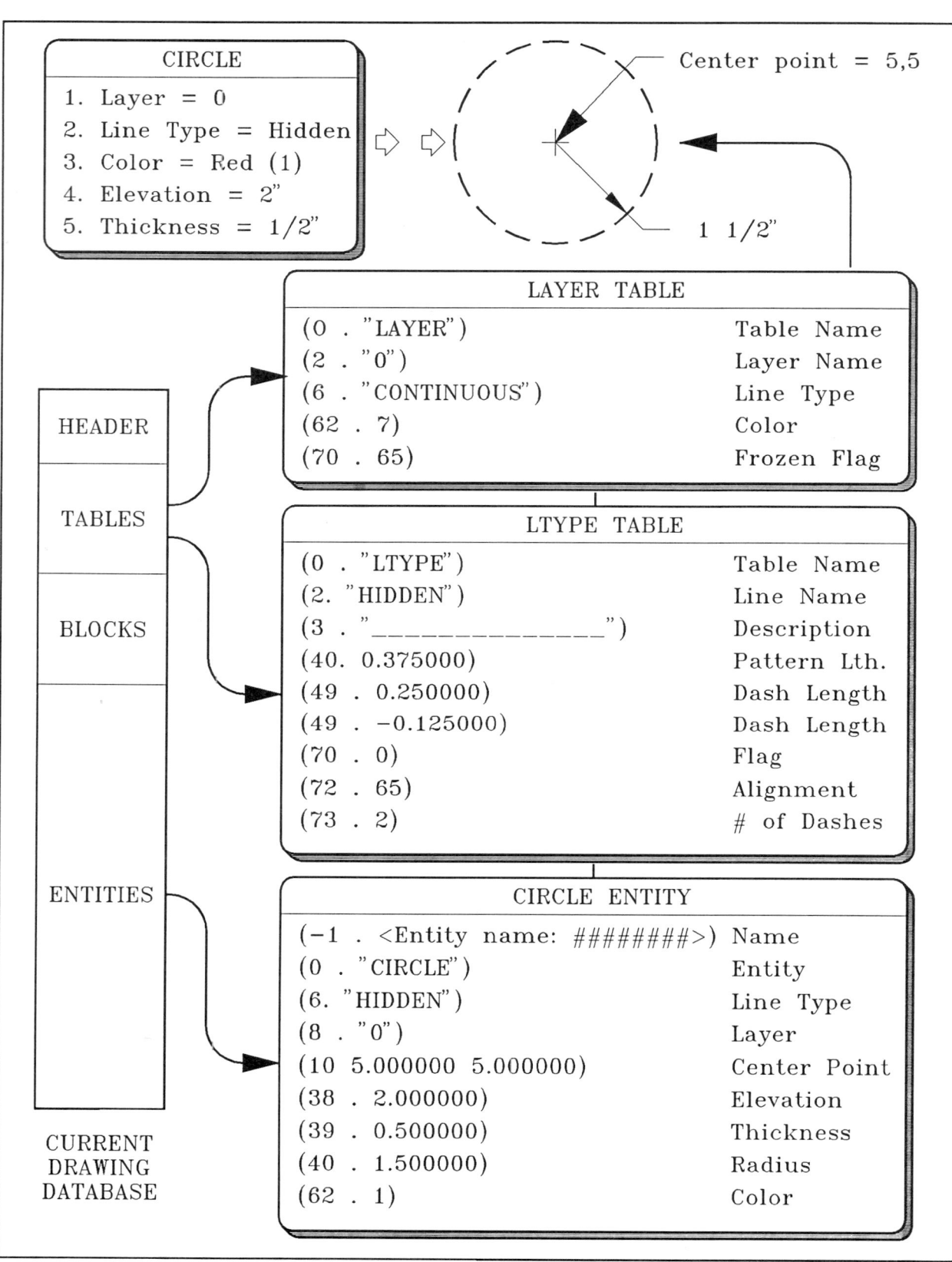

Accessing the Drawing Database

CHAPTER 12

The Drawing Database

While you are used to seeing AutoCAD draw lines, arcs and circles on your monitor, AutoCAD also functions "behind the scenes" as a database manager. Like any other good database manager, AutoCAD stores and sorts, adds and deletes, lists and reports records of information. Each record of the AutoCAD database stores information about geometric entities, reference tables, and the AutoCAD drawing environment. This chapter will show you how to use AutoLISP to access and manipulate AutoCAD's drawing database.

The Benefits of Accessing AutoCAD's Database

By accessing AutoCAD's database through AutoLISP, you gain almost complete control over the AutoCAD drawing editor. This control gives you three important benefits in customizing AutoCAD.

- You can modify, tally or delete drawing objects using your own editing commands that AutoCAD doesn't provide.

- You can make your custom programs more "intelligent," searching the drawing to mix and match entities. This automatic search makes it easier for your users to select groups of drawing entities.

- Using entity access, you also can make material counts for processing outside AutoCAD.

How To Skills Checklist

Using this chapter, you will learn how to:

- Get entity names from AutoCAD, and step through the drawing database using ENTLAST, ENTNEXT, ENTSEL, and ENTDEL.

- Make a program using AutoCAD's OFFSET command to automate multiple offsets.

- Work with Selection Sets, using SSGET, SSLENGTH, SSNAME, SSDEL, SSADD, and SSMEMB to group and process sets of drawing entities.

- Create a new set of Selection Set tools, called SSTOOLS, to merge selection sets, delete one set from the other, and find the common entities between two sets.

- Retrieve entity data, using ENTGET.

- Use ASSOC to extract entity data for AutoLISP calculations.

- Use DXF codes to identify entity data.

❏ Modify entities using ENTMOD, ENTUPD and SUBST, developing commands that are not available from AutoCAD.
❏ Process complex entities, like polylines, using SSGET and ENTSEL.

Macros, AutoLISP Tools and Programs

MACROS
[RCLOUD] puts the Revision Cloud function in a macro that handles settings.
[ENTGET] is a testing macro to get the entity name and entity data list of an entity.
[MARK] marks the database for [CATCH], using the MARK function.
[CATCH] selects all entities created since last MARKed, using the CATCH function.
[LAST-N] selects the last n number of entities created.

AutoLISP TOOLS
DXF is a simple function to retrieve the data element of an association list.
SSINTER builds a selection set of entities common to two selection sets.
SSUNION adds two selection sets together.
SSDIFF creates a selection set by taking the difference between two sets.
MARK puts a mark in the database for CATCH to use as a starting point.
CATCH selects all entities from the MARKed position through the end of the database.
LASTN retrieves a given number of entities starting from the end of the database.

PROGRAMS
MOFFSET is a program that creates multiple offsets of a single selected object. It eliminates the need to repeatedly select an entity during the AutoCAD Offset command.
APLATE is an area calculator that subtracts the area of holes and cutouts for any plate-like surface with punched or cut holes.
C:CSCALE automatically changes the drawing scale of selected entities relative to plot scale. It resizes text and block entities and aligns them.
C:BSCALE is a program to respecify block insertion scale factors after the block is present in the drawing. It allows independent rescaling of X, Y and Z factors.
C:RCLOUD is a program that lets you sketch a crude revision cloud that is reprocessed into a neatly reformatted cloud.

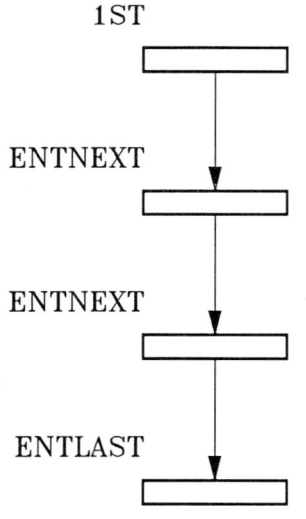

Drawing Database Order

Entity Names

Every entity within an AutoCAD drawing has a unique identification number. This number, commonly called the entity name, is the only means you have to retrieve data for the entity using AutoLISP. The entity name points to the data defining the entity in the drawing database. The entity name lets your programs go directly to the entity information that you need. AutoLISP has three functions to get an entity name directly from the AutoCAD database. They are:

Entity Names   12—3

ENTLAST   ENTNEXT   ENTSEL

You can use ENTSEL to prompt the user to select a single entity, like the Break and Offset commands do. ENTSEL can take an optional prompt to pass along to the user, but if you do not supply one, ENTSEL defaults to the familiar "Select objects:" prompt.

We also will look at ENTDEL, an entity data function that deletes entities. When you specify an entity name, ENTDEL erases the entity. It erases one entity at a time.

AutoCAD maintains its database in the order in which you create entities. This order is maintained throughout the entire editing session. Deleted entities remain in the database and are just "turned off." The order is changed when you reload the drawing because deleted entities are omitted from the database each time the drawing is called up for editing. AutoLISP can always tell you what the first entity is in the current database, as well as the last.

Selecting an Entity with ENTSEL, ENTNEXT and ENTLAST

First setup, then try the entity name functions.

```
Enter selection:                    Begin a NEW drawing named CA12
```

    Copy CA12.MNU to TEST.MNU.

    Copy MAIN.MNU to TEST.MNU.

```
Select [TEST] [24 x 36] [FULL] [INITIATE]          It should bring up TEST.MNU.

Command: ERASE                      Erase the border.
Command: ZOOM                       Center to height of 8.

Command: PLINE                      Draw the plate with punch holes as shown.

Command: (setq 1st (entnext))                      Gets the first entity.
Lisp returns: <Entity name: 60000014>              Your name will differ.

Command: (setq lent (entlast))                     Gets the last entity.
Lisp returns: <Entity name: 6000085C>

Command: (setq plate (entsel "Select the plate outline: "))  Supply the optional prompt:

Select the plate outline:                          Pick the outline
Lisp returns: (<Entity name: 60000474> (160.093700 141.593700))   Your data will differ.
```

The ENTSEL function returns both the entity name and the point that was picked on the screen. The point is returned in the form of a LIST. Unless you use OSNAP, don't assume that the point you picked actually lies on the selected object. The AutoCAD pickbox permits object selection even if the point is not

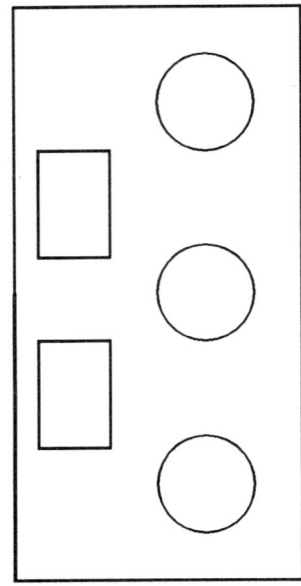

Plate with Holes and Cutouts

exactly on the entity. You can supply the list returned by ENTSEL to any AutoCAD command selection prompt that allows point picking, except the Fillet and Chamfer commands.

You can use ENTNEXT to step through the AutoCAD database, entity by entity. ENTNEXT, called without an argument, returns the first entity of the database. If you pass an entity name as the optional argument to ENTNEXT, AutoLISP returns the next entity that follows the entity name that you supply.

You can pass entity names to any AutoCAD command requesting object selection where Last is allowed as an option. You can't pass entity names if selection by point picking is required, but you can use an ENTSEL type list. AutoCAD commands accept the entity names regardless of the way you retrieve them.

Try Erasing the second entity of the database by entity name, then delete the first entity using ENTDEL.

```
Command: ERASE
Select objects: (entnext 1st)      Get the next (2nd) entity and <RETURN> to erase.

Command: OOPS                       Bring it back.

Command: (entdel 1st)              Now the 1st is gone.

Command: OOPS                       Bring it back. OOPS can't...
*Invalid*                           because AutoCAD didn't erase it.

Command: (entdel 1st)              And it's back
```

This last example shows an important difference between using AutoCAD and using AutoLISP. You can perform some functions in AutoLISP without full support from AutoCAD. In this case, AutoCAD leaves AutoLISP with the duty to restore the entity after it is erased. You can toggle entities between being erased and being visible. To bring a deleted entity back, you ENTDEL it again, even if the AutoCAD Erase command deleted the entity.

Let's explore entity names some more by making a program called MOFFSET. MOFFSET offsets a selected entity any number of times without your having to reselect the entity.

Enter the header comments, and the beginning of the function. When requesting input, use UFUNS, the user interface functions that you developed earlier. Supply defaults. Note that the NUM and DIS variables are global. They are not on the local variable list, so their defaults carry over to repeated uses.

Making a Multiple Offset Command MOFFSET

 Copy MOFFSET.LSP to TEST.LSP.

 Create a new LISP function file called TEST.LSP.

Select [EDIT-LSP]   Edit TEST.LSP, and enter:

```
;* C:MOFFSET is a multiple offset command. It requires the user to select one
;* object, tell how many times to offset it and give a direction for the offset.
;* It requires the User GET functions.
(defun C:MOFFSET( / ent spt intval)
   (setq ent (entsel "\nSelect object to offset: ")    ;get the object
         num (uint 5 "" "How many times" num)          ;# times
         dis (udist 1 "" "Offset distance" dis nil)    ;the distance to offset
         spt (upoint 1 "" "Select side" nil (cadr ent));get which side
   );setq
```

Next, MOFFSET sets a variable that is used to store the incremented distance value. This value is incremented by the original offset distance each time the program goes through the loop.

Save what you've done and continue adding to TEST.LSP:

```
   (setq intval dis)                      ;set a variable to increment interval

   (repeat num                            ;program loop
      (command "offset" intval ent spt "");run the offset command
      (setq intval (+ intval dis))        ;increment the offset distance
   );repeat
   (princ)         ;clean ending
);defun
(princ)            ;clean loading
;*
```

Save, Exit and reenter AutoCAD. TEST.LSP should load.

Command: **(load "ufuns")**   You MUST load the user interface functions from the previous chapter. MOFFSET needs to use them.

Command: **MOFFSET**   Try the function.
Select object to offset: Select any Line, Pline, Circle or Arc.
How many times: **5**
Offset distance: **1**
Select side: Pick a side.

```
Command: MOFFSET
Select object to offset:
How many times <5>:
Offset distance <1">:
Select side:
```

Try it again and see the defaults.
Select any object.
<RETURN> for the default.

 Delete TEST.LSP.

 Rename TEST.LSP to MOFFSET.LSP.

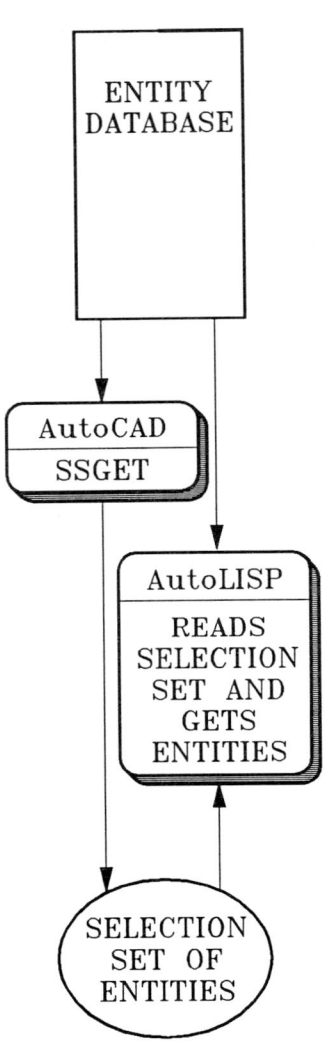

AutoLISP Selection Sets

Your screen should look like the Multiple Offset Command.

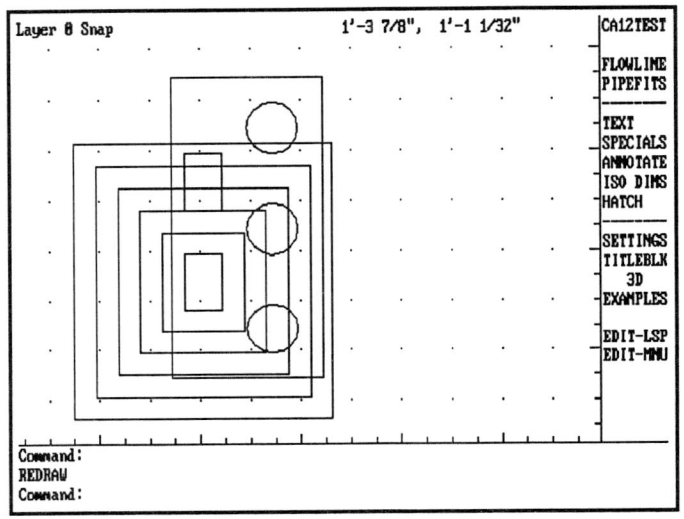

Multiple Offset Command

Entity Selection Sets

Selection Sets are nothing more than groups of AutoCAD entities. All AutoCAD edit commands operate on groups of entities. At the start of each edit command, AutoCAD asks for the set of objects to edit. The selection mode used is independent of the objects in the set. The user chooses the most convenient manner of selecting objects.

AutoLISP provides six functions to work with selection sets. These are:

SSGET   SSLENGTH   SSNAME   SSDEL   SSADD   SSMEMB

The SSGET (short for Selection Set GET) function simulates the selection routine of the AutoCAD program. You use SSGET to group entities in a set. You can use the function with or without formal arguments. If called with no arguments, SSGET issues the standard "Select objects:" prompt of AutoCAD. SSGET returns the name of a selection set. This name must be SETQed to a variable if you expect to process the set.

Creating Selection Sets with SSGET

Command: **(setq ss1 (ssget))** Select objects: **W** First corner: Other corner: 16 Found. Select objects: Lisp returns: <Selection set: 1>	Window everything, including the plate. And "ss1" says it is selection set number 1.

The selection set number has no real significance to your program. AutoLISP simply receives a handle from AutoCAD referencing that set of objects. The message <selection set #*n*> is a reference to the selection set *n* created during the current drawing session. Let's try the other SSxxx functions, starting with SSLENGTH.

Command: **(sslength ss1)** Lisp returns: 16	Find out how many are in there.
Command: **(ssdel (car plate) ss1)** Lisp returns: <Selection set: 1>	Remove one from the set. PLATE was set to the plate earlier. Just the last atom.
Command: **SELECT** Select objects: **!SS1** 15 found.	Check which one is not included. AutoCAD tells us one less.
Command: **(ssmemb (car plate) ss1)** Lisp returns: nil	See if entity is a member. Because you just subtracted it.
Command: **(ssadd (car plate) ss1)** Lisp returns: <Selection set: 1>	Add it back to the set.
Command: **SELECT** Select objects: **!SS1** 16 found.	Which ones are included. Back to the full set.
Command: **(ssmemb (car plate) ss1)** Lisp returns: <Entity name: 60000474>	See if entity is a member. AutoLISP tells us it's back in.
Command: **(ssname ss1 0)** Lisp returns: <Entity name: 60001B6C>	Gets first entity from the set.

NOTE. When you add or remove entities from an existing Selection Set using SSADD or SSDEL, you do not need to again SETQ the variable bound to that set. The variable remains set to the Selection Set and the addition or deletion is made to the set itself. If you use no arguments, SSADD creates a null (empty) set.

AutoLISP provides the SSNAME function to get the names of individual entities. You give SSNAME the name of the selection set and the index number of the entity you want. AutoLISP returns the name of the entity. SSNAME is sort of an "NTH" function of selection sets. However, the first entity is really the 0 offset element from the beginning of the set. When you use SSNAME, remember to start counting with 0! For example, count 0 to 15 for 16 elements.

Selections sets are a powerful tool. Let's make a command, called APLATE, to calculate the area of a plate with punched holes removed from it. To help you get started, the first part of the example reviews the Add and Subtract features of AutoCAD's Area command.

To review, try passing the "plate" entity to the Area command and then subtract the first entity of selection set "ss1." Your areas will vary from the example's.

Making a Plate Area Calculator APLATE

```
Command: U                                          Use Undos to get rid of the Moffset circles.
Command: AREA
<First point>/Entity/Add/Subtract: A
<First point>/Entity/Subtract: E
(ADD mode) Select circle or polyline: !PLATE        PLATE was SETQed earlier.
Area = 13.50 square in. (0.0938 square ft.), Perimeter = 1'-3"
Total area = 13.50 square in. (0.0938 square ft.)
(ADD mode) Select circle or polyline:
<First point>/Entity/Subtract: S
<First point>/Entity/Add: E
(SUBTRACT mode) Select circle or polyline:          Select a circle.
Area = 0.44 square in. (0.0031 square ft.), Circumference = 0'-2 23/64"
Total area = 13.06 square in. (0.0907 square ft.)
(SUBTRACT mode) Select circle or polyline:          <RETURN> twice.
```

Write the function APLATE, getting the name of an entity using the ENTSEL function. Recall that the ENTSEL function can pass a prompt to the user. Tell the user what they are to select, and to get the set of objects.

 Copy APLATE.LSP to TEST.LSP.

 Create a new LISP function file called TEST.LSP.

Select **[EDIT-LSP]**   Edit TEST.LSP, and enter:

```
;* APLATE calculates the area of a plate, automatically subtracting
;* the area of any included holes or openings.
(defun C:APLATE( / plate ss1 count emax)
   (setq plate (entsel "\nPick the plate outline: "))    ;get the plate outline
   (prompt "\nSelect all the holes...")                  ;Tell them what to select
   (setq ss1 (ssget))                   ;get the selection set
   (ssdel (car plate) ss1)              ;Make sure they didn't include the plate
```

Start the Area command, add the first entity and begin the subtraction of entities. Leave the command sent to AutoCAD via the COMMAND pipeline unfinished at the point where AutoCAD expects an entity name. Then, start a counter to loop through the selection set so that each entity in the set may be

More on Selection Sets 12—9

sent to the Area command and subtracted from the running total. Use SSLENGTH to obtain the number of entities in the selection set so that you can maintain an index counter of your location in the set.

Save what you've done and continue adding to TEST.LSP:

```
(command "area" "a" "e" plate "" "s" "e")    ;start the command
(setq count 0                                ;set the initial count value
      emax (sslength ss1)                    ;find the max count value
);setq
```

Loop through the selection set one at a time and pass the extracted name to the command processor. Increment your counter at the end of each cycle.

Save what you've done and continue adding to TEST.LSP:

```
(while (< count emax)                        ;start the program loop
   (command (ssname ss1 count))              ;pass the entity name to command
   (setq count (1+ count))                   ;increment the counter
);while
(command "" "")                              ;exit the command
```

When the loop is finished, complete the Area command with two <RETURN>s. Then, report the last area count. AutoCAD keeps the final area total in the AREA system variable.

```
   (prompt (strcat "\nFinal area is " (rtos (getvar "AREA"))"sq."))    ;inform of last area
   (princ)     ;clean living
);defun
(princ)        ;clean ending
;*
;*end of C:APLATE
```

Save, Exit and Reenter AutoCAD. TEST.LSP should load. Test it:

Command: **U**   If you forgot, get rid of the Moffset circles.
Command: **APLATE**   It prompts:
Pick the plate outline: Pick it.
Select all the holes...
Select objects: Select them.
Select objects: <RETURN>, it scrolls by and you get:
Final area is 10 21/64"sq.

 Delete TEST.LSP.

 Rename TEST.LSP to APLATE.LSP.

More on Selection Sets

You can use the SSGET function to perform selection by one of the typical AutoCAD selection methods. You can use a single point as an argument, like '(0 5), selecting an entity passing through that point. You can use the Last entity and Previous selection set options as arguments "L" and "P" respectively. To select entities by windows "W," or crossing windows "C," use the command option letter ("W" or "C") followed by the two points that form the box. Using an "X" argument (Release 9 or later releases) lets you make a selection by entity data. The various arguments that SSGET accepts are shown in the table below.

SSGET FORM	PURPOSE
(ssget '(0 5))	selects entity passing thru 0,5.
(ssget "L")	selects last entity.
(ssget "P")	selects previous selection set.
(ssget "W" '(0 0) '(5 5))	objects within window 0,0 to 5,5.
(ssget "C" '(0 0) '(5 5))	objects within/crossing box 0,0 to 5,5.
(ssget "X" *entity-data*)	retrieves entities that match data.

When you use SSGET with either the "W" or "C" option, you must supply the two corner points. You can't pick the corner points. If you don't use the optional entity data argument, SSGET "X" returns all non-deleted entities in the drawing. Optional data arguments filter the selection by its entity data.

You can construct a function selecting all the objects within the limits of the drawing using SSGET and one of the System Variables as window arguments. This function is shown in the following example.

Using SSGET

```
Command: (ssget "C" (getvar "LIMMIN") (getvar "LIMMAX"))     All within and crossing limits
Lisp returns: <Selection set 2>

Command: (ssget "X")                                          Get all entities in database
Lisp returns: <Selection set 3>

Command: (ssget "P")                                          Get the Previous set
Lisp returns: <Selection set 4>
```

NOTE. AutoLISP can't keep more than six AutoLISP selection sets "open" at one time. Open means the selection set is still set to a variable. Be sure to reset your variables to NIL, declare them local, or reuse the names, or you may run out of sets. The book uses SS1, SS2, SS3, and SS4 over and over again to help keep track of the number of sets open. If you run out, SSGET returns nil.

AutoLISP doesn't provide all the tools you need to work with entity selection sets. Some of your more advanced programs will require the ability to merge (append) selection sets, find common entities in selection sets (a union), or subtract selection sets. In the following section, we will help you develop a general set of AutoLISP tools, called SSTOOLS, to manipulate selection sets.

Developing a Selection Set Toolkit SSTOOLS

SSUNION merges (forms a union of) all entities in two selection sets, using the SELECT command. SSDIFF subtracts one set from another using Object Selection's Remove mode. SSDIFF is similar to SSUNION in its code. SSINTER finds the intersection of entities common to two sets. The SSTOOLS are shown in the following diagrams.

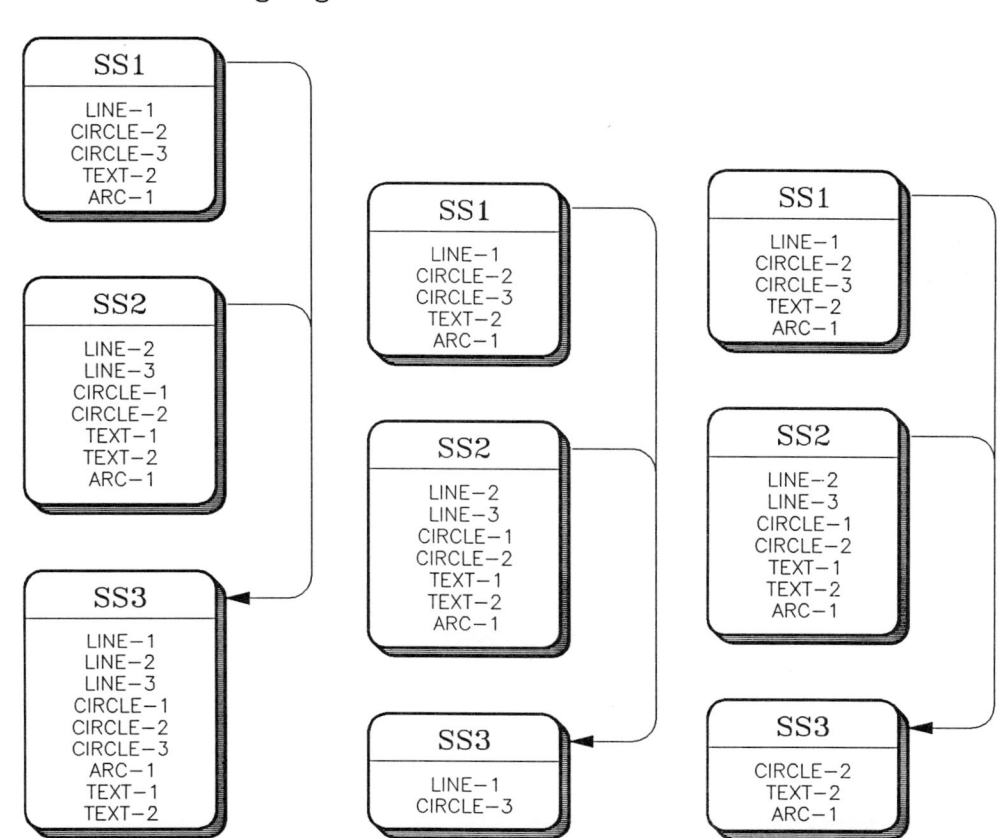

Creating SSTOOLS

To create SSDIFF, copy SSUNION, change it's name and description, and add a "R" to the "(command..." line.

 Copy SSTOOLS.LSP to TEST.LSP.

Create a new LISP function file called TEST.LSP.

Select **[EDIT-LSP]**   Edit TEST.LSP, and enter:

```
;* SSTOOLS.LSP is a kit of selection set tools. They all return ss3, a global.

;* SSUNION adds two selection sets using the Select command & creates new ss3
;* It must be used prior to the command which is to receive its selection
```

```
(defun ssunion (ss1 ss2 / hilite ss3)
  (setq hilite (getvar "HIGHLIGHT"))      ;speeds up selection
  (setvar "HIGHLIGHT" 0)
  (command "SELECT" ss1 ss2 "")            ;uses the select command to combine set
  (setq ss3 (ssget "P"))                   ;combined set is the previous
  (setvar "HIGHLIGHT" hilite)
  ss3                                      ;returns ss3
);defun
;*
;* SSDIFF subtracts ss2 from ss1 using the Select command & creates new ss3
(defun ssdiff (ss1 ss2 / hilite ss3)
  (setq hilite (getvar "HIGHLIGHT"))      ;speeds up selection
  (setvar "HIGHLIGHT" 0)
  (command "SELECT" ss1 "R" ss2 "")        ;uses the select command Remove mode
  (setq ss3 (ssget "P"))                   ;combined set is the previous
  (setvar "HIGHLIGHT" hilite)
  ss3                                      ;returns ss3
);defun
;*
```

SSINTER finds the intersection of entities common to both sets. The function uses entity names within one selection set and determines the membership of each name in the other set. Use SSMEMB to determine the membership.

Save what you've done and continue adding to TEST.LSP:

```
;* SSINTER takes the common entities (intersection) of ss1 and ss2
;* and returns new ss
(defun ssinter (ss1 ss2 / count count1 count2 smax smax1 smax2 more less name / hilite ss3)
   (setq count 0                           ;set counters
         count1 0
         count2 0
         smax1 (sslength ss1)              ;get number of entities in sets
         smax2 (sslength ss2)
         ss3 (ssadd)                       ;start a new empty set
   );setq
   (if (>= count1 count2)                  ;find out which has more entities
      (setq more ss1 less ss2 smax smax1)  ;flip names around
      (setq more ss2 less ss1 smax smax2)
   );if
   (while (< count smax)                   ;start a program loop
      (setq name (ssname more count))      ;get the entity name
      (if (ssmemb name less)               ;see if it's in the other set
         (ssadd name ss3)                  ;if it is, add it to the output set
      );if
      (setq count (1+ count))              ;increment counter
   );while
   ss3                                     ;pass out new set
);defun
;*
;*end of SSTOOLS.LSP
```

Save, Exit and Reenter AutoCAD. TEST.LSP should load. To test, you need to make some sets:

```
Command: (setq ss1 (ssget))         Select the plate.
Lisp returns: <Selection set: 5>

Command: (setq ss2 (ssget))         Select the rectangular holes.
Lisp returns: <Selection set: 6>

Command: (setq ss4 (ssget))         Yes, SS4. Select the circles.
Lisp returns: <Selection set: 7>

Command: (setq ss3 (ssunion ss2 ss4))   Test SSUNION.
Select objects:   2 found.
Select objects:   3 found.
Select objects:
Lisp returns: <Selection set: 8>

Command: SELECT                     Check it.
Select objects: !SS3                The new set. All the holes.
5 found.                            Hit <RETURN> to finish.

Select objects: (setq ss3 (ssdiff ss1 ss2))     Test SSDIFF. It scrolls by.
Command: SELECT                     Enter !SS3 to check it. All but the rectangles.

Command: SELECT                     You can use SSINTER in a command.
Select objects: (ssinter ss1 ss2)   Only the rectangles.
2 found.
```

Save the TEST.LSP file for the next exercise.

Traversing The Database

You often will find it necessary to skip around in the database, but AutoLISP does not provide a direct method to do this.

In the next example, you will create the LASTN function. LASTN lets you select more than the Last entity, it lets you work backwards in the database, specifying how far back you want to go. To move backwards, you use ENTDEL to delete the last entity in the drawing; then use the ENTLAST function and add the name to a selection set. The next time you call ENTLAST you get a "new" last entity. Of course, you reinstate the entities by processing the selection set and calling ENTDEL for each entity after you complete the selection. There is a second version of this function on the CA DISK, named LASTN-R9.LSP. It is for use with release 9 only, and uses SSGET "X."

Making the Function LASTN

Select **[EDIT-LSP]**   Continue adding to TEST.LSP:

```
;* lastn retrieves the last n entities in the drawing
(defun lastn (numb / ss3 count smax)
  (setq ss3 (ssadd))            ;create a new set
  (repeat numb                  ;loop the number of times
    (ssadd (entlast) ss3)       ;add the last entity to the set
    (entdel (entlast))          ;delete the last entity
  );repeat
  (setq count 0                 ;initialize counter
        smax (sslength ss3)     ;get number of entities in set
  );setq
  (while (< count smax)         ;loop to reinstate the deleted entities
    (entdel (ssname ss3 count)) ;brings them back
    (setq count (1+ count))     ;increment the counter
  );while
  ss3                           ;return set
);defun
```

Save, Exit and Reenter AutoCAD. TEST.LSP should load.

```
Command: SELECT                 Test LASTN.
Select objects: (lastn 3)
3 found.
```

To help you traverse the database, create two AutoLISP functions, called CATCH and MARK, to mark the database. The CATCH function advances along the database from the entity name you feed to it to the last entity of the database. You must store the entity name as the temporary global variable #MARK. The C:MARK command presets the variable and marks the database by inserting and deleting an entity, but MARK remembers the entity's name in the global #MARK. All entities created (or Exploded) after MARK and before CATCH are selected.

Select **[EDIT-LSP]**   Continue adding to TEST.LSP:

```
;* C:MARK is used by CATCH. It marks the database.
(defun C:MARK ( / pt val)
  (setq val (getvar "cmdecho"))         ;Saves echo setting
  (setvar "cmdecho" 0)                  ;Echo off
  (if (setq #mark (entlast))            ;If dwg not empty
    nil                                 ;then use last entity
    (progn                              ;else:
      (command nil nil nil "POINT" "@" ) ;^Cs & Creates mark entity
      (setq #mark (entlast))            ;Gets entities name
      (entdel #mark)                    ;Deletes entity
    )
  )
  (prompt
"\n#MARK SET -- (CATCH) will select all subsequent entities... ")
  (setvar "cmdecho" val)                ;Restores echo
  (princ)                               ;Forces nothing to be returned
)
```

```
;*
;* CATCH starts at MARK and retrieves to the end of the database.
;* It's good to use with the Explode command. Use MARK first, prior to EXPLODE.
(defun catch ()
  (if #mark
    (progn                            ;then
      (setq ss3 (ssadd))        ;start a new set
      (while (setq #mark (entnext #mark))   ;test/get next, while there's an entity
        (ssadd #mark ss3)       ;add it to the set
      );while
      ss3                       ;pass out the set
    );progn then
    (prompt "\n#MARK not set. Run MARK before (CATCH).\n")    ;else
  );if
);defun
;*
(princ)
;* end of SSTOOLS.LSP
```

Save, Exit and reenter AutoCAD. TEST.LSP should load.

Command: **MARK** MARK SET -- (CATCH) will select all subsequent entities...	Test them. Set a mark.
Command: **LINE**	Draw some entities.
Command: **SELECT** Select objects: **(catch)**	Test Catch.
5 found.	More or less.
Command: **(catch)** MARK not set. Run MARK before (CATCH).	Again to test the error prompt.
Select **[EDIT-MNU]**	Edit TEST.MNU.

Change the [CA00TEST] header to [CA12TEST] and add the items shown to the EXAMPLES page:

```
***SCREEN
**ROOT
[CA12TEST]

**EXAMPLES
[Examples]
[]
[MARK     ](mark)
[]
[CATCH    ](catch)
[]
[LAST-N   ](lastn (getint "^MLast n? entities: "))
```

Save and Exit to AutoCAD. TEST.MNU reloads. Test them.

   Delete TEST.LSP.

   Rename TEST.LSP to SSTOOLS.LSP.

Command: **QUIT**

Retrieving Entity Data

Your use of selection sets would be limited if you were not able to access individual entity data. You retrieve entity data by using AutoLISP's ENTGET function. Recall that ENTGET peeks into the AutoCAD database at the entity you request and returns its data. ENTGET takes an entity name as its only argument. The data is returned in an association list. The process is shown in the Entity Data Retrieval diagram.

You need to create an entity set to work with. This entity set is shown in the screen shot called AutoCAD's Entities. The entity set is included in the drawing called ENTITY.DWG on the CA DISK.

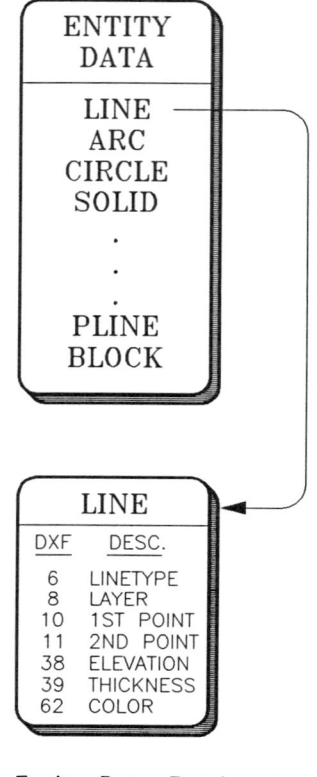

Entity Data Retrieval

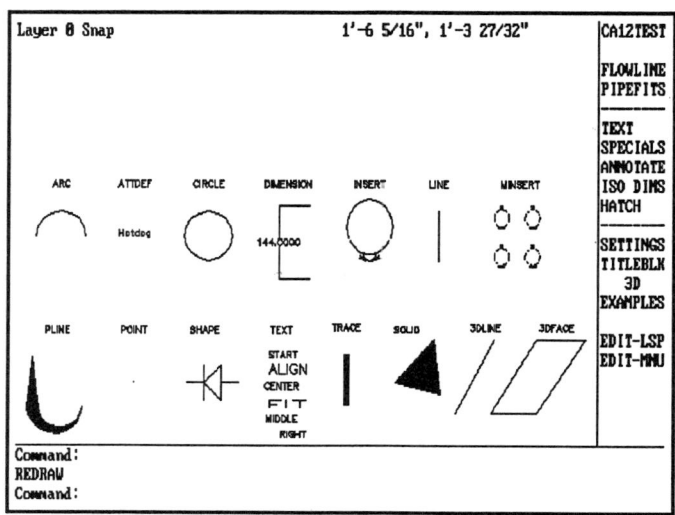

AutoCAD's Entities

Even if you have the CA DISK, you also need to find your ES.SHX file, or compile ES.SHP at the AutoCAD Main Menu. Copy ES.SHX to the \CA-ACAD directory. ES.SHP and ES.SHX come on the AutoCAD Support disk(s).

Examining Entities with ENTGET

 Edit an EXISTING drawing named ENTITY.DWG. Start the drawing, load TEST.MNU. Skip the drawing creation in this sequence.

 Begin a NEW drawing named ENTITY and draw.

Select **[TEST] [24 x 36] [FULL] [INITIATE]**   It should bring up TEST.MNU.

Command: **ZOOM**   Center, height 9".
Command: **DUP**   Copy ES.SHX to the \CA-ACAD directory.
Command: **LOAD**   Load ES. Then type SHAPE ? for listing.

Command: **ARC**   Draw the first entity.
Command: **SHAPE**   Enter one of the shapes.

Create each of the entities shown. They only need to be similar to the screen shot.
Both Insert and Minsert a block named LADYBUG.
Use each of the text alignment types shown.

If you've created your own drawing, your point coordinates and entity names will vary in rest of the chapter's examples.

Let's review the entity data format. Select the text that says "Center" in the drawing using the ENTSEL function. Then, use the ENTGET function to retrieve the data.

Command: **(setq en (entsel))**
Select object: Pick the "CENTER" Text

Lisp returns: (<Entity name: 600001E0> (5.483078 3.971583))

Command: **(setq ed (entget (car en)))**
Lisp returns: ((-1 . <Entity name: 600001E0>) (0 . "TEXT") (8 . "0") (10 4.825000 3.875000)
(40 . 0.150000) (1 . "CENTER") (50 . 0.000000) (41 . 1.000000) (51 . 0.000000)
(7 . "STD1-8") (71 . 0) (72 . 1) (11 5.250000 3.875000))

Since ENTSEL returns a list of entity name and point, you used CAR to extract the first atom, the entity name. ENTGET returned the entity data as an association list, ED. Your entity names may vary from the book's examples.

Entity Association Lists

An association list is a special two element list. The first element is a key item, word or code used in retrieving the second portion of the list, the data element. AutoLISP's ASSOC function returns the sublist (key word and data) that starts with the key word, from the list. You construct an association list with a dot "." separating the key and data elements. This is called a dotted pair. The dotted pair lets you use the CDR function to retrieve the data, regardless of the number of data items.

To see how ASSOC works, try getting a group of information using the ASSOC function with the key word "8," the code for layer. Then, make your own dotted pair list using AutoLISP's CONS function.

```
Command: (assoc 8 ed)         Give the code and entity list.
Lisp returns: (8 . "0")       It returns the association list.

Command: (cdr (assoc 8 ed))   And the data itself with CDR of the group.
Lisp returns: "0"             Returns "0", the layer name.

Command: (cons 8 "0")         Make a layer association list.
Lisp returns: (8 . "0")       A dotted pair just like AutoLISP's.
```

By definition, the dotted pair can have only two elements. You create a dotted pair using the basic list CONStructor function. A CONS created list is memory efficient because space is allocated for only two items. When working with entity data, AutoLISP uses the dotted pair form of a list.

DXF Group Codes and AutoLISP

AutoCAD uses the DXF format (Drawing eXchange Format) codes to report entity data. A DXF code precedes each type of data, whether a point, layer name, elevation etc. The DXF code enables programmers to manage AutoCAD's entity data without needing to know the order of the data. In fact, the order you get in your examples may vary from the order the book shows!

The complete DXF codes are shown in two tables: The Entity DXF Groups; and the Table DXF Groups. These tables are reproduced again in Appendix C for easy access.

The following exercise shows the listings for some common entities and their DXF codes. Try getting the entities shown. Continue on your own to look at the remaining entities. First, make a menu item to make the whole process easy.

Using DXF Codes

Select **[EDIT-MNU]**   Edit TEST.MNU. Add the following to the EXAMPLES page:

```
[ENTGET  ]+
(setq ed (entget (setq en (car (entsel "^MED - Select entity to get data:^M")))));
[]
```

Save and Exit to AutoCAD. TEST.MNU reloads.

Select **[EXAMPLES]**   Select some entities:

Select **[ENTGET]**   "ED -" reminds us that it SETQed ED to the data.
ED - Select entity to get data: Select the LINE.
Lisp returns:
((-1 . <Entity name: 6000003C>)   The entity name.
 (0 . "LINE")   The entity type.
 (8 . "0")   Its layer name.
 (10 810.000000 624.000000)   First endpoint.
 (11 810.000000 720.000000))   Second endpoint.

DXF Group Codes and AutoLISP 12—19

```
Command: !EN                                    It also SETQed ED to the entity name.
Lisp returns: <Entity name: 6000003C>

Select [ENTGET]
ED - Select entity to get data:                 Try the block INSERT.
Lisp returns:
((-1 . <Entity name: 60000258>)                 The entity name.
 (0 . "INSERT")                                 The entity type.
 (8 . "0")                                      Its layer.
 (2 . "LADYBUG")                                The block name.
 (10 675.000000 624.000000)                     Insertion point.
 (41 . 96.000000) (42 . 96.000000) (43 . 96.000000)    X, Y and Z scale factors.
 (44 . 0.000000) (45 . 0.000000)                Minserted Column and Row distance are none.
 (50 . 0.000000)                                Block rotation angle.
 (70 . 0) (71 . 0))                             Minserted Column and Row counts are none.

Select [ENTGET]
ED - Select entity to get data:                 Pick the CENTER text.
Lisp returns:
((-1 . <Entity name: 600001E0>)                 Entity name.
 (0 . "TEXT")                                   Entity type.
 (1 . "CENTER")                                 Text string.
 (7 . "STD1-8")                                 Style name.
 (8 . "0")                                      Layer name.
 (10 471.714300 372.000000)                     AutoCAD's text alignment point.
 (11 504.000000 372.000000)                     Optional alignment point, like centered.
 (40 . 12.000000)                               Text height, preset style or not.
 (41 . 1.000000)                                Width factor.
 (50 . 0.000000)                                Rotation angle.
 (51 . 0.000000)                                Obliquing angle.
 (71 . 0)                                       Generation flag, if mirrored, upside down.
 (72 . 1)                                       Justification flag. 1 means centered.

Select [ENTGET]
ED - Select entity to get data:                 Pick the Face.
Lisp returns:
((-1 . <Entity name: 600002BC>)                 Entity's name.
 (0 . "3DFACE")                                 Entity type.
 (8 . "0")                                      Layer name.
 (10 1002.000000 324.000000 0.000000)           1st 3D point.
 (11 1096.500000 468.000000 0.000000)           2nd 3D point.
 (12 1002.000000 468.000000 0.000000)           3rd 3D point.
 (13 912.000000 324.000000 0.000000))           4th 3D point.
```

Try them all and see if you can recognize the codes. Refer to the DXF Tables.

Not all key elements of an entity's association data list are DXF codes. The special "-1" and "-2" key element codes represent the primary and secondary entity names. These minus key codes are not DXF codes because the information associated with these key codes is not stored as part of the drawing file. These key codes are used to return the temporary entity names created during the drawing session.

ENTITY DXF GROUP CODES

LN = LINE	PT = POINT	CI = CIRCLE	AR = ARC	TR = TRACE	SD = SOLID	TX = TEXT
SH = SHAPE	BK = BLOCK	IN = INSERT	AD = ATTDEF	AT = ATTRIB	PL = POLYLINE	VT = VERTEX
DM = DIMENSION	3L = 3DLINE	3F = 3DFACE	SQ = SEQEND			

ENTITIES

CODE	DESCRIPTION	LN	PT	CI	AR	TR	SD	TX	SH	BK	IN	AT	AD	AR	PL	VT	DM	3L	3F	SQ
0	Primary Text Value							1												
1	Name: Shape, Block, Tag								2	2	2	2								
2	Prompt Strg,												3							
3	Line Type Name	6	6	6	6	6	6	6	6	6	6	6	6	6	6	6	6	6	6	6
7	Text Style Name							7				7	7							
8	Layer Name	8	8	8	8	8	8	8	8	8	8	8	8	8	8	8	8	8	8	8
10	X - Start or Insert Point	10				10	10	10	10	10	10	10	10	10		10		10	10	
	X - Center Point			10	10																
	X - Corner Point						10														
	X - Definition Point																10				
11	X - End or Insert Point	11																			
	X - Corner Point					11	11														
	X - Alignment Point											11	11								
	X - Middle Point Of Dim.																11				
12	X - Corner Point					12	12														
	X - Insert Point																12				
13	X - Corner Point					13	13														
	X - Definition Point																13				
14	X - Definition Point																14				
15	X - Definition Point																15				
16	X - Definition Point																16				
20	Y - Start or Insert Point	20				20	20	20	20	20	20	20	20	20		20		20	20	
	Y - Center Point			20	20																
	Y - Corner Point						20														
	Y - Definition Point																20				
21	Y - End or Insert Point	21																			
	Y - Corner Point					21	21														
	Y - Alignment Point											21	21								
	Y - Middle Point Of Dim.																21				
22	Y - Corner Point					22	22														
	Y - Insert Point																22				
23	Y - Corner Point					23	23														
	Y - Definition Point																23				
24	Y - Definition Point																24				
25	Y - Definition Point																25				
26	Y - Definition Point																26				
30	Z - Corner Point																	30	30	
31	Z - Corner Point																	31	31	
32	Z - Corner Point																		32	
33	Z - Corner Point																		33	
38	Entity Elevation	38	38	38	38	38	38	38	38	38	38	38	38	38	38	38	38			38
39	Entity Thickness	39	39	39	39	39	39	39	39	39	39	39	39	39	39	39	39			39

ENTITIES DXF GROUP CODES

CODE	DESCRIPTION	LN	PT	CI	AR	TR	SD	TX	SH	BK	IN	AD	AR	PL	VT	DM	3L	3F	SQ
0																			
40	Radius			40	40														
	Height, Size or Width							40	40										
	Leader Length															40			
41	X Scale Factor or Width							41	41		41	41	41	41	41		41		
42	Y Scale Factor or Buldge										42				42				
43	Z Scale Factor										43								
44	Column Spacing										44								
45	Row Spacing										45								
50	Rotation Angle							50			50	50							
	Start Angle												50			50			
51	Curve Fit Tangent																		
	End Angle											51	51						
	Obliquing Angle							51	51										
62	Color	62	62	62	62	62	62	62	62	62	62	62	62	62	62	62	62	62	62
66	Entities follow flag										66			66					
70	Dimension Type															70			
	Vertex or Polyline Flag											70	70	70	70				
	Attribute Flag										70								
	Column Count																		
	Block Type Flag									70									
71	Text Generation Flag							71				71	71						
	Row Count										71								
72	Text Justification							72				72	72						
73	Field Length											73	73						

TABLE DXF GROUP CODES

LT = LINETYPE   LY = LAYER   ST = STYLE   VW = VIEW

CODE	DESCRIPTION	LT	LY	ST	VW
	TABLES				
0	Table Name	2	2	2	2
2	Descriptive Text	3		3	
3	Font File Name			4	
4	Bigfont File Name				
6	Line Name		6		
10	X - View Center Point				10
11	X - View Direction				11
20	Y - View Center Point				20
21	Y - View Direction				21
31	Z - View Direction				31
40	View Height				40
	Pattern Length	40			
	Fixed Text Height			40	
41	View Width		41		41
42	Text Width			42	
49	Last Height Used	49			
50	Dash length			50	
62	Obliquing Angle		62		
70	Color	70	70	70	70
71	Number of Table Entries			71	
72	Text Generation Flag	72			
73	Alignment Codes	73			
	Number of dash items				

To simplify the retrieval of entity data, let's make a simple function called DXF.LSP. DXF's purpose is to accept a DXF group code and an entity list. It returns the data item as the CDR of the group codes association list. You can use the DXF function to simplify entity data retrieval. DXF is used by other functions throughout the rest of the book.

Making a DXF Extraction Function

 Copy DXF.LSP to TEST.LSP.

 Create a new LISP function file called TEST.LSP.

```
Select [EDIT-LSP]                    Edit TEST.LSP, and enter:

;* DXF takes an integer dxf code and an entity data list.
;* It returns the data element of the association pair.
(defun dxf(code elist)
   (cdr (assoc code elist))   ;finds the association pair, strips 1st element
);defun
;*
```

Save, Exit and reenter AutoCAD. TEST.LSP should load. Test it.

```
Select [ENTGET]                      First make sure you have same data in ED variable.
ED - Select entity to get data:      Pick the CENTER text, then enter:
Command: (dxf 7 ed)                  Gets the Text Style name.
Lisp returns: "STD1-8"               It works.
```

 Delete TEST.LSP.

 Rename TEST.LSP to DXF.LSP.

Entity Properties and Default Values

Every entity resides on a layer and has a layer name (DXF code 8). However, there are four kinds of entity data that are reported only if the value is not the default setting. The four kinds of optional data and their default values are:

DEFAULTABLE ENTITY PROPERTIES

OPTIONAL PROPERTY	DEFAULT VALUE	DXF CODE
ELEVATION	0.0	38
THICKNESS	0.0	39
LINETYPE	BYLAYER	6
COLOR	BYLAYER	62

Tips and Techniques

Here are some tips and techniques.

- ❑ To avoid running out of selection sets, declare your sets as Locals, reuse the same names or reset to NIL when you are through.
- ❑ When you build a selection set with SSADD in a loop, initialize it first to a null set by calling SSADD without arguments.
- ❑ Using AutoLISP's entity access commands is usually faster than using AutoCAD's editing commands to manipulate entities.
- ❑ Use the MARK and CATCH functions when Exploding dimensions, polylines, or blocks. Then you can easily select the resulting entities for editing.
- ❑ When you create polylines and other entities for area calculations, use a unique layer, linetype or color to make them easy to select with SSGET "X" filters.

Here are some suggestions for extending the routines provided in the chapter.

- ❑ If you use SSGET "X" with one or more of its optional entity data filters, you can search the entire drawing and return the entities that match the filters. You can filter for Entity type, Block name of Inserts, Linetype, Style etc. Use a DXF coded association data list in the form returned by ENTGET to specify your filters.
- ❑ You can extend the techniques used in C:APLATE to other area calculation programs, like doing floor plan calculations.
- ❑ You can modify C:RCLOUD to create a program to smooth polyline contours generated when tracing contours with Sketch.

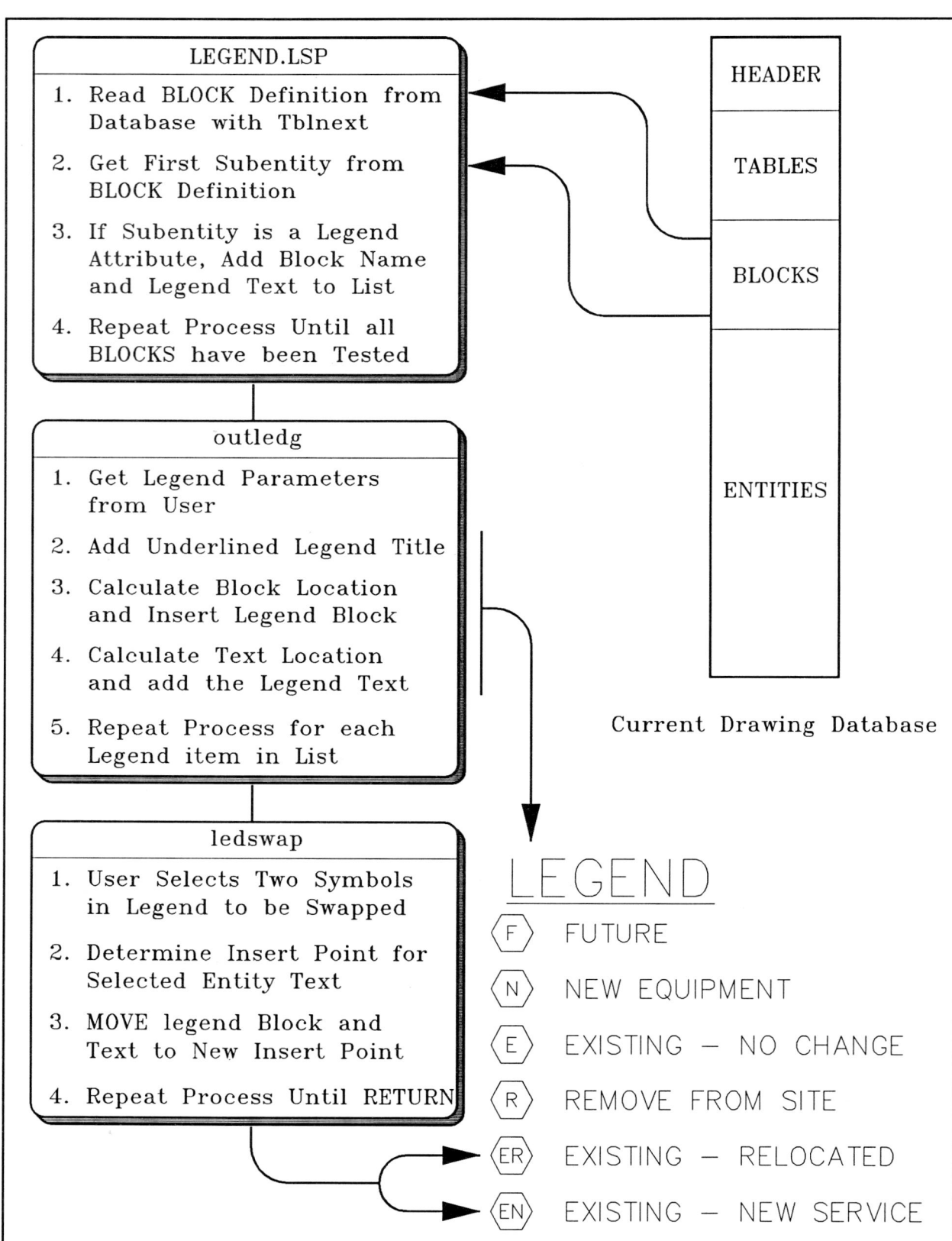

The LEGEND Program

CHAPTER 13

AutoLISP Table Access and More!

SYMBOL Table Access is your avenue to read what is in an AutoCAD drawing. This chapter introduces you to some everyday uses of table access, showing you how to use AutoLISP to read the LAYER, LTYPE, VIEW, STYLE, and BLOCK reference tables. The chapter shows how to use this table access to generate symbol reference charts. The chapter also shows how to use the LTYPE table to store and update personal variables in an AutoCAD drawing.

The Benefits of AutoLISP Table Access

Using Symbol Table access gives you the following benefits in your custom applications.

- Table access provides information about what is defined in the drawing reference tables. You can use this access in your programs to control drawing choices and prevent user errors.

- By accessing block definitions in the BLOCK table, you can determine a block's characteristics to control its use in your application.

- You can use access to the Linetypes table to store string or other types of user variable information in the drawing, and automatically update this information.

How To Skills Checklist

In this chapter, you will see how to:

- Scan the AutoCAD's drawing reference tables, LAYER, LTYPE, VIEW, STYLE, and BLOCK, using AutoLISP's TBLSEARCH and TBLNEXT functions.
- Search block data to automatically generate legends.
- Access the LTYPE table, imbed information in blocks, store "personal variables" in drawings, protect them from purging, and update their information.

Macros, AutoLISP Tools and Programs

MACROS
[SWAP] and [LEGEND] call the legend programs from the screen menu.
[WLAYER] loads and calls the WLAYER program.

AutoLISP TOOLS

ETOS is a function that takes any expression and converts it to a string.
OUTLEDG is a formatting routine that takes a double element list of data and puts it in the AutoCAD drawing as text.
PUTVAR can be used inside other AutoLISP programs to update a personal variable.

PROGRAMS
LEGEND is a program that creates a legend by scanning the BLOCK table, extracting all specially marked legend blocks. It builds a legend and places the legend in the drawing at a user specified location.
C:LEGEND collects legend blocks and formats them.
OUTLEG draws the legend.
LEDSWAP lets the user swap items in the legend easily.
C:WLAYER scans the drawing's layer list and erases, skips or wblocks each layer to a drawing file name of the first 8 characters of the layer name.
PVAR provides the function to create a personal variable system to retain personal values in a drawing's Ltype table when you end it. You can use numeric or string variables.
C:PVAR is used to query, view or update the personal variables.
PUTVAR stores the variable as a linetype definition.
RETVAR retrieves the variable from the linetype definition.
PVCHK verifies the personal variables block is defined or inserts it
GETPVAL is used to get a new personal variable value from the user

Symbol Tables

AutoCAD and AutoLISP refer to the "tables of named stuff" that are stored with each drawing as the Symbol Tables. Do not confuse this usage of "symbol" with graphic elements commonly called symbols, like blocks, Inserts and Shapes. In Symbol Tables, "symbol" means the name assigned to the named stuff that AutoCAD keeps track of. This includes named "things," like Block Definitions (not Inserts), named "places," Views and Layers, and named properties, Linetypes and Text Styles. Unlike entity access, AutoLISP's built-in table access functions only let you retrieve information. AutoLISP cannot modify the tables in the database.

Named Blocks

TBLSEARCH and TBLNEXT are the AutoLISP Symbol Table functions. TBLSEARCH scans any table, LAYER, LTYPE, VIEW, STYLE, or BLOCK. TBLSEARCH simply returns the table data list for the symbol name you feed to it. A table data list is similar to an entity data list. You commonly use TBLSEARCH to search for a View or Layer name, a Linetype or Text Style.

Although entity access gives you information on block Inserts, sometimes you need access to entity information within the Block Definition. This access provides the key to generating the legend example shown below. Each Block in the drawing, containing a special Attribute for the legend, receives a legend entry. You need to define the Attribute as Constant, with the attribute tag as the key to determining if the Block is a "legend symbol."
Let's see how TBLSEARCH works with some blocks in an example of legend generation. The legend symbols that you will work with are shown in the Legend Symbol Blocks drawing.

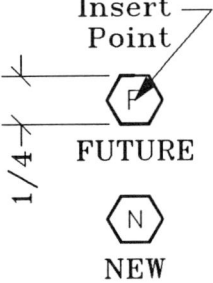
FUTURE

NEW

EXIST

REMOVE

EX—RELOC

EX—NEW—S

Legend Symbol Blocks Library

Named Blocks 13—3

Make the six blocks with constant attribute definitions and the values shown.

LEGEND BLOCK CHARACTERISTICS

BLOCK NAME	ATTRIBUTE TAG	ATTRIBUTE VALUE
FUTURE	LEGEND	FUTURE
NEW	LEGEND	NEW EQUIPMENT
EXIST	LEGEND	EXISTING - NO CHANGE
REMOVE	LEGEND	EXISTING - REMOVE FROM SITE
EX-RELOC	LEGEND	EXISTING - RELOCATE
EX-NEW-S	LEGEND	EXISTING - NEW SERVICE

Initializing a Legend Generator

Edit the EXISTING drawing named LEGEND. You can Zoom in, Insert and Explode one of the blocks to examine it.

Create a NEW drawing named LEGEND.

Copy CA13.MNU to TEST.MNU.

Copy MAIN.MNU to TEST.MNU.

```
Select [TEST] [24 x 36] [FULL] [INITIATE]           It brings up TEST.MNU.
Command: ERASE                              Erase the border.
Command: ZOOM                               Left, corner -1,-1 with height 3.

Command: POLYGON
Number of sides: 6                          at 0,0 Circumscribed with 0.125-inch radius.

Command: TEXT                               Set style STD3-32, use Middle at 0,0 and rotation 0:
Start point or Align/Center/Fit/Middle/Right/Style: S
Text: F                                     Give it an F for FUTURE

Command: ATTDEF                             Change modes to Invisible and Constant:
Attribute modes -- Invisible:Y Constant:Y  Verify: N
Attribute tag: LEGEND
Attribute value: FUTURE                     Place it next to symbol.

Command: BLOCK
Block name (or ?): FUTURE
Insertion base point: 0,0
Select objects: L                           You MUST select the Attdef first.
Select objects: W                           Then the rest.

Command: OOPS                               Recycle it to make the others.
Command: CHANGE                             Change Attdef tag and legend Text to make other symbols.
Command: BLOCK                              Repeat Oops, Change, Block for each.
```

The Attdef must be the first entity selected in the block. The LEGEND program will rely on its being first to speed up searching.

You can get information about the block with TBLSEARCH. You tell AutoLISP the table name and the name of the block to search for. The list returned contains information about the Block definition. This header includes the insertion point, Block name and the original insertion base point.

```
Command: (setq tbdata (tblsearch "BLOCK" "FUTURE"))
Lisp returns: ((0 . "BLOCK")                The block table.
(2 . "FUTURE")                              Block name.
(70 . 64)                                   Flag value.
(10 0.000000 0.000000 0.000000)             Insertion base point.
(-2 . <Entity name: 40000FBC>))             First subentity.
```

You also can retrieve table references without knowing the names of blocks. AutoLISP provides the TBLNEXT function to step through the reference tables one entry at a time. The first time TBLNEXT is called in a drawing session, it returns the first item of the table. Each additional call returns a subsequent item. It automatically advances itself to the next item. You can force TBLNEXT to restart at the first item of the table by including an optional (non-NIL) rewind argument, like T.

Let's get the first item of the block table, using the rewind argument to assure a start from the top. Then use TBLNEXT again to get the second Block.

```
Command: (tblnext "BLOCK" T)                Returns the "FUTURE" list as above.

Command: (tblnext "BLOCK")                  "NEW" will be next if you created them in our order.
Lisp returns: ((0 . "BLOCK") (2 . "NEW") (70 . 0) (10 0.000000 0.000000 0.000000)
(-2 . <Entity name: 40000997>))
```

Notice that the table entries do not list the data for subentities. If you want to find the entities contained in the Block definition, you have to search a little deeper.

The table data list for the Block contains a -2 code followed by an entity name. This is the name of the first subentity of the Block definition. Treat subentity names as you would any other entity name. Use the ENTGET function to get the data associated with the entity. The data format is exactly like normal entity data.

```
Command: (setq edata (entget (cdr (assoc -2 tbdata))))        TBDATA was set above.
Lisp returns: ((-1 . <Entity name: 40000FBC>) (0 . "ATTDEF") (8 . "0")
(10 0.203125 -0.046875) (40 . 0.093750) (1 . "FUTURE") (3 . "")
(2 . "LEGEND") (70 . 3) (73 . 0) (50 . 0.000000) (41 . 1.000000)
(51 . 0.000000) (7 . "STD3-32") (71 . 0) (72 . 0) (11 0.000000 0.000000))
```

This entity data list begins with the same entity name (ours is 40000FBC), but now it has a normal -1 code. To distinguish between table header data and subentity data, AutoLISP codes the table data's pointer to the first subentity returned by the TBLSEARCH function with a -2, and uses a -1 group to identify the entity data list of each subentity in the Block definition.

ENTGET found your special LEGEND Attdef first because you selected it first when you created the Block. Since you specified a fixed order during object selection, you can depend on the LEGEND attribute being first in the block definition. This fixed order avoids the need to search the entire Block definition. This is not always the case. Unless you have complete control over the original block definitions, don't plan on any specific order of entities within blocks.

In the next few exercises, we will show you how to pick your way through the table data, build a list of this data, make use of it and place it in the drawing to form a legend. You need to develop a group of functions that automate table searching, get the needed data, prompt for the location and build the legend, using the global drawing scale and default text.

First, you need to write a function, called LEGEND, to retrieve the legend entries and store them in a list. The function begins by searching through the block table. It looks for the ATTDEF tag of LEGEND. It ignores any symbol Blocks that were not created with the LEGEND Attdef first. Each time it finds a legend entry, it uses the APPEND function to build a list of lists, with each sublist being the Block name and Attribute value. When TBLNEXT reaches the end of the Blocks table, it returns nil. The LEGEND program then calls two other functions, OUTLEDG and LEDSWAP, to output and format the list. We'll show you those in a moment. For the moment, concentrate on creating the LEGEND program.

Developing a Legend Generator Program LEGEND

Select **[TEST]** **[24 x 36]** **[1/8"=1'0"]** **[INITIATE]**   Re-setup to reset your scale.

  Copy LEGEND.LSP to TEST.LSP.

  Create a new LISP function file called TEST.LSP.

Select **[EDIT-LSP]**   Edit TEST.LSP, and enter:

```
;* C:LEGEND scans the BLOCKs table and extracts all specially marked symbol
;* Blocks, then builds a legend in the drawing listing all such Blocks.
;* Mark symbol Blocks with a constant invisible ATTDEF tagged LEGEND with
;* the symbol's description as its value.
;* The Attdef MUST be the first selected entity when making the Block.
;* The LEGEND functions consist of C:LEGEND, OUTLEDG and LEDSWAP.
;* Legend must have the DXF function preloaded, and expects a
;* global dwg scale variable named #DWGSC set to plot scale.

(defun C:LEGEND ( / frst tbdata elist edata llist)
   (prompt "\nBuilding legend list...\n")       ;tell user what is going on
   (setq frst t)                                ;flag to rewind to first item in table
   (while (setq tbdata (tblnext "BLOCK" frst))  ;Until all Blocks are tested - retrieve data
      (setq edata (entget (dxf -2 tbdata)))     ;Get 1st subentity's data.
      (if (and (= (dxf 0 edata) "ATTDEF") (= (dxf 2 edata) "LEGEND")) ;is it a legend symbol
   (if llist                                    ;then, If llist exists,
      (setq llist                               ;then append Block name and legend descr,
             (append llist (list (list (dxf 2 tbdata) (dxf 1 edata)))))
);setq
(setq llist (list (list (dxf 2 tbdata) (dxf 1 edata))))
                                                ;else create new list w/ name and descr.
      );if llist
);if and =
     (setq frst nil) ;resets flag to disable rewind
   );while
   (outledg llist)                              ;outputs legend, using the list llist.
   (ledswap)                                    ;Allows user to swap entries.
   (princ)                                      ;end cleanly
);defun C:LEGEND
;*
```

If you ran LEGEND now, it would crash because OUTLEDG and LEDSWAP aren't done yet. To test LEGEND, you need to temporarily disable their lines (outledge llist) and (ledswap) with a leading ;** and change (princ) to (princ llist). Do this even if you DO have the CA-DISK, so that you can see what the list looks like.

Even if you have the CA DISK, change the end of TEST.LSP to read:

```
   ;** (outledg llist)                          ;outputs legend, using the list llist.
   ;** (ledswap)                                ;Allows user to swap entries.
   (princ llist)                                ;end cleanly
);defun C:LEGEND
;*
```

Save, Exit and reenter AutoCAD. TEST.LSP should load.

Command: **(load "UFUNS")**   UFUNS and DXF are used by LEGEND.
Command: **(load "DXF")**

Command: **LEGEND**   Test it.
Building legend list...
Lisp returns: (("FUTURE" "FUTURE") ("NEW" "NEW EQUIPMENT") ("EXIST" "EXISTING - NO CHANGE") ("REMOVE" "REMOVE FROM SITE") ("EX-RELOC" "EXISTING - RELOCATED") ("EX-NEW-S" "EXISTING - NEW SERVICE"))

If you got the example list, you're ready to try the OUTLEDG function. Entries for the legend are automatically drawn using predefined Text styles, scaling the symbols by the #dwgsc global scale factor.

First the function figures the number of legend entries to guide its positioning. OUTLEDG prompts you with the number of symbols in the legend and temporarily Inserts the first Block to give you visual help in selecting the Text position and vertical spacing. You pick the insertion point for the first legend symbol, the horizontal offset to the descriptive text, and the vertical step down to the next symbol. This process lets you adjust the format to any proportion of symbols to text.

Then the block is erased, and the "LEGEND" title is placed. FOREACH steps through each sublist of legend entries, inserting each symbol block (car item), and descriptive text (cadr item). The insertion point for the next symbol is calculated according to text height. Text is offset by the horizontal distance that you indicated. After stepping down a vertical step, the process repeats until all the symbols are placed.

Outputting the Legend

Select **[EDIT-LSP]**   Edit TEST.LSP.

Even if you have the CA DISK, change TEST.LSP and remove the ;** from the OUTLEDG line, but leave it on (ledswap). Delete the "llist" variable from the PRINC expression:

```
(outledg llist)                    ;outputs legend, using the list llist.
;** (ledswap)                          ;Allows user to swap entries.
(princ)                            ;end cleanly
```

If you don't have the CA DISK, add to TEST.LSP:

```
;* OUTLEDG is used by LEGEND to draw the legend. It uses STD1-8 and STD1-4
;* height text styles which must be defined.
;* Legend must have the DXF and UFUNS functions preloaded, and expects a
;* global dwg scale variable named #DWGSC set to plot scale.
(defun outledg (llist / pt1 txtht txdist step)
  (setq $ortho (getvar "ORTHOMODE"))       ;Save
  (setvar "ORTHOMODE" 1)                   ;Reset
  (prompt                                  ;number of entries
    (strcat "\nThere are " (itoa (length llist)) " entries for the legend.")
  ) ;prompt
  (setq pt1 (upoint 1 "" "Starting point of first legend symbol" nil nil))
  (command "insert" (caar llist) pt1 #dwgsc "" "")   ;Temp. inserts 1st symbol.

  (setq txtht (dxf 40 (tblsearch "STYLE" "STD1-8"))  ;current text style has variable height
        txdist (udist 1 "" "Distance to first legend text" txdist pt1)   ;Horiz space.
        step (udist 1 "" "Distance to next legend symbol" step pt1)  ;Vert space.
  ) ;setq                                  ;Get relative positions.

  (command "erase" "l" ""                  ;Erase temp insert.
    "text" "S" "STD1-4"                    ;switch text styles
    (polar pt1 (* 0.5 pi) (- step txtht))  ;Enter LEGEND title above
    0.0 "%%uLEGEND%%u"                     ;pt1, %%u underscores.
    "text" "s" "STD1-8" nil                ;resets the style
  ) ;command
```

13—8 Customizing AutoCAD

```
    (foreach item llist                                  ;Insert ea symbol & descr.
      (command "insert" (car item) pt1 #dwgsc "" "")    ;Inserts symbol.
      (setq bpt (polar pt1 (* pi 1.5) (/ txtht 2.0)))    ;Calc vert base pt of text.
      (command "text" (polar bpt 0 txdist) 0.0 (cadr item)) ;Puts in text.
      (setq pt1 (polar pt1 (* 1.5 pi) step))             ;Steps to next line.
    );foreach
    (setvar "ORTHOMODE" $ortho)                          ;Restores
);defun OUTLEDG
;*
```

Save, Exit and reenter AutoCAD. TEST.LSP should load.

Command: **!#DWGSC**	Check it. If NIL, select [EXAMPLES] to set it.
Command: **LEGEND** Building legend list... There are 6 entries for the legend. Starting point of first legend symbol: Distance to first legend text: Distance to next legend symbol:	**LEGEND** appears! Tells you what's happening. Pick a point. Give about 24 inches. About 36 inches and it draws.

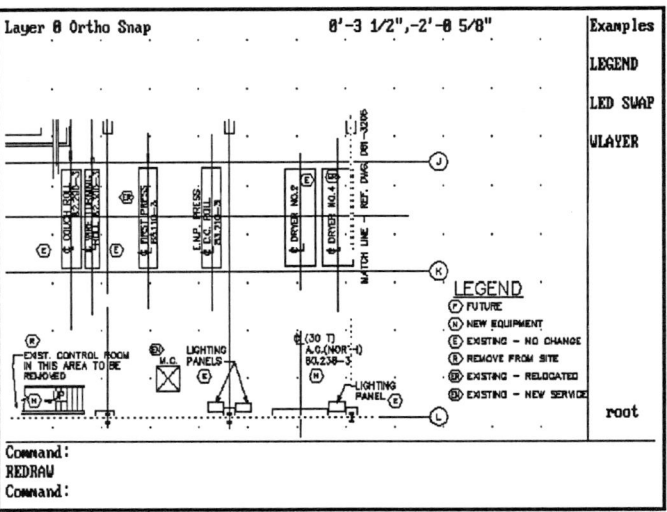

The Output Legend

Since the legend probably was not built in the order you would have chosen, you may wish to automate swapping its parts around. The LEDSWAP tool, shown next, lets you pick a pair of legend symbols to swap on the screen. It automatically finds the associated descriptive text string and swaps the pair.

Although the entity names of objects will never be the same from one drawing session to the next, the relative order of the entities generally remains unchanged. However, certain commands like EXPLODE will rearrange entities. Since LEDSWAP is called by the LEGEND command, the order of entities will be predictable. Let's look at the entity selection it uses.

Adding a SWAP Utility LEDSWAP

```
Command: (setq s1name (entsel))
Command: Select object:                                     Select a symbol.
Lisp returns: (<Entity name: 6000120C> (2890.000000 676.000000))

Command: (setq t1name (entnext (car s1name)))               Find out the next one.
Lisp returns: <Entity name: 60001220>
```

Note that ENTSEL returned the symbol name to S1NAME as an ENTSEL style set of entity name and point. ENTNEXT applies CAR to get the bare entity name of the text, set to T1NAME. You will see the difference in the following treatment of the symbol and text entity names. Add the following LEDSWAP to your TEST.LSP.

Select **[EDIT-LSP]**   Edit TEST.LSP.

Even if you have the CA DISK, change TEST.LSP and remove the ;** from the LEDSWAP line at the end of the LEGEND function.

If you don't have the CA DISK, add to TEST.LSP:

```
;* LEDSWAP is used by LEGEND to rearrange the entries.
(defun ledswap ( / redrw s1name s2name t1name t2name bp1 bp2 )
  (while (setq s1name (entsel "\nSelect first symbol or RETURN to exit: "))
    (if (setq s2name (entsel "\nSelect second symbol or RETURN to skip: "))
      (progn
        (setq t1name (entnext (car s1name))        ;Get 1st descr. ename.
              t2name (entnext (car s2name))        ;Get 2nd descr. ename.
              bp1 (osnap (cadr s1name) "INS")      ;Get 1st INSert pt.
              bp2 (osnap (cadr s2name) "INS")      ;Get 2nd INSert pt.
        )
        (command "MOVE" s1name t1name "" bp1 bp2   ;Move 1st to 2nd.
                 "MOVE" s2name t2name "" bp2 bp1   ;Move 2nd to 1st.
        )
        (entupd (car s1name))                      ;"redraw" 1st.
        (entupd t1name)
      );progn
    );if s2name
  );while s1name
  (princ)                                          ;finish cleanly
);defun LEDSWAP
(princ)
;*
;* end of LEGEND.LSP
```

Save, Exit and reenter AutoCAD. TEST.LSP should load.

Note that the Move commands issued by LEDSWAP make no distinction between the ENTSEL style entity lists and the bare entity names of the Text entities. You can feed either kind to AutoCAD's "Select Objects:" prompt. However, the ENTUPD function that redraws the first pair must make a distinction and extract the bare entity name from the S1NAME list to avoid a bad argument type. The LEDSWAP function buffers its action with a WHILE and an IF to give you a chance to exit or change your mind by hitting a <RETURN> or <SPACE>.

Select **[EDIT-MNU]**

Change the [CA00TEST] label to [CA13TEST] and add these two item to the EXAMPLES page:

```
***SCREEN
**ROOT
[CA13TEST]
[]

**EXAMPLES
[Examples]
[]
[LEGEND  ]^C^C^C(if C:LEGEND nil (load "legend")) (load "dxf");+
(if upoint nil (load "ufuns")) SNAP ON ORTHO ON LEGEND
[]
[LED SWAP]^C^C^C(if ledswap nil (load "legend")) SNAP ON ORTHO ON (ledswap);
[]
```

Save and Exit to AutoCAD. TEST.MNU reloads.

Command: **ERASE**	Erase the previous test legend and test it all:
Select **[EXAMPLES] [LEGEND]**	It loads if needed, sets up and:
Command: LEGEND Building legend list... There are 6 entries for the legend. Starting point of first legend symbol: Distance to first legend text: Distance to next legend symbol: Select first symbol or RETURN to exit: Select second symbol or RETURN to skip: Select first symbol or RETURN to exit:	The LEGEND appears, with the same questions as before. Pick point. Show it. Show it. This is new. Pick the FUTURE symbol. Pick the NEW symbol. Keep swapping if you like.

 Delete TEST.LSP.

 Rename TEST.LSP to LEGEND.LSP.

The menu item won't load the LEGEND file until you rename or copy TEST.LSP to LEGEND.LSP, but TEST.LSP was automatically reloaded by [EDIT-LSP].

OUTLEDG and LEDSWAP were made as separate functions. This modularity makes testing easier, and also makes it easier for you to cannibalize them for other uses.

You've used the Symbol Table functions, TBLSEARCH and TBLNEXT. You've taken the table data, used it to find entity data and modified your drawing. As these examples show, table access is simple and straight forward, but it gives you access to information that wields great power over AutoCAD's Drawing Editor.

This table access will enable you to a put personal variables in the Linetype table. But before we look at Linetypes let's discuss accessing Views, Layers and Styles.

Named Views, Layers and Styles

Views, Layers and Styles are even simpler to work with than block definitions. They have no subsets of data, like the entities in a block definition.

If you use standard view names, but they do not have constant locations, you can use VIEW table access to check them. Then, your setup routine only needs to prompt the user to make Views, if they don't exist.

Programs that modify text also can check the STYLE table to determine if text is fixed height or variable height so the programs can deal with the differences.

You should still be in the LEGEND drawing. It has the same View A and text styles as your prototype. Let's take a look at these tables.

Table Access to Views, Layers and Styles

```
Command: (tblnext "view" t)              Give the rewind argument.
((0 . "VIEW")                            Type of table.
 (2 . "A")                               The entries reference name.
 (70 . 0)                                Flag value .
 (40 . 2982.000000)                      View height.
 (10 1728.000000 1342.000000)            View center point.
 (41 . 0.000000)                         View width.
 (11 0.000000 0.000000)                  View direction from origin X,Y coordinates.
 (31 . 1.000000))                        Z View direction coordinate.

Command: (tblnext "view")                Try again.
nil                                      There are no more views in drawing.

Command: (tblnext "style" t)             Get the first style.
((0 . "STYLE")                           Type of table.
 (2 . "STANDARD")                        Entry reference name.
 (70 . 0)                                Flag value.
 (40 . 0.000000)                         Style fixed height.
 (41 . 1.000000)                         Style width factor.
 (50 . 0.000000)                         Obliquing angle.
 (71 . 0)                                Generation flag.
 (42 . 0.200000)                         Last height used.
 (3 . "txt")                             Regular font shape file.
 (4 . ""))                               Bigfont shape file, if any.
```

LAYER table searches are useful for determining if a layer exists. If it doesn't exist, you can define it. Otherwise, you can let whatever the user has set up override your programs. Look up the table entry for layer "0."

```
Command: (tblsearch "layer" "0")
Lisp returns: ((0 . "LAYER")        The type of table.
(2 . "0")                           Entries reference name.
(70 . 64)                           Flags for layer entries.
(62 . 7)                            Layer's color.
(6 . "CONTINUOUS"))                 The layer's linetype.
```

If you ever want to erase or Wblock a drawing by layers, here's a simple program, called WLAYER, that does just that. WLAYER provides an example of using SSGET "X" filtering.

If you Wblock, WLAYER sends each layer to a separate file by layer name. It gives you the options of skipping layers, or erasing the layers from the drawing as they are Wblocked. You should still be in the LEGEND drawing. Use the LEGEND drawing to try WLAYER.

WLAYER Wblocking by Layer with SSGET X Filtering

```
Command: SELECT                         Try SSGET "X" for all entities on layer name "0".
Select objects: (ssget "X" '((8 . "0")))
214 found.
```

 Copy WLAYER.LSP to TEST.LSP.

 Create a new LISP function file called TEST.LSP.

 Make layers named CL, PH, COL and RAIL. Draw a few entities on each layer. Then:

```
Select [EDIT-LSP]                       Edit TEST.LSP, and enter:

;* C:WLAYER  scans the drawing's layer list and Erases, skips or Wblocks each
;* layer to a drawing file name of the first 8 character of the layer name.
;* It assumes that the files on disk do not exist. It will bomb if one does exist
;* It prompts with each layer name, then Erases, Wblocks (and optionally erases),
;* or Skips it.
;* It requires the DXF and UFUNS function be preloaded.
(defun C:WLAYER( / prmpt frst bpt tbdata ents lname)
  (setvar "CMDECHO" 0)
  (setq frst t                                     ;set a flag to wind table to top
    bpt   (getvar "INSBASE")                       ;get the drawing's insertion basepoint
  );setq
  (while (setq tbdata (tblnext "layer" frst))      ;start a while loop, get each layer
    (setq ents (ssget "X" (list (cons 8 (dxf 2 tbdata))))) ;make set of all entities on layer
      lname (substr (dxf 2 tbdata) 1 8)            ;parse the string to max of 8 char
      frst nil                                     ;reset rewind flag
    );setq
    (setq prmpt (strcat "Erase or Wblock layer " lname "? E/W or RETURN to skip? "))
    (if (and ents (setq prmpt (ukword 0 "Wblock Erase" prmpt nil)))
;check for entities and choice
```

```
        (if (= prmpt "Wblock")
          (if (= "Yes" (ukword 1 "Yes No" "Erase also? Y/N" nil))
            (command "wblock" lname "" bpt ents "")        ;wblock & erase
            (command "wblock" lname "" bpt ents "" "Oops") ;wblock & restore
          )
          (command "ERASE" ents "")         ;wblock them out if exist
        )
      );if
    );while
    (setvar "CMDECHO" 1)
    (princ)                                 ;finish cleanly
);defun
;*
```

Save, Exit and reenter AutoCAD. TEST.LSP should load.

Select **[EDIT-MNU]**   Add the following to the EXAMPLES page:

```
[WLAYER  ]^C^C^C(if C:WLAYER nil (load "wlayer")) (load "dxf");+
(if ukword nil (load "ufuns")) WLAYER
```

Save and Exit to AutoCAD. TEST.MNU reloads.

Select **[EXAMPLES] [WLAYER]**   It loads if needed, then:

```
Command: WLAYER
Erase or Wblock layer 0? E/W or RETURN to skip? :          <RETURN> to skip it.
Erase or Wblock layer CL? E/W or RETURN to skip? : E       Erase.
Erase or Wblock layer PH? E/W or RETURN to skip? : W       Wblock it.
Erase also? Y/N: Y                                         And erase it.
Erase or Wblock layer COL? E/W or RETURN to skip? : W
Erase also? Y/N: N
Erase or Wblock layer RAIL? E/W or RETURN to skip? :
```

  Delete TEST.LSP.

  Rename TEST.LSP to WLAYER.LSP.

Command: **QUIT**

The menu item won't load WLAYER until you rename or copy TEST.LSP to WLAYER.LSP, but [EDIT-LSP] automatically reloaded the function.

Named Properties: Linetypes and More

Rather than go through a routine LTYPE table access example, we want to show you how to use the Ltype table to store your own drawing variables. A Symbol Table is just a data structure. The Ltype table's data is routinely interpreted as a linetype definition, but you can use it to store other data if you want. Most customizers of AutoCAD wish they had a greater number of User Variables (user definable System Variables) than the USERR1-5 and USERI1-5 that AutoCAD makes available. The next exercise shows you how to store your own Personal Variable data in the Ltype table in the drawing file itself.

13—14 Customizing AutoCAD

Let's review what a linetype really is. First, use the Linetype command to see how AutoCAD lists the description and contents of its ACAD.LIN file. Then, create your own Linetype. Call it PERVAR, for personal variable.

PVARs Using Linetypes for Personal Variables

```
Enter selection:                                 Begin a NEW drawing named PVARS.

Select [TEST] [24 x 36] [FULL] [INITIATE]        It brings up TEST.MNU.

Command: ZOOM                                    Zoom Center with height 14".
Command: LINETYPE                                First check an existing one.
?/Create/Load/Set: C
Name of linetype to create: DASHDOT
File for storage of linetype <ACAD>:
Wait, checking if linetype already defined...
DASHDOT already exists in this file. Current definition is:
*DASHDOT,__ . __ . __ . __ . __ . __ . __ . __ . __
A,0.5,-0.25,0,-0.25
Overwrite? <N> <^C>                              *Cancel* to not change it.
?/Create/Load/Set: C                             Create your own.
Name of linetype to create: PERVAR
File for storage of linetype <ACAD>: TEST        Keep the ACAD.LIN file clean.
Creating new file
Descriptive text: This is my personal variable.
Enter pattern (on next line):
A,1.5,-2.5                                       That's minus 2.5
New definition written to file.
```

Like any other linetype, AutoCAD requires you to load it before using it. Load it and load DASHDOT for comparison. Then, set PERVAR current and draw a Line. It should look like a big dashed line. Use TBLSEARCH to see how AutoCAD stores information in the Ltype symbol table. List DASHDOT for comparison, then get the table data for PERVAR.

```
?/Create/Load/Set: L
Name of linetype to load: PERVAR
File to search <TEST>:
Linetype PERVAR loaded.
?/Create/Load/Set: L                             Load Linetype DASHDOT.
Name of linetype to load: DASHDOT
File to search <TEST>: ACAD                      From the ACAD.LIN file so you can look at it.
Linetype DASHDOT loaded.
?/Create/Load/Set: S                             And Set PERVAR current to use it.
New entity linetype (or ?) <BYLAYER>: PERVAR

Command: LINE                                    Draw a long Line.

Command: (tblsearch "LTYPE" "DASHDOT")           Check it out.
Lisp returns: ((0 . "LTYPE")                     The linetype table name.
(2 . "DASHDOT")                                  Table entry name.
(70 . 64)                                        Maximum number of items that may follow.
                                                 The visual description:
(3 . "__ . __ . __ . __ . __ . __ . __ . __ . __")
(72 . 65)                                        Alignment code, ASCII value for "A", the only allowed code.
(73 . 4)                                         Number of dash length items.
```

(40 . 1.000000)	Absolute total of dash pattern.
(49 . 0.500000)	First dash length.
(49 . -0.250000)	Pen up dash length.
(49 . 0.000000)	Dot.
(49 . -0.250000))	Pen up dash length.

Command: **(setq tbdata (tblsearch "LTYPE" "PERVAR"))**   Your linetype is next:

Lisp returns: ((0 . "LTYPE") (2 . "PERVAR") (70 . 64) (3 . "This is my personal variable.") (72 . 65) (73 . 2) (40 . 4.000000) (49 . 1.500000) (49 . -2.500000))

The 2 group Linetype name can be any string value you like up to 31 characters. The 70 is for DXF processing and irrelevant here. The 3 group description is any string up to 47 characters. The 73 value tells how many 49 groups will follow. The 49 groups are the actual linetype pattern values: 0 denotes a point; positive values are pen down segment lengths; and negative values are pen up (space) lengths. These 49 values are simply REAL numbers, but the leading one must be greater then zero to satisfy AutoCAD's alignment, and at least two must exist. The 40 group is the unit pattern length, the sum of the absolute values of the 49 group segments.

You can treat this table data the way you treated Block table data. Let's take one of the pieces of information, retrieve it and bind it to a variable. Start by getting the description.

Lisp returns: Command: (setq pervar (cdr (assoc 3 tbdata)))
Lisp returns: "This is my personal variable."   A STRing.

So we see that the same data which formed an innocent Linetype named PERVAR, also can be retrieved as an AutoLISP variable. Let's call Linetypes that we treat in this manner PVAR Linetypes.

The ability to retrieve a value from the PVAR Linetype has little value in real applications, unless you also can update the value. You cannot update the tables directly with AutoLISP. But AutoCAD can handle the updates with commands like the Linetype command. To modify the LTYPE, BLOCK, LAYER, VIEW or STYLE table data, use the AutoCAD command related to the table.

To update the Ltype table, recreate PERVAR and then reload its definition.

```
Command: LINETYPE
?/Create/Load/Set: C
Name of linetype to create: PERVAR        Enter the name
File for storage of linetype <ACAD>: TEST
Wait, checking if linetype already defined...
PERVAR already exists in this file.  Current definition is:
*PERVAR,This is my personal variable
A,1.5,-2.5
Overwrite? <N>:  Y                        Change the description.
```

```
Descriptive text: (10.0 12.0)
Enter pattern (on next line):
A,0,0
New definition written to file.
?/Create/Load/Set:L                         Now load the new definition.
Linetype was loaded before.  Reload it? Y

File to search <TEST>:                      OK. That's the one.
Linetype PERVAR loaded.
?/Create/Load/Set:                          <RETURN>

Command: (setq tbdata (tblsearch "LTYPE" "PERVAR"))    Look at the new table data:
Lisp returns: ((0 . "LTYPE") (2 . "PERVAR") (70 . 64) (3 . "(10.0 12.0)") (72 . 65) (73 .
2) (40 . 0.000000) (49 . 0.000000) (49 . 0.000000)
```

You changed both the description of the variable as well as the values. Since you can format any kind of data as a string, use the 3 group description to store your PVAR data. The 49 groups can store only numbers. Set the first two to 0. Recall that linetypes must have two or more in the 49 groups.

```
Command: LINE From point:                   Pick any point.
To point: (setq pervar (read (cdr (assoc 3 tbdata))))   It draws a line to 10,12.
To point: <^C>                              Cancel.
Command: !PERVAR                            Check the variable.
Lisp returns: (10.000000 12.000000)
```

The READ function returns the first list, or atom of a string, as long as the string contains no <SPACE>s outside of any list. Use READ to strip the outer pair of ""'s from the string. In the example above, (read "(10.0 12.0)") returned (10.000000 12.000000). The following table shows the data types, typical values, the conversion string values you need for storage, and what READ yields:

DATA TYPE VALUE TABLE

TYPE	TYPICAL VALUE	STRING VALUE	READ VALUE
REAL	12.3456789	"12.3456789"	12.345680
INT	12345	"12345"	12345
STR	"String value"	""String value""	"String value"
POINT	(1.23 3.21)	"(1.23 3.21)"	(1.230000 3.210000)

Using the table's format, you can store and update any string that is less than 47 characters, store and update any point (list) or number as a personal variable in any AutoCAD drawing. If you automate this storage and update, you can maintain critical information, like the elevations of piping platforms, floor to floor heights, and stadia points in surveys in a drawing.

You can automate the process several ways. You can treat each personal variable as a small table, or one line of a large table of information where the 49 group codes store keys to how the variable table is formatted. You also could set up a code to string together linetype PVARs to form a longer string than 47 characters.
The next example shows how to treat personal variables as individual pieces of data, one variable name per value.

Converting Data Types to Strings ETOS

Since the book's method requires storing REALs, INTegers, POINTs (really LISTs), and STRings as string values, you need a way to indiscriminately convert them to the forms in the Data Type Value Table.

This is a chicken and egg problem. And the chicken comes first. It lays an egg called ETOS (Expression TO String) a function that uses AutoLISP to write directly to files. ETOS is hatched in the next chapter on AutoLISP I/O. For now, accept ETOS as a "black box." Any expression that you feed ETOS is returned as a string in the Data Type Value Table format. Create the conversion ETOS function.

Creating ETOS

   Copy ETOS.LSP to TEST.LSP.

   Create a new LISP function file called TEST.LSP.

Select **[EDIT-LSP]**   Edit TEST.LSP, and enter:

```
;* ETOS (Expression TO String) takes any Expression and converts to a string.
;* "STRings" are returned as double ""STRings""
(defun etos (arg / file)
  (if (= 'STR (type arg)) (setq arg (strcat "\"" arg "\"")))  ;format ""STRings""
  (setq file (open "$" "w"))           ;Open temp file.
  (princ arg file)                     ;Write it
  (close file)
  (setq file (open "$" "r"))           ;Reopen to read
  (setq arg (read-line file))          ;Set to string read from file.
  (close file)
  (close (open "$" "w"))               ;Scrunch file down to 0 bytes.
  arg                                  ;Returns as string
) ;defun ETOS
;*
```

Save, Exit and reenter AutoCAD. TEST.LSP should load.

Command: **(etos "test string")**   Test it.
Lisp returns: ""test string""

Command: **(setq x '(test list 123))**   Make a list.
Lisp returns: (TEST LIST 123)
Command: **(etos x)**
Lisp returns: "(TEST LIST 123)"

   Delete TEST.LSP.

   Rename TEST.LSP to ETOS.LSP.

Defining and Protecting Your PVARS

Unlike the standard AutoCAD User System Variables, you can give personal variables unique names. The example PVAR system uses predefined linetype names which are recognized by the first 4 characters "PVAR." You can make the following characters anything that you like. The next section shows you how to define a PVAR set, and how to store them in a linetype file as a block for easy insertion into a drawing. It also shows how to store them invisibly in a drawing so that they are hard to delete accidentally.

The best way to assure the safety of personal variables is to use them somewhere in the drawing, making use of the PVAR linetype. The simplest way to do this is to make and Insert a tiny Block that contains entities using each PVAR as a linetype. To avoid having to force a regen if you want to add a variable to an existing group, the Block can reserve as many PVAR names as you want. These can be dummy PVARs, with nil, "" (0 0) or 0 values. You can use the Rename command to change their names, or you can use the PVAR and associated functions to change any value. Let's define ten PVARs and make a protective Block.

Enhancing the PVAR's Function

```
Command: LINETYPE
?/Create/Load/Set: C
Name of linetype to create: PVAR1
File for storage of linetype <TEST>:     Take the default.
Wait, checking if linetype already defined...
Descriptive text: 1.0                    A real value.
Enter pattern (on next line):
A,0,0
New definition written to file.          Don't load it yet.
```

Linetypes are stored in a disk file, like the standard ACAD.LIN or the PVAR.LIN file. A Linetype file is an ASCII text file. The easiest way to make nine more PVARs is to use your text editor.

   Copy PVAR.LIN to TEST.LIN.

   Edit the TEST.LIN file that the Linetype command has created.

```
Command: ED                              Edit the TEST.LIN file. It looks like this:

*PERVAR,(10.0 25.0)
A,0,0
*PVAR1,1.0
A,0,0
```

Delete the *PERVAR line, the A,0,0 line following it, and any blank lines. Copy the PVAR1 and A,0,0 lines down nine times, renaming each PVAR2, PVAR3, through PVAR10, changing the data description as follows:

```
*PVAR1,1.0
A,0,0
*PVAR2,2.0
A,0,0
*PVAR3,3.0
A,0,0
*PVAR4,"string four"
A,0,0
*PVAR5,"5"
A,0,0
*PVAR6,6
A,0,0
*PVAR7,7
A,0,0
*PVAR8,
A,0,0
*PVAR9,(9.00 9.00)
A,0,0
*PVAR10,(10.0 10.0)
A,0,0
```

Be SURE you enter a <RETURN> after the last 0.
Save, Exit and reenter AutoCAD.

 Delete TEST.LIN.

 Rename TEST.LIN to PVAR.LIN.

You want to use each PVAR, creating ten zero length Line entities. Then, make them into a Block. Use AutoCAD's entity Linetypes, not Layer Linetypes, so the drawing won't have to carry a layer for each PVAR.

You could load PVAR1 with the Linetype command, set it current, draw a Line at 1,1 and repeat the process nine times for the other nine PVARs. But let's use a quick AutoLISP routine to load 10 linetypes, set each, and draw 10 Lines with them.

If you have the CA DISK, you have the drawing file PERVARS.DWG so you can skip to the INSERT command at the bottom of this sequence.

Skip to the INSERT.

 Create the linetypes.

```
Command: LAYER                                  Make name PERVARS-DONOT-DELETE and leave it current.

Command: (foreach x '(1 2 3 4 5 6 7 8 9 10)
1>(setq x (strcat "PVAR" (itoa x)))
1>(command "LINETYPE" "L" x
1>"PVAR" "S" x "" nil "LINE" "1,1" "@" ""))     And it creates all ten.

Command: LIST                                   Select Window over 1,1. See if all 10 PVAR Lines drew.

Command: BLOCK                                  Block all 10 Points to name PERSONAL-VARS at base 1,1.
Command: INSERT                                 PVARS-DONOT-DELETE block at 1,1.
Command: SELECT                                 Select Last to make it the Previous selection.
Command: LAYER                                  Set 0 current and Freeze PERVARS-DONOT-DELETE.
Command: WBLOCK                                 Name PERVARS.
Block name:                                     <RETURN> to select, enter basepoint 1,1 then P to select Prev.

Command: INSERT                                 Insert PERVARS, Osnap INT to corner of titleblock.
```

You Wblocked out a "data block" on a frozen layer. Whenever you insert the "data block," it will bring the frozen layer with it. Now, all you have to do to put the PVAR template into a drawing is to insert the data block PERVARS. The best place to put the data block is in the corner of a standard titleblock on a frozen layer, or the lower left limits corner, say at -0.001,-0.001.

The PVAR Functions

To have functions to update and get the values from stored PVARs, you need to be able to check that they are in the drawing. The PVCHK (PVar CHecK) function does just that. It checks the BLOCK table for the PVARS-DONOT-DELETE block. If it can't find it, PVCHK tries to insert PERVARS and checks again before returning T or NIL. The complete PVAR program and its supporting functions is shown in the PVAR Program drawing.

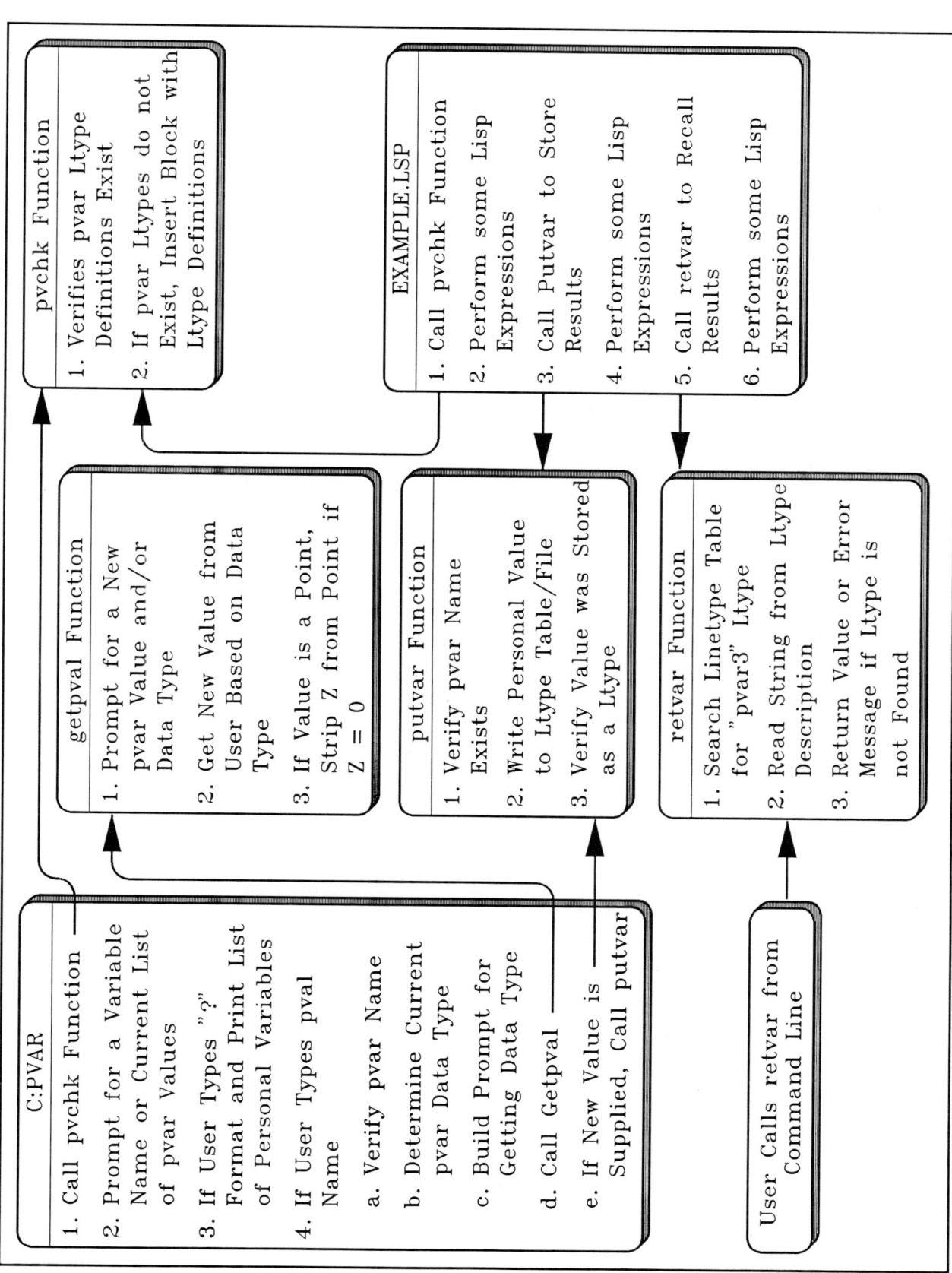

The PVAR Program

Enter the PVAR.LSP header information and the PVCHK function.

The Complete Personal Variables Function PVAR

 Copy PVAR.LSP to TEST.LSP.

 Create a new LISP function file called TEST.LSP.

```
Select [EDIT-LSP]                        Edit TEST.LSP,
```

Enter the PVARS header info and PVCHK function:

```
;* PVAR.LSP is the Customizing AutoCAD Personal Variable system. Included are:
;* RETVAR to retrieve a the value of a personal variable inside a LISP function.
;* PUTVAR to store away a value.
;* PVCHK checks if PVARS-DONOT-DELETE data block is defined, else it tries insert.
;* GETPVAL is used by C:PVAR to get new PVAR value from user.
;* C:PVAR, is similar to the AutoCAD SETVAR command. It lists all pvars, or
;* displays the value of one pvar and updates it if changed. Sorry, PVAR
;* is not an AutoCAD transparent command because it used the LINETYPE command.
;* PUTVAR and RETVAR work from within other functions if no command is pending.
;* ETOS, DXF, and UFUNS must be loaded.
;* A valid pvar is a special linetype, previously loaded in the drawing.
;* Pvars are defined in PVAR.LIN file in the form:    *PVARxxxxx,value
;*                                                    A,0,0
;* Valid data types are STRing, REAL, INTeger, POINT (actually a list) and NIL.

;* PVCHK (PVar CHecK) checks if definition exists, if not adds it to block
;* definition table which also adds PVARs to LTYPE table, w/o actual insertion.
(defun pvchk ()
  (or (tblsearch "block" "PVARS-DONOT-DELETE")   ;if found, exit and return T
      (command "insert" "pervars" nil)           ;otherwise insert, continue
      (tblsearch "block" "PVARS-DONOT-DELETE")   ;if found, exit T, otherwise error:
      (prompt "\nERROR: PERVARS data block and .DWG file not found.")   ;nil
  )                                              ;returns T if OK, else nil
);defun
;*
```

Save, Exit and reenter AutoCAD. TEST.LSP should load.

```
Command: (pvchk)                     Test it.
Lisp returns: T                      It's there.
Command: RENAME
Block/LAyer/LType/Style/View: B      Rename blocks so you can test if not there.
Old block name: PERVARS
New block name: XXX                  Any new name!

Command: RENAME                      And rename PVARS-DONOT-DELETE to XX.

Command: (pvchk)                     Test again. It Inserts PERVARS and:
Lisp returns: T
```

To make sure the block is in the drawing, use PVCHK at the start of the other PVAR functions. The functions depend on predefined PVAR linetypes.

Now, that you know the PVARs are in the drawing, you need a RETVAR (RETurn VARiable) function to return their existing values. RETVAR is the equivalent of AutoLISP's built-in GETVAR. It's simple and takes one argument, the PVAR name. It checks PVCHK, gets the named PVAR from the Ltype table, extracts the stored string value with DXF and converts it to its real data value with READ.

Select **[EDIT-LSP]**   Edit TEST.LSP, and add:

```
;* RETVAR return the value of a stored personal variable. It uses PVCHK and DXF.
(defun retvar (pvname / tb)
  (if (and (pvchk)                            ;if PVARS-DONOT-DELETE data block present
       (setq tb (tblsearch "LTYPE" pvname))   ;and pvar name is in LTYPE table
      )
     (read (dxf 3 tb))                        ;then read stored string to return value
     (prompt "ERROR: variable not found")     ;else pvar not defined, return nil
) );defun & if
;*
```

Save, Exit and reenter AutoCAD. TEST.LSP should load.

Command: **(load "DXF")**   It must be loaded.
Command: **(retvar "pvar1")**   Returns: 1.000000.
Command: **(retvar "pvar4")**   Returns: "string four".
Command: **(retvar "pvar7")**   Returns: 7.
Command: **(retvar "pvar8")**   Returns: nil.
Command: **(retvar "pvar9")**   Returns: (9.000000 9.000000).

Now, you need to update the values. Do this just as you did before. Use the Linetype command to write a new description to the file and reload it from the COMMAND pipeline.

Linetypes are stored internally, like a Block definition. The values, once they are placed in the drawing, are independent of the .LIN file containing their original definitions. The values travel with the drawing whether or not they are in any .LIN file. Use a scratch file named $.LIN for updating, and reserve the named .LIN files, like PVAR.LIN, as insertable master default sets of values for specific applications.

PUTVAR (PUT VARiable) is the equivalent of AutoLISP's SETVAR. It takes two arguments, the PVAR name and a value. PUTVAR checks PVCHK and the Ltype table for the PVAR name, uses ETOS to convert the value to a string and writes the linetype with COMMAND. The (close (open "$.LIN" "w")) expression is another egg that you will find in the next chapter. It insures the file $.LIN is empty.

The only complication that you encounter in PUTVAR is that you must check the EXPERT System Variable to know whether the Linetype command is going to ask if you want to overwrite an existing value. Then, PUTVAR checks RETVAR to see if the value was successfully updated.

13—24 Customizing AutoCAD

```
;* PUTVAR updates a PVAR in LTYPE table. It uses RETVAR, which uses PVCHK and DXF.
(defun putvar (pvname nval / ex)
   (if          ;if PVARS-DONOT-DELETE data block present and pvar name is in LTYPE table
     (and (pvchk) (setq tb (tblsearch "LTYPE" pvname)))
     (progn                                        ;then
       (if (= 'STR (type nval)) (setq nval (substr nval 1 45)))  ;truncates if too long
       (setq nval (etos nval))                     ;converts to string to store
       (close (open "$.LIN" "w"))                  ;nulls out temp file
       (command "LINETYPE" "C" pvname "$"          ;writing to LTYPE table/file
                nval "0,0" "L" pvname              ;and reload
       )
       (if (< (getvar "EXPERT") 3)
         (command "Y" "" "")                       ;if expert then "Y"
         (command "" "")
       )
       (equal (retvar pvname) (read nval))         ;verify, returns T if OK, nil if fails
   ) );if & progn
);defun
;*
```

Save, Exit and reenter AutoCAD. TEST.LSP should load.

Command: **(load "DXF")**	It must be loaded.
Command: **(load "ETOS")**	It must be loaded.
Command: **(putvar "pvar1" 1.23)**	Test it.
LINETYPE	And the linetype command scrolls by.
Lisp returns: T	Try again with a bad name, and it will return nil.

That's all you absolutely need for your AutoLISP programs to make use of PVARs. You can define them, retrieve them, and update them. The last two PVAR.LSP functions, called C:PVAR and GETPVAL, provide a friendly interface, like the AutoCAD SETVAR command.

C:PVAR is the command function that displays a list of all defined PVARs. It can set new values. The new value setting is done by a second function, GETPVAL (GET Pvar VALue). GETPVAL is an interface function and uses the UFUNS.LSP functions to get input. It takes the arguments DTYPE, KWD, PRMPT, and VAL, all supplied by C:PVAR. The allowable DTYPEs (DataTYPEs) are "REAL," "INT," "STR," "NIL" and "POINT." These are strings not 'quoted symbols like the TYPE command returns. The normal KWD (KeyWorD) is "\\" to allow respecifying data type. The prompt PRMPT is a string, and the VAL is the current data value.

The first half of GETPVAL is a series of COND tests and expressions, one for each data type. The second half is executed only if the user enters a \ backslash to reset a data type. The second half of the function then gets the new data type from the user and recursively calls the GETPVAL function with a new prompt and a null "" KWD keyword. The last part reformats points as 2-D or 3-D and returns the new value. Let's enter and test GETPVAL first.

Select [EDIT-LSP]   Edit TEST.LSP, and add:

```
;* GETPVAL (GET Pvar VALue) is used by C:PVAR to get new PVAR value from user
```

```
;* and return it. If user specifies new data type, it calls itself recursively.
;* If you wish to prevent the user from setting to nil or changing data type,
;* disable the lines marked with ;!comments by adding leading ;! asterisks and
;* remove the leading ;$ from the line marked with ;$ (or opposite to reverse).
;* It uses the UFUNS.LSP and ETOS functions.
(defun getpval (dtype kwd prmpt val / nval)
  (if (= kwd "\\")                                                           ;!
      (prompt "\nEnter new value, or a \\ to specify new data type.") ;!
  )                                                                          ;!
  (setq nval
    (cond                              ;issue appropriate GET for data type
        ((= dtype "NIL") nil)          ;if user specs new data type as NIL
        ((= dtype "REAL") (udist 1 kwd prmpt val nil))    ;get new real
        ((= dtype "STR")  (substr (ustr prmpt val T) 1 45)) ;get new string
        ((= dtype "POINT") (upoint 17 kwd prmpt val nil)) ;get new point
        ((= dtype "INT")  (uint 1 kwd prmpt val))         ;get new integer
        (T (prompt prmpt) "\\")        ;otherwise invalid or nil
    );cond
  );setq
  (if (= nval ".") (setq nval ""))
  (if (= nval "\\")                              ;!if user entered \ to reset data type
;$    (setq nval val)    ;$ alt then to prevent users changing data types
     (progn                             ;! alt then to allow set new data type
       (setq dtype                                                  ;!
         (strcase                                                   ;!
           (ukword 1 "Real Str Point Int Nil"                       ;!bit & keywords
             "Specify data type Real/Str/Point/Int/Nil" nil   ;!prompt & defaults
       ) ) )                                                        ;!
       (setq prmpt (strcat "Enter a " dtype))                 ;!and reset prompt
       (getpval dtype "" prmpt nil)                           ;!call recursively
     );progn
     (if                              ;else, check if point
       (and (= dtype "POINT") (= 0 (caddr nval))) ;if point & has a "null" Z
         (list (car nval) (cadr nval))            ;then return w/ Z stripped
         nval                                     ;else return new value
     );else, if point
  );if = "\\"
);defun getpval
;*
```

Save, Exit and reenter AutoCAD. TEST.LSP should load.

Command: **(load "DXF")**   If not loaded, it must be loaded.
Command: **(load "ETOS")**   If not loaded, it must be loaded.
Command: **(load "UFUNS")**   If not loaded, it must be loaded.

Command: **(getpval "REAL" "\\" "\nEnter: " 12.3)**
Enter new value, or a \ to specify new data type.
Enter: <1'-0 19/64">: **32.1**   Note. C:PVAR will add a decimal display.
Lisp returns: 32.100000

Command: **(getpval "REAL" "\\" "\nEnter: " 12.3)**
Enter new value, or a \ to specify new data type.
Enter: <1'-0 19/64">: ****   Enter \ to respecify.
Specify data type Real/Str/Point/Int/Nil: **S**   String.
Enter a STR: **A new string**
Lisp returns: "A new string"

Try the other data types if you like.

If you don't want to allow your users the ability to set PVARs to nil, or you want to prevent their changing the data types of PVARs, GETPVAL has an alternate else statement that is currently disabled by a leading ;$. If you want to restrict your users, follow the instructions in the GETPVAL header information, then test it again.

When the user enters a ? to the Personal variable name prompt, the first half of the command function C:PVAR reads the entire Ltype table and lists the PVAR name, data type and value for each PVAR. It checks for a valid PVAR name, extracts the description with DXF, READs it to get its data value and uses TYPE to get its data type.

If it is a LIST, C:PVAR assumes it is a point and uses MAPCAR to map the RTOS function to each element of the list to display in current units. It also uses MAPCAR to map PRINC and display the list because each line consists of several pieces of data. You can't use PROMPT and STRCAT because one bit of data is the DTYPE in SYMbol form.

If the user enters a PVAR name, the second half of PVAR is executed. It verifies the name and gets the values, then sets the arguments to send to GETPVAL to get the new value. As it sets the arguments for GETPVAL, it converts the symbol format of DTYPE to a string format with ETOS. It changes "LIST" to "POINT," since a point isn't a true data type in AutoLISP. It formats a custom prompt, PRMPT, appropriate to each data type (REAL and INT get the same prompt). This lets GETPVAL display the numbers and points in both decimal and current units. When this is done, it calls GETPVAL to set NVAL, the new value. It then compares the value returned by GETPVAL to the old value, and calls PUTVAR to update the PVAR, if the PVAR was changed.

Select [EDIT-LSP]   Edit TEST.LSP, and add:

```
;* C:PVAR can query, view or update the personal variables.
;* It uses the UFUNS.LSP, ETOS, DXF, GETPVAL, PVCHK, and PUTVAR functions.
(defun C:PVAR ( / tb dtype val strval kwd nval prmpt first)  ;VNAME floats global
  (if (pvchk)                               ;if PVARS-DONOT-DELETE data block present
    (if                                     ;get pvar name, if user ? for list of names
      (= "?" (setq vname (ustr "\nPersonal variable name (or ?)" vname nil)))  ;which pvar
      (progn                                               ;then list...
        (textscr)                                          ;flip to text screen
        (prompt "\nPersonal Variables Library\n")          ;Print a header to the list
        (prompt "\nName \tType \tValue")                   ;use tabs to align columns
        (setq first T)                                     ;set rewind flag on
        (while (setq tb (tblnext "LTYPE" first))           ;while there is a linetype
          (setq vname (dxf 2 tb) first nil)     ;store name, set rewind flag off
          (if (= (substr vname 1 4) "PVAR")     ;if name is a personal var.
            (progn                              ;then print name, type, value.
              (setq val (read (dxf 3 tb))       ;get description, cvt to data value
                    dtype (type val)            ;get data type as 'symbol
              )
              (if (= 'LIST dtype)               ;if it's a so called POINT
                (setq dtype 'POINT              ;then adjust dtype &
                      val (mapcar 'rtos val)    ;convert to current units
                ) )
              (mapcar 'princ (list "\n" vname "\t" dtype "\t" val)) ;list it
) ) );while linetype, if, progn
```

```
          (setq vname "")                          ;clear default
       );progn then list
       (if                                         ;else they entered name to update. If...
         (and (= (strcase (substr vname 1 4)) "PVAR") ;...valid PVAR name and
              (setq tb (tblsearch "LTYPE" vname))    ;look up the name ok...
         )
         (progn                                    ;...then
           (setq strval (dxf 3 tb)                 ;description str - stored value
                 val (read strval)                 ;the actual data value
                 dtype (etos (type val))           ;data type in string form for prompt
                 kwd "\\"                          ;set keyword for new data type
           );setq
           (cond                                   ;build prompt & format if point
             ( (= "LIST" dtype)
               (setq dtype "POINT"                                       ;set data "type"
                     prmpt (strcat vname " is " strval ", " dtype)       ;POINT prompt
             ) )
             ( (numberp val)
               (setq prmpt (strcat vname " is " strval ", " dtype)) ;REAL or INT
             )
             ( (= "STR" dtype)
               (setq prmpt                                               ;prompt for STRing
                 (strcat "Enter string (45 char max), or a period . for an empty string\n"
                   vname " is a" (if (= "" val) "n empty " " ") dtype
             ) ) )
             (T (setq prmpt (strcat "\n" vname " is invalid or nil.")))
           );cond
           (setq nval (getpval dtype kwd prmpt val))   ;get new value
           (if (/= val nval)                           ;if new then update LTYPE table
             (if (putvar vname nval) nil (prompt "\nReset failed."))
           )
           (mapcar 'princ (list "\n" vname " is " nval "\n"))
         );progn then
         (progn                                    ;else error unknown variable
           (setq vname "")                         ;clear default
           (prompt (strcat "\nPersonal variable " vname " doesn't exist."))
         );progn else
       );if entered a name, valid pvar name
     );if user queries ?
   );if pvchk
   (princ)
);defun C:PVAR
;*

(princ)
;*end of PVAR.LSP
```

Save, Exit and reenter AutoCAD. TEST.LSP should load.

Command:	**(load "DXF")**	If not loaded, it must be loaded.
Command:	**(load "ETOS")**	If not loaded, it must be loaded.
Command:	**(load "UFUNS")**	If not loaded, it must be loaded.
Command:	**PVAR**	Test it.
Personal variable name (or ?): **?**		Ask for a list. You get:

Personal Variables Library

```
Name     Type    Value

PVAR1    REAL    1.230000
PVAR2    REAL    2.000000
PVAR3    REAL    3.000000
PVAR4    STR     string four
PVAR5    STR     5
PVAR6    INT     6
PVAR7    INT     7
PVAR8    nil     nil
PVAR9    POINT   (9" 9")
PVAR10   POINT   (10" 10")

Command: PVAR
Personal variable name (or ?): PVAR8
Enter new value, or a \ to specify new data type.
PVAR8 is invalid or nil.
Specify data type Real/Str/Point/Int/Nil: P          Point data type.
Enter a POINT: 10,20
LINETYPE                                Command scrolls by as it updated.
PVAR8 is (10.000000 20.000000)

Command: PVAR
Personal variable name (or ?) <PVAR8>: PVAR1
Enter new value, or a \ to specify new data type.
PVAR1 is 1.230000, REAL <1 15/64">: 3.21          A new real.
LINETYPE                                Command scrolls by as it updated.
PVAR1 is 3.210000
```

Test several more combinations.

  Delete TEST.LSP. Delete TEST.MNU.

  Rename TEST.LSP to PVAR.LSP. Rename TEST.MNU to MY13.MNU.

This implementation requires that you predefine the PVAR names. But you can easily change the portions of code that check the names in the Ltype table to create new PVARs automatically. You also could write a function to take all currently defined PVARs in the drawing and write them to a linetype file, borrowing the code from the first half of C:PVAR and from the PUTVAR function. You could add another useful function to Wblock out all current PVARs to a "data block" file. We are sure you will find many uses for PVARs in your programs.

Integrating the Example Menu

You do not need to integrate this chapter's menu into the CA-MENU. If you find any of these macros useful, add them to your own menus. Here is a diagram of the menu file.

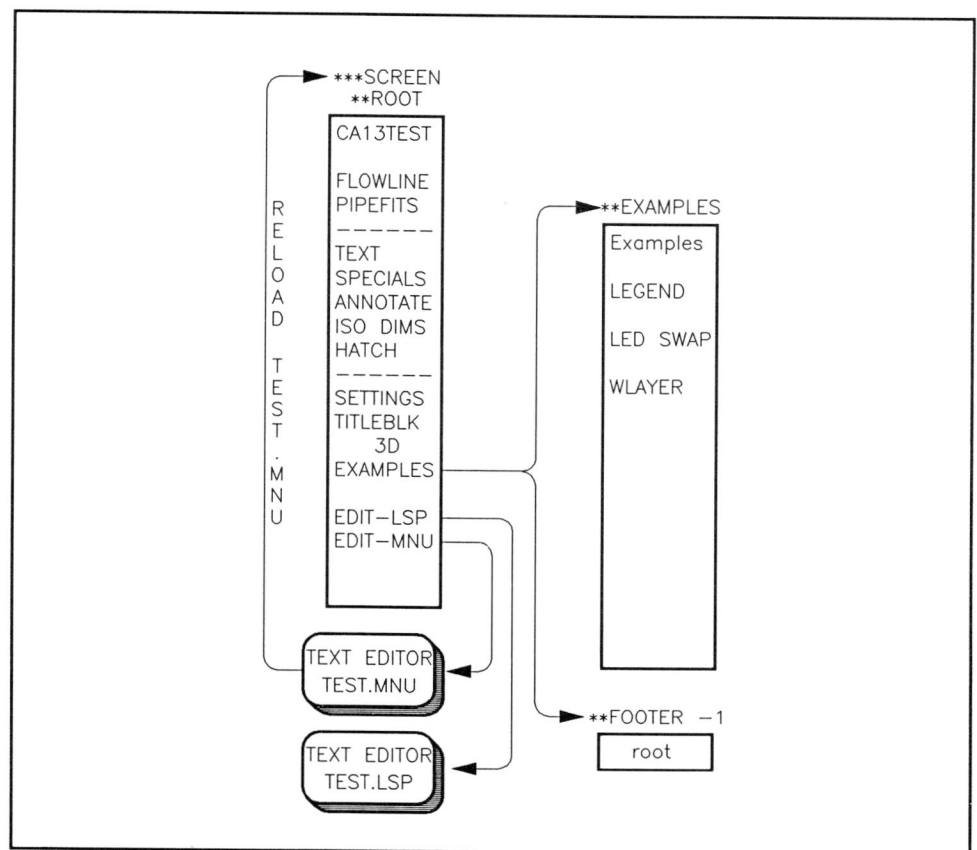

The Chapter 13 Menu

Summary Tips and Techniques

Here are some summary tips and suggestions for extending this chapter's routines to other applications.

- ❑ The Ltype table is suited to PVARs. If you used the Block table, the drawing would automatically regenerate with each update of the variable. Entries from the other tables also are easily wiped out by the PURGE or WBLOCK command, making these tables more difficult to deal with. An alternative method for implementing PVARs is to use attributes to store them, and Release 9's SSGET "X" to access them.

- ❑ Whether you use Ltype table access, or attributes with SSGET "X" to implement PVARs, we recommend working through the basic idea of personal variables in your application. Like many AutoCAD users, you may be frustrated with AutoCAD's failure to store plotting data. Think about building a PLOT routine that uses PVARs to store all drawing specific plotting requirements in the drawing so they will be accessible with the drawing!

- ❑ When your programs require a value setting that cannot be saved as System Variables, use PVARs. Put expressions in your ACAD.LSP file to automatically initialize their values.

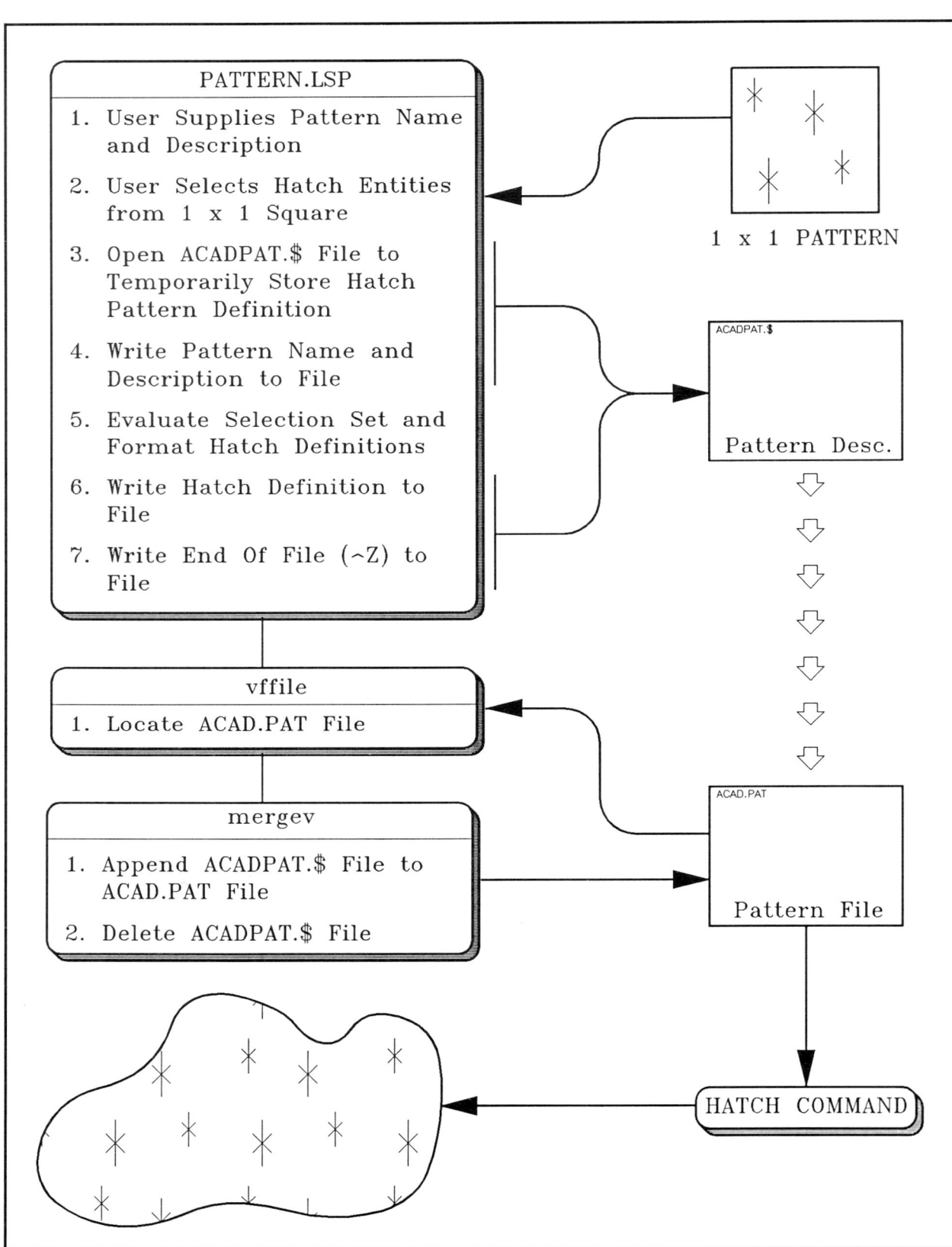

Automatic Hatch Generator

CHAPTER 14

AutoLISP Input/Output

AutoLISP I/O is your only AutoCAD "link to the outside world." AutoCAD has limited import modes. Using AutoLisp I/O lets you import external data, communicating with and accessing data files. This chapter shows you how to read and write files using AutoLISP, and how to manage your files using a combination of DOS and AutoLISP controls. The chapter also shows you how to integrate file I/O in two applications, an automatic hatch generator program called PATTERN, and REFDWG, a program that generates a drawing schedule and inserts the schedule in your drawing.

The Benefits of AutoLISP I/O

There are two main benefits to using AutoLISP I/O in your customized applications. These are:

- You can process large amounts of data, and centralize your data.

- You gain efficiencies in storage and processing. If you store data internally in AutoCAD, you encounter a significant overhead in drawing file storage. Using external files gives you more efficient data storage and better maintenance and control over your data.

How To Skills Checklist

In this chapter, you will learn how to:

- ❑ Read and write data to the screen. The screen is the default AutoLISP write device. To get started in writing data, the chapter shows you how to write different formats to the screen.

- ❑ Read and write an AutoLISP data file.

- ❑ Handle files at the DOS level, using AutoLISP and Shell.

- ❑ Back up files within AutoLISP.

- ❑ Merge files using DOS copy instead of using AutoLISP's append.

- ❑ Use a single Shell command to execute multiple commands at the DOS level.

- ❑ Verify execution of your SHELL commands from AutoLISP.

- ❑ Get your current directory.

- ❑ Verify file merging in a single AutoLISP function.

- ❑ Combine AutoLISP I/O and DOS level controls to read and write files while running an AutoLISP program.

- ❏ Write to AutoCAD's text screen using ANSI formats. For your applications, you may want to take over the text screen to present information to your users.
- ❏ Evaluate file formats for efficiency. Your application program's efficiency depends, in part, on selecting the appropriate file format.

Macros, AutoLISP Tools and Programs

MACROS
[FINDFILE] gets input and makes a menu item out of the VFFILE function.
[MERGFILE] gets input and executes the MERGEV function.
[BACKUP] gets input and backs up a file with the BACKUP function.
[PATTERN] loads the necessary function files and executes the PATTERN command function.
[REFDWG] loads all needed functions and runs the REFDWG command function.

AutoLISP TOOLS
FILELIB.LSP includes AutoLISP utilities that retrieve the current DOS directory path, verify a file's existence, format a file name, make a backup up file with an extension .BAK, and merge two files.
ANSILIB.LSP is a library of functions used to create text screen formatting using the ANSI escape calls. These functions clear the text screen, allow cursor movement control, and allow control of text color and effects such as blinking.

PROGRAMS
C:PATTERN is an automatic hatch generator. This program automatically generates most hatch patterns, creating a hatch pattern definition from selected entities. It appends the pattern to the ACAD.PAT file.
C:REFDWG is a drawing schedule program. This program generates a drawing schedule and inserts it in your drawing. C:REFDWG displays a list of project reference drawings on the text screen. Then, using ANSI library functions, the user indicates the reference drawings to list in the current drawing. After the selections are made, the program places the text in the drawings at a point specified by the user.

Device Input/Output

You've used the PROMPT function, which writes only to the screen. You also have used the PRIN1, PRINC and PRINT functions, which can write to files and devices, and GETSTRING, which gets input from the user. AutoLISP provides several more functions to read and write data to and from the screen, to and from a file, or any other device. The AutoLISP functions that write to devices are:

WRITE-LINE   WRITE-CHAR   PRINT   PRINC   PRIN1

The reading functions supplied by AutoLISP are:

READ-LINE   READ-CHAR

WRITE-LINE and READ-LINE write and read strings of data. WRITE-CHAR and READ-CHAR work with the integer equivalents of ASCII characters. PRINT, PRINC, and PRIN1 accept any expression, STRING, LIST, numbers etc., and write out a string.

Reading and Writing

Your default writing device is the "CONSOLE." It works like COPY CON: in DOS. AutoLISP reads input from the CONSOLE (keyboard) and writes output to the CONSOLE (screen), unless you tell it to do something different. You can use other "devices," like a printer (PRN or LPT1), or a serial port (COM1, COM2). AutoLISP always "returns" values to the screen.

Let's review the 4 Ps, PROMPT, PRINT, PRINC, and PRIN1. Then, look at WRITE-LINE and WRITE-CHAR.

Reading and Writing Data to the Console

```
Enter selection:                        Begin a NEW drawing named FILES.

Command: (prompt "PROMPT only writes to the screen, without quotes. ")
Lisp prints and returns: PROMPT only writes to the screen, without quotes. nil

Command: (princ "PRINC is similar, but returns string ")
Lisp prints and returns: PRINC is similar, but returns string "PRINC is similar, but returns
string "

Command: (write-line "WRITE-LINE forces a return. ")
Lisp prints: WRITE-LINE forces a return.
Lisp returns: "WRITE-LINE forces a return. "
```

PROMPT is the recommended method for prompting because it writes to both text and graphics screens. PROMPT always returns nil. It puts no line feed before or after the prompt, unless you imbed "\n" codes, or ^M in menu items. PRINC is similar to PROMPT, but returns the string value instead of NIL. Don't use PRINC in place of PROMPT, because PRINC doesn't print to the command prompt line of dual screen systems. WRITE-LINE is like PRINC except that it adds a <RETURN> whether you like it or not. You may have noticed the book's habit of putting a <SPACE> at the end of the prompt string to avoid having input, or more output, run into the prompt.

The other writing functions are PRINT, PRIN1 and WRITE-CHAR.

```
Command: (princ "While PRINC doesn't quote,\nor expand control characters,\n")
Lisp prints: While PRINC doesn't quote,
or expand control characters,
Lisp returns: "While PRINC doesn't quote,\nor expand control characters,\n"

Command: (prin1 "PRIN1 quotes and expands control \ncharacters.")
Lisp prints and returns: "PRIN1 quotes and expands control \ncharacters.""PRIN1 quotes and
expands control \ncharacters."
```

```
Command: (print "PRINT adds a leading \nnewline & trailing space. ")
Lisp prints and returns: "PRINT adds a leading \nnewline & trailing space. " "PRINT adds a
leading \nnewline & trailing space. "

Command: (write-char "A")                  Oops, we forgot to warn you!
Lisp returns: error: bad argument type
(WRITE-CHAR "A")
Command: (write-char 65)                   WRITE-CHAR wants an ASCII value, not a string character.
Lisp prints and returns: A65
```

"Expanding control characters" means printing them in a string like "...\n..." instead or actually issuing a new line or other code to the screen or device. Note that the "\n" forced a real new line in PRINC, but remained an expanded string in the others. WRITE-CHAR is very simple, it prints unquoted like PRINC, and returns its ASCII value.

General File Handling

AutoLISP can process only one type of DOS file, an ASCII text file. Some programming languages can jump back and forth in the file. This is called random access. AutoLISP does not have random access. AutoLISP only reads a file in a sequential order. It reads from top to bottom.

Given this constraint, AutoLISP provides the basic tools to read and write data either one character at a time, or a full line at a time. Before you can read and write to a file, you have to "open" a file to talk to it, and "close" a file to store in it. It will come as no surprise to find that the functions to open and close files are called:

OPEN   and   **CLOSE**

File processing is simple. You OPEN a file, process it by reading or writing. Then, you CLOSE it. Each time you open a file, AutoLISP returns a file handle to you. This handle is a name reported from DOS that lets you tell AutoLISP which file you want to work with. You can work with several files at once, each with a handle. Set the handle to a variable and supply it as an optional argument to the read, print and write functions to direct your data to the correct file. File handles are always changing and their actual names are meaningless, just like entity names, so work with the variables.

You have been working with files all along, but you may not have realized it. A file is really a device! It is a device with the added feature of being able to store and retrieve the information on disk. The screen/keyboard (CONsole), printers and disk files are all devices. DOS, AutoCAD and AutoLISP treat the CONsole as the default device. If not set to other devices, data is sent to the screen and received from the keyboard. DOS reserves names for all devices except files, they are:

DEVICE NAMES	HARDWARE COMPONENTS
CON	Screen (video monitor) and keyboard.
PRN	Printer hooked to the 1st parallel port.
COM1	The 1st serial port. Can be a printer, modem etc.
COM2	Second serial port with same purposes as COM1.
LPT1	The first parallel port.
LPT2	The second parallel port.
NUL	A null device used to suppress output data.

You can use as many COM and LPT ports as you have. You need to open devices for communication, just like files. However, since you can both write to and read from a device, you must tell which access mode you need by passing a key letter to OPEN. You can open a device to read "r," write "w," or append "a," but not all devices support all modes. You cannot open the PRN and CONsole devices to read. PRN and CONsole are write only. Append adds to an existing file. If a files does not exist, append creates a new file. Your access mode is passed along to AutoLISP at the time the file is opened. Your access mode must be in lower case for AutoLISP to use it. Here is a list of Device Access Modes:

DEVICE ACCESS MODES

MODE	ACCESS CONTROL MEANING
r	Read the contents, only accesses existing files.
w	Write to the file, creates a new file each time.
a	Append data to an existing file, or opens new one.

Let's try some file processing. Set the variable FP to store the file handle returned by AutoLISP. Then, send some data to a file called "TEST." You can use any valid DOS filename. Write a few lines and close the file. You MUST close the device when finished, or DOS withholds the file from access by other programs and calls. The unclosed files become lost clusters when you exit AutoCAD. The filename is only entered on the disk directory listing after it has been closed!

Opening and Closing Files

Command: **(print "Without a file handle, the data goes to the screen")**
Lisp prints and returns: "Without a file handle, the data goes to the screen" "Without a file handle, the data goes to the screen"

Command: **(setq fp (open "TEST" "w"))**   OPEN a file for writing.
Lisp returns: <File: #C136>   The file handle FP.

Command: **(write-line "It's all in the lines..." fp)**   Write data to the file with handle FP.

Lisp returns: "It's all in the lines..."   But it writes to the file.

Command: **(write-line "That we write and read." fp)**
Lisp returns: "That we write and read."

```
Command: (write-char 79 fp)          Output a character, ASCII "O".
Lisp returns: 79

Command: (write-char 75 fp)          And a "K".
Lisp returns: 75

Command: (close fp)                  Now close it.
Lisp returns: nil
```

To read a file you do just the opposite. You open the file with "r" and extract data using the READ-LINE and READ-CHAR functions. Remember to supply the file handle FP variable. Don't worry about moving to the next line in the file, AutoLISP automatically advances with each READ-LINE. When the end of the file is reached, READ-LINE returns NIL. Try these:

```
Command: (setq fp (open "TEST" "r"))   Open it again to READ.
Lisp returns: <File: #BF7E>

Command: (read-line fp)                Read the first line.
Lisp returns: "It's all in the lines..."

Command: (read-line fp)                And the second.
Lisp returns: "That we write and read."

Command: (read-char fp)                Now the next character.
Lisp returns: 79                       It's the ASCII number.

Command: (chr (read-char fp))          Read and convert it.
Lisp returns: "K"

Command: (read-line fp)
Lisp returns: nil                      There's nothing left.

Command: (close fp)                    Close it.
Lisp returns: nil
```

Writing to the Printer and Other Devices

Printers are almost like files. They are addressed as PRN or LPTn, where n is the port number. But you can't read from a printer, at least not through AutoLISP. Printers can cause problems. Unlike true DOS files, printers can become "unavailable" after they are opened. If you disconnect the printer, or it runs out of paper, AutoLISP doesn't like this and causes the computer to lock up!

If you need to, you can write to, and read from devices like the COM ports. We won't ask you to do it because we assume your digitizer or plotter is sitting on your COM port(s). If you wanted to write to the PRN and COM ports, it would look like:

```
Command: (setq fp (open "PRN" "w"))
Lisp returns: <File: #E64E>

Command: (write-line "Wake up Mr. Printer" fp)
Lisp returns: "Wake up Mr. Printer"

Command: (setq ser1 (open "COM1" "w"))
Lisp returns: <File: #D60A>

Command: (write-line "This goes to the first serial port" ser1)
Lisp returns: "This goes to the first serial port"

Command: (setq nowher (open "NUL" "w"))
Lisp returns: <File: #D574>

Command: (write-line "This goes to nowhere..." nowher)
Lisp returns: "This goes to nowhere..."

Command: (close fp) (close ser1) (close nowher)
```

Instead, we recommend that you write everything to a file. This avoids hanging your system by writing to the printer. Then, you can copy the file to the printer using the PRN device name. DOS handles printer errors when you use it to copy. You can use AutoCAD's SHELL to issue the DOS command COPY. If your printer is on the second parallel port, use LPT2 in place of PRN.

Make sure your printer is ready.

Command: **SHELL**
Dos command: **COPY TEST PRN**   The little TEST file should print.

Testing for Files and Paths

Now that you know the basics of reading and writing, you need to know how to manage files. As you just saw with printers, AutoLISP lacks some file and device management capabilities. To get around these limitations, you will need to integrate AutoLISP and DOS to handle typical file management tasks. These include checking if a file exists before trying to read from it, finding it in a directory on the disk using a path, and making sure you're getting a valid file name from the user.

You can check the existence of a file by using OPEN to read. You can do directory checking by creating a temporary file in DOS via SHELL, then reading the file in AutoLISP. You check for a valid file name by examining the string entered by the user. The next exercise develops three functions to help you with these tasks. These functions are used throughout the rest of the book.

The CPATH (Current PATH) function uses AutoCAD's SHell command and the DOS CD Change Directory command to return the full directory path. If used without a directory name, CD displays the current directory name. CPATH uses a DOS redirection symbol to send the current directory path name to a temporary file. Then, you return to AutoCAD, use AutoLISP file access to read the temporary file and return the contents of the file. You should still be in the FILES drawing. If not, restart the FILES drawing.

Making Functions to Check Files

 Copy CA14.MNU to TEST.MNU.

 Copy MAIN.MNU to TEST.MNU.

```
Select [TEST] [24 x 36] [FULL] [INITIATE]            It brings up TEST.MNU.
```

 Copy FILELIB.LSP to TEST.LSP.

 Create a new AutoLISP function file called TEST.LSP.

```
Select [EDIT-LSP]                Edit TEST.LSP, and enter:

;* CPATH returns current DOS dir path in form "d:\\path\\...\\"
;* If SH (shell) fails w/ insufficient memory, increase size in ACAD.PGP
(defun cpath (/ path fp)
                                  ;Creates 0 byte file, overwrite if existing...
  (close (open "$" "w"))          ;...allows READ-LINE to test SHell's success.
  (command "SH" "CD > $" )              ;Redirect current dir to temp file $.
  (setq fp (open "$" "r"))
  (if (setq path (read-line fp))       ;Nil only if SHell failed.
    (setq path (strcat path "\\"))     ;Puts path in form "d:\\path\\...\\"
                                       ;Sets path to "" if SHell failed.
    (progn (prompt "\nSHell FAILED!") (setq path ""))
  )
  (close fp)
  path                            ;Returns path, which is "" only if SHell failed.
)
;*
```

Save, Exit and reenter AutoCAD. TEST.LSP should load.

```
Command: (cpath)                  Test it.
SH
DOS Command: CD > $
Lisp returns: "C:\\CA-ACAD\\"
```

Note that the format that CPATH returns is the same format as (getvar "ACADPREFIX") or (getvar "DWGPREFIX"). CPATH is useful if your program needs to identify what directory it is in, or needs to switch directories. DOS has no provision to automatically restore directories if you change, but AutoLISP can get the name with CPATH, switch in SHELL, do something, then restore the saved name.

The VFFILE (VeriFy FILE) function verifies the existence of a file. Use it in cases where the user gives a file name and your programs must insure the name supplied by the user actually exists. VFFILE determines if the file exists by using the file handle returned by OPEN. If a file doesn't exist, then NIL is returned instead of a file handle. VFFILE reprompts the user, giving the user the option of giving up.

First, type in an expression to see how easy it is to check a file name. Then, enter the full-blown function.

```
Command: (if (setq fp (open "stuff" "r")) (not (close fp)))
Lisp returns: nil             It doesn't exist. FP is NIL so CLOSE isn't evaluated.
Command: (if (setq fp (open "test" "r")) (not (close fp)))
Lisp returns: T               CLOSE always returns NIL, so NOT makes it T.
```

Select **[EDIT-LSP]**   Edit TEST.LSP, and add:

```
;* VFFILE verifies a file's exist, reprompts for path till found, returns fspec or NIL.
;* WARNING * DOS 3.3 APPEND may cause FALSE UNPREDICTABLE results.
;* May be called w/ fname in form: "\\path\\filename.ext" or just "filename.ext"
(defun vffile (fname / fspec fp path prmpt)
  (setq path "")                  ;Starts w/ current dir, or fname with path.
  (while                          ;While file can't be opened, reprompt...
    (and (/= "Q" path) (not (setq fp (open (setq fspec (strcat path fname)) "r"))))
    (setq prmpt                                   ;Build prompt
      (strcat "\n\nFile " fname " not found in "
        (if (= "" path) "current directory," path)  ;Current dir if path=""
        "\nEnter path to search, or Q to quit: "
      );strcat
      path (strcase (getstring prmpt))
  ) );while & setq
  (if fp (close fp))
  (if (= "Q" path) (progn (prompt "File not found. ") nil) fspec) ;returns nil or fspec
);defun vffile
;*
```

Save, Exit and reenter AutoCAD. TEST.LSP should load.

Select **[EDIT-MNU]**
Change [CA00TEST] to [CA14TEST] on the root page, then add to the EXAMPLES page:

```
***SCREEN
**ROOT
[CA14TEST]

**EXAMPLES
[Examples]
[]
[FINDFILE](vffile (getstring "^MEnter filename to find: "));
[]
```

Save and Exit to AutoCAD. TEST.MNU reloads.

```
Select [EXAMPLES] then [FINDFILE]
Command: (vffile (getstring "
2> Enter filename to find: "))
Enter filename to find: ACAD.LIN
File acad.lin not found in current directory,
Enter path to search, or Q to quit: \ACAD\
Lisp returns: "\\ACAD\\ACAD.LIN"
```

Try again, entering Q to quit. It will return NIL.

You must use a trailing "\" in VFFILE's path input. VFFILE also accepts a full "\\path\\filename.ext" as its filename argument, but the path can't be reentered.

Your programs also must manage file extensions. On occasion, the user may enter a file extension along with the file name even though your program assumes a file type. You frequently will have to determine if they included the extension, and add an extension if they did not.

The FFNAME (Format File NAME) function checks for a period "." in the file name by parsing the string. Notice that FFNAME starts at the end of the string and works to the beginning. If no extension exists, FFNAME adds one.

Select [EDIT-LSP]   Edit TEST.LSP, and add:

```
;* FFNAME formats a filename as "filename.ext" given input FNAME w/ or w/o
;* extension, and EXT as "EXT".
(defun ffname (fname ext / inc lngth)
  (setq inc -1  lngth (strlen fname))      ;Initialize, lngth is filename length
  (while
    (not                                    ;loops until OR is non-NIL
      (or                                   ;eval 2nd AND only if 1st is NIL
        (and                                ;SETQ FNAME only if "." = non-NIL
          (= "." (substr fname (- lngth (setq inc (1+ inc))) 1))   ;find "."
          (setq                             ;strip last char and append EXT
            fname (strcat (substr fname 1 (- lngth inc)) ext)
          )
        )
        (and                                ;SETQ FNAME only if...
          (or (= inc 3) (= inc lngth))      ;...if "." not found in last 3 char
          (setq fname (strcat fname "." ext)) ;then append EXT to whole fname
) ) ) )
  fname
);defun FFNAME
;*
```

Save, Exit and reenter AutoCAD. TEST.LSP should load.

Command: **(ffname "new" "txt")**   Append extension.
Lisp returns: "new.txt"

Command: **(ffname "new.saf" "txt")**   Forces correct extension.
Lisp returns: "new.txt"

Take a moment to dissect the FFNAME function. It uses a WHILE-NOT-OR-AND structure as a conditional loop. Think of it as an exercise in writing a function without using deeply nested IF statements. Write out a filename and extension. Step through the function logically to help you understand the AND and OR.

Now that you have verified files, you need a way to merge the files and back them up. The next exercise provides you with three more file handling functions, called BACKUP, MERGEF and MERGEV. Then, it shows you how to use the functions.

AutoLISP and DOS File Handling

The BACKUP function show how to use AutoCAD's FILES command and SHell command to manage files.

BACKUP uses the AutoCAD FILES command. BACKUP uses the FFNAME program to format the filename. If you need verification, have your program attempt to test for a filename.BAK after BACKUP.

Select **[EDIT-LSP]**   Edit TEST.LSP, and add:

```
;* BACKUP copies file fspec to its path\name.BAK, using
;* the FFNAME function to format the filename.
(defun backup (fspec)
   (command "FILES" "5" fspec      ;Copy filespec.EXT to
     (ffname fspec "BAK")          ;filespec w/ EXT stripped & repl w/ BAK
     "" "" nil                     ;NIL cancels error if file didn't exist.
) );*defun BACKUP
;*
```

MERGEF puts two files together without verification, using the DOS "+" symbol. MERGEF deletes the second file. You can add a Yes/No option if you like. In DOS, a plus symbol placed between successive file names causes DOS to combine the files using the last file name. Many text files have a <^Z> at the end. This tells programs to stop reading them. If you use the "a" append mode of OPEN to write additions to a file, the <^Z> will still be in the file above your additions. Using DOS's COPY with + handles the <^Z> for you.

Save what you've done and continue adding to TEST.LSP:

```
;* MERGEF appends file2 to file1, strips file1's ^Z if present, deletes file2
;* If SH (shell) fails w/ insufficient memory, increase size in ACAD.PGP
;* It uses the VFFILE function to verify the filenames existence.
(defun mergef (file1 file2)
   (setq file1 (vffile file1)  file2 (vffile file2))      ;verifies files
   (command "SH" (strcat "COPY " file1 " + " file2 " " file1)
           "SH" (strcat "DEL " file2) nil ;NIL cancels error if SH failed
) )
;*
```

MERGEV combines two files and verifies the copy procedure. It deletes the second file. You can add a Yes/No option. MERGEV uses an interesting technique of creating a small batch file using AutoLISP. It runs the batch file through AutoCAD's SHell facility. The last statement of the batch file causes the batch file to delete itself. If the batch file has not been deleted, it means that the COPY was never run and/or the SHell command failed. MERGEV checks if the batch file exists after the combine process. If it is not there, the function returns NIL.

Save what you've done and continue adding to TEST.LSP:

```
;* MERGEV copies file1 to its path\name.BAK then appends file2 to file1,
;* strips file1's ^Z if present, and deletes file2, returning NIL if SHell fails.
;* If SH (shell) fails w/ insufficient memory, increase size in ACAD.PGP
;* It uses the FFNAME function to format the filename and VFFILE to verify existence.
(defun mergev (file1 file2 / file)
  (setq file1 (vffile file1)  file2 (vffile file2))     ;verifies files
  (setq file (open "$.bat" "w"))             ;create temp $.BAT file
  (princ                                      ;write ea string to file as a...
    (strcat                                   ;...separate command line w/  "\n"
      "COPY " file1 " "                       ;Copy filespec.EXT to filespec w/...
      (ffname file1 "BAK")                    ;... EXTension repl w/ BAK
      "\n"                                    ;newline
      "COPY " file1 " + " file2 " " file1 "\n"   ;merge to name of file1
      "DEL " file2 "\n"                       ;DEL file2
      "DEL %0.bat"                            ;DEL temp file $.BAT itself
    ) file                                    ;close strcat & send to file
  );princ
  (close file)
  (command "SH" "$")                          ;execute the above $.BAT
                                              ;Tests if SHELL worked - if...
  (if (setq file (open "$.bat" "r"))          ;...file deleted itself OK.
    (progn (close file) (prompt "\nSHell FAILED!"))  ;Then returns nil if failed
    T                                         ;Else returns T if OK
  )
);defun mergev
;*
```

Save, Exit and reenter AutoCAD.TEST.LSP should load. Test it:

Select [EDIT-MNU]   Change and add the following to the EXAMPLES page:

```
[MERGFILE](mergev (getstring "^MFilename to append to: ");+
(getstring "^MFile to add, and then delete: "));
[]
[BACKUP  ](backup (getstring "^MFilename to backup to .BAK: "));
[]
```

Save and Exit to AutoCAD. TEST.MNU reloads. Test them.

Select [EXAMPLES] [BACKUP]
Command: (backup (getstring "
2> Filename to backup to .BAK: "))
Filename to backup to .BAK: **TEST**   Flips through the FILES menu and creates TEST.BAK.

Command: **DIR**   Enter TEST.* to see if it worked. Note the size.

Command: **(mergef "test" "test.bak")**
SH
DOS Command: COPY test + test.bak test   TEST
TEST.BAK
1 File(s) copied
Command: SH
DOS Command: DEL test.bak
Lisp returns: nil

Command: **DIR**   Enter TEST.* If OK, its size is doubled and TEST.BAK gone.

Select **[EXAMPLES] [BACKUP]**   Backup TEST again and try MERGEV.

Select **[MERGFILE]**
Command: (mergev (getstring "
2> Filename to append to: ")
1> (getstring "
2> File to add, and then delete: "))
Filename to append to: **TEST**
File to add, and then delete: **TEST.BAK**   SHell and DOS COPY commands scroll by, then:
Lisp returns: T   True. It worked. Check with DIR.

  Delete TEST.LSP.

  Rename TEST.LSP to FILELIB.LSP.

You can try these functions on other files to see how they handle errors.
Now it's time to create a program, named PATTERN. PATTERN puts these file handling functions to good use.

Automatic Hatch Pattern Generator

We hate calculating hatch patterns. It is such a time consuming job and the process is exactly the same, regardless of the pattern. This obviously is a job for an AutoLISP function. PATTERN works by examining the entities of a sample pattern created in a 1x1 box. PATTERN operates under the following conventions when you create the sample pattern:

- All dots are drawn using the POINT entity.

- All line segments are drawn using the LINE entity.

- All lines are drawn at angle increments of 45 degrees (0, 45, 90, 135, 225 etc.).

- All other entity types are ignored by the program.

If you don't remember how the hatch pattern file works, please refer back to the chapter on Linetypes and Hatches.

Write the function PATTERN. It starts by gathering input from the user. It gets the name of the pattern, the pattern description and the entities of the sample pattern. It opens a temporary file and writes the name and description to the file.

Creating PATTERN an Automatic Hatch Generator

```
Enter selection:                                Begin a NEW drawing named PATTERN

Select [TEST] [24 x 36] [FULL] [INITIATE]       It brings up TEST.MNU.
Command:ERASE                                   Erase the border.
```

 Copy PATTERN.LSP to TEST.LSP.

 Create a new LISP function file called TEST.LSP.

```
Select [EDIT-LSP]                               Edit TEST.LSP, and enter:
```

```lisp
;* C:PATTERN writes a HATCH pattern from selection set of LINES and/or POINTS.
;* LINES must be at 0 or angular multiples of 45 degrees. Other angles are NOT
;* filtered out and will become irregular in alignment. Selection set should
;* also be created inside a 1-unit square box for correct alignment.
;* Calls DXF, SHELL and VFFILE (FILELIB.LSP) functions.
(defun C:PATTERN ( / hname hdes ss1 fp count en ed et pt1 pt2 ang dlen
                     deltax deltay skip olin fspec)
           ;INPUT
  (setq hname ""   hdes "")                     ;init for input, WHILEs insist on string
  (while (= "" (setq hname (getstring "\nName of pattern: "))))  ;Force a hatch name
  (while (= "" (setq hdes  (getstring "\nDescription: " T))))    ;For ACAD.PAT file
  (prompt "\nSelect unit pattern entities: ")
  (while (not (setq ss1 (ssget))))              ;Get entities.
  (setq fp (open "acadpat.$" "w"))              ;Open temp pattern file.
  (textscr)
           ;WRITE HEADER                        ;Write * to file to mark start...
  (princ (strcat "*" hname) fp)                 ;...of new pattern.
  (write-line (strcat "," hdes) fp)             ;Write ,description to file.
```

Here's the sticky part, calculating the geometry. Fortunately, it is already done for you in this example. PATTERN retrieves the entity name, data and type from the AutoCAD database. It tests the type using a COND structure, calculates the hatch entry for the entity and sets it to a variable "olin" that gets written to the file. If the entity is a POINT, then the calculation is simple. If it's a LINE, the function tests the angle of the line and makes the pattern calculations based on that angle. If you have a valid entity, the function writes the hatch entry to the file and the screen. It helps to keep the user amused!

Save what you've done and continue adding to TEST.LSP:

```
              ;CALCULATE & WRITE BODY OF PATTERN
  (setq count 0   emax (sslength ss1))   ;EMAX=number of entities selected.
  (while (< count emax)                  ;Examine ea entity selected.
    (setq en (ssname ss1 count)          ;Entity name
          ed (entget en)                 ;Entity data
          et (dxf 0 ed)                  ;Entity type
          count (1+ count)
    )
    (cond
      ((= et "POINT")                    ;cond-1 - If it's a POINT...
        (setq olin                       ;...calc & format the hatch "line".
          (strcat "0," (rtos (car  (dxf 10 ed)) 2 6) ","
                       (rtos (cadr (dxf 10 ed)) 2 6) ",0,1,0,-1"
        ) )
        (prompt (strcat "\n" olin))      ;Display "line" for amusement.
        (write-line olin fp)             ;Write it to file.
      );cond-1
      ((= et "LINE")                     ;cond-2 - If it's a LINE...
        (setq pt1 (dxf 10 ed)            ;Endpt 1
              pt2 (dxf 11 ed)            ;Endpt 2
              ang (angle pt1 pt2)        ;Angle
              dlen (distance pt1 pt2)    ;Length
        );setq
        (if (= "1.00"                    ;Test whether 90 deg multiple.
               (rtos                     ;If so, abs of sin or cos...
                 (+ (setq deltax (abs (cos ang)))   ;...will be 1 and other 0.
                    (setq deltay (abs (sin ang)))
                 ) 2 2
            ) );=
          (setq deltax 0.0               ;Then set offset along line &
                deltay 1.0               ;Offset to parallel line &
                skip (- dlen deltay)     ;Dash length
          )
                                         ;Else assume to be 45 deg family &
                                         ;use offsets set in IF test above &
          (setq skip (- dlen (* deltay 2.0)))     ;set Dash length.
        );if
        (setq olin                       ;Format LINE hatch "line".
          (strcat (angtos ang 0 6)    "," (rtos (car pt1) 2 6)
             "," (rtos (cadr pt1) 2 6) "," (rtos deltax 2 6)
             "," (rtos deltay 2 6)    "," (rtos dlen 2 6)
             "," (rtos skip 2 6)
        ) );setq&strcat
        (prompt (strcat "\n" olin))
        (write-line olin fp)
      );cond-2
      (T (prompt                         ;cond-3 - not LINE or POINT
           (strcat "\nInvalid entity " et " skipped.")
         )
      )
    );cond
  );while
```

14—16 Customizing AutoCAD

The final part of PATTERN manages the files. It looks for the ACAD.PAT file. If it does not find the file, PATTERN asks the user where the file is using the VFFILE function. Once ACAD.PAT has been found, the program asks the user if they want to merge the temporary file with the ACAD.PAT file.

Save what you've done and continue adding to TEST.LSP:

```
            ;CLOSE FILE & APPEND FOR USE
  (write-char 26 fp)                      ;Write ^Z EOF char.
  (close fp)                              ;Close temp ACADPAT.$ file.
  (setq fspec (vffile "ACAD.PAT"))        ;Verify path of ACAD.PAT
  (initget "Yes No")
  (if
    (and                         ;Test - do you want to merge and did it work?
      (/= "" fspec)                       ;Did VFFILE find ACAD.PAT
      (/= "No"                            ;Do you want to append for use
        (getkword
          (strcat
            "\nTemp pattern file must be appended to ACAD.PAT for use."
            "\nDo you want to append pattern to " fspec "? Yes/No/<Y>: "
  ) ) )
      (mergev fspec "ACADPAT.$")          ;Then merge {backup ACAD.PAT,...
                                          ;...append & del acadpat.$}
    )
    nil                                   ;If above OK, then nil
    (prompt "\nTemp file ACADPAT.$ left in current directory. ")   ;Else
  );if
  (graphscr)
  (princ)
);defun
(princ)
;*
```

Save, Exit and reenter AutoCAD. TEST.LSP should load.

Select [EDIT-MNU]   Add the following to the EXAMPLES page:

```
[PATTERN ]^C^C^C(load "DXF") (if VFFILE nil (load "FILELIB"));+
(if C:PATTERN nil (load "PATTERN")) PATTERN
```

Save and Exit to AutoCAD. TEST.MNU reloads.

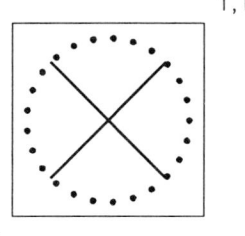

The Initial Hatch Pattern

The [PATTERN] menu item is designed to load any necessary .LSP files. Note that the PATTERN.LSP file is still named TEST.LSP, unless you have the CA DISK. It doesn't matter right now. C:PATTERN was loaded automatically by your [EDIT-LSP] menu item.

Draw a sample unit pattern similar to that shown in the Initial Hatch Pattern. Test the PATTERN program.

Command: **ERASE**   Erase the border.
Command: **ZOOM**   Left -1,-1 with Height 6".
Command: **UNITS**   Set to decimal.
Command: **SNAP**   Set to 0.1
Command: **PLINE**   Draw a 1x1 box.
Command: **DUP**   Copy your original ACAD.PAT file to \CA-ACAD*.*

Command: **CIRCLE 3P/2P/TTR/<Center point>:**   Pick center point of square.
Diameter/<Radius>: **0.42426**

Command: **DIVIDE**   Divide the Circle into 25 segments to create POINTs.

Command: **ERASE**   Select circle.

Command: **LINE**   Draw 2 Lines at 45 deg., watching the COORDS.

                                                 Try PATTERN on the X and circle of dots you just made:

Select **[EXAMPLES] [PATTERN]**   The DXF and FILELIB files load, then:

Command: **PATTERN**
Name of pattern: **XDOTS**
Description: **Dots in a circle pattern with a large X**
Select unit pattern entities...
Select objects: Window the X and points you drew.
Select objects: Return and 27 lines of numbers scroll by.

Temp pattern file must be appended to ACAD.PAT for use.
Do you want to append pattern to ACAD.PAT? Yes/No/<Y>:   <RETURN> to append it.
SH
DOS Command: $

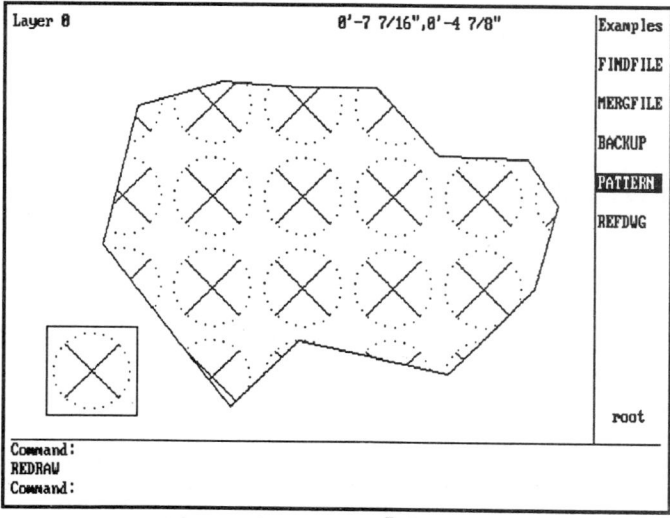

The Final Hatch Pattern

Command: **ED**

Look at ACAD.PAT with your text editor. The order and numbers may vary.

```
*XDOTS,Dots in a circle pattern with large X
135.000000,0.800000,0.200000,0.707107,0.707107,0.848528,-0.565685
45.000000,0.200000,0.200000,0.707107,0.707107,0.848528,-0.565685
0,0.910931,0.394491,0,1,0,-1
0,0.871782,0.295611,0,1,0,-1
0,0.809272,0.209574,0,1,0,-1
0,0.727330,0.141785,0,1,0,-1
0,0.631104,0.096505,0,1,0,-1
0,0.526640,0.076577,0,1,0,-1
0,0.420502,0.083255,0,1,0,-1
0,0.319359,0.116118,0,1,0,-1
0,0.229566,0.173102,0,1,0,-1
0,0.156766,0.250626,0,1,0,-1
0,0.105533,0.343819,0,1,0,-1
0,0.079085,0.446826,0,1,0,-1
0,0.079085,0.553174,0,1,0,-1
0,0.105533,0.656181,0,1,0,-1
0,0.156766,0.749374,0,1,0,-1
0,0.229566,0.826898,0,1,0,-1
0,0.319359,0.883882,0,1,0,-1
0,0.420502,0.916745,0,1,0,-1
0,0.526640,0.923423,0,1,0,-1
0,0.631104,0.903495,0,1,0,-1
0,0.727330,0.858215,0,1,0,-1
0,0.809272,0.790426,0,1,0,-1
0,0.871782,0.704389,0,1,0,-1
0,0.910931,0.605509,0,1,0,-1
0,0.924260,0.500000,0,1,0,-1
```

Command: **PLINE**

Make hatch boundary of irregular shape.

Command: **HATCH**

Enter XDOTS and try it on the Pline.

 Delete TEST.LSP.

 Rename TEST.LSP to PATTERN.LSP.

Try using PATTERN a few more times, including giving it wrong input. You are at the point where you have developed enough tools so that the programs that you write are robust.

If you use PEDIT to fit the polyline, AutoCAD will take longer to hatch the area because the pattern must conform to PEDIT's curve fitting. This is typical of the Hatch command and not the pattern.

Using ANSI Formatting Codes

ANSI is a system level device driver. Like the AutoCAD tablet, printer and plotter drivers, it contains a set of instructions that control the operations of a device. The ANSI system device driver is not part of the AutoCAD program. It comes along with your DOS operating system. The driver has a standard name of ANSI.SYS with all DOS versions. As you may recall from the Getting Started chapter, the ANSI driver is loaded by including it in your system CONFIG.SYS file. When the system is booted, the driver is automatically loaded.

The remainder of the book requires that you have the ANSI driver loaded. If the screen formatting effects are not working, your driver is not loaded. Go back to the Getting Started chapter for instructions on how to load this device. Remember to reboot your computer after editing the CONFIG.SYS file.

Let's look at the ANSI codes. The CA DISK includes a help screen showing the common ANSI formatting codes together with their effects on the screen. You can stay in your current drawing, or reload the last drawing.

Sample Text Screen Formatting

```
C:\CA-ACAD> TYPE                    Only if you have the CA DISK.
File to list: ANSI.HLP              It displays a help screen.
```

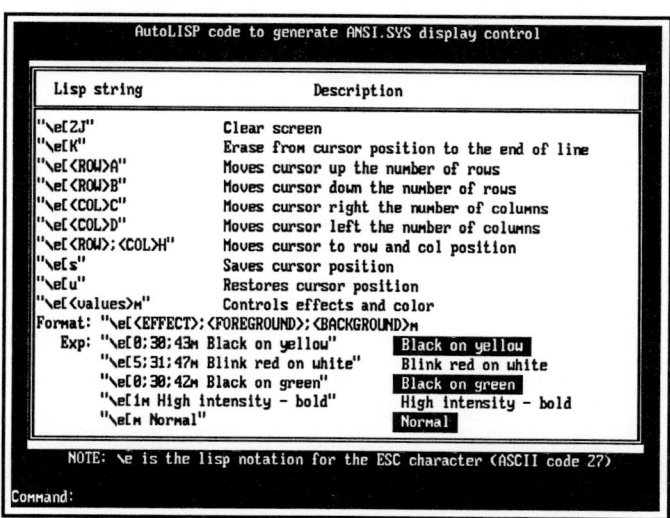

ANSI Formatting Effects

Your screen should look like the illustration, but in color if your screen supports color. You can apply the same screen formatting control through the AutoLISP language. You're familiar with the new line "\n" character in a prompt string. The backslash character acts like an AutoLISP escape character to send a formatting sequence through AutoLISP.

The ANSI device driver works with an ASCII 27 <ESC> escape character. The escape character must precede any instruction to the ANSI driver device. The escape character is written as "\e" in an AutoLISP prompt. It must be lowercase. The escape character is only a code to invoke a formatting instruction, it does not cause any formatting effect itself. To get the formatting effect you need to include additional codes. Those codes are shown in the illustration, and in the help screen. A typical formatting code looks like:

`"\e[2J"`

Please note that upper case and lower case codes are not equivalent.
The formatting codes are dependent on the text screen controller and not on the graphics screen. Life is simple and carefree if you have a dual screen system. If you have a single screen system, apply the codes according to the following rules:

- Only apply the codes to the TEXT screen. Flip to the text screen first.
- Do not FLIP to the graphics screen while formatting.
- Always return the screen to the original mode.

Try a little interactive formatting with AutoLISP. Refer back to the illustration for the formatting codes.

Command: **<F1>**	First flip to the text screen.
Command: **(prompt "\e[2J")**	Clears the screen.
Command: **(prompt "\e[1m")**	Bolds the characters.
Command: **(prompt "\e[m")**	Resets the screen.
Command: **(prompt "\e[s")**	This saves the current screen cursor position.

You also can make the cursor move to an exact place on the text screen. Cursor positioning is determined according to a row and column map. The typical screen has 25 rows and 80 columns. When designing programs to format a screen of text, many programmers use a Screen Position Chart, like the 25 x 80 chart shown.

They plan the screen by filling in the map with the information they need to present. We'll do this with our next program. You can make a map with grid paper. If you have the CA DISK, the chart shown is the SCR-MAP.DWG file.

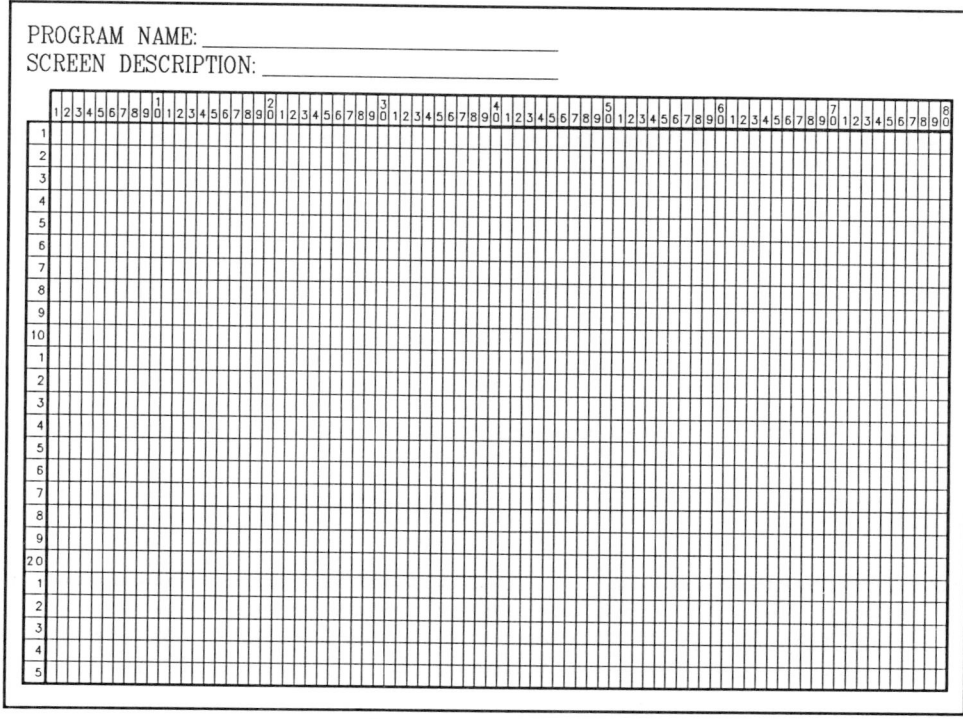

25 x 80 Screen Position Chart

You tell ANSI where to put the cursor by supplying the desired row and column.

Command:	`(prompt "\e[10;40H")`	Go near the screen center.
Command:	`(prompt "\e[u")`	Restore the position saved before.

Now, you can apply what you know to make a library of ANSI/AutoLISP functions. Most of the functions are simple one-line PROMPT commands. Use PRINC to keep the functions clean. This exercise also provides two functions, called CENTER and DOWNROW. CENTER takes a string of text and calculates the text indent to achieve a centered look. DOWNROW moves the cursor down a specified number of rows.

Making an ANSI Library of Screen Control Functions

 Copy ANSILIB.LSP to TEST.LSP.

 Create a new LISP function file called TEST.LSP.

Select **[EDIT-LSP]**   Edit TEST.LSP, and enter:

```
;*;WARNING: ANSILIB.LSP uses ; semicolons in strings -Do Not run LSPSTRIP/S
;* ANSILIB.LSP library of functions consist of the most common
;* tasks required for screen formatting using the ANSI escape calls

(defun cls()            ;this function clears the text screen
    (prompt "\e[2J") ;issues ansi code
    (princ)
);defun

(defun bold()           ;this function sets high intensity character mode on
    (prompt "\e[1m") ;issues ansi code
    (princ)
);defun

(defun normal()         ;this function resets character mode to normal intensity
    (prompt "\e[m")   ;issues code
    (princ)
);defun

(defun center(str / hleng)      ;this function takes a string and centers it on text scr.
    (prompt "\e[;1A") ;omit row number causes cursor to move in left col
    (setq hleng (/ (- 80 (strlen str)) 2)) ; gets half the string length
    (prompt (strcat "\e[" (itoa hleng) "C")) ; move half distance of cols
    (prompt str) ; prints the string
    (princ)
);defun

(defun goto(row col)   ;moves cursor to a position on the screen
    (prompt (strcat "\e[" (itoa row) ";" (itoa col) "H"))
    (princ)
);defun

(defun downrow(numb)   ;moves cursor down a number of rows
    (prompt (strcat "\e[" (itoa numb) "B"))  ;issues ansi code
    (princ)
);defun

(defun savepos()        ;saves the current position of the cursor
    (prompt "\e[s")    ;issues ansi code
    (princ)
);defun

(defun restpos()        ;function restores previously saved cursor position
    (prompt "\e[u")    ;issues ansi code
    (princ)
);defun
```

Save, Exit and reenter AutoCAD. TEST.LSP should load. It's Test Time:

```
Command: <F1>              Flip to the text screen.
Command: (cls)             Clears the screen.
Command: (bold)            Makes the characters bold.
Command: (normal)          Makes them normal.
Command: (center "HELLO")  Centers the string.
Command: (goto 10 20)      Goes to a screen position.
Command: (savepos)         Saves the position.
Command: (downrow 5)       Moves down 5 rows.
Command: (restpos)         Restores the position.
```

 Delete TEST.LSP.

 Rename TEST.LSP to ANSILIB.LSP.

If they all work, move on and process your first external file.

File Formats: The Basis of External Data Handling

Files can have many formats. The important part of creating and reading external files is to decide what format your file should be. Plan your data file formats with your entire system in mind. If possible, standardize on one format and write one function to process that type of format.

The simplest files may have a series of entries formatted as they are to appear on the screen. These files are read one line at a time and the full line is processed. A typical series of lines might look like:

```
REFERENCE DRAWINGS - PAPER PLANT
D78-1042 SECONDARY FIBER SYSTEM PIPING
D81-3003 NO.3 PAPER MACHINE - WET END PROCESS FLOW DIAGRAM
D81-3888 SHT.6 MAIN CONTROL PANEL-BACK PANEL & BILL OF MAT'L
```

Other files might have several pieces of information on a single line. When this is the case, you use a file record delimiter to separate the data. A common delimiter is the comma ",." This is the CDF format. A comma delimited file looks like:

```
PIPE-1,12.5,3000,1800,0,134.0
PIPE-2,14.5,2400,1200,0,122.0
```

Comma delimited files are processed by parsing the string and searching for the commas. You use a WHILE NOT EQUAL "," expression in AutoLISP to do this.

```
(while (/= "," testchar)
   ... get the next character
);while
```

You also can use a space as a delimiter, but only for files where spaces have no significance. You can use space delimiters exactly like a commas. You use AutoLISP to process the file and look for the spaces with a WHILE NOT EQUAL expression.

You also can use space delimiting to separate entries by one or several spaces to arrange your data in a column format. This is the SDF format, the old Fortran language standard for data files. Each entry is expected to start in one column and end some number of columns afterward. A column format file having fields that start at specific column locations looks like:

```
PIPE-1     12.5     3000 1800 0   134.0
PIPE-2     14.5     2400 1200 0   122.0
```

The file is processed by parsing the string at these predetermined tab spots. You put this in a FOREACH loop with a list of all the "tab" positions. Each time through the loop, you can chop the string by SUBSTR. Think of it as making the string eat itself.

Use the previous drawing. Try this exercise at the command line.

Handling File Formats

Type a string for variables substituting a <SPACE> for each • shown:

```
Command: (setq s "PIPE-1•••••12.5••••••3000•1800•0••134.0")

Command: (foreach tab '(11 10 5 5 3 5) (princ (read (substr s 1 tab))) (terpri)
1> (setq s (substr s (1+ tab)))
1> (prompt s) (terpri))
PIPE-1                              PROMPTs the first parsed value, a symbol.
12.5       3000 1800 0   134.0      The remaining string.
12.5                                Second value, a real.
3000 1800 0   134.0                 Remaining string.
3000                                Third value... etc.
1800 0   134.0
1800
0   134.0
0
134.0                               The string.
134.0                               And the last value.
Lisp returns: nil                   Returned from FOREACH
```

Other variations include using an asterisk "*" as the first character on a line to mark the data entry. An asterisk combined with a comma delimited file would appear in the data file as:

```
*PIPE-1
12.5,3000,1800,0,134.0
```

You have a choice of putting all the data on the next line after the key, or on the same line. If you put the data on two lines, AutoLISP will process it faster. You may recognize this format as the ACAD.PAT and ACAD.LIN file formats.

A format uniquely suited to AutoLISP is to structure your data as a LIST inside the string you read from a file. The READ function reads the first atom or expression in a string, so you can use it to convert the string to a list and then get its data directly with FOREACH. This is fast because it avoids parsing the individual string characters. It is simple because it avoids the need to count columns. Make a string as if you had read it from a file, then READ it and extract the data.

```
Command: (setq lst "(PIPE-1 12.5 3000 1800 0 134.0)")
Lisp returns: "(PIPE-1 12.5 3000 1800 0 134.0)"
Command: (setq lst (read lst))                          Convert to a list.
Lisp returns: (PIPE-1 12.500000 3000 1800 0 134.000000)

Command: (foreach s lst (princ s) (terpri))             Extract the data:
PIPE-1
12.500000
3000
1800
0
134.000000
Lisp returns: nil
```

Using the example list and the SET function, you can set symbols like PIPE-1 to variables in the list.

```
Command: (set (car lst) (nth 1 lst))    Set PIPE-1 to 12.500000.
Lisp returns: 12.500000
Command: !pipe-1                         Check it.
Lisp returns: 12.500000
```

In the next series of exercises, you will develop a Reference Drawing Command, called REFDWG. It presents a list of drawings for a project, asks the user to select which ones they want. Then, it enters the drawing names as text in the current AutoCAD drawing.

The example uses a simple pre-formatted data file. However, REFDWG gives special significance to the piece of information of the file. The project name must be the second line of the file. The first line is left blank to conform to database output restrictions of dBASE III. The first drawing name starts on the third line of the file. Subsequent drawing names follow on successive lines, one per line. Let's create this file now.

Creating a Reference Drawing Data File

 Use the REFDWG.TXT file.

 Create a data file called REFDWG.TXT

```
Command: ED
File to edit: REFDWG.TXT          Enter the following:

(Leave the first line blank, start on the second line.)
REFERENCE DRAWINGS - PAPER PLANT
D78-1042    SECONDARY FIBER SYSTEM PIPING
D78-1060    SECONDARY FIBER SYSTEM - SECTIONS
D81-3003    NO.3 PAPER MACHINE - WET END PROCESS FLOW DIAGRAM
D81-3204    NO.3 PAPER MACHINE - GENERAL ARRANGEMENT
D81-3211    NO.3 PAPER MACHINE - PROFILE VIEW
D81-3862    MOTOR AND CONTROL CONDUIT LAYOUT
D81-3862    NO.3 PAPER MACHINE MOTOR AND CONTROL CONDUIT LAYOUT
D81-3887    MAIN CONTROL PANEL STEEL LAYOUT & CUTOUTS
D81-3888 SHT. 1   MAIN CONTROL PANEL PUSHBUTTON WIRING
D81-3888 SHT. 2   MAIN CONTROL PANEL PUSHBUTTON WIRING cont.
D81-3888 SHT. 3   MAIN CONTROL PANEL I/O LAYOUT RACK #MC1
D81-3888 SHT. 4   MAIN CONTROL PANEL I/O LAYOUT RACK #MC2
D81-3888 SHT. 5   MAIN CONTROL PANEL FACE LAYOUT
D81-3888 SHT. 6   MAIN CONTROL PANEL-BACK PANEL LAYOUT & BILL OF MAT'L
```

Save, Exit and reenter AutoCAD.

You can generate this type of file easily with common database programs. In the chapter on LOTUS and dBASE file processing, we will show you a robust project drawing manager written in dBASE that generates a reference drawing text file.

The REFDWG Function

In the next part of the exercise, you make the function to read a file of reference drawings, list the drawings on the text screen, prompt the user to select the drawing names they want, then place the names in the current AutoCAD drawing in a Reference Drawing Schedule.

The main function name is REFDWG. This prints the drawing names to the screen and then calls two other functions, REFSEL to select the drawing names, and REFOUT to put the names as text in the AutoCAD drawing.

REFDWG begins by asking for the reference drawing list file name. It enters a giant COND structure. It tests if the file exists, or if it can be opened for reading. Then, REFDWG Flips to the text screen with TEXTSCR and clears it with the CLS function. It reads the first line in the input file. The first line is your header for the screen. REFDWG sets the cursor at the top left screen position 1 1, turns on BOLD mode and CENTERs the label on the first line. It reset the screen to normal and starts putting the reference names on the screen.

REFDWG initializes the start row, column and count, (65 is the letter A). Instead of numbers, the function uses capital letters because you can get more than 10 selections with only one character from the user. There are only 25 lines, but there are 26 letters. The program loops until the last line of the file is read. As each item is read, it is appended to a master list of drawings. A letter is placed before each name on the screen so that you can tell the user how to pick the drawing names. The function uses STRCAT with the CHR to convert a number to the letter equivalent. The function starts at 65 to get an uppercase "A." Each time through, the "row" and "count" variables are incremented. When it is done reading the file, REFDWG closes the file, moves the cursor down 2 rows and stores the current position.

Creating REFDWG

Copy REFDWG.LSP to TEST.LSP.

Create a new LISP function file called TEST.LSP.

Select [EDIT-LSP]   Edit TEST.LSP, and enter:

```
;* C:REFDWG is a command which displays a file of project reference drawings,
;* permits the user to make any number of selections from the drawing list
;* and then formats and enters the drawings as text in AutoCAD. There is a
;* max limit of 15 drawings per input file. It requires a fixed height text style.
;* The FILELIB.LSP, ANSILIB.LSP and UFUNS.LSP files must be loaded.
;*
(defun C:REFDWG ( / fp label row col count item itlist)
   (setq file (ustr "Reference drawing file name" nil nil) ;get the input file
         file (ffname file "TXT"))       ;deletes extension adds ours
   );setq
   (cond
      ((setq fp (open file "r"))         ;get the file handle
       (textscr)                         ;flip to the text screen
       (cls)                             ;clear it - cursors at upper left cell

       (read-line fp)                    ;skip first line (for dBASE)
       (setq label (read-line fp))       ;read second line of file
       (bold)                            ;turn on high intensity characters
       (center label)                    ;center and print project label to screen
       (normal)                          ;reset character attribute to normal

       (setq row 3                       ;initialize cursor positioning variables
             col 2
             count 65                    ;and the first letter for entry ids
       );setq

          (while (setq item (read-line fp))  ;start a loop, read each line of file
             (goto row col)                  ;move to screen position
                (prompt (strcat (chr count) ". " (substr item 1 75))) ;print letter and entry
```

```
            (setq itlist (append itlist (list item))   ;make a list of all entries
                  row (1+ row)                ;increment row and column
                  count (1+ count)
            );setq
        );while
        (close fp)                            ;close the input file
        (bold)
        (savepos)
        (goto 24 1)
        (center "DO NOT FLIP THE SCREEN!")    ;add warning
        (restpos)
        (normal)
        (goto (+ row 1) 2)                    ;move down 1 row
        (savepos)                             ;save the current cursor position
        (setq slist (refsel itlist))          ;call the function to select entries
        (refout slist label)                  ;send to output function
      );only if the file was found!
      (t (prompt "\nFile not found."))
    );end of condition
    (princ)                                   ;finish cleanly
);defun
```

Save what you've done.
Even if you have the CA DISK, temporarily disable two lines with ";*" so they look like:

```
;*        (setq slist (refsel itlist))       ;call the function to select entries
;*        (refout slist label)               ;send to output
```

Save, Exit and reenter AutoCAD. TEST.LSP should load.

Command: **(load "ANSILIB")**   If not loaded, it must be loaded.
Command: **(load "FILELIB")**   Make sure it's loaded.
Command: **(load "UFUNS")**   If not loaded, it must be loaded.

Command: **REFDWG**
Reference drawing file name: **REFDWG**   A bold title and an alphabetically coded list appears.

REFSEL, the next part of the function, loops, waiting for the user to identify each drawing they need. The user enters a letter. The program converts the letter to an integer and tests if it is within the range of "A" through the last item.

If the input is valid, the program puts a bold asterisk "*" to the left of item in the 1st column. The program checks to see if it is a duplicate entry. If it is not, the selected name is put on the selected list. After each selection, the prompt is reissued. The prompt first restores the cursor to its previous place, saved in the REFDWG function, and it deletes to end of line with "\e[K."

The REFDWG Function 14—29

If the input is invalid, the program gives a warning message. The program uses GETSTRING to pause for input and clears the error. The "\e[K" clears out the line before the prompt. After all drawings are marked, the program sets the screen to NORMAL and clears it with CLS.

Select **[EDIT-LSP]**
Even if you have the disk, remove the first ";*" from the (setq slist...) line:

```
        (setq slist (refsel itlist))        ;call the function to select entries
```

Now add the following:

```
(defun refsel(itlist / ltr ltrs elemno mnulist pt txtst)
   (cond                                    ;test for list of entries
      (itlist
         (bold)                             ;set high intensity on
         (while (/= 0                       ;test for null input from user
            (setq ltr (ascii (strcase       ;converts to upcase and then to integer
               (getstring "\e[u\e[KEnter letter or RETURN: "))))
;restore cursor position and prompt
            );setq
            (if (and (> ltr 64) (< ltr count))   ;check range of letter from user
               (progn
                  (setq elemno (- ltr 65))       ;find which one they selected
                  (goto (+ elemno 3) 1)          ;goto that element line
                  (prompt "*")                   ;mark it with asterisk
                  (setq addelem (nth elemno itlist))    ;extract marked element from list
                  (if (not (member addelem mnulist))   ;check for duplicates
                     (setq mnulist (append mnulist (list addelem))) ;add it to selected list
                  );if
               );progn
               (getstring "\e[u\e[KInvalid selection. RETURN to continue. ")
;not in selection range
            );if
         );while
         (normal)                           ;reset character attributes
         (cls)                              ;clear screen
      );itlist exists
   );cond
   mnulist                                  ;return selected list of entries
);defun
```

Save, Exit and reenter AutoCAD. TEST.LSP should load.

```
Command: REFDWG
Reference drawing file name: REFDWG

Enter letter or RETURN:          The list and this prompt appear. Enter several letters.
```

Your screen should look like the Marked Reference drawing.

14—30 Customizing AutoCAD

```
           REFERENCE DRAWINGS - PAPER PLANT
    A. D78-1042  SECONDARY FIBER SYSTEM PIPING
   *B. D78-1068  SECONDARY FIBER SYSTEM - SECTIONS
    C. D81-3003  NO.3 PAPER MACHINE - WET END PROCESS FLOW DIAGRAM
    D. D81-3204  NO.3 PAPER MACHINE - GENERAL ARRANGEMENT
   *E. D81-3211  NO.3 PAPER MACHINE - PROFILE VIEW
   *F. D81-3862  MOTOR AND CONTROL CONDUIT LAYOUT
    G. D81-3862  NO.3 PAPER MACHINE MOTOR AND CONTROL CONDUIT LAYOUT
    H. D81-3887  MAIN CONTROL PANEL STEEL LAYOUT & CUTOUTS
    I. D81-3888 SHT. 1  MAIN CONTROL PANEL PUSHBUTTON WIRING
   *J. D81-3888 SHT. 2  MAIN CONTROL PANEL PUSHBUTTON WIRING cont.
    K. D81-3888 SHT. 3  MAIN CONTROL PANEL I/O LAYOUT RACK #MC1
   *L. D81-3888 SHT. 4  MAIN CONTROL PANEL I/O LAYOUT RACK #MC2
   *M. D81-3888 SHT. 5  MAIN CONTROL PANEL FACE LAYOUT
    N. D81-3888 SHT. 6  MAIN CONTROL PANEL-BACK PANEL LAYOUT & BILL OF MAT'L

   Enter letter or RETURN:



                        DO NOT FLIP THE SCREEN!
```

Marked Reference Drawing

The final function, REFOUT, asks for the starting point of text and the style name. The style must be a fixed height style. The function loops through each element of the selected drawing list and puts the text on the screen.

Select **[EDIT-LSP]**
Even if you have the disk, remove the first ";*" from the (refout slist...) line:

      (refout slist label)   ;send to output

Now add the following:

```
(defun refout (slist label / pt txtst $hlite)
   (cond
      (slist
         (graphscr)                                       ;flip to graphics
         (setq pt (upoint 1 "" "Starting point" nil nil)  ;get text alignment point
               txtst (ustr "\nText style" "STD1-8" nil)   ;and the style
         );setq
         (setq $hlite (getvar "HIGHLIGHT"))               ;save setting
         (setvar "HIGHLIGHT" 0)                           ;turn off highlight
         (command "text" "s" txtst pt 0.0                 ;change text style
            (strcat "%%u" label "%%u") "text" ""          ;put on project label
         );command
         (foreach elem slist                              ;step through each entry
            (command elem "text" "")                      ;place text, restart command
         );foreach
         (command nil)                                    ;finish command - cancel
         (setvar "HIGHLIGHT" $hlite)                      ;reset highlight mode
      );slist is not nil
   );cond
);defun
(princ)
```

Save, Exit and reenter AutoCAD. TEST.LSP should load.

Select **[EDIT-MNU]**   Add the following to the EXAMPLES page:

[REFDWG]^C^C^C(if bold nil (load "ANSILIB")) (if upoint nil (load "UFUNS"));+
(load "DXF") (if C:REFDWG nil (load "REFDWG")) REFDWG

Save and Exit to AutoCAD. TEST.MNU reloads. Test it.

Select **[EXAMPLES] [REFDWG]**   The files load if needed. Then:

```
Command: REFDWG
Reference drawing file name: REFDWG
Starting point:                          Pick a point.
Text style <STD1-8>:                     Stuff scrolls by, and text goes in.

Command: QUIT
```

 Delete TEST.MNU. Delete TEST.LSP.

 Rename TEST.MNU to MY14.MNU. Rename TEST.LSP to REFDWG.LSP.

Your screen should look like Reference Drawing as AutoCAD Text.

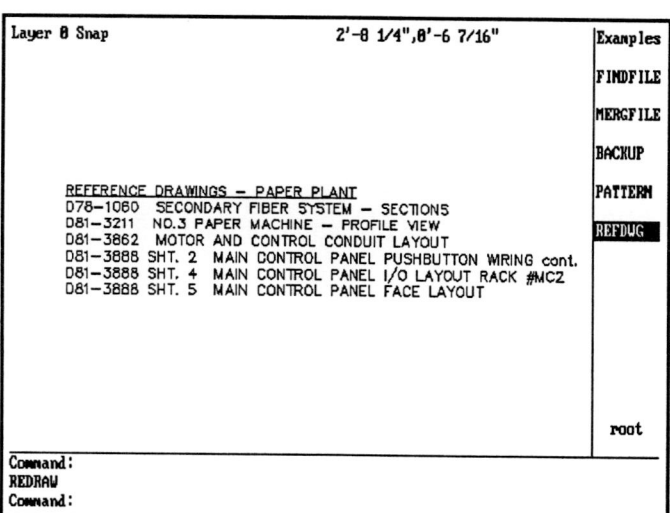

Reference Drawing as AutoCAD Text

Integrating the Example Menu

You do not need to integrate this chapter's menu into the CA-MENU. If any of these macros are useful, add them to your own menus. Here is the chapter's menu.

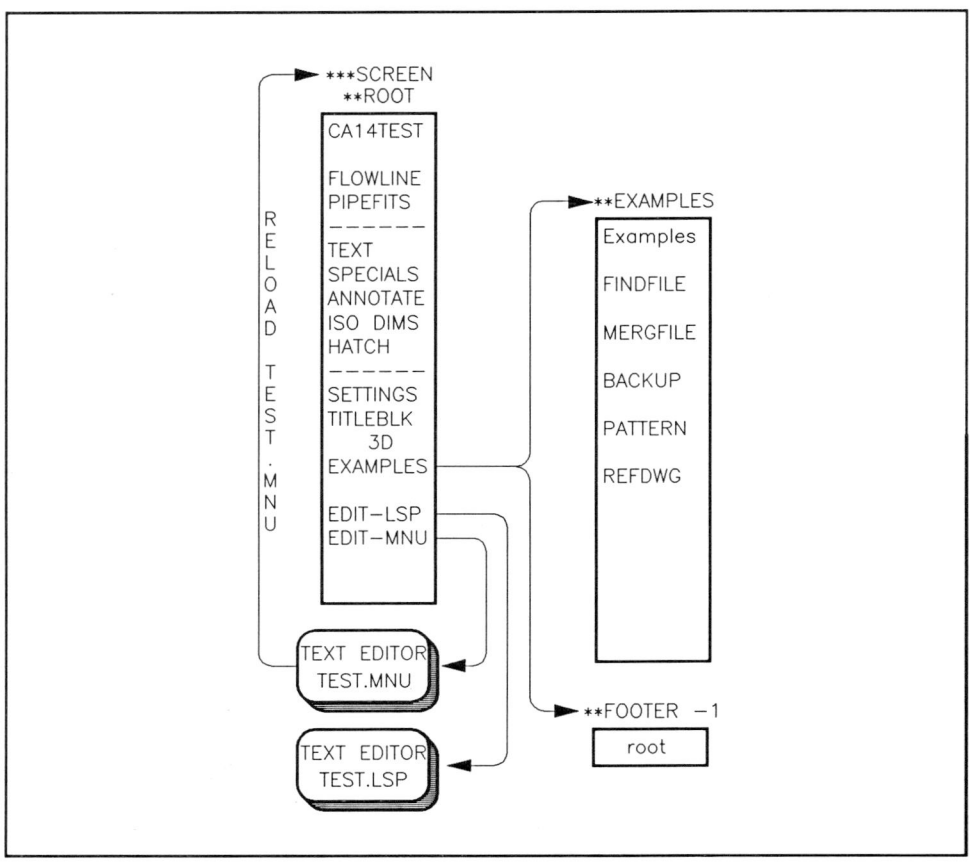

The Chapter 14 Menu

Tips and Techniques
Here are some tips and techniques for reading and writing files.

Writing to the screen

- Watch the FLIP Screen. It will destroy your ANSI formats.
- Consider getting input with GRREAD, then control the screen to give you more control over reading and writing to the screen.

Writing to the printer

- You can write to the printer like any other device, but we recommend copying to a file, then letting DOS handle the printing.

AutoLISP file access

- The formula is open, process and close.
- Use temporary files to clean out the process.
- Use $ to identify your temporary files so that you can find them.
- It is better to process small files.
- Remember to close your files.

DOS level controls

- ❑ Watch out for AutoLISP's append. Remember to use DOS Copy + when you are dealing with <^Z>s.
- ❑ Use CPATH to control directories. If your users change directories via SHELL, they can crash AutoCAD.

Automatic hatch generators

- ❑ You can expand the hatch generator program to a general purpose unit hatch generator.
- ❑ Consider adding continuous lines.
- ❑ Think about adding 60 degree angle increments.

File Formats

- ❑ There is no single best format for data files. Choose the format that fits your application best.

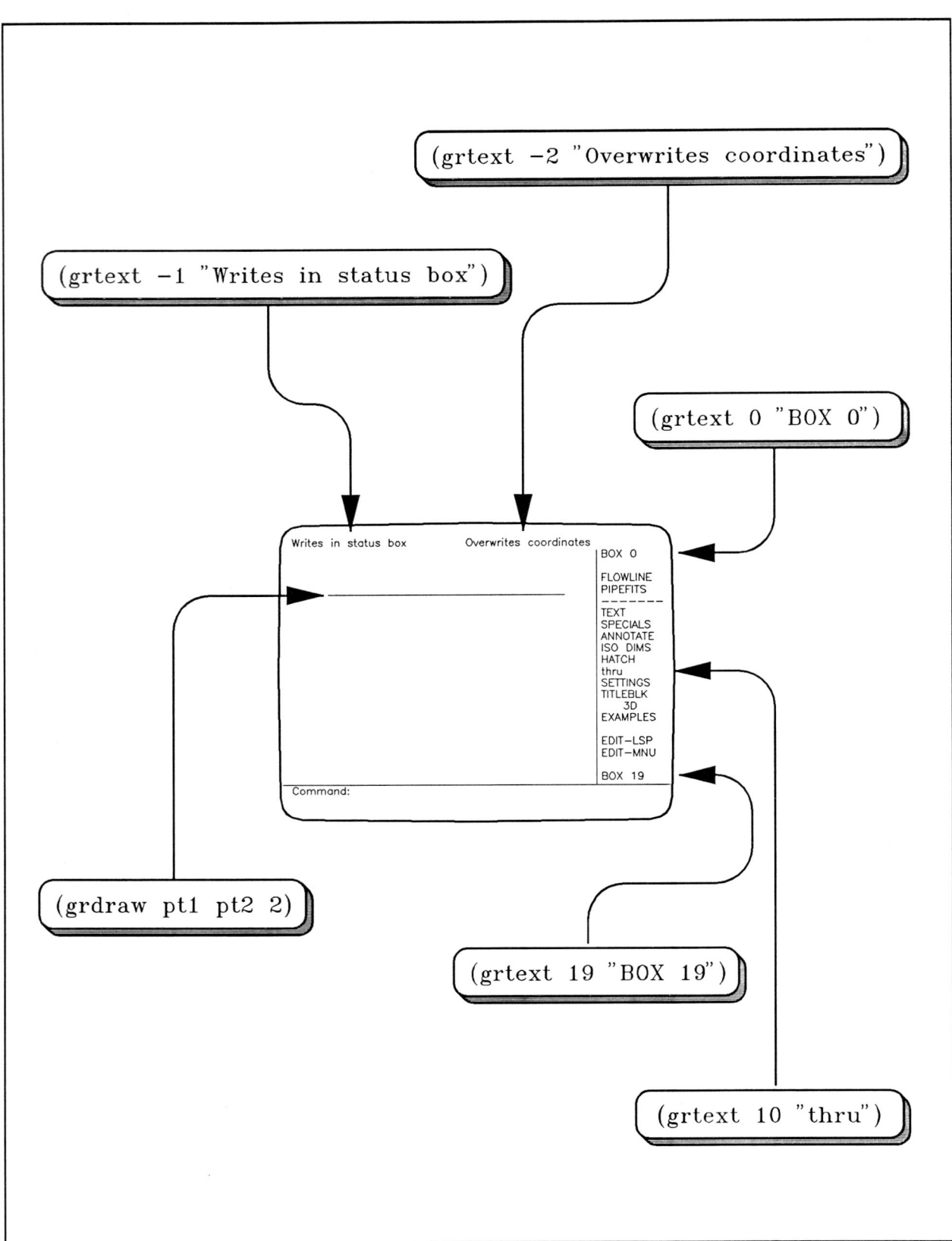

GRTEXT and GRDRAW Functions

CHAPTER 15

AutoLISP Device Access

AutoLISP gives you tools to take direct control of AutoCAD's text screen, graphics screen and other devices. Controlling these devices gives you greater control over interactions with your users. This chapter provides a program called DVIEW that shows you how to display a box around views defined in a drawing, and to display the view name on the status line. The chapter provides a second program called ETEXT, a single line text editor, that shows you how to integrate keyboard device access with ANSI screen formatting.

The Benefits of Using AutoLISP's Device Access

Using AutoLISP to access the graphics screen and other devices gives you more control over user input and output from your programs. Using device access is an alternative to command prompt responses, but you can use device access only under total AutoLISP control. This device access lets you:

- Control the graphics screen, screen menu labels and status line boxes to provide information to your user.

- Provide a dynamic way to label the screen for your program, quickly update labels, or to use temporary labels for your application.

- Augment 8 character screen menu labels by using status line boxes with 25 to 40 characters to provide more information to your users.

- Monitor and control user input by filtering the input devices.

How To Skills Checklist

This chapter shows you how to:

- ❑ Use GRTEXT to put custom labels in status boxes.
- ❑ Use GRDRAW to draw temporary lines on the screen.
- ❑ Use AutoLISP to make a window-box on the graphics screen, using GRTEXT and GRDRAW.
- ❑ Use GRREAD to monitor and control user input to your applications program.
- ❑ Use GRREAD to continuously track the digitizing cursor.

Macros, AutoLISP Tools and Programs

MACROS
[DVIEW] provides menu access to the DVIEW command.
[EditTEXT] provides menu access to the ETEXT command.
[DDRAW] provides menu access to the DDRAW command.

AutoLISP TOOLS
STRIP is a function that removes a specified entity association list from an entity data list.

PROGRAMS
C:DVIEW creates a dynamic view window by drawing a temporary view window on the graphics screen using GRTEXT and GRDRAW. It uses GRDRAW to display a box around the views defined in a drawing. It uses GRTEXT to display the view name on the status line. Each view is displayed sequentially in a continuous loop until the user selects a view or quits the program.
C:ETEXT is a single line text editor that uses GRREAD and the ANSI.SYS codes. The program lets you select a single line of text and edit it by moving the cursor within the text, adding or deleting characters.
C:DDRAW is a drawing aid that automates drawing lines at specified angles. It switches the snap rotation angle, based on the closest increment to the user's selected point. The program displays a compass icon to assist the user in selecting the angle for each line.
C:XPOINT is a function to demonstrate alternative methods of marking key points on the graphic screen.

Dynamic Screen Labeling Using GRTEXT

GRTEXT is an AutoLISP function that displays strings of information in selected boxes of the screen device. AutoCAD uses two status line boxes and the screen menu boxes to provide information to the user. The AutoLISP GRTEXT function give you access to those boxes. The boxes are:

BOX NAME	BOX NUMBER	MAXIMUM CHARACTERS	HIGHLIGHT FEATURE
STATUS	-1	40	NO
COORDS	-2	25	NO
MENU	0 to 19+	8	YES

GRTEXT takes two arguments: the box number, and the prompt string you want to display. The status box numbers listed in the table are typically fixed but vary with the type of video monitor that you use. The number of screen menu boxes is dependent on the number of boxes available for the screen type. The are numbered starting at 0. Be careful if you can't anticipate what screen your users will use. The Color Graphic Adapter card (CGA standard) and many other monitors only have 20 or 21 menu boxes.

Let's start by writing a message in each box.

Using GRTEXT to Place Text on the Graphics Screen

```
Enter selection:                                    Begin a NEW drawing named CA15.

Command: (grtext -1 "Writes in status box")         A -1 writes to the status box.
Lisp returns: "Writes in status box"
Command: <Ortho off><Ortho on>                      Any toggle wipes it.

Command: (grtext -2 "Overwrites coordinates")
Lisp returns: "Overwrites coordinates"              Move the cursor to wipe it.

Command: (grtext 0 "Box 0")                         Writes to first menu box.
Lisp returns: "Box 0"

Command: (grtext 10 "thru")                         Writes to 11th menu box.
Lisp returns: "thru"

Command: (grtext 19 "Box 19")                       Writes to 20th line. The last line on a CGA screen.
Lisp returns: "Box 19"
```

Your screen should look like the AutoCAD Graphics Screen Boxes.

```
┌─────────────────────────────────────────────────────────────────────┐
│ Writes in status box        Overwrites coordinates    │Box 0        │
│                                                       ├─────────────┤
│                                                       │FLOWLINE     │
│                                                       │PIPEFITS     │
│                                                       ├─────────────┤
│                                                       │TEXT         │
│                                                       │SPECIALS     │
│                                                       │ANNOTATE     │
│                                                       │ISO DIMS     │
│                                                       │HATCH        │
│                                                       │thru         │
│                                                       │SETTINGS     │
│                                                       │TITLEBLK     │
│                                                       │   3D        │
│                                                       │EXAMPLES     │
│                                                       ├─────────────┤
│                                                       │EDIT-LSP     │
│                                                       │EDIT-MNU     │
│                                                       ├─────────────┤
│                                                       │Box 19       │
│                                                       │             │
├───────────────────────────────────────────────────────┴─────────────┤
│ Command: (grtext 19 "Box 19")                                       │
│ "Box 19"                                                            │
│ Command:                                                            │
└─────────────────────────────────────────────────────────────────────┘
```

AutoCAD Graphics Screen Boxes

AutoLISP can only print one STRING of data to each screen box. AutoCAD will truncate any string that is too long to fit. You may find that the COORDS box overwrites the tail of the STATUS box.

The life span of a GRTEXT label is short and unstable, particularly a label that you put in the COORDS box. AutoCAD actively uses the STATUS and COORDS boxes. Each time a user changes layers, flips a toggle, picks a point or moves the cursor, AutoCAD updates the respective boxes. There is no sensible way to control the STATUS box, but you can freeze the COORDS box by turning AutoCAD's coordinate settings OFF. We don't recommend doing this as standard practice. We recommend using these boxes only for short-lived information.

It is safer to use the screen menu boxes, numbers 0 thru 19+, for dynamic screen labeling. AutoCAD doesn't use these boxes during the drawing process. The application developer has nearly total control over screen menu labels. Screen menu boxes are good for informing the user of current program settings. For example, a parametric program might tell the user the current pipe fitting size, or the current snap rotation angle.

Screen menu boxes are updated only by GRTEXT, or when the screen page is changed. Since screen menus are changed only by your programs, or your users, you can control screen menu boxes.

Some video monitors refresh the screen boxes each time the graphics screen is redrawn. Others refresh individual boxes after they are highlighted. This causes a loss of the label information. You need to test your screen to see if your screen labels get wiped. Test your screen by running the cursor along the screen menu and then redraw your screen. If the labels get wiped, you will have to plan on using labels more sparingly. We will show you how you can still use them.

Dynamic labels are easily removed from the screen and the original contents restored. To restore the screen to original labels, call GRTEXT with no arguments. Test REDRAW and highlight, then clear with GRTEXT:

Command: **REDRAW** Move the highlight menu bar up and down. Still OK?

Command: **(grtext -1 "Writes in status box")** Write it again.
Lisp returns: "Writes in status box"

Command: **(grtext)** Restores all to original.
Lisp returns: nil

You can highlight text sent to status and screen boxes in the same way that AutoCAD does. Your can use highlighting for tutorial purposes, for example, instructing the user to "follow the bouncing box."

You do highlighting by including an optional argument in the GRTEXT expression. There are two modes:

    0 is de-highlight
    1 is highlight

Another inconsistency among video monitors, particularly with ADI drivers, affects how you highlight screen boxes. Because monitors use various ways to handle the reversing of text, you need to use a two step process to highlight text in screen boxes. First, send the string in the unhighlighted state, then send the same string with highlight turned on.

Command: **(grtext 0 "Brighter")** With no highlight.
Lisp returns: "Brighter"

Command: **(grtext 0 "Brighter" 1)** Try it with a highlight.
Lisp returns: "Brighter"

The effect is shown in the screen shot of Highlighted and Unhighlighted Menu Boxes.

```
Layer 0 Ortho Snap                320'-0",0'-0"            Brighter
                                                           FLOWLINE
                                                           PIPEFITS

                                                           TEXT          Brighter
                                                           SPECIALS
                                                           ANNOTATE      FLOWLINE
                                                           ISO DIMS      PIPEFITS
                                                           HATCH
                                                                         TEXT
                                                           SETTINGS      SPECIALS
                                                           TITLEBLK      ANNOTATE
                                                              3D         ISO DIMS
                                                           EXAMPLES      HATCH

                                                           EDIT-LSP      SETTINGS
                                                           EDIT-MNU      TITLEBLK
                                                                            3D
                                                                         EXAMPLES

                                                                         EDIT-LSP
Command: (grtext 0 "Brighter" 1)                                         EDIT-MNU
"Brighter"
Command:
          Command: (grtext 0 "Brighter")
          "Brighter"
          Command:
```

Highlighted and Unhighlighted Menu Boxes

It may work on your screen in one step, but it is better to play safe to ensure your programs work on other screens.

Using AutoLISP's GRDRAW, you can draw directly on the screen, just like AutoCAD.

GRDRAW: Drawing on the Screen

You will often find drawing temporary straight vectors helpful in informing your user, like putting temporary markers on the screen. AutoLISP can draw vectors between any two points on the screen, in any of your monitor's supported colors. The points are in real world 2D coordinates. Don't try to draw complex images with GRDRAW. Look for alternatives.

Like screen box highlighting, AutoCAD can highlight a temporary vector. You've seen the effects each time you make a selection set of entities. AutoCAD indicates which objects are in the selection set by dashing them. (Some monitors use other methods.) Entity highlighting occurs only if the HIGHLIGHT mode is enabled. The GRDRAW function has an optional argument to invoke the highlighting effects of the graphics monitor. To highlight the temporary vector, give the fourth argument a number greater than zero.

Set two points near the top left and top right of the screen, along the same horizontal vector. Draw the temporary vector in yellow with GRDRAW. Then, try drawing the same vector green and highlighted.

Using GRDRAW to Draw Vectors

`Command: <Ortho on>`	
`Command: (setq pt1 (getpoint "Point: "))`	Set a point.
`Point:`	Pick upper left.
`Lisp returns: (300.000000 2000.000000)`	
`Command: (setq pt2 (getpoint "Point: " pt1))`	Set another point, using basept PT1.
`Point:`	Pick upper right.
`Lisp returns: (3000.000000 2000.000000)`	
`Command: (grdraw pt1 pt2 2)`	Give GRDRAW points and a color (2 is yellow).
`Lisp returns: nil`	
`Command: (grdraw pt1 pt2 3 1)`	Draws it green and highlighted.

You've seen AutoCAD "reverse" the color of rubberband and window lines as they overlap other entities in the drawing. GRDRAW also can reverse colors. Color reversal occurs only if an entity is overdrawn, and if GRDRAW is given a -1 color argument. The reversed color pairs depend on your video system. Try it.

`Command: LAYER`	Set color to BLUE.
`Command: LINE`	Trace over part of GRDRAW's vector.
`From point: !PT1`	First point.
`To point:`	Pick point with Ortho On.
`Command: REDRAW`	To clear previous GRDRAW vector.
`Command: (grdraw pt1 pt2 -1)`	Turns the BLUE part to some other color.
`Lisp returns: nil`	

You can seemingly "erase" and "undraw" GRDRAW vectors and real lines by drawing over them using black, color 0.

GRCLEAR is the last, and the least of AutoLISP's graphics screen functions. It clears the graphics screen like a blank slide. It doesn't take any arguments. Try a GRDRAW black vector, then GRCLEAR.

`Command: (grdraw pt1 pt2 0)`	Black it out.
`Command: REDRAW`	Brings the blue line back.
`Command: (grclear)`	All's clear.
`Lisp returns: nil`	
`Command: REDRAW`	Brings it back.
`Command: QUIT`	

It is time to use GRDRAW and GRTEXT in a program called DVIEW.

Making a Dynamic View

DVIEW (Dynamic VIEW) uses GRDRAW to display the bounds of defined Views, and uses GRTEXT to display their names. It rotates dynamically through the views until you pick one, or quit.

DVIEW uses symbol table access to look up each view name in the drawing. It gets the view name, height and width. Then, it takes half the values because Views are stored with their center values. It determines the aspect ratio, ARATIO, of the screen based on the number of pixels along the X and Y directions. The SCREENSIZE system variable reports the pixel limits for the monitor. Our screen size is (570.000000 410.000000) and our aspect ratio is 410.0/570.0, or 0.719298. Your screen size and aspect ratio may differ. The program applies the aspect ratio to either the height or width so that the GRDRAW window box is drawn in the correct proportion.

The program retrieves the center point of the view and calculates the four corners of the box. It then draws the box, using white vectors and writes the name of the view in the layer status box.

DVIEW asks the user if they want to "Quit," see the "Next" view or "Go" to the current choice. If "Go" is selected, the drawing view is switched to their current choice. If another view is requested, the program overwrites the previous window box by overlaying another box drawn in black.

At the end of the program, it overwrites the last window box and calls GRTEXT to clean up the status and menu box areas.

A Dynamic VIEW Command Using GRDRAW and GRTEXT

Enter selection: Edit any of your own EXISTING drawings.

 Copy CA15.MNU to TEST.MNU.

 Copy MAIN.MNU to TEST.MNU.

Command: **MENU**   Load TEST.MNU.

 Copy DVIEW.LSP to TEST.LSP.

 Create a new LISP function file called TEST.LSP.

Select [**EDIT-LSP**]   Edit TEST.LSP, and enter:

```
;* C:DVIEW draws a window box around each view in a drawing and
;* displays its name in the layer status box. It prompts for Next view,
;* Quit, or Go to the currently indicated view.
;* Requires DXF and UFUNS.LSP to be loaded.
(defun C:DVIEW ( / tbdata vname Y X CP P1 P2 P3 P4 V)
   (while (and (/= V "Quit")(/= V "Go"))
     (if (setq tbdata (tblnext "VIEW" T))           ;checks to see if any view exists
        (while (/= tbdata nil)                       ;starts the loop
          (setq vname (dxf 2 tbdata)                ;extracts view name
              Y (* (dxf 40 tbdata) 0.5)              ;view height
              X (* (dxf 41 tbdata) 0.5)              ;view width
              $screen (getvar "SCREENSIZE")          ;screen pixel resolution
              aratio (/ (cadr $screen) (car $screen)) ;aspect ratio based on pixels
          );setq
          (if (> X (* Y aratio))                     ;corrects for proportions of
             (setq Y (* X aratio))                   ;the screen. Either X or Y
             (setq X (* Y (/ 1.0 aratio)))           ;will determine window maximum.
          );if
          (setq CP (dxf 10 tbdata)                   ;center point of view
              P1 (list (- (car CP) X) (- (cadr CP) Y)) ;lower left corner
              P2 (list (- (car CP) X) (+ (cadr CP) Y)) ;upper left
              P3 (list (+ (car CP) X) (+ (cadr CP) Y)) ;upper right
              P4 (list (+ (car CP) X) (- (cadr CP) Y)) ;lower right
          );setq
          (grdraw P1 P2 7 1)                         ;draws left vertical vector
          (grdraw P2 P3 7 1)                         ;draws top horizontal
          (grdraw P3 P4 7 1)                         ;draws right vertical
          (grdraw P4 P1 7 1)                         ;draws bottom horizontal
          (grtext -1 (strcat "View name: " vname))   ;prints name in layer status box
          (setq V (ukword 1 "Next Go Quit" "Display another VIEW? Quit/Go/Next" "Next"))
          (cond
             ((= "Next" V) (setq tbdata (tblnext "VIEW"))) ;gets next view from table
             ((= "Go" V)                             ;2nd cond, execute VIEW
                (progn
                   (command "VIEW" "R" vname)        ;change view
                   (setq tbdata nil)                 ;nils variable causing while loop termination
             ) );progn & 2nd cond
             ((= "Quit" V) (setq tbdata nil))        ;nils variable causing while loop termination
          );cond
          (grdraw P1 P2 0)    ; each of these draws
          (grdraw P2 P3 0)    ; a black vector over
          (grdraw P3 P4 0)    ; the current screen
          (grdraw P4 P1 0)    ; vectors. It's the only way short of a full redraw
        );while
     );if
  );while or
  (grtext)                      ;clears out the temporary screen labels
  (princ)
);defun C:DVIEW
;*
```

Save, Exit and reenter AutoCAD. TEST.LSP should load.

Select **[EDIT-MNU]**

Change the top to [CA15TEST] and add the following to the EXAMPLES page:

[CA15TEST]

```
**EXAMPLES
[Examples]
[]
[DVIEW    ]^C^C^C(if C:DVIEW nil (load "dview")) (load "dxf");+
(if ukword nil (load "ufuns")) DVIEW
[]
```

Save and Exit to AutoCAD. TEST.MNU reloads.

Command: **VIEW**	Make several named views, make some overlapping.
Select **[EXAMPLES] [DVIEW]**	Test it. The AutoLISP files load, then:

```
Command: DVIEW
Display another VIEW? Quit/Go/Next <Next>:          <RETURN> to look at a few views.
Display another VIEW? Quit/Go/Next <Next>:          The box draws and names display.
Display another VIEW? Quit/Go/Next <Next>: G        G to Go, and it restores:
VIEW ?/Delete/Restore/Save/Window: R
View name to restore: B                             Shows the view name here.
```

 Delete TEST.LSP.

 Rename TEST.LSP to DVIEW.LSP

Command: **QUIT**

The [DVIEW] didn't load DVIEW.LSP because it was already loaded from TEST.LSP. It will auto-load, since TEST.LSP has been renamed DVIEW.LSP. If the views are off screen as DVIEW rotates through views, DVIEW displays their names on the status line so that you can still pick the views.

Alternatives for Displaying Screen Information

AutoLISP device access isn't always the best solution for displaying information. You can sometimes use existing AutoCAD commands to make good temporary tools in AutoLISP programs. Plines are good temporary marker lines because they are continuous and you can erase them easily. Points are good entities to place marks. Adjust the point mode PDMODE and size PDSIZE to the style and size of the image you need. Look for alternatives when you program. If a simple Pline shows it, or X marks the spot, use them.

The following XPOINT command function is an example illustrating how to use AutoCAD drawing commands to inform your users in an AutoLISP program. XPOINT builds a simple set of points, and processes those points, using a temporary Pline to construct a frame for the user. The Pline command is started and passed each point from the user. As this process is going on, the functions builds a list of points.

15—10 Customizing AutoCAD

Having a point list, XPOINT resets the point display modes and processes each point PT. It uses PDMODE 3 for an X and size -8 (8% screen size). Each time an X is required, XPOINT places it at PT, do THE STUFF. It then deletes the point. The X is a quick way to keep users visually informed. THE STUFF in our example is a circle. In your own application, it can be anything you need.

If your display supports popups, you can load CA-MENU and try the POINTS icon menu under popup menu 1 to review the point modes. Then, reload TEST.MNU.

Let's make XPOINT.

X Marks the Spot XPOINT

Enter selection: Begin a NEW drawing named XPOINTS.

  Copy XPOINT.LSP to TEST.LSP.

  Create a new LISP function file called TEST.LSP.

Command: **MENU**   Load TEST.MNU.
Select **[EDIT-LSP]**   Edit TEST.LSP, and enter:

```
;* C:XPOINT is an illustrative function to demonstrate alternative methods
;* of marking key points on the graphics screen. It illustrates how core tools
;* of AutoCAD can assist the AutoLISP programmer. Both the Pline and Point
;* commands are used to mark the users progress during a custom command.
;* The "STUFF" that this function does can be whatever you want to accomplish.
;* It requires the UFUNS.LSP file be loaded.

(defun C:XPOINT(/ pt ptlist ent)
   (command "pline")                          ;start the pline to mark path of points
   (while (setq pt (getpoint "\nPoint: "))    ;start loop - issue prompt, get input
      (command pt)                            ;send point to command processor
      (setq ptlist (append ptlist (list pt))) ;add point to list
   );while
   (command "")                               ;end pline command
   (setq ent (entlast))                       ;save pline ename to erase later
   (setq $pdsize (getvar "PDSIZE")            ;store user's point settings
         $pdmode (getvar "PDMODE"))
   );setq
   (setvar "PDSIZE" -8)                       ;set point size to 8% screen
   (setvar "PDMODE" 3)                        ;mode to draw points as an X
   (foreach pt ptlist                         ;loops through each point
      (command "point" pt)                    ;X marks the point on the screen
      (if
         (= "Y" (ukword 1 "Y N" "Do you want a circle?" "Y")) ;If you want STUFF
            (command "erase" (entlast)        ;Then erase point and do stuff
               "" "circle" pt "d"             ;THIS IS THE STUFF
               (fix (/ (getvar "VIEWSIZE") 12));THE STUFF IS HERE
            )
            (entdel (entlast))                ;Else erase point & continue
      );if
```

```
  );foreach
    (entdel ent)                    ;ENTDEL is faster than Erase
    (setvar "PDSIZE" $pdsize)       ;reset point size and mode
    (setvar "PDMODE" $pdmode)       ;to previous values
    (princ)
);defun
;*
```

Save, Exit and reenter AutoCAD. TEST.LSP should load. Test it:

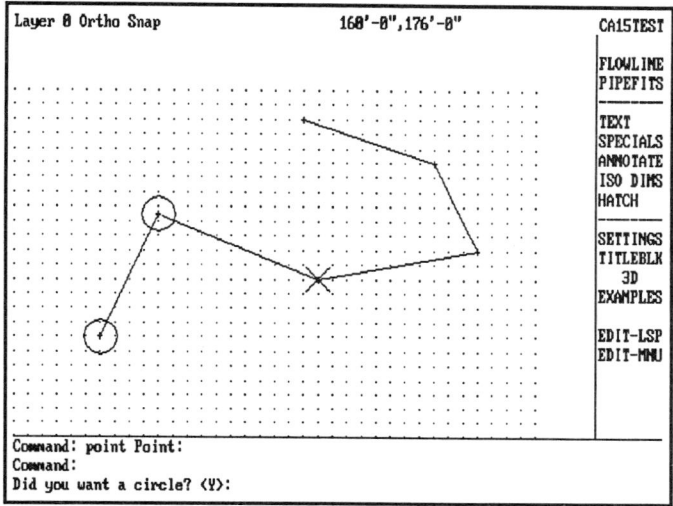

XPOINT in Progress

Command: **XPOINT**	
Command: Point:	Enter a few points.
Command: Do you want a circle <Y>: **Y**	Commands scroll. Circle is drawn.
Command: Do you want a circle <Y>:	Repeat with Y or N until through.
	The Pline should erase, leaving only circles.

  Delete TEST.LSP.

Rename TEST.LSP to XPOINT.LSP.

Command: **QUIT**

When you are done, your screen should look like XPOINT Completed.

```
Layer 0 Ortho Snap                 160'-0",176'-0"                    CA15TEST

                                                                      FLOWLINE
                                                                      PIPEFITS

                                                                      TEXT
                                                                      SPECIALS
                                                                      ANNOTATE
                                                                      ISO DIMS
                                                                      HATCH

                                                                      SETTINGS
                                                                      TITLEBLK
                                                                         3D
                                                                      EXAMPLES

                                                                      EDIT-LSP
                                                                      EDIT-MNU


Command:
Command: redraw
Command:
```

XPOINT Completed

GRREAD: Device Access for Input

AutoLISP can read input directly from any supported input device and tell you what the user is doing. GRREAD can detect when a user selects a point, enters a key from the keyboard, picks a screen menu item or picks something from the tablet menu. AutoLISP returns a list of a coded device number and the data entered by the user.

To see how this works, enter something at the keyboard, pick a point and select a menu item. Then, let's look at what the device codes mean.

Using GRREAD for Accessing Device Data

Enter selection: Begin a NEW drawing named CA15.

Select **[TEST] [24 x 36] [FULL] [INITIATE]**   It brings up TEST.MNU.

Command: **(setq char (grread))**   Enter an upper case "A".
Lisp returns: (2 65)   Notice no <RETURN> was needed.

Command: **(setq pt (grread))**   Pick a point. Your coords will vary.
Lisp returns: (3 (200.000000 2400.000000))

Command: **(setq mnu (grread))**   Select a screen item with your pointer.
Lisp returns:(4 10)

In each case, a LIST is returned. The first element of the list is the device code number. It provides the means for AutoLISP programs to determine the device from which the input originated. The device codes and returned data are:

DEVICE	CODE	DATA RETURNED
Keyboard	2	The ASCII character
Point Pick	3	Drawing coordinates
Screen menu Pick	4	Box number
Tracked Point	5	Tracked drawing coordinates
Button menu item	6	Button item
Tablet area 1	7	Box number
Tablet area 2	8	Box number
Tablet area 3	9	Box number
Tablet area 4	10	Box number
AUX1 menu	11	Box number
Pointer Button	12	Coord when button is picked
Menu by INS key	13	Box number

As you saw, the keyboard input "A" returned the ASCII number for the character A, the digitizer point pick returned a list of 2 reals, and the screen menu pick returned the box number selected. Each list is formatted so that CAR returns the device code, and CADR returns the data.

```
Command: (car char)
Lisp returns: 2                              The keyboard code.

Command: (cadr pt)
Lisp returns: (200.000000 2400.000000)       A point.
```

Here is a useful application of GRREAD that uses ANSI screen formatting codes. The ETEXT program is a single line text entity editor, done of course, in AutoLISP.

ETEXT An AutoCAD Text Editor

ETEXT begins with the user selecting the text entity to edit. It gets the data for the entity and determines the number of characters in the string. It formats the text screen using ANSI escape strings. It uses the ANSI codes in-line here instead of library functions from ANSILIB. This is a bit faster and more compact. ETEXT prints a headline, the cursor control key instructions, then the text string selected for editing.

The editing part of the program starts a loop using a WHILE that terminates when the user enters a <RETURN>. A COND test filters for the ASCII number of the key that has been pressed. For each directional key, a cursor position variable is updated and the cursor is repositioned on the screen. For the delete key, the edit string is parsed, the character removed and the string reprinted to the screen. Insertion of characters, for any valid printable character, does just the opposite. It adds characters to the string and reprints it.

Making ETEXT

 Copy ETEXT.LSP to TEST.LSP.

 Create a new LISP function file called TEST.LSP.

```
Select [EDIT-LSP]               Edit TEST.LSP, and enter:

;*;WARNING: ETEXT.LSP contains ; semicolons in strings - Do Not LSPSTRIP/S
;* ETEXT is a single line text editor allowing users to move the cursor
;* along the text and insert or delete character as required. ANSI.SYS
;* is required for program to work. DXF function must be loaded.
;*The screen cursor is moved via the keys:
;*
;*      Cursor Movements  KEY         Character Insert/Delete
;*      <CTRL LEFT> Arrow             Insert - just start typing
;*      <CTRL RIGHT> Arrow            Delete - <DEL> key
;*
;* DO NOT USE THE FLIP SCREEN <F1> KEY

(defun C:ETEXT ( / entity txt txtlen curpos key)
  (setq entity (entget (car (entsel "\nPick text to edit: "))))   ;Get text to edit
  (if (= "TEXT" (dxf 0 entity))                         ;Test if text was selected
    (progn                                              ;If text...
      (setq txt (dxf 1 entity)                          ;Get text string
            txtlen (1+ (strlen txt))                    ;Get text length
            curpos txtlen                               ;initial cursor position = text length
      );setq
      (textscr)                                         ;Flip to text screen

      (prompt "\e[2J")                                  ;clear screen and puts header info
(prompt
"\e[7;1;37;44m            Customizing AutoCAD Line Editor
"
)
(prompt
"\e[2;1H    <CTRL LEFT> moves left, <CTRL RIGHT> moves right, <DEL> deletes character    "
)
      (prompt "\e[0;37;40m")                            ;sets ansi screen attributes to normal

      (prompt "\e[4;1HEdit text:")                      ;Print title at line four
      (prompt "\e[1m")                                  ;bold
      (prompt (strcat
        "\e[10;1H       DO NOT HIT <F1> flipscr <INS> <HOME> <END> <PG UP> <PG DN>"
        "\n                    or Unshifted CURSOR KEYS!"
        )     )
      (prompt "\e[m")                                   ;normal
      (prompt (strcat "\e[6;1H" txt))                   ;Print text at line six
```

```
        (while (/= key 13)                          ;While key is not RETURN...
          (prompt (strcat "\e[6;" (itoa curpos) "H")) ;Position cursor
          (setq key (last (grread)))                ;Get key from keyboard
          (cond                                     ;Test key
            ((= key 243)                            ;If key was <CTRL LEFT> ARROW...
              (setq curpos (1- curpos))             ;Subtract 1 from cursor position
              (if (< curpos 1) (setq curpos txtlen))
;If cursor is less than 1 set cursor to text length
            );back arrow
            ((= key 244)                            ;If key was <CTRL RIGHT> ARROW...
              (setq curpos (1+ curpos))             ;Add 1 to cursor position
              (if (> curpos txtlen) (setq curpos 1))
;If cursor is greater than text length set cursor to 1
            );forward arrow
            ((or (= key 46) (= key 211))            ;If key was DEL...
              (setq txt (strcat (substr txt 1 (1- curpos)) (substr txt (1+ curpos))))
;Create new text string less the character at the cursor position
              (setq txtlen (1+ (strlen txt)))       ;Get new text length
              (prompt (strcat "\e[6;1H" txt " "))   ;Display new text string
            );delete mode
            ((and (> key 31) (< key 127))           ;If valid key...
              (setq txt (strcat (substr txt 1 (1- curpos)) (chr key) (substr txt curpos)))
;Create new text string including the new character at the cursor position
              (setq txtlen (1+ (strlen txt)))       ;Get new text length
              (setq curpos (1+ curpos))             ;Get new cursor position
              (prompt (strcat "\e[6;1H" txt " "))   ;Display new text string
            );entered a text character
            (T nil)
          );cond
        );while
        (setq entity (subst (cons 1 txt) (assoc 1 entity) entity))  ;Change entity list
        (entmod entity)                             ;Update entity
      );progn
      (prompt "\nEntity was not text")              ;Print bad pick message
    );if
    (prompt "\n\t\n\t\n\n\n\n\n\n\n\n\n\n\n\n\n\n\n\n\n\n\n\n\n") ;clear screen
    (princ)
);defun C:ETEXT
;*
```

Save, Exit and reenter AutoCAD. TEST.LSP should load.

Select **[EDIT-MNU]**   Add the following to the EXAMPLES page:

[EditTEXT]^C^C^C(load"dxf") (if C:ETEXT nil (load "etext") ETEXT

Save and Exit to AutoCAD. TEST.MNU reloads.

Command: **ZOOM**   Center, height 6".

Command: **TEXT**   Type "THE LAZY COW." as text to test.

Select **[EXAMPLES] [EditTEXT]**   Test it:
Hold down control key and hit left/right arrows as noted:

15—16 Customizing AutoCAD

```
Command: <CTRL LEFT>          Several to get to start of "COW"
Command: BROWN                Type "BROWN " -- DON'T TOUCH <INSERT> key.
Command: <CTRL RIGHT>         To delete "LAZY" move cursor on each character...
Command: <DEL>                ...and hit <DEL> key.
Command: JUMPED OVER THE FENCE.   To add, <CTRL RIGHT> to end and type.
Command: <RETURN>             A <RETURN> exits and updates the text in your drawing.
```

  Delete TEST.LSP.

  Rename TEST.LSP to ETEXT.LSP

A screen shot of the ETEXT line editor is shown below.

```
              Customizing AutoCAD Line Editor
    <CTRL LEFT> moves left, <CTRL RIGHT> moves right, <DEL> deletes character

Edit text:

THE BROWN COW JUMPED OVER THE FENCE



              DO NOT HIT <F1> flipscr <INS> <HOME> <END> <PG UP> <PG DN>
                          or Unshifted CURSOR KEYS!
```

The ETEXT Line Editor

GRREAD doesn't recognize the unshifted cursor keys which will cause a flip screen, so ETEXT uses control shifted cursor keys. If you feel this procedure is error prone and you hit the unshifted cursors accidently, you could substitute <TAB> (ASCII 9) for <CTRL RIGHT> (ASCII 244) and <BACKTAB> (shift-tab, ASCII 143) for <CTRL LEFT> (ASCII 243) in the code.

Continuous Coordinate Tracking

One unique capability of GRREAD is the ability to track the motion of the digitizer or mouse pointer as the user moves it. AutoLISP samples the motion and returns the current coordinates each time a call for a tracked point is made. You can request tracked points by including a NON-NIL argument in the GRREAD expression.

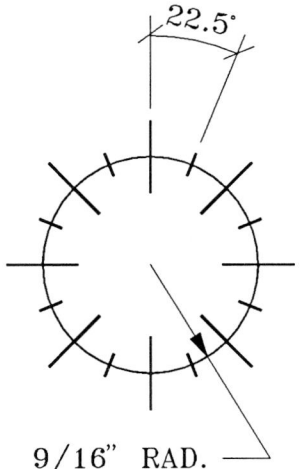

A Drawing Compass

Let's look at a program that tracks the digitizer coordinates and automatically rotates the snap angle to draw at preset angles. The program also illustrates that ENTMOD is faster to use than AutoCAD's Move command for modifying single entities. Try GRREAD at the keyboard first.

DDRAW An AutoRotate Snap

```
Command: (grread T)                                    Supply the argument True.
Lisp returns: (5 (1530.000000 1830.000000))            Immediately!
```

In programs that work with continuous tracking, a continuous WHILE loop is set up. The tracking task is made part of the test condition. It continuously samples the data. What is done with the data afterwards depends on your application, but the technique of continuous sampling remains the same.

DDRAW is a command function that automates drawing at preset angles. It is an angular snap function. There are two functions in the file, STRIP and DDRAW. STRIP is a simple support function, like DXF, but it removes an association sublist from the entity data list. DDRAW uses a simple block COMPASS as a visual indicator at each point when you draw. Use STRIP to remove the 10 dxf code group, the insertion base point of the COMPASS symbol. Each time a point is picked, the function APPENDs back a revised insertion base point and ENTMODs it to move the COMPASS symbol. The compass is shown in the diagram of a Drawing Compass.

 Copy DDRAW.LSP to TEST.LSP.

 Create a new LISP function file called TEST.LSP.

 You have the COMPASS.DWG for a block.

 Draw a compass as illustrated and Wblock it, base point at center, to COMPASS.DWG

```
Select [EDIT-LSP]                       Enter the STRIP function:
;* STRIP removes a specified association list from an entity data list.
(defun strip (code data) ;code is dxf code number, data is association list of data
   (cdr (subst (car data) (assoc code data) data))   ;strips an ASSOC group from entity list
);defun
;*
```

The main DDRAW function starts out by setting the angular resolution. In our case it is 22.5 degrees. DDRAW calculates a percentage of the screen height and uses the number to scale the COMPASS symbol. It inserts the symbol at the point first selected by the user and waits for the user to move the cross hairs outside of the compass boundary. Then, it sets the angle.

Two WHILE loops are set up. One to keep the command running and the other to allow sampling and point selection to alternate. The tracked value is compared to the center point of the compass. The "inside" loop continues until the distance condition has been satisfied. If the distance is greater than a certain amount, the program switches the snap rotation angle and prompts for a point to be entered.

The snap angle, snap base and ortho modes are adjusted for the new direction. A length of line is requested, but the user can hit a <RETURN> to reset the angle, or Quit. No matter how the user enters a length, the line is drawn in the direction of the selected rotation angle. The compass is moved using the ENTMOD and ENTUPD functions. If you try the Move command, you'll see that the function is amazingly swift compared to Move. It circumvents the selection set operations of AutoCAD and regenerates the entity in a different place.

Save what you've done and continue adding to TEST.LSP, above STRIP:

```
;* DDRAW automates drawing line segments at specific angles.
;* It flips the rotation between snap increments. DXF and UFUNS.LSP must be loaded.

(defun C:DDRAW ()
  (setq prmpt "\nQuit, RETURN to reset angle, or enter length.") ;a one-time prompt
  (setq anginc 0.392699)                           ;angle increment
  (setq maxd (* (getvar "VIEWSIZE") 0.15)          ;symbol height & max
        pt1 (upoint 1 "" "Starting point" nil nil)) ;critical distance
  );setq
  (command "insert" "compass" pt1 maxd "" 0.0 "line" pt1) ;puts in compass, starts line
  (setq ed (entget (entlast))                      ;gets the entity name
        esub (strip 10 ed)                          ;strips ent data of pt
  );setq
  (setvar "ORTHOMODE" 0)                           ;turn off ortho mode
  (setq more T)
  (while more                                      ;loop
    (setq inside T)                                ;inside of compass flag

    ;* get the dynamic rotation angle...
    (while inside                   ;loops until flag is nil
      (setq gd (grread t))          ;samples coordinates
      (if (= (car gd) 5)            ;if a real sampling (not a key press...)
        (if (> (distance pt1 (cadr gd)) maxd) (setq inside nil))) ;check if outside
    );if                                            ;circle, calc distance
    );while

    ;*set up the environment for the new angle, calc the angle...
    (setq ang (* (fix (+ (/ (angle pt1 (cadr gd)) anginc) (/ anginc 2.0))) anginc))
;constant increment
    (setvar "SNAPANG" ang)    ;change rotation angle , specific angle list
    (setvar "SNAPBASE" pt1)   ;reset base point
    (setvar "ORTHOMODE" 1)    ;turn on ortho mode

    ;*get length of line, draw it and move the compass symbol
    (if prmpt (progn (prompt prmpt) (setq prmpt nil)))   ;issue one-time prompt
    (setq d1 (udist 0 "Q" "Q/<RETURN>/<length>" nil pt1)) ;find out the length
    (cond
      ((numberp d1)                                 ;cond1 see if it is nil
        (setq pt2 (polar pt1 ang d1))               ;calc the new point
        (entmod (append esub (list (cons 10 pt2)))) ;changes compass
        (command (setq pt1 pt2))                    ;pass point to Line command
      );point cond1
      ((= d1 "Q")                                   ;restore snap & erase compass
        (command "" "snap" "r" "0,0" "0" "erase" (dxf -1 ed) "")
        (setq more nil inside nil)
      );quit
```

```
    );cond
     (setvar "ORTHOMODE" 0)                          ;turn off the ortho
  );while more
   (princ)
);defun
;*
```

Save, Exit and reenter AutoCAD. TEST.LSP should load.

Select **[EDIT-MNU]**   Add the following to the EXAMPLES page:

```
[DDRAW    ]^C^C^C(if C:DDRAW nil (load "ddraw")) (load "dxf");+
(if ukword nil (load "ufuns")) DDRAW
```

Save and Exit to AutoCAD. TEST.MNU reloads.

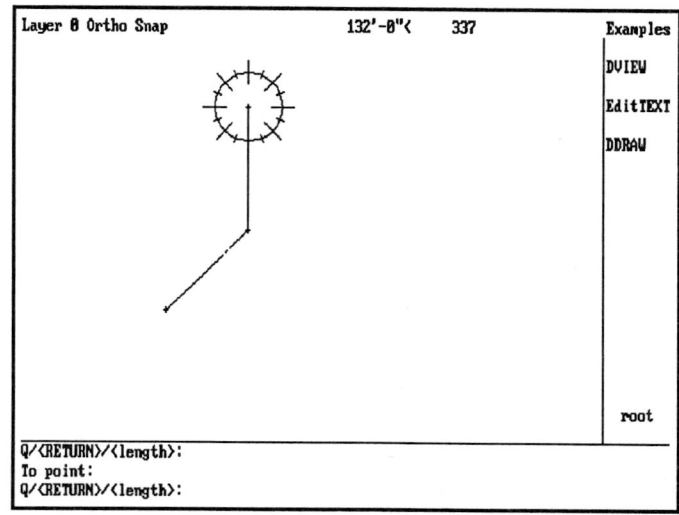

DDRAW Routine

Command: **ZOOM**   Center, height 12"

Select **[EXAMPLES] [DDRAW]**
Command: DDRAW   Test it. The COMPASS block must exist.
Starting point: Select a point.
Quit, RETURN to reset angle, or enter length.   A one-time prompt.
Q/<RETURN>/<length>: **10'**   Enter distances, and pick a few points.
To point:
Q/<RETURN>/<length>:   Try a <RETURN> to reset the angle.
To point:
Q/<RETURN>/<length>: **Q**   Q to Quit, and the COMPASS erases.

 Delete TEST.LSP and TEST.MNU.

 Rename TEST.LSP to DDRAW.LSP. Rename TEST.MNU to MY15.MNU.

Command: **QUIT**

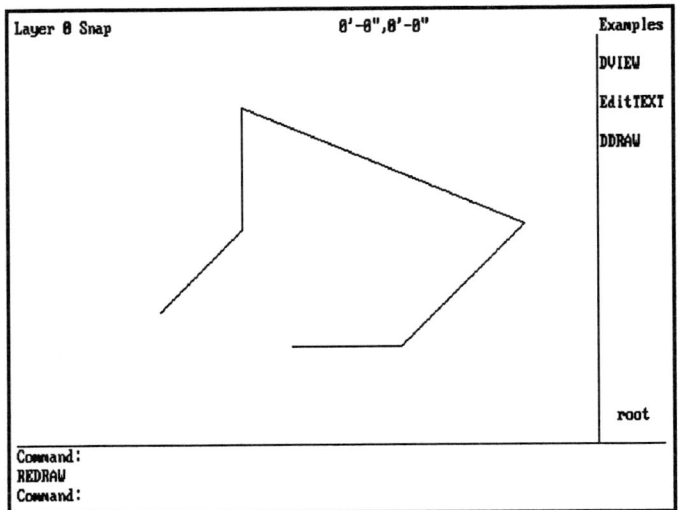

DDRAW Completed

If you prefer Plines, you can change the "line" to "pline" on the 6th line, or you could add a choice. You also could change the preset angle increments. Practice to get smooth with the function. It is like tightrope walking. Don't look down at the compass, look where you are going.

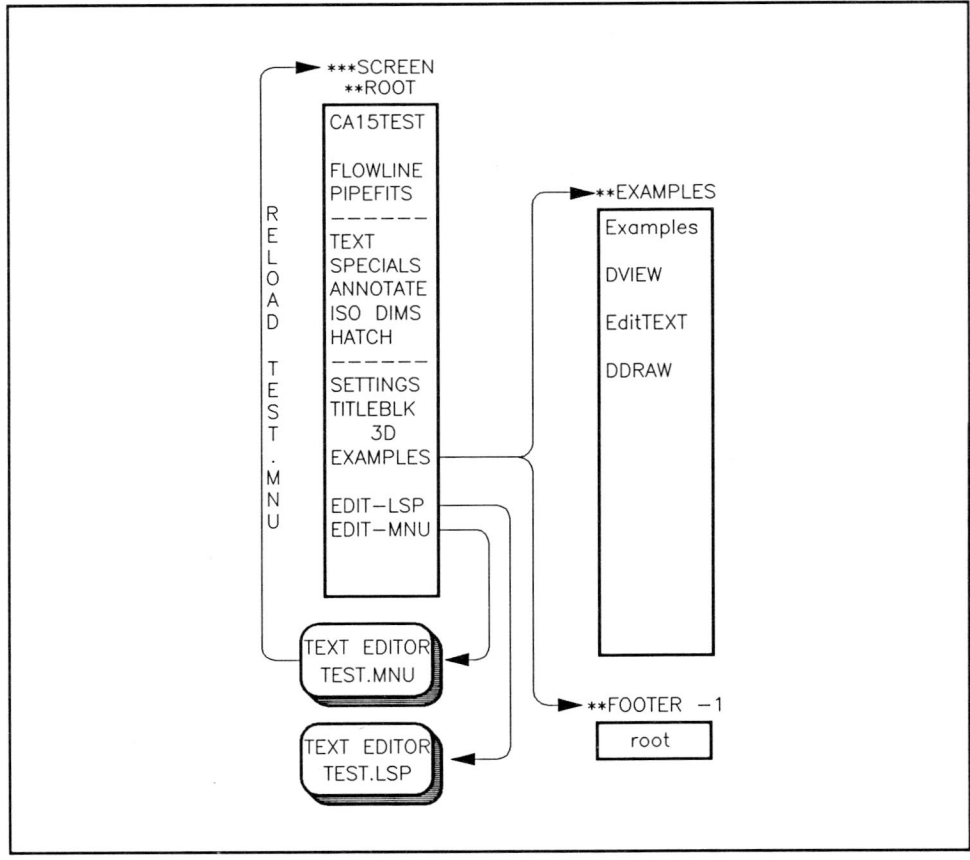

The Chapter 15 Menu

Integrating the Example Menu

You do not need to integrate this chapter's menu into the CA-MENU. If any of these macros are useful, add them to any of your own menus. The chapter's menu is shown on the previous page.

Tips and Techniques

Here are some summary tips and techniques for using GRTEXT, GRDRAW, and GRREAD.

GRTEXT

- ❏ You can't use ANSI formatting codes. You are accessing the graphics screen!
- ❏ Remember to control your normal status mode and coordinate display so that your user can't blow out your GRTEXT.

GRDRAW

- ❏ Use GRDRAW for simple temporary lines. It doesn't clutter up the drawing database.
- ❏ Use it for graphics marks like lines around a selection set, or marking a temporary grid.

Remember to consider combining GRTEXT and GRDRAW, like the chapter's window box program, DVIEW. You also can use these AutoLISP commands as screen refreshers in a multi-tasking environment. They restore the current graphics screen.

GRREAD

- ❏ DO NOT HIT <F1> flipscr <INS> <HOME> <END> <PG UP> <PG DN> or CURSOR KEYS when using ANSI.SYS. GRREAD does not control them. Control C <^C> is the only other exception to its control.
- ❏ Is the only way (currently) to get single stroke input.
- ❏ Is good for Yes/No input.
- ❏ Is a good candidate for writing tutorial programs with direct access.

Think about adding to the ETEXT program to make a full, multi-line text editor. You have the core tools.

This completes the book's basic tour through AutoLISP. The next section, PART V., provides a set of advanced AutoLISP applications, starting with developing an AutoLISP toolkit. These advanced applications use many of the routines that you have developed in this basic tour, and cover more AutoLISP functions.

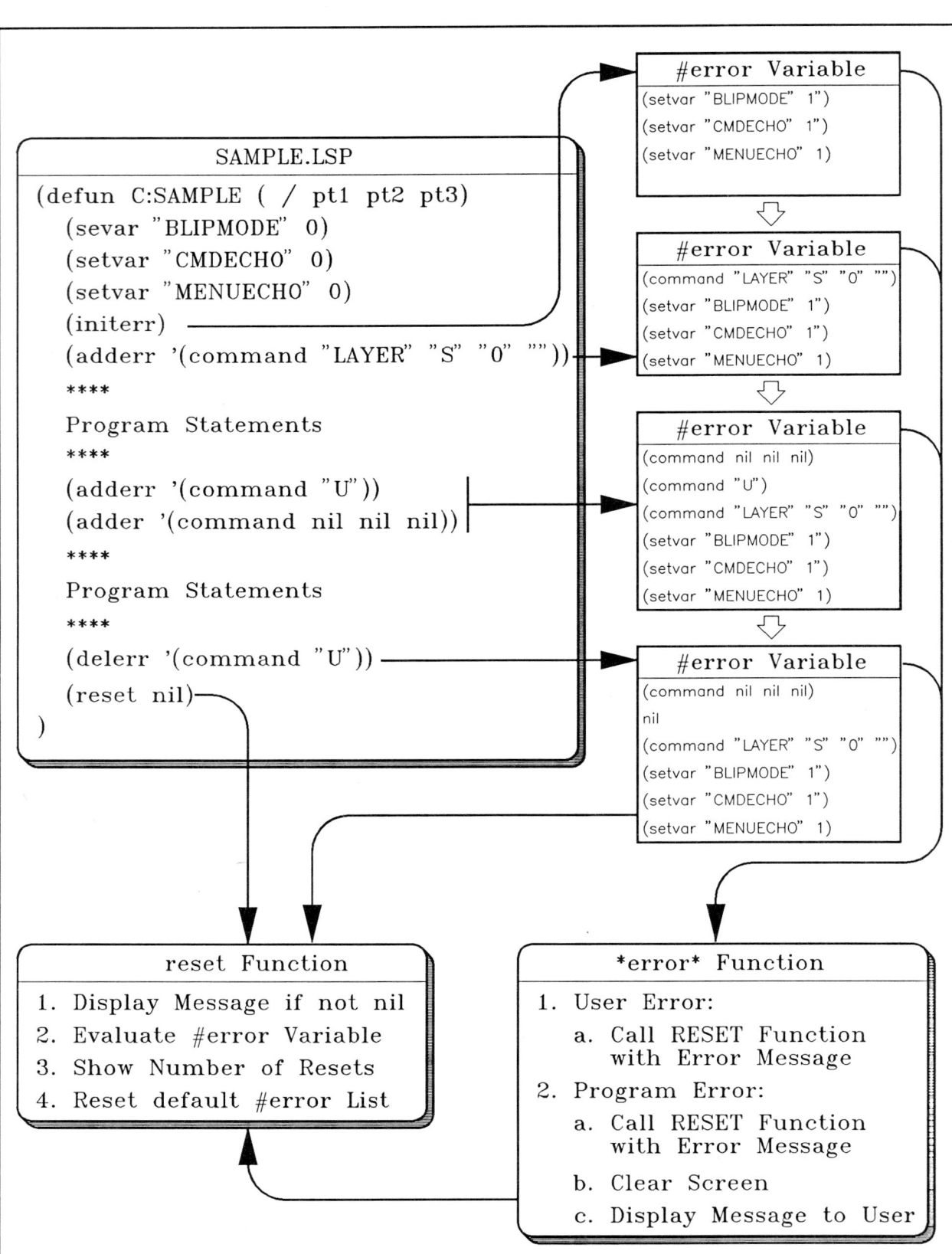

Error Recovery System

PART V. ADVANCED APPLICATIONS

CHAPTER 16

Developing an AutoLISP Toolkit

Every customizer eventually realizes the need for a standard set of routines. This need emerges from performing the same tasks over and over again. In the previous chapters, you have developed new tools and routines that you will continue to use in the remainder of the book. You are probably tired of loading DXF and UFUNS! This chapter shows you how to organize these routines and automate their loading.

Part of the management of these standard routines is knowing where to store them. AutoCAD provides the ACAD.LSP file to help you with this task. This chapter shows you how to use the ACAD.LSP file.

The Benefits of Building a Standard Set of Subroutines

There are several benefits to you, and to your users, in standardizing your AutoLISP routines.

- A common set of routines simplifies your AutoLISP programming. It pays to standardize your AutoLISP routines in a library, and not "re-invent the wheel" for each new routine.

- It pays to use VMON to manage AutoLISP memory. This lets you run larger programs. Part of developing a standard toolkit is to standardize your procedures, including memory management.

- A standard toolkit lets you maintain control of your applications environment. It lets you direct user errors and helps keep your users out of an uncontrolled environment.

- Using ACAD.LSP lets you provide a seamless integration for your user. It benefits your users to work in a cohesive system. Don't be content to just "get it to work." There is less tendency for your users to get lost if you put in extra effort to integrate your AutoLISP routines.

How To Skills Checklist
In this chapter, you will learn how to:

❑ Organize your AutoLISP functions into and use the ACAD.LSP file.

❑ Use VMON for AutoLISP memory management.

❑ Build error trapping functions, and use them in your AutoLISP programs.

Macros, AutoLISP Tools and Programs

MACROS
[S-LEADER] loads LEADTEXT.LSP and executes the S-LEAD leader and text command function. It shows how to handle errors in a function.

AutoLISP TOOLS
CA-ACAD.LSP is Customizing AutoCAD's ACAD.LSP file. It contains a collection of functions and expressions that are used in many of the book's application programs. These functions include expressions to set global variables to establish and control the editing environment. It also includes functions from DXF.LSP, DDRAW.LSP, ETOS.LSP, ANGTOC.LSP, and the User Interface Functions from the UFUNS.LSP file. Other functions are:
CALOAD manages the loading of files.
MLAYER checks to make sure a layer is defined.
ERROR evaluates errors and displays the appropriate message. It also evaluates a list of functions to restore the drawing environment when an error occurs. A variety of functions support the error handling to maintain the environment.
RESET is a function that uses the error function modules to restore the drawing environment after normal program completion.

PROGRAMS
LEADTEXT.LSP is a program that draws a straight line leader with an unlimited number of segments. The program calculates the leader text placement and justification. DTEXT lets the user add multiple lines of text to the leader. The program demonstrates the use of error handling.
C:S-LEAD draws a pline leader with as many segments as desired. The arrow head is incorporated as part of the pline.
LTEXT determines the text justification and initiates a DTEXT command to let the user place text.

 If you have the CA DISK, you need to copy CA-ACAD.LSP to ACAD.LSP for use in subsequent chapters. If you do not have the disk, you need to create ACAD.LSP.

The ACAD.LSP File
ACAD.LSP is a file used for AutoLISP programs and expressions. Each time AutoCAD starts up a new or existing drawing, it loads the ACAD.LSP file. AutoLISP automatically evaluates each item of the ACAD.LSP file. By using the ACAD.LSP file, you can insure that your drawing environment's initial

settings are established for your programs and menus to operate correctly. ACAD.LSP files commonly include:

- AutoLISP defuns.
- Immediately executed AutoLISP expressions.
- Setting of global variables and AutoCAD system variables.
- Initialization of error routines.
- Function Loader.
- Layer manager/definer.

AutoLISP cannot execute AutoCAD commands during the automatic initial load of ACAD.LSP. AutoLISP cannot execute commands because the AutoLISP environment and the ACAD.LSP file are loaded before the AutoCAD Command: prompt and commands are present in memory. Therefore, AutoCAD ignores any COMMAND functions, and gives an error, "error: incorrect request for command list data." Strangely, the error message may not appear until you execute your first command or function after the Command: prompt appears. However, the drawing database is available during this initial load. You can use entity and table access, and set system variables. Many developers use this access to retrieve setup information stored within the drawing.

Before gathering your standard routines into ACAD.LSP, let's take a look at AutoLISP memory management. It is vital to efficiency and to how many routines you put in the file.

Memory Management

Since the very early days of AutoCAD, it has been able to "page" drawing data out of memory to the disk. This "virtual memory" is the basic reason why AutoCAD drawings have never been limited by the amount of the computer's RAM. AutoCAD pages part of the drawing to disk each time memory is in short supply. AutoCAD extends its "paging" abilities to AutoLISP functions. It can page to EXTended or EXPanded RAM as well as to disk. There are several memory management functions in AutoLISP. The functions and what they do are:

```
VMON    activates the "paging" system for AutoLISP.
MEM     reports the amount of AutoLISP type memory available.
ALLOC   request more memory for use by AutoLISP.
EXPAND  increases the working space of AutoLISP into new memory.
GC      collects any "garbage", unused portions of memory.
```

ALLOC, EXPAND, and GC are functions used by AutoLISP to manage its use of memory, and are not for our use. GC reclaims small portions of memory, but AutoLISP automatically runs this function when it determines memory is tight, so you don't have to.

The only functions we will use are VMON and MEM. MEM is a simple function that tells you how much AutoLISP memory is available and what it is used for. Let's try MEM.

A Look at MEM

```
Enter selection:
```

Begin a NEW drawing named CA16.

 Copy CA16.MNU to TEST.MNU.

 Copy MAIN.MNU to TEST.MNU.

```
Command: (mem)
Lisp returns: Nodes:     2560
Free nodes:   145
Segments:     5
Allocate:     512
Collections:  24
Swap-ins:     0
Page file:    504
Free heap:    17480
```

```
       HEAP
  AutoLISP FUNCTIONS
   USER FUNCTIONS
      VARIABLES
        LISTS

       STACK
  PROGRAM POINTERS
  PROGRAM COUNTERS
   TEMPORARY VALUES
  FUNCTION ARGUMENTS
```

Heap and Stack Use

AutoLISP is a separate dedicated program that shares the computer with AutoCAD. Unfortunately, AutoLISP is restricted to 64K of "data." To AutoLISP "data" includes all your programs, variables and their data. You have control over up to 45K. Without virtual memory paging, you could easily run out of RAM.

VMON (Virtual Memory ON) extends the AutoCAD paging system to AutoLISP. To VMON, "data" means your variables and their values, and it does NOT page them. VMON recognizes only functions created by DEFUN as "programs." It swaps out infrequently used function definitions when more space is needed. A function eligible for page swapping is assigned a page ID number that AutoCAD and AutoLISP use to keep track of the function's state. Data, whether used by a function or just SETQed in an AutoLISP expression, is never swapped out. The benefit of the function paging is that it lets you run larger programs, and it makes room for more programs. Once VMON is on, it can't be turned off during the current session.

Unfortunately, good memory management is not as simplistic as just turning VMON on and forgetting it. One problem is that VMON won't help with memory that is used by variables and strings. Variable values and strings are data, not program, and are not paged. VMON paging is for user defined functions only.

The CLEAN Alternative

Some developers use a technique called CLEAN as an alternative to using VMON. The CLEAN technique involves manipulating the ATOMLIST. CLEAN is a user defined function that you use to reset the ATOMLIST to a subset of itself. All function and variable names not included in the reset ATOMLIST are discarded, releasing their memory. Typically, systems using CLEAN will call the function to clean the ATOMLIST before loading a new program for use. This means that the next time you want to use a function that CLEAN has wiped from the ATOMLIST, you have to wait for it to reload. You also must be careful to not clean out your standard subroutine functions and global variables. CLEAN is incompatible with VMON because access to the ATOMLIST is blocked when paging occurs. An example using a CLEAN function is shown in the Memory Management section of the AutoLISP Programmers Reference.

Memory Management Practices

Whether you use VMON or CLEAN, you need to use good management to get the most out of your memory. Good practices, include:

- Keep your variables local whenever possible. Reset variables to NIL when you do not need them. Make sure to CLOSE file handle variables first, or you will leave open files.

- Standardize your variable names and reuse them throughout all your programs.

- Keep variable and function names to six characters or less.

- Make one-shot routines as .LSP files that execute as they load. Reset their variables to NIL at the end of the file.

- Load large functions only when you use them. DEFUN the same name, such as ROUTINE, for all your large functions, executing them from a menu item.

- If you reuse function names, like ROUTINE, AutoLISP must have enough memory to hold for both the old and new definitions for a short time. SETQ ROUTINE to NIL at the top of each file before defining the new function for the name ROUTINE.

- If you set new values to existing variables that hold large data lists, or strings; set these variables to NIL first, then set the new values.

- Use LAMBDA for recursive operations.

We recommend activating VMON to give AutoCAD the task of managing your memory. VMON turns on the paging system as soon as it is called. Once activated, any function that you define with DEFUN is assigned a page table ID. As a team, AutoCAD and AutoLISP work together to manage function swaps and retrieval. VMON affects only those functions loaded or defined after it is turned on, so we recommend putting it the top of your ACAD.LSP file.

Other than VMON, you can make up your ACAD.LSP file in any order. The book groups the functions according to their purpose. There is nothing special about the ACAD.LSP file format. Think of it as any other external .LSP function file. The only difference is that AutoCAD automatically loads this one, like DOS loads an AUTOEXEC.BAT.

Up to now, AutoCAD has been loading the "empty" ACAD.LSP file that you created in chapter 2, or that you loaded from the CA DISK. Let's turn to making a real ACAD.LSP file.

AutoLISP Setup and Initialization

For setup and initialization, you have a group of AutoLISP expressions that need to be evaluated as the file is loaded. These fall into several tasks: turning VMON on, setting the global scale factor and giving the user a welcome message. In this setup group, you also want to establish any constant variables that you will use in other programs. Some common constant values are values like angular rotations, PI/2 or PI/4 etc. Let's add this information.

Creating an AutoLISP Toolkit with ACAD.LSP

  Copy CA-ACAD.LSP to TEST.LSP.

  Create a new LISP function file called TEST.LSP.

Select **[TEST] [24 x 36] [FULL] [INITIATE]**   It brings up TEST.MNU.

Select **[EDIT-LSP]**   Edit TEST.LSP, and enter:

```
;* ACAD.LSP file for use with Customizing AutoCAD
;* This file contains the routines used in many of the application programs
;* presented in the book.
;*

(vmon)                              ;*turn on virtual memory pager for AutoLISP
(prompt "\nLoading Customizing AutoCAD LISP Tools")   ;give them a message

(setq DEBUG T)                      ;allows global control of error and program
                                    ;diagnostics. Make True during development
                                    ;and NIL after programs are completed.

;* Set global variables
(setq #dwgsc (getvar "DIMSCALE")   ;drawing scale
      #pi45    (* pi 0.25)          ;45 deg = quarter of PI
      #pi90    (* pi 0.5)           ;half of PI
      #pi270   (* pi 1.5)           ;1-1/2 of PI
      #pi360   (* pi 2.0)           ;twice PI
);setq
;*
```

General Utility Functions

You have previously defined several general utility functions that you must be tired of loading. We are! These functions include:

```
DXF      Extracts data from dxf group. It is from DXF.LSP.
STRIP    Removes an association sublist. It is from DDRAW.LSP.
ETOS     Converts any expression to a string. It is from ETOS.LSP.
ANGTOC   Converts an angle to decimal degree format for commands.
         It is from ANGTOC.LSP.
```

Write a GENERAL LIBRARY header. Extract the four functions from their files and merge them into the TEST.LSP file.

Save what you've done and add this header to TEST.LSP:

```
;********************** GENERAL LIBRARY ********************************
; Converting angles, calculating points, managing layers,
; loading functions, general DXF function to extract dxf group data, etc.
;_____
;*
```

Merge the following four functions as shown below:

```
;* DXF is the standard function for retrieving the CDR of a dxf pair
(defun dxf (code list / a)
  (cdr (assoc code list))
);defun
;*
;* STRIP removes a specified association list from an entity data list.
(defun strip (code data) ;code is dxf code number, data is association list of data
  (cdr (subst (car data) (assoc code data) data))   ;strips an ASSOC group from entity list
);defun
;*
;* ETOS (Expression TO String) takes any Expression and converts to a string.
;* "STRings" are returned as double ""STRings""
(defun etos (arg / file)
  (if (= 'STR (type arg)) (setq arg (strcat "\"" arg "\"")))  ;format ""STRings""
  (setq file (open "$" "w"))           ;Open temp file.
  (princ arg file)                     ;Write it
  (close file)
  (setq file (open "$" "r"))           ;Reopen to read
  (setq arg (read-line file))          ;Set to string read from file.
  (close file)
  (close (open "$" "w"))               ;Scrunch file down to 0 bytes.
  arg                                  ;Returns as string
);defun ETOS
;*
;* ANGTOC Converts an angle to unit independent format of decimal degrees
(defun angtoc (ang)
  (setq ang
    (rtos (atof (angtos ang 0 8)) 2 6)   ;makes strings like "<<45.000000"
  )
  (strcat "<<" ang)
)
;*
```

Let's take the opportunity to add a few new functions. You will use the following functions in some future examples.

```
DTOR      Converts any decimal degree angle to radians.
MIDOF     Calculates the mid point between two points.
CALOAD    Manages the loading of function files.
MLAYER    Automates the setup of layers.
```

CALOAD manages the loading of function files. It is good for development menu use. It keeps track of which files have been loaded, using a global list. It only loads files if they are not already loaded. If a file is not found, CALOAD informs the user and cancels the macro or command.

MLAYER automates the creation and setup of layers using table access. It will assign default color and linetypes if the layer has not already been created. The function checks the layer symbol table to make sure the layer exists. If the layer exists, MLAYER makes it current. Otherwise, MLAYER creates a new layer and assigns the color and linetype. MLAYER is a handy way to do all your layer creations, and to check for correct layers before issuing draw commands. Predefined layers in the prototype drawing will override the settings passed to MLAYER.

Save what you've done and continue adding to TEST.LSP:

```
;* DTOR converts any decimal degree angle to radians
(defun dtor (deg) (* deg 0.017453292519943))     ;deg times PI/180
;*

;* MIDOF Returns the midpoint between 2 points
(defun midof (pt1 pt2)
   (polar pt1 (angle pt1 pt2) (/ (distance pt1 pt2) 2.0))
);defun
;*

;* CALOAD is our function manager.  It manages the loading of files by determining
;* if the file has previously been loaded. When loaded the file name is placed on
;* a name list. Prior to loading it checks this list to see if it's been loaded,
;* & checks to see if file exists. If it can't find it, it cancels and prompts.
;* It assumes and adds a .LSP extension to allow file checking.
(if DEBUG                                         ;conditionally define function
  (defun caload (subdir fname)                    ;alt. definition if DEBUG set T
    nil
);defun
  (defun caload (subdir fname / fn fspec)         ;standard definition
    (if (not (member (strcase fname) #cafuns))    ;if not prev loaded
      (progn                                      ;then load it
        (setq fspec (strcat subdir fname ".LSP")) ;assumes .LSP
        (if (setq fn (open fspec "r"))            ;if exists
          (progn
            (prompt "\nInitial load.  Please wait...")   ;give a message
            (setq #cafuns (cons (strcase fname) #cafuns)) ;then put it on list &
            (close fn)                            ;close test "r" filename &
            (load fspec)                          ;load it
          )
```

```
            (command nil nil        ;else ^Cs cancel cmd or macro which needed file
          (prompt (strcat "\n" fspec " not found. "))  ;prompt=NIL=^C
        )
      );if
    );progn then
    nil                 ;else already loaded, so NIL
  );if
);defun CALOAD
);if CALOAD is activated
;*

;* MLAYER Checks to make sure layer is defined.
;* It defines it if not and makes it current. If it exists, it just makes it current.
(defun mlayer (name color ltype)
  (if (tblsearch "LAYER" name)        ;of prev layer. Can be passed to the error routine.
    (command "layer" "s" name "")
    (command "layer" "m" name "c" color name "lt" ltype name "")
  );if
  (princ)
);defun
;*
```

For testing, temporarily set the (setq DEBUG T) line near the top of the file to:

(setq DEBUG **nil**)

Save, Exit and reenter AutoCAD. TEST.LSP should load.

```
Loading Customizing AutoCAD LISP Tools     This is the prompt at the file beginning.
Lisp returns: MLAYER                        Now test the new routines:

Command: (caload "\\ca-acad\\" "dxf")      Test file manager. Takes 2 "\" to get one.
Initial load.  Please wait...               This is its 1st time loaded by CALOAD.
Lisp returns: DXF                           The last evaluated.

Command: LINE                               Try it transparently:
To point:    (caload "/ca-acad/" "dxf")    Try to load it again. See, "/" works too.
Lisp returns: nil                           It's already loaded.
To point:

To point: (caload "\\ACAD\\" "STUFF")      And if it can't load it:
\ACAD\STUFF.LSP not found.
Lisp returns: Command: nil
Command:                                    And LINE is cancelled.

Command: (mlayer "CEN32" "RED" "CENTER")
                                            Command scrolls, CEN32 is created and current.

Command: (mlayer "CEN31" "RED" "CENTER")
                                            Test on existing layer. It sets it current.

Command: (midof '(0 0) '(1 1))             Prove that MIDOF works.
Lisp returns: (0.500000 0.500000)           As expected.
```

CALOAD didn't need the "\\CA-ACAD\\" path since it was the current directory, but it tested the path handling.

NOTE. Leave the DEBUG set to NIL for the remaining exercises in this chapter. Set it to True before going on to other chapters.

User Interface Functions

You have worked with the UFUNS.LSP set of user interface functions that buffer GET functions, handle keywords and prompts with defaults and check for correct input from the user. You need to merge these into your ACAD.LSP standard file. Edit the header as shown below. The exercise doesn't reprint the full functions below. It shows the first two lines of each function that you need to merge.

Select **[EDIT-LSP]**   Merge UFUNS.LSP into TEST.LSP. Edit the header. The file should include these functions. ONLY THE 1ST TWO LINES ARE SHOWN:

```
;*********************** USER INTERFACE FUNCTIONS ***************************
; These functions take a full complement of prompting arguments, input options
; and INITGET control bits. A message, default and other options are passed
; to the GETxxx functions
;_____
;*

;* UDIST Userinterface distance function
(defun udist (bit kwd msg def bpt /...

;* UINT Userinterface integer function
(defun uint (bit kwd msg def /...

;* UREAL Userinterface real function
(defun ureal (bit kwd msg def /...

;* UPOINT Userinterface point function
(defun upoint (bit kwd msg def bpt /...

;* UANGLE Userinterface angle function
(defun uangle (bit kwd msg def bpt /...

;* USTR Userinterface string
(defun ustr (msg def spflag /...

;* UKWORD Userinterface key word
(defun ukword (bit kwd msg def /...
```

Save, Exit and Reenter AutoCAD -- TEST.LSP should load.

ERROR Error Handling When Functions Crash

All programs experience errors, whether they are caused internally in the program or by the user. It is important to control a program after an error has occurred. You can repair the drawing editor environment with a few simple steps after an error happens.

AutoLISP has one predefined error trapping function called *ERROR*. Yes, the name includes the asterisks. AutoLISP executes this function each time it detects an error. You've seen *ERROR*'s standard response many times. It prints the cause and the guilty code from inside out in increasingly deep nests, then returns the function name.

Not only does AutoLISP have automatic error handling, it lets you redefine the steps that AutoLISP performs in an error. To do this, you redefine the *ERROR* routine with DEFUN.

NOTE. This is the ONLY function that we recommend redefining on the AutoLISP atomlist. *ERROR* has been written for your use and it is not required for the general execution of AutoLISP programs.

ERROR has a fixed number of arguments, one. *ERROR* expects a STRING type error message as its argument. Normally, you don't call the *ERROR* function, or pass it an argument. AutoLISP calls the function and passes the argument. It is up to you to decide what to do with the message and what to make the *ERROR* function do. Let's look at a simple hands-on example.

Adding Error Handling to the Toolkit

```
Command: (defun *error* (msg)              A new error function.
1> (prompt (strcat "Error: " msg))
1> (prompt "\nDon't do that!\n"))
Lisp returns: *ERROR*                      The function name.

Command: (defun test ()                    Make a test function.
1> (repeat 1000 (prompt "Cancel this stuff.")))
Lisp returns: TEST

Command: (test)                            Test it.
Cancel this stuff.
^Cancel this stuff.                        Hit a <^C> to cancel it.
Error: console break                       The error prompt with MSG.
Don't do that!                             The 2nd error function prompt.
Lisp returns: nil                          Unless you change it, *ERROR* returns NIL.
```

The MSG argument "console break" was supplied by AutoLISP. You can put any AutoLISP expression in the error routine.

An Error Trap System

On the simplest level, the *ERROR* function cleanly exits from a cancelled or flawed program. However, you can do much more with error trapping. To protect the state of the drawing editor environment, you need a recovery method. If you don't have a recovery procedure, your settings and system variables are left scrambled. The error recovery system can and should do the following:

- Reset SETVAR system setting variables.
- Reset the previous current layer.
- Undo any entities or sequence of tasks left incomplete.
- Instruct the user to contact someone and report the problem.
- Print messages to an external error file, like STDERR.TXT.

Programs also need a way to update and alter the set of tasks to execute when an error occurs. The following exercise helps you develop a system to keep track of errors and update your executable tasks. The system keeps a list of tasks, adding executable AutoLISP expressions to a global #ERROR list. ADDERR adds tasks to, and DELERR removes individual tasks from the master error list #ERROR.

The main function *ERROR* processes the error list by evaluating each expression of the list. An error list initialization routine, INITERR, establishes the starting contents of this error list. A RESERR routine reestablishes these settings back to the default values after a program is completed.

Since we encourage the use of error trapping, we recommend centralizing this code in the ACAD.LSP file. The ACAD.LSP file is your present TEST.LSP file. Add these routines to the TEST.LSP file. Then, we will show you how to integrate them into some functions.

Creating *ERROR* Trap Routines

Select **[EDIT-LSP]**   Edit TEST.LSP, and add:

```
;*************************** ERROR TRAP ROUTINES ***************************
; The *ERROR* function is designed to maintain the system's integrity. The main
; function executes each item on the #error list. The list must be a valid LISP
: expression. The error list can be modified throughout the course of a
; program using ADDERR and DELERR. INITERR resets the #error list to your
; standard settings. RESERR resets the environment. It can be called at the end
; of a program.
;_____
;*

(if #err nil (setq #err *error*))  ;* stores the original standard error function

;* *ERROR*
(if DEBUG
  nil                        ;let standard error routine stand
  (defun *error* (msg)
    (command nil nil nil)    ;cancels any pending command
    (reset msg)      ;calls RESET to process #ERROR list, restoring environment
    (grtext)                             ;clear the graphics text boxes
                ;* Check the type of error. If user did not cancel program,
                ;* use standard message. Otherwise, it's a program bug.
    (if (and (/= msg "console break") (/= msg "Function cancelled"))  ;compare messages
      (progn
        (grclear)                                      ;clear the graphics screen
```

```
              (grtext -1 "Pgm ERROR, call supervisor!") ;prompt the user in status box
        ) ;progn
    ) ;if
    (princ)
  ) ;defun *ERROR*
) ;if *ERROR* is activated.
;*

;* RESET function resets the standard environment. You can call it at the end
;* of a function to put the settings back in order, including any tasks you
;* ADDERR but do not DELERR. If you call it from a function for normal
;* termination, pass it a NIL argument.
(defun reset (msg / x)
  (if msg
    (prompt (strcat "Error: " msg "\n"))       ;print out error message
    (prompt "\nResetting environment... ")
  )
  (setq x 1)                                    ;sets the counter x
  (foreach func #error                          ;loops through error items
    (eval func)                                 ;evaluates each one
    (grtext -2 (strcat "Reset " (itoa x)))      ;Shows user a running count of
    (setq x (1+ x))                             ;the number of items reset
  ) ;foreach
  (initerr)                                     ;resets default #ERROR list
  (princ)
) ;defun RESET
;*

;* INITERR sets default #ERROR list for resetting environment and/or error recovery.
(defun initerr ()
  (setq #error (append
    '((setvar "BLIPMODE" 1))                    ;blipmode should be on
    '((setvar "CMDECHO" 1))                     ;command echo on
    '((setvar "MENUECHO" 1))                    ;menu echo on
  )) ;setq
) ;defun

;* ADDERR adds one item to the #ERROR list. ITEM must be quoted expression,
;* or in form: (list 'FUNCTION arg arg arg...) where arguments are quoted or
;* not depending on whether you pass current value or literally.
(defun adderr (item)
  (if (not (member item #error))                ;if it's not a duplicate item
    (setq #error (cons item #error))            ;add it to the list
  ) ;if
) ;defun
;*

;* DELERR deletes one item from the #ERROR list. ITEM is same format as ADDERR.
(defun delerr (item)
  (setq #error (subst '() item #error))         ;replace item with a null list
) ;defun
;*

;* Disable following SETQ w/ leading ; to enable above custom *ERROR* function.
;* The RESET, ADDERR, DELERR and INITERR functions will still work, regardless.
(setq *error* #err) ;*remove leading ; if any to enable the standard *ERROR* function.

(prompt "...done.\n") (princ)
;* end of ACAD.LSP
```

Even if you have the CA DISK, make sure the DEBUG line is NIL at the top of the file, and disable the (setq *error* #err) line above with a leading ; semicolon:

(setq DEBUG **nil**)

;(setq *error* #err) ;*remove leading ; if any to enable the standard *ERROR* function.

Save, Exit and reenter AutoCAD. TEST.LSP should load.

Command: **!#ERROR**   Check the global list.
Lisp returns: nil   It's empty.

Command: **(initerr)**   Creates our default global #ERROR list:
Lisp returns: ((SETVAR "BLIPMODE" 1) (SETVAR "CMDECHO" 1) (SETVAR "MENUECHO" 1))

Command: **!#ERROR**   Check again, now it returns task list:
Lisp returns: ((SETVAR "BLIPMODE" 1) (SETVAR "CMDECHO" 1) (SETVAR "MENUECHO" 1))

Command: **(adderr '(command "ERASE" "Last" ""))**   Add a task to the #ERROR list.
Lisp returns: ((COMMAND "ERASE" "Last" "") (SETVAR "BLIPMODE" 1) (SETVAR "CMDECHO" 1) (SETVAR "MENUECHO" 1))

Command: **(adderr '(do stuff))**   Adds task to #ERROR list.
Lisp returns: ((DO STUFF) (COMMAND "ERASE" "Last" "") (SETVAR "BLIPMODE" 1) (SETVAR "CMDECHO" 1) (SETVAR "MENUECHO" 1))

Command: **(delerr '(do stuff))**   Deletes task from #ERROR list.
Lisp returns: (nil (COMMAND "ERASE" "Last" "") (SETVAR "BLIPMODE" 1) (SETVAR "CMDECHO" 1) (SETVAR "MENUECHO" 1))

Command: **LINE**   Draw two lines for the test.
Command: **SETVAR**   Set BLIPMODE from <1> to 0 for testing.

Command: **(getpoint a b c)**   Oh Oh, an error:
Error: bad argument type
ERASE   *ERROR* erased Last and evaluated each #ERROR list
Select objects: expression. It blanked the screen to call attention to
Command: the status line message: "Pgm ERROR call supervisor!".

Command: **REDRAW**   Yes, the last line is gone.

Command: **!#ERROR**   Check again, it's back to your defaults:
Lisp returns: ((SETVAR "BLIPMODE" 1) (SETVAR "CMDECHO" 1) (SETVAR "MENUECHO" 1))

Command: **SETVAR**   Check BLIPMODE again. It is reset back to 1.

  Copy CA-ACAD to ACAD.LSP. This replaces the "empty" ACAD.LSP file that has been loading until now.

  Copy TEST.LSP to replace ACAD.LSP. Copy ACAD.LSP to CA-ACAD.LSP for a backup.

Command: **QUIT**   You must exit and restart to restore the original *ERROR*.

Now ACAD.LSP will automatically load in your Customizing AutoCAD environment. Be careful not to mix the file with any other ACAD.LSP file, since they must all share the same name in different directories for autoloading. We suggest you have backups of all your ACAD.LSP files. Store them under other names.

If you want normal *ERROR* reporting for debugging when you write programs, type:

```
(setq *error* #err)
```

To restore the CA-customized *ERROR* for testing, make sure DEBUG is set to NIL in ACAD.LSP, and reload ACAD.LSP.

If you want to disable the custom *ERROR* for an extended time, remove the ";" from the line:

```
;(setq *error* #err) ;*remove leading ; if......
```

at the end of the ACAD.LSP file, or set DEBUG to True in ACAD.LSP, and reload ACAD.LSP. You saved the original *ERROR* function as variable #ERR in the ACAD.LSP.

Now, let's see how error recovery works by making a program called LEADTEXT.

LEADTEXT.LSP has two parts. S-LEAD is a command function that draws a leader line with any number of segments. It then calls L-TEXT. L-TEXT is a function that determines text orientation and places Dtext at the end of the leader line. LEADTEXT provides an example of error recovery, and combines MLAYER layer setting and RESET use.

LEADTEXT A Leader/Text Program with Error Recovery

```
Enter selection:                              Begin a NEW drawing named LEADER
Loading Customizing AutoCAD LISP Tools        The new ACAD.LSP loads.
```
                                              Make sure DEBUG is set to NIL and the custom *ERROR* is enabled.

  Copy LEADTEXT.LSP to TEST.LSP.

  Create a new LISP function file called TEST.LSP.

```
Select [TEST] [24 x 36] [FULL] [INITIATE]              It brings up TEST.MNU.

Select [EDIT-LSP]                       Add the S-LEAD function first.
```

16—16 Customizing AutoCAD

```
;* LEADTEXT contains 2 functions. S-LEAD draws leaders. L-TEXT adds text to
;* leaders.

;* C:S-LEAD (Straight LEADer) draws a pline leader using as many line segments
;* as desired. The start point is the arrow head. L-TEXT is called to provide
;* correctly justified DTEXT.

(defun C:S-LEAD ( / pt1 pt2 pt3)
  (setvar "BLIPMODE" 0)                                 ;turn blipmode off
  (setvar "CMDECHO" 0)                                  ;turn command echo off
  (setvar "MENUECHO" 0)                                 ;turn menu echo off
  (initerr)                                             ;initialize error list
  (adderr                          ;adds command function to reset layer name
    (list 'command "LAYER" "S" (getvar "CLAYER") "")    ;getvar is eval here to put layer
  )                                                     ;name in #ERROR list
  (mlayer "ANN01" "G" "")                               ;creates/sets annotation layer
  (setq pt1 (upoint 1 "" "Pick arrow head point" nil nil)    ;get start point of arrow head
        pt2 (upoint 1 "" "To point" nil pt1)            ;get point for 1st line segment
        pt3 (polar pt1 (angle pt1 pt2) (* 0.125 #dwgsc)) ;calc arrow head size
  );setq

  (adderr '(command "U"))                               ;add undo to error list
  (adderr '(command nil nil nil))                       ;add cancel to error list
  (command "PLINE" pt1 "W" "0" (* 0.0625 #dwgsc) pt3 "W" "0" "0" pt2)  ;draw arrow head
                                                        ; and first line segment
  (while (setq pt2 (getpoint "\nTo point: " pt2))       ;loop for additional line segments.
    (command pt2)                                       ;until ENTER is pressed
  );end while
  (command "")                                          ;end pline
  (delerr '(command "U"))                               ;the pline can stay now if user cancels
  (l-text)                                              ;call l-text function
  (reset nil)                                           ;resets environment and #ERROR list
);end defun
```

Even if you have the CA DISK, add a leading ; semicolon to disable the (l-text) line near the end of S-LEAD to test it:

```
;   (l-text)                                            ;call l-text function
```

Save, Exit and reenter AutoCAD. TEST.LSP should load.

Command: **ZOOM**	Center, height of 6".
Command: **S-LEAD**	Run the command. It makes a new layer.
Pick arrow head point:	Pick a couple of points.
To point:	<RETURN> when finished picking. It restores old layer.
Command: **S-LEAD**	Try again, but cancel after two points.
Pick arrow head point:	Pick one point.
To point:	Pick one more point.
To point: **<^C>**	Now CANCEL.

The leader line is erased because the program "knows" that the entire leader has not been completed. The program includes several error recovery statements that modify the recovery actions stored in #ERROR, based on the point at which the cancel occurred. Let's complete the leader command by adding the text subroutine.

Select **[EDIT-LSP]**
Even if you have the CA DISK, remove the leading ; to re-enable the (l-text) line near the end of S-LEAD.

```
  (l-text)                                    ;call l-text function
```

Add the L-TEXT function:

```
;* L-TEXT (Leader TEXT) starts a DTEXT command following any leader. The text
;* justification (Left or Right) is determined by the last point and mid point
;* of the last line/arc of the leader.

(defun l-text ( / pt1 pt2)
   (setq pt1 (getvar "LASTPOINT")          ;get last point of leader
         pt2 (osnap pt1 "quick,midp")      ;get midpoint using pt1, use quick
         txtht (dxf 40 (tblsearch "STYLE" (getvar "TEXTSTYLE")))) ;current height
   );setq
   (prompt "\nEnter text: ")
   (if (> (car pt1) (car pt2))             ;determine X point position
      (progn                               ;if X of pt1 is greater, Left text
         (setq pt1 (polar pt1 5.759 txtht)) ;based on text size
         (command "DTEXT" pt1 0)           ;start DTEXT command - leave user there
      );end progn
      (progn                               ;if X of pt2 is greater use Right text
         (setq pt1 (polar pt1 3.665 txtht)) ;based on text size
         (command "DTEXT" "R" pt1 0)       ;start DTEXT command - leave user there
      );end progn
   );end if
);defun L-TEXT
;*
(princ)

;*end of LEADTEXT.LSP
```

Save, Exit and reenter AutoCAD. TEST.LSP should load.

```
Command: S-LEAD                    Run the command.
Pick arrow head point:             Pick a couple of points.
To point:                          <RETURN> when finished picking.
Enter text: ENTER                  Enter some text.
Text: A NOTE
Text: HERE...
Text:                              <RETURN> when finished entering text.
```

Select **[EDIT-MNU]**   Make a menu item.
Change the top page label to [CA16TEST] and add the following to the EXAMPLES page of TEST.MNU.

```
***SCREEN
**ROOT
[CA16TEST]
[]

**EXAMPLES
[Examples]
[]
[S-LEADER]^C^C^C(caload "" "LEADTEXT") S-LEAD
[]
```

Save and Exit to AutoCAD. TEST.MNU reloads.

  Delete TEST.LSP

  Rename TEST.LSP to LEADTEXT.LSP.

Select **[EXAMPLES] [S-LEADER]**   Test it. It should load the functions and execute.

Try canceling or forcing an error.

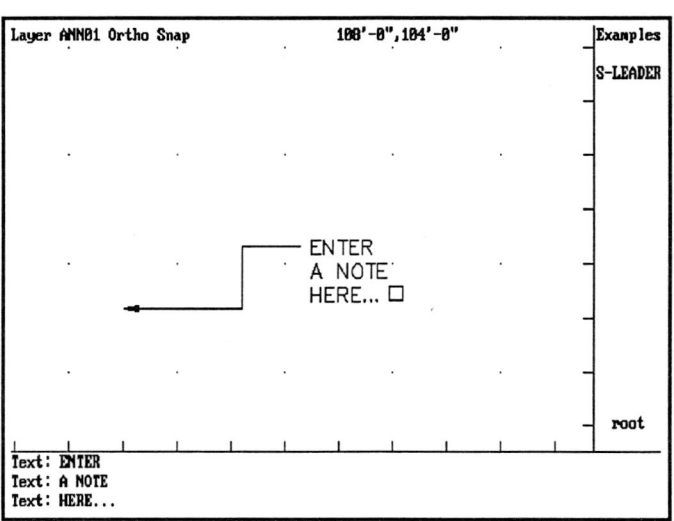

Leader and Text Using Error Recovery

Change the error trapping function back to the default *error* routine to continue working with the book exercises.

Command: **ED** Edit ACAD.LSP.
Change the DEBUG back to T near the top of the file.
Turn the custom error function back off by removing the leading ";". The DEBUG and last lines should be:

(setq DEBUG **T**)

(setq *error* #err) ;*remove leading ; if any to enable the standard *ERROR* function

(prompt "...done.\n") (princ)
;* end of ACAD.LSP

Save, Exit, and reenter AutoCAD.

 Delete TEST.MNU.

 Rename TEST.MNU to MY16.MNU.

Command: **QUIT**

In the last chapter you will pull the book's menu together and put some finishing touches on your application environment, including enabling the CA LOAD and the custom *ERROR* functions. Leave both functions disabled while you are still in an exercise and test environment.

Working with Other Application Programs

Some users encounter conflicts when they attempt to run other programs developed for AutoCAD with their own customized programs. Most of the conflicts between personal applications and other programs arise from their conflicting use of AutoLISP.

The type of problems that you may encounter include encrypted or undocumented programs and menus dominating the AutoCAD environment, AutoLISP files, function lists and variables. AutoCAD has limitations on how it handles files. The AutoCAD program has standard named files, containing libraries of information that it automatically uses. AutoCAD also can have only one menu file loaded at one time. Using other programs may limit your access to AutoCAD's open architecture if they:

16—20 Customizing AutoCAD

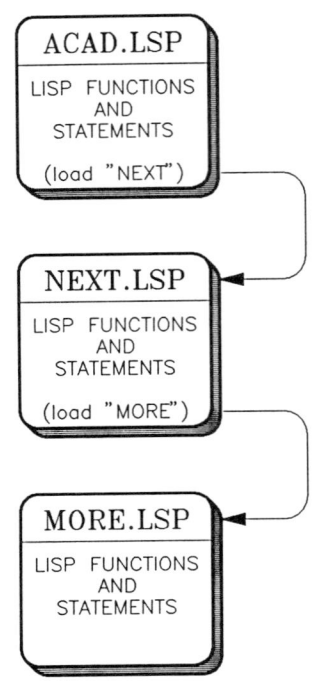

Chain Loading

- Have a protected ACAD.LSP file.
- Have encrypted menu systems.
- Have hard coded disk drives and/or directories.
- Do not incorporate a scheme for user customization.
- Do not document the AutoLISP function names and variables reserved by the program.
- Use CLEAN routines and you use VMON.

Many of these limitations do not have reasonable work arounds. Here are some ideas on integrating your personal programs with other programs.

To determine if a AutoLISP file is protected, just TYPE the contents using DOS. If you get "AutoLISP PROTECTED file" then you don't have access to the developer's file.

If you want to use the program with your own personal program, you can chain load both programs. To work around the protected ACAD.LSP file, you can rename the protected ACAD.LSP file, say to MOREACAD.LSP. Give your standard file the ACAD.LSP name. Load the MOREACAD.LSP file in the last line of your ACAD.LSP file.

Chain Loading ACAD.LSP Files

`Enter selection:`	Begin a NEW drawing, again named CA16.
`Loading Customizing AutoCAD LISP Tools`	ACAD.LSP loads. Make sure you previously set DEBUG to T and disabled the custom *ERROR*, as described in the previous sequence.

 You have a MOREACAD.LSP file and a TESTACAD.LSP file.

 Create the files MOREACAD.LSP and TESTACAD.LSP.

`Command:` **ED**   Create MOREACAD.LSP, and enter the following in it:

```
(prompt "\n\nThis is the MOREACAD.LSP file.")
(prompt "\nIts contents are assumed to be ")
(prompt "\nprotected in our example.")
```

Save, Exit and reenter AutoCAD. Start a new file:

Command: **ED** Create TESTACAD.LSP, and enter the following:

```
(prompt "\nThis is the TESTACAD.LSP file being loaded.")
(prompt "\nIt loads the MOREACAD.LSP file.")
(prompt "\nand shows an error message.")
(prompt "\nBut it's harmless...")
(prompt "\nYou could change *ERROR* to trap")
(prompt "\nthe message and suppress the printing of it.")
(load "moreacad")
```

Save, Exit and reenter AutoCAD, and test this:

Command: **(load "testacad")**

```
This is the TESTACAD.LSP file being loaded.
It loads the MOREACAD.LSP file.
and shows an error message.
But its harmless...
You could change *ERROR* to trap
the message and suppress the printing of it.
This is the MOREACAD.LSP file.             The second file is loaded.
Its contents are assumed to be
protected in our example.
Lisp returns: Error: invalid character     It generated an error.
```

The AutoCAD User Reference does not recommend chain loading files, but after loading the one file, AutoLISP will load the next file in the series. If you do this, please be aware that the entire first file will load before the second file, no matter where the call to the second file is located.

Chain loading causes an "invalid character" error, but you can work around that if you wish by modifying *ERROR* to suppress the printing of the error message. You should put the chaining LOAD function at the end the file.

If you want compatibility for your own applications, the best solution is to leave your ACAD.LSP unprotected. It is a simple matter to set a variable to the path or configuration of other programs. A line that you can use is:

(setq #mydir "C:\\PATH\\")

From this, an application program can find its own configuration file that another developer can use for setup information. The configuration file should contain the names of drives and directories where the user has installed the program. The file could be a simple comma delimited file with a key word and then the path. Your programs could make an association of key words and related paths. Each time a symbol or AutoLISP file is needed from disk, the key word could be used to retrieve the path name to the file. It sounds a bit slower than it really is. An example file would look like:

```
ACAD,C:\ACAD
LISP,C:\ACAD\PROGRAMS
SYMBOLS,C:\ACAD\SYMB
```

Other compatibility problems that you may encounter are more difficult. They are summarized here.

Encrypted Menu Systems. You have to treat encrypted menus as part of the system. By now, you know how to create your own menu. Have your menu call the other menu that you wish to use. You'll have to manually change back to your personal menu.

Hard Coded Disk Drives and/or Directories. Unfortunately, if the programs that you want to use are intertwined, there is little hope of finding a way to manage them. Some programs insist on being installed in a directory called C:\>ACAD. Try putting your programs in another directory.

Integrating the Example Menu

You do not need to integrate this chapter's menu into the CA-MENU. If you find the macros helpful, add them to any your own menus.

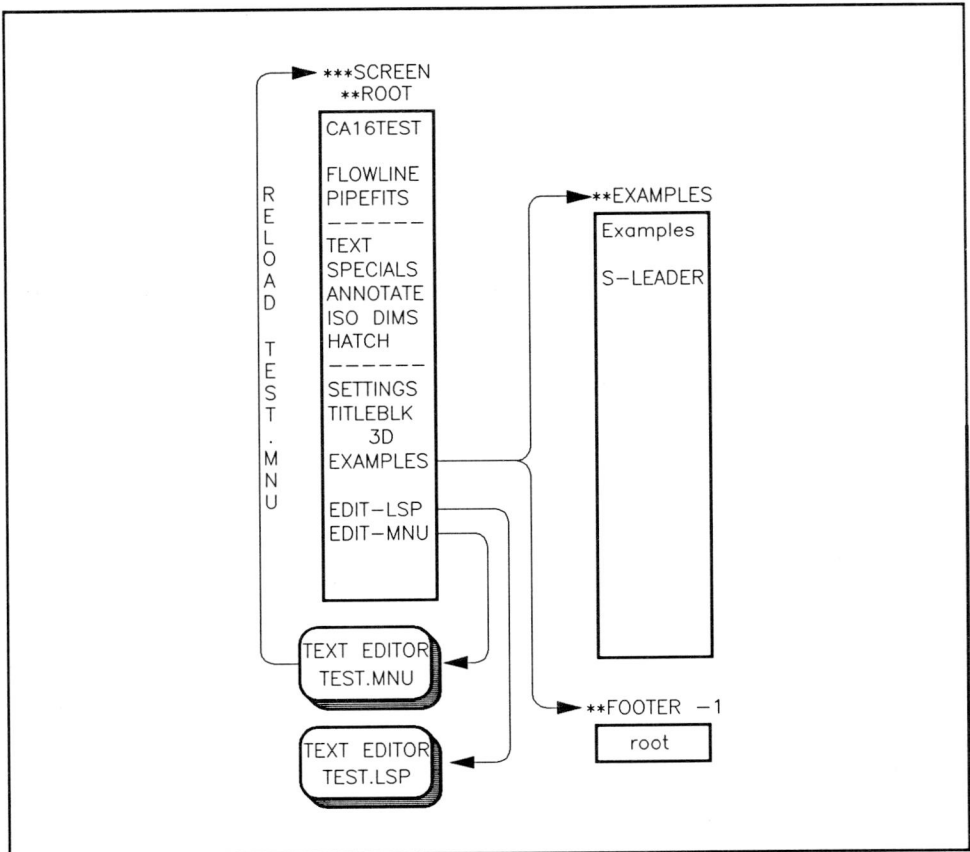

The Chapter 16 Menu

Summary Tips and Techniques

Here are some summary tips and techniques for managing your AutoLISP toolkit.

- Keep ACAD.LSP small. You don't want your users to wait each time a new drawing is called. You want them "to get in the right line at the bank."

- During setup, give your users status information about their application. For example, give them the current scale of the drawing.

- Use VMON to conserve AutoLISP memory. EXPANDED/EXTENDED memory makes VMON a lot more efficient.

- Use the CALOAD function loading manager in your menu macros.

- Filter your user's inputs and present data defaults to your user in your interface functions. This helps with the program maintenance of keeping a common interface.

- Use INITGET and good input filters to filter types and ranges of data to help your users avoid errors.

- Start error trapping when you start writing your routines. This helps protect your application's environment.

- Use RESET to restore the environment when your program terminates.

- If you need to, try to chain load other programs' ACAD.LSP files from your ACAD.LSP file.

Because this book is a teaching guide, we've haven't made programs as complete as you would in a production applications environment. This means sacrificing some error checking, resetting, subdirectory usage and program coordination. This chapter and the chapter on AutoLISP File I/O give you many of the tools that you need to develop "complete" programs. When you develop your own applications, refer to these chapters for the tools to coordinate your applications and to do error checking.

It is time to move on to some advanced applications, starting with AutoLISP 3-D tools.

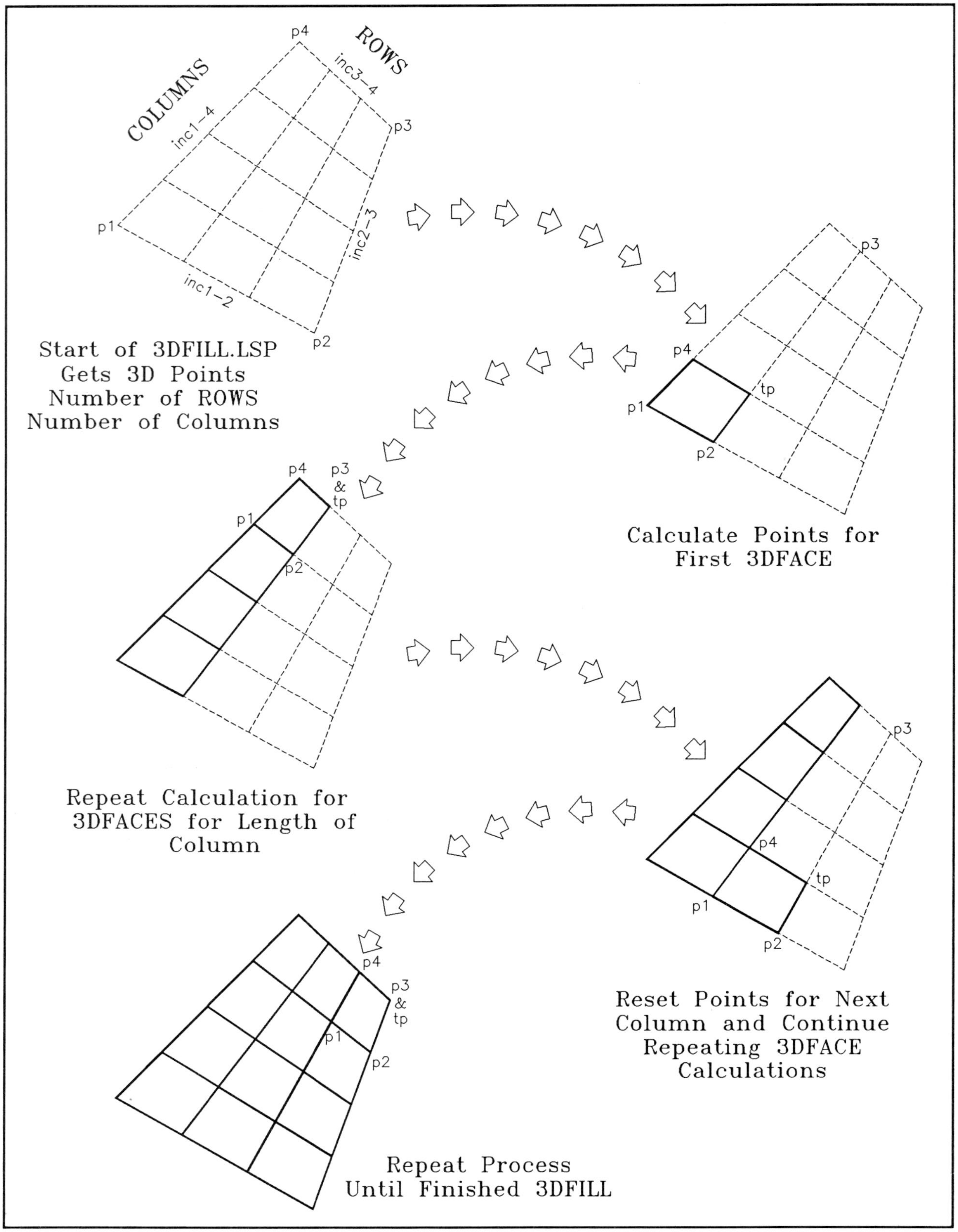

p4
ROWS
COLUMNS
inc3-4
inc1-4
p3
p1
inc2-3
inc1-2
p2
Start of 3DFILL.LSP
Gets 3D Points
Number of ROWS
Number of Columns
p3
p4
tp
p1
p2
Calculate Points for
First 3DFACE
p4
p3
&
tp
p1
p2
Repeat Calculation for
3DFACES for Length of
Column
p3
p4
tp
p1
p2
Reset Points for Next
Column and Continue
Repeating 3DFACE
Calculations
p4
p3
&
tp
p1
p2
Repeat Process
Until Finished 3DFILL

Automatic Mesh Generator

CHAPTER 17

3-D Manipulations with AutoLISP

This chapter shows you how to use AutoLISP to draw 3-D objects. AutoCAD gives you little direct help in 3-D drafting. If you need 3-D to generate parts assemblies, rendering backgrounds or mechanical drawings, you need to develop your own tools and commands. This chapter develops a basic set of AutoLISP 3-D tools and shows you how to create 3-D tools that are unavailable in AutoCAD.

The Benefits of Making 3-D Tools

If you have the tools, it is often easier to write a simple AutoLISP routine to create an object than to "draw" the object in AutoCAD's 3-D.

The chapter provides two key application benefits. It shows how to:

- Use AutoLISP to make meshes in 3-D that subdivide areas into segments and then pass the images into AutoShade for color renderings.
- Use AutoLISP to convert plans to elevation drawings.

In addition, you can use DXB files to get 3-D visualizations with varying viewpoints. You can place several view points of an image in a single "flat" drawing.

How to Skills Checklist

In this chapter you will learn how to:

- ❏ Extract 3-D points.
- ❏ Control 3-D elevation of points.
- ❏ Calculate points, angles, distances, arcs and altitudes in 3-D space.
- ❏ Generate meshes.
- ❏ Create elevation drawings.
- ❏ Project surface faces.

Macros, AutoLISP Tools and Programs

MACROS
[MAKEWALL] is a parametric macro that converts Plines into center lines and draws 2-sided walls with height (THickness) suitable for a 3-D view.

[WINDOW] draws a 3-D window in an opening, with wall above and below the head and sill.
[DOOR] draws a 3-D door opening and inserts a 2-D plan view door block.
[MAKE ELV] creates an elevation drawing from a plan drawn with ELEVation and THickness.

AutoLISP TOOLS
3DTOOLS.LSP is a collection of functions for determining and manipulating points, lines and curves in 3D space, including:
3DDIST is a 3-D distance function.
BANGLE is a 3-D angle function.
MID3D calculates the mid of 3-D points.
RADIUS calculates the radius of an arc from its chord and altitude.
ARCLEN calculates arc length given the radius and included angle.
INCANG calculates included angle given chord and radius.
ALTITD calculates altitude from radius and included angle.
CIRXY calculates a point along the circumference of an arc.
C:PICKVPT is a pick and view command. It lets you look at standard viewpoints, or pick target and eye points. PICKVPT lets the user select a predetermined view by entering a key letter. It also provides the option of selecting angles in the xyz planes, or entering a target specification.
C:ZARC creates an arc in the z axis by representing the arc with 3DLINEs.
3DPOLAR provides the 3D equivalent to the AutoLISP polar function.
C:3DPOLAR is a command with a user interface for the 3DPOLAR function. It is equivalent to using the ID command with relative polar coordinates.

PROGRAMS
C:3DFILL is a mesh generator. It fills a 3D area, defined by 4 points, with 3DFACEs. The program divides the surface of an area into a user supplied number of rows and columns.
C:MELEV makes an elevation from a plan drawing. The program draws an elevation view of a 3D wall plan, including window and door openings. The user can specify the view point for the elevation. C:MELEV prompts the user for 3D plane selection for drawing an elevation. Supporting functions evaluate the entities and draw the elevation.

2-D vs. 3-D ENTITIES

This exercise assumes that you are familiar with the use of ELEVation and THickness in adding 2 1/2-D extrusions to most AutoCAD entity types. In addition to 2 1/2-D extrusions, AutoCAD has four 3-D entities:

```
3DFACE   3DLINE   POINT   INSERT
```

These are the only entities whose points are stored with 3 coordinate values. 3DFACE, 3DLINE and POINT are fully independent in 3-D space. INSERT's insertion point is fully independent, but its rotation angle is still in the XY plane.

In the database, a 2-D point looks like:

```
(10.000000 20.000000)
```

2-D vs. 3-D ENTITIES 17—3

A 3-D point looks like:

(10.000000 20.000000 30.000000)

If you supply 2-D AutoCAD commands with a 3-D point in the form 1,2,3 it will cause an error. However, if you supply a 2-D command with an AutoLISP point list in the form (1 2 3), it accepts the point list and ignores the Z coordinate. In the same way, 2-D AutoLISP functions and entity data lists accept and ignore the Z of 3-D point lists. Unfortunately, functions like DISTANCE, POLAR and GETDIST are strictly 2-D through Release 9. If you feed them 3-D points, you won't get what you want.

Draw a normal Line, then a 3DLINE. View them. Then, look at how AutoCAD stores the two entities.

Examining a 3-D Entity

```
Enter selection:                         Begin a NEW drawing named CA17.
Loading Customizing AutoCAD LISP Tools...done.    Make sure ACAD.LSP loaded.
```

 Copy CA17.MNU to TEST.MNU.

 Copy MAIN.MNU to TEST.MNU.

```
Select [TEST] [24 x 36] [FULL] [INITIATE]              It brings up TEST.MNU.

Command: ERASE                           Erase the border.
Command: ZOOM                            Left, corner 0,0 and height 12"
Command: UNITS                           Set decimal and decimal angles, each with 4 decimal places.

Command: LINE                            From 0,0 to 1,1
Command: CHANGE                          Change Last to color Green.

Command: 3DLINE                          From 0,0,0 to 1,1,1
Command: CHANGE                          Change Last to color Yellow.

Command: VPOINT                          0,-1,0
                                         It's a front view of the XZ plane looking straight up the Y axis.

Command: (entget (entnext))              The line is first, with 2-D points.
Lisp returns: (-1 . <Entity name: 60000050>) (0 . "LINE") (8 . "0") (62 . 3)
(10 0.000000 0.000000) (11 1.000000 1.000000))

Command: (entget (entlast))              The 3DLINE is last. It has 3-D points.
((-1 . <Entity name: 60000078>) (0 . "3DLINE") (8 . "0") (62 . 2)
(10 0.000000 0.000000 0.000000) (11 1.000000 1.000000 1.000000))

Command: LIST                            Select the two and compare.
```

When you listed them, AutoCAD did not report an angle for the 3DLINE. A single angle would not be meaningful. Later, we will help you develop tools to report angles in the XY plane and the YZ plane.

Let's see how AutoLISP works with 3-D coordinates. AutoLISP does not GET 3-D points, unless you specifically ask for them with the INITGET function. INITGET has a bit code (16) that instructs AutoCAD to send 3-D points to AutoLISP. Both the AutoLISP GETPOINT and GETCORNER functions can return a 3-D point since they request a user point. Try INITGET and retrieve 3-D points.

```
Command: (initget 16)
Lisp returns: nil
```

```
Command: (setq pt1 (getpoint "Point: "))
Point:
(4.407000 7.626000 0.000000)
```
Pick a point.
Z is zero because your elevation is set to 0.

```
Command: (initget 16)
Lisp returns: nil
```

```
Command: (setq pt2 (getcorner "Other corner: " pt1))
Other corner:
(8.932000 10.862000 0.000000)
```
Pick it.
Same with the corner point.

If a Z coordinate is not supplied by the user, 3-D initialized GET functions take the current ELEVATION setting as Z. Now, look at how the DISTANCE and ANGLE functions work with 3-D points. Set the two points given, and try the DISTANCE and ANGLE functions.

```
Command: (setq pt1 '(0 0 0) pt2 '(1 1 1))
Lisp returns: (1 1 1)
```

```
Command: (distance pt1 pt2)
Lisp returns: 1.414214
```
Try the DISTANCE function on the 2 points.

```
Command: (angtos (angle pt1 pt2))
Lisp returns: "45.0000"
```
We'll also convert it to degrees.

AutoLISP ignores the Z coordinate of points passed to all the functions like DISTANCE, POLAR and INTERS. It calculates based on 2-D points. You can confirm the results with the List command.

The first simple tools that you need are 3-D distance and polar routines.

3-D Distances

The shortest distance between two points is a straight line, even when the points are 3-D points. The standard distance formula is "the square root of the sum of the squares of the differences between the Xs, Ys and Zs." As a matter of fact, it is written the same way in AutoLISP.

3DDIST is the 3-D distance function that you will make. 3DDIST calculates distance by manipulating lists of points. It uses MAPCAR to perform the tasks of subtracting the XYZ differences. It squares the differences by multiplication, then tallies the result using APPLY. MAPCAR and APPLY are two built-in AutoLISP functions that apply a single AutoLISP function to a group of data.

APPLY "applies" a function to the elements of the list. For example, the add function "+" can take multiple arguments, but they must be individual number arguments. If NUMLIST is (1 3 5), then (apply '+ numlist) would return 9, the sum of 1, 3 and 5. It takes the list as arguments to the "applied" add function.

MAPCAR "maps" a function to each element of one or more lists and returns a corresponding list. If NUMLIST is (1 3 5) and NUM2 is (2 4 6), then (mapcar '+ numlist num2) returns 1+2, 3+4 and 5+6 in the form (3 7 11). It "maps" the + function to the CAR of the lists, then to their second elements, their third elements, and so on, and returns a list of the results.

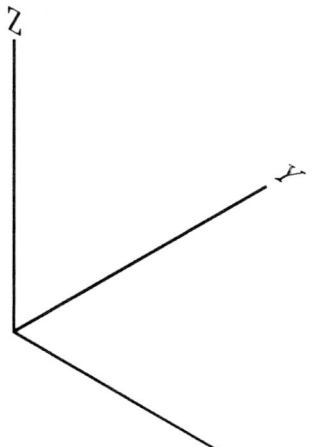

Three Dimensional Axis System

You must quote the applied or mapped function names with a ' single quote. If not, AutoLISP attempts to evaluate the function itself.

Making A 3-D Distance Function 3DDIST

   Copy 3DTOOLS.LSP to TEST.LSP.

   Create a new LISP function file called TEST.LSP.

Select **[EDIT-LSP]**   Edit TEST.LSP, and enter:

```
;* 3DTOOLS.LSP is a collection of subroutines for dealing with points,
;* lines and curves in 3-D space, plus the PICKVPT command for easy Vpoints.

;* 3DDIST calculates the distance between two - 3D points.
(defun 3ddist(pt1 pt2 / deltas sqrs sums)
   (setq deltas (mapcar '- pt2 pt1)            ;calculates the change in X, Y and Z
         sqrs (mapcar '* deltas deltas)        ;squares them by multiplication
         sums (apply '+ sqrs)                  ;adds them all together - for a sum
   );setq
   (sqrt sums)                                 ;takes the square root of the sum
);defun
;*
```

Save, Exit and reenter AutoCAD. TEST.LSP should load.

Command: **(3ddist pt1 pt2)**   Test with the (0 0 0) and (1 1 1) points from earlier.
Lisp returns: 1.732051   The true 3-D distance.

You can write a 3-D midpoint function in the same manner. Use the list '(2.0 2.0 2.0) as one of MAPCAR's arguments to divide each coordinate of the 3-D point list in half.

Edit the TEST.LSP file and enter, or look at, the MID3D function.

```
;* MID3D calculates the mid point of two 3D points.
(defun mid3d(pt1 pt2)
   (mapcar '+ pt1                  ;adds the 1st point to
      (mapcar '/                   ;half of the difference
         (mapcar '- pt2 pt1)       ;between the XYZ coordinates.
         '(2.0 2.0 2.0)
      );mapcar
   );mapcar
);defun
```

Save, Exit and reenter AutoCAD. TEST.LSP should load.

Command: **(mid3d pt1 pt2)**
Lisp returns: (0.500000 0.500000 0.500000)

Try it.
Half way to 1,1,1.

MID3D is a very compact function. First, it MAPCARs '- to PT2 and PT1 to get a delta list of X, Y and Z differences. It then MAPCARs '/ to the delta list and the 2's list to halve the deltas. Finally it MAPCARs '+ to the original PT1 and the half delta list to get the midpoint list that it returns.

Try some polar functions next.

3-D Polar Functions

With AutoCAD, you've become accustomed to polar coordinates in 2-D. AutoCAD commands accept a polar coordinate in the format of DISTANCE<ANGLE. However, there is no current equivalent in 3-D.

Polar notation in a 3-D coordinate system involves 2 rotation angles. You know about the first angle. It is the angle from the X axis in the XY plane and is usually requested as "Rotation angle <90>: ." In 3-D polar calculations, you also need to determine the angle from the XY plane toward the Z axis. We will call the XY plane rotation angle "Alpha" and the angle toward the Z axis "Beta."

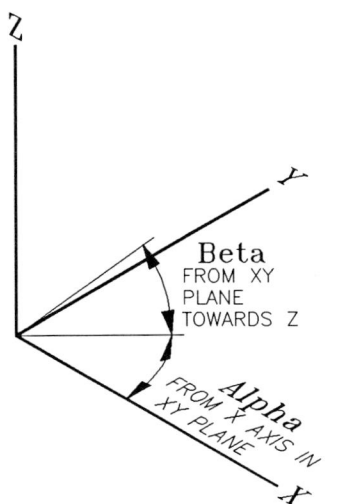

Alpha and Beta Angles

Let's write a function that calculates the beta angle. In 3-D polar functions, the distance along the XY plane will be the ADJacent distance and the OPPosite leg will be the rise in Z elevation. The equation is TAN(angle)=OPP/ADJ where OPPosite is Z1 - Z2 (the vertical Z difference), and ADJacent is the base distance from PT1 to PT2 in the XY plane.

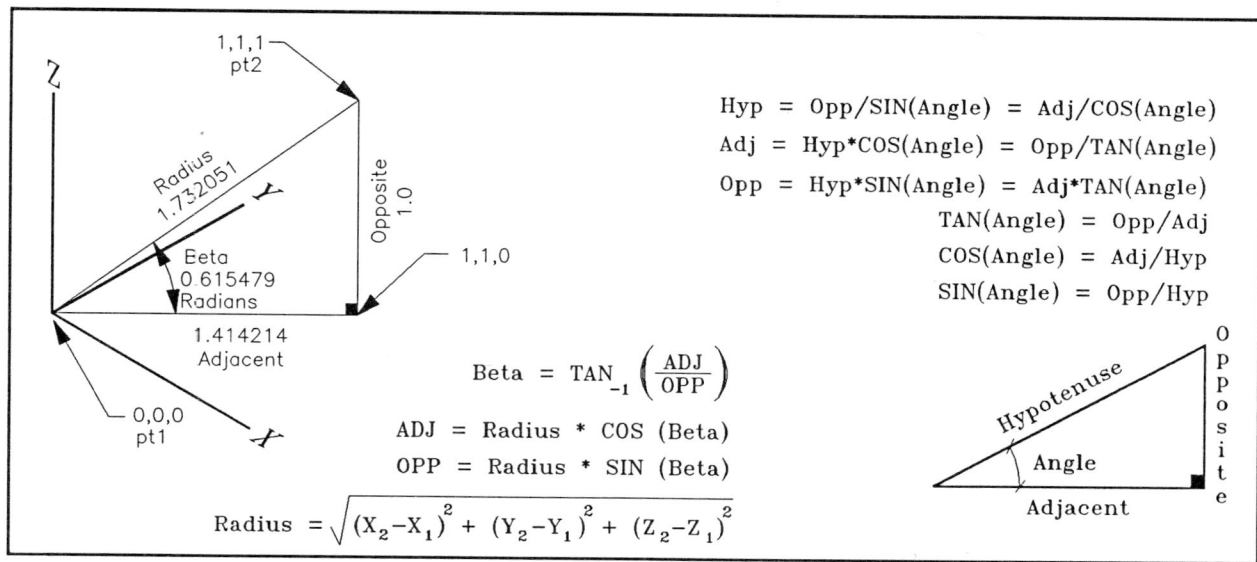

Formulas and Triangles

To determine the angle, use the AutoLISP arctangent function ATAN. Try the ATAN function at the keyboard, using equal numbers for the OPPosite and ADJacent distances. Then write a function called BANGLE for Beta ANGLE.

Making a 3D Polar Function BANGLE

Command: **(atan 1 1)**	Test an equal leg triangle.
Lisp returns: 0.785398	45 degrees.
Select **[EDIT-LSP]**	Edit TEST.LSP, and add:

```
;* BANGLE calculates the inclined angle from the XY plane between two points.
(defun bangle(pt1 pt2)                      ;takes arctangent.
   (if (zerop (setq adj (distance pt1 pt2))) ;if dist is 0, angle is 90 or all Z
      1.57079633                            ;half of pi
      (atan (/ (- (caddr pt2) (caddr pt1))) ;divides the rise - delta Zs
            (distance pt1 pt2)))            ;by the run - length in XY plane
   );atan
  );if
);defun
;*
```

Save, Exit and reenter AutoCAD. TEST.LSP should load.

Command: **(bangle pt1 pt2)**	Test it on (0 0 0) and (1 1 1).
Lisp returns: 0.615479	35.2643 degrees.

You can use this simple function to write a command for easy Vpoints.

PICKVPT is a command to buffer input and calculate angles and view points for you. It adds "look at" and "look from" options, and several standard single letter view options to the standard AutoCAD Vpoint command. The UPOINT function is used for the key word options. PICKVPT requires Release 9.

A Pick and View Function PICKVPT

Select **[EDIT-LSP]**   Edit TEST.LSP, and add:

```
;* PICKVPT buffers the Vpoint command allowing the user to select a view based
;* on the entry of key letters. A target and eye option is also included,
;* although it is not compatible with AutoCAD 2.6
(defun C:PICKVPT( / pt eye alpha beta)
   (setq pt
      (upoint 17 "L R F B P" "Target point or Left/Right/Front/Back/Plan view" nil nil)
   )
   (cond
      ((= pt "L")                          ;left view
         (command "vpoint" "-1,0,0")
      );l
      ((= pt "R")                          ;right view
         (command "vpoint" "1,0,0")
      );r
      ((= pt "F")                          ;front view
         (command "vpoint" "0,-1,0")
      );f
      ((= pt "B")                          ;back view
         (command "vpoint" "0,1,0")
      );b
      ((= pt "P")                          ;plan view
         (command "vpoint" "0,0,1")
      );p
      (T                                   ; a point was entered
         (setq vpt (upoint 17 "" "Location point of eye" nil pt)   ;need eye point
               alpha (angle pt vpt)        ;rotation angle in XY plane
               beta  (bangle pt vpt)       ;rotation angle in XZ plane
         );setq
         (command "vpoint" "r" (angtoc alpha) (angtoc beta)) ;AutoCAD release 9 option
      );T
   );cond
(princ)
);defun
;*
```

Save, Exit and reenter AutoCAD. TEST.LSP should load.

Test it:

Command: **PICKVPT**   Try a key view point. Back.
Target point or Left/Right/Front/Back/Plan view: **B**
vpoint Rotate/<View point> <0'-0",-0'-1",0'-0">: 0,1,0 Regenerating drawing.
Command: **PICKVPT**   Try a target and eyepoint:
Target point or Left/Right/Front/Back/Plan view: **0,0,0**
Location point of eye: **1,1,1**
vpoint Rotate/<View point> <0'-0",0'-1",0'-0">: r Enter angle in X-Y plane from X axis <90>: <<45.000000 Enter angle from X-Y plane <0>: <<35.264390 Regenerating drawing.

Polar Rotations in Space

Another useful 3D tool is 3DPOLAR. 3DPOLAR works like POLAR. In addition to the arguments of POLAR, 3DPOLAR takes the "beta" angle and a 3-D point instead of a 2-D point. It calculates the projection of the 3-D point on the XY plane, then takes the Z difference (the rise) between the two points. The X and Y components are shortened by the "beta" angle. Again, the function MAPCAR is used for the point calculations.

Making More 3D Tools

Select **[EDIT-LSP]** Edit TEST.LSP, and add:

```
;* 3DPOLAR calculates a point in 3D space given rotation from X axis in XY plane,
;* rotation from XY plane toward Z, base point, and offset distance.
(defun 3dpolar (pt alpha beta dist / dx dy dz)
    (setq dx (* dist (cos alpha) (cos beta))   ;shortened by X and Z projection
          dy (* dist (sin alpha) (cos beta))   ;shortened by Y and Z projection
          dz (* dist (sin beta)))              ;only Z projection causes shortening
    );setq
    (mapcar '+ (list dx dy dz) pt)             ;adds delta values to base point
);defun
;*

;* C:3DPOLAR is a friendlier interface for the above 3DPOLAR.
;* To use transparently, enter (c:3dpolar) in menu macros.
(defun C:3DPOLAR ( / bpt)
  (3dpolar
    (setq bpt (upoint 17 "" "3D Basepoint" nil nil))
    (uangle 1 "" "Angle from X axis in XY plane" nil bpt)
    (uangle 1 "" "Angle from XY plane toward Z" nil bpt)
    (udist 1 "" "Distance" nil nil)
) );defun
;*
```

Save, Exit and reenter AutoCAD. TEST.LSP should load.

```
Command: 3DPOLAR
Basepoint: 0,0,0
Angle from X axis in XY plane: 45
Angle from XY plane toward Z: 35.2643     The previously measured angle.
Distance: 1.732051                        The previously measured distance to 1,1,1.
Lisp returns: (1.000001 1.000001 0.999997)         Off a squinch, we rounded off the angle.
```

Now that you have 3DPOLAR, it's easy to write a quick vertical arc generator, called C:ZARC:

Select [EDIT-LSP]   Edit TEST.LSP, and add:

```
;* C:ZARC is a simple vertical arc, using UFUNS for input and 3DPOLAR.
(defun C:ZARC ( / bpt rad alpha beta1 beta2 anginc )
  (setq bpt (upoint 17 "" "Centerpoint" nil nil)
        rad (udist 1 "" "Radius" nil bpt)
        alpha (uangle 1 "" "Angle from X axis in XY plane" nil bpt)
        beta1 (uangle 1 "" "Start angle from XY plane toward Z" nil bpt)
        beta2 (uangle 1 "" "End angle from XY plane toward Z" nil bpt)
        anginc (ureal 1 "" "Resolution (parts per 360 deg.)" nil)
        anginc (/ 6.28318531 anginc)               ;2PI/anginc
  );setq
  (command "3DLINE" (3dpolar bpt alpha beta1 rad))
  (while (< (setq beta1 (+ beta1 anginc)) beta2)   ;while < end angle
    (command (3dpolar bpt alpha beta1 rad))       ;draw another segment
  )
  (command (3dpolar bpt alpha beta2 rad) "")      ;finish 3dline "Arc"
);defun C:ZARC
;*
```

Save, Exit and reenter AutoCAD. TEST.LSP should load.

Command: **ZARC**   Test it.
Centerpoint: Pick a point.
Radius: Pick or enter.
Angle from X axis in XY plane: Line at which the arc plane intersects XY.
Start angle from XY plane toward Z: Show or type.
End angle from XY plane toward Z: Show or type.
Resolution (parts per 360 deg.): **32**
3DLINE From point: Draws arc with 3Dlines. Use PICKVPT to see it.

Let's make a few calculation tools that you can apply to arcs, circles and spheres.

Other Polar and Curve Formulas

The following functions are helpful when you work with curved surfaces, arcs, and circles in both 2-D and 3-D. We haven't used them elsewhere in the book, but you will find them useful. With the exception of CPT, for center point, all functions use the basic curve formulas shown in the diagram.

The formula for a circle calculates a point on the circumference of the circle given an angle and radius:

```
Coord = Radius*COS(angle),Radius*SIN(angle)
```

In AutoLISP you write it:

```
(list (* rad (cos ang)) (* rad (sin ang)))
```

This formula and derivations appear in most of the following functions. The functions' starting angles are similar to how AutoCAD draws arcs.

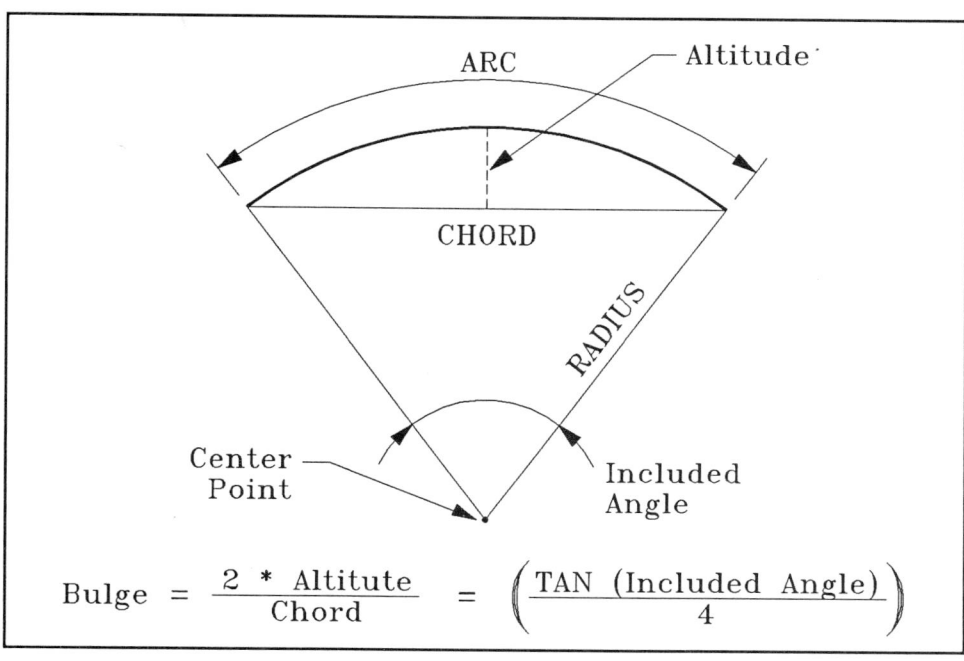

Trigonometry of a Curve

Making 3D Curve Functions

Select **[EDIT-LSP]**   Edit TEST.LSP, and add:

```
;* RADIUS calculates the radius of an arc given the chord distance and
;* altitude dimensions.
(defun radius (chord altitude) ;ok
   (/ (+ (expt (/ chord 2.0) 2.0) (* altitude altitude)) (* 2.0 altitude))
);defun
;*

;* ARCLEN calculates the arc length given the radius and included angle
(defun arclen (rad ang / ang)
   (while (> ang #pi360) (setq ang (- ang #pi360)))    ;assures angle is less than full circle
   (* rad ang)                                          ;multiplies radius by included angle
);defun
;*

;* INCANG calculates the included angle given a chord distance and radius
(defun incang (chord radius)
   (* 2.0 (atan                                          ;twice the arc tangent
      (/ chord 2.0                                       ;of opposite over adjacent
         (sqrt (- (expt radius 2) (expt (/ chord 2.0) 2)))) ;calcs adjacent leg of triangle
   ));divide & atan
   );
);defun
;*

;* ALTITD calculates the altitude of an arc given the radius and included angle.
(defun altitd (radius angle)
   (- radius (* radius (cos (/ angle 2.0))))   ;radius minus the adjacent leg of
);defun                                         ;the triangle
;*
```

Save, Exit and reenter AutoCAD. Test the functions against AutoCAD's Circle command.

Command: **PICKVPT**	Go to plan view.
Target point or Left/Right/Front/Back/Plan view: **P**	
Command: **ARC**	Center 0,0, Start 1,0 and End 0,1.
Command: **LINE**	Draw a line between two endpoints of the arc.
Command: **LIST**	Both line and arc: LINE reports: Length = 1.4142, Angle = 135 ARC reports: center point X=0.0000 Y=0.0000 radius 1.0000 start angle 0 end angle 90
Command: **(setq dst (sqrt 2))**	Returns: 1.414214 -- a distance equal to lines length.
Command: **(setq ia (incang dst 1))**	Returns: 1.570796, 90 degrees. IA saves included angle.
Command: **(setq alt (altitd 1 ia))**	Returns 0.292893, the altitude distance. Compare using Dist command and Osnap MID to MID, then:
Command: **(arclen 1 ia)**	Returns: 1.570796, the arc length. Compare to a Pline:
Command: **PEDIT**	Select the arc. It changes to a pline arc.
Command: **LIST**	Reports: Length = 1.5708
Command: **(radius dst alt)**	Returns: 1.000000

 Delete TEST.LSP.

 Rename TEST.LSP to 3DTOOLS.LSP.

Mesh Generation

It is hard to visualize some 3-D surfaces in wireframes when they are represented by a single large 3Dface. Other surfaces may actually curve although their sides are straight. You can visualize such surfaces more accurately if they are divided into a number of smaller segments. Mesh generation involves filling a bounded surface with a pattern. Mesh generation often is used to subdivide an element into smaller elements to generate more sets of data, permitting a finer degree of analysis. Meshes also are used to add texture to a drawing. Regardless of the application, the technique is the same.

The 3DFILL function works with a simple mesh, one that uses 3DFACEs to fill the area between any four points. The points are true 3-D points, at any X,Y,Z coordinate. 3DFILL is a command function that gets 4 points from the user, asks for the number of rows and columns in the mesh, then calculates the points while simultaneously drawing the 3DFACEs. Let's look at bit closer at the problem.

Making A Mesh Generator 3DFILL

  Copy 3DFILL.LSP to TEST.LSP.

  Create a new LISP function file called TEST.LSP.

Select **[EDIT-LSP]**   Start a new TEST.LSP.
Enter the following, then we'll discuss how it works.

```
; calc-inc returns a list of XYZ point offsets between two points & divided by
; row or columns
(defun calc-inc (p1 p2 inc)
  (mapcar '/                          ;divides the delta in XYZ by increment number
    (mapcar '- p1 p2)                 ;gets delta of coordinates
    (list inc inc inc)                ;build a dummy list to serve as the dividing numbers
  );mapcar
);defun
;*
```

Save, Exit and reenter AutoCAD. TEST.LSP should load.

Command: **(calc-inc '(1.0 1.0 1.0) '(0.0 0.0 0.0) 4)**   The 1/4 increment for 1,1,1 to 0,0,0.
Lisp returns: (0.250000 0.250000 0.250000)

CALC-INC uses a few MAPCARs. It calculates the X, Y and Z component differences between two points and divides each by the increment number, returning the three increment values as a list. The increment value is the number of columns (or rows) of the current face being drawn.

The main body of the function gets the four points from the user and calculates the unit slope of a line from the 1st to 2nd and 3rd to 4th points. The unit slope is the distance by which the program will increment the starting point of the 3DFACE each time through the loop. The first part gets the input and uses CALC-INC to calculate the ROW increment values.

Select **[EDIT-LSP]**   Edit TEST.LSP, and add:

```
;* 3DFILL fills four 3-D points with rows and columns of 3DFACE's
(defun C:3DFILL ( / p1 p2 p3 p4 tp row col inc1-2 inc3-4 inc1-4 inc2-3)
   (setq p1 (upoint 17 "" "Enter 1st point" nil nil)      ;get 1st point
         p2 (upoint 17 "" "Enter 2nd point" nil p1)       ;get 2nd point
         p3 (upoint 17 "" "Enter 3rd point" nil p2)       ;get 3rd point
         p4 (upoint 17 "" "Enter 4th point" nil p3)       ;get 4th point
         row (uint 1 "" "Number of rows (---)" 1)         ;get number of rows, 1 is default
         col (uint 1 "" "Number of columns (|||)" 1)      ;get number of columns, 1 is default
         inc1-2 (calc-inc p2 p1 row)   ;calculate XYZ increment from 1st to 2nd point
         inc3-4 (calc-inc p3 p4 row)   ;calculate XYZ increment from 4th to 3rd point
   );setq
```

Next, a double looping segment of the program goes through each row and draws all the mesh elements for that row. For each row, an inner loop is run for the columns of the mesh. In the row statements, a unit slope along the 1 to 4 points is calculated. The constant unit slopes for 1 to 2 and 3 to 4 are applied in calculating P2 and P3. Then, the slope along the P2 and P3 line is calculated. This slope is used in the column loop part of the program. Take another look at the illustration at the start of the chapter.

The column loop determines the points for each column mesh element and draws the 3DFACE. Before entering the column loop the base point for the next loop cycle is set and then recalled at the end of the column loop.

Save what you've done and continue adding to TEST.LSP:

```
  (repeat row                             ;repeat 3DFACE for each row
    (setq inc1-4 (calc-inc p4 p1 col)     ;calculate temp XYZ inc from 4th to 1st pt
          p2 (mapcar '+ p1 inc1-2)        ;calculate new 2nd point
          p3 (mapcar '+ p4 inc3-4)        ;calculate new 3rd point
          inc2-3 (calc-inc p3 p2 col)     ;calculate temp XYZ inc from 3rd to 2nd pt
          bp p2                           ;save the base point.
    );setq
    (repeat col                           ;repeat 3DFACE for each column
      (setq p4 (mapcar '+ p1 inc1-4)      ;calculate new 4th point
            tp (mapcar '+ p2 inc2-3)      ;calculate temporary point
      );setq
      (command "3DFACE" p1 p2 tp p4 "")   ;draw 3DFACE
      (setq p1 p4 p2 tp)                  ;set 1st to 4th point, set 2nd to temp point
    );repeat                              ;end column repeat
    (setq p4 p3 p1 bp)                    ;set 1st & 4th point to 3rd point
  );repeat                                ;end row repeat
(princ)
);defun C:3DFILL
;*
```

Save, Exit and reenter AutoCAD. Try the routine:

```
Command: ERASE                      Clean up the screen.
Command: ZOOM                       Center at 0,0 and height 7.
Command: 3DFILL
Enter 1st point: 0,0,0
Enter 2nd point: 0,0,1.5
Enter 3rd point: .Z
of 0,0,1
(need XY): @3<22.5
Enter 4th point: 3,0,-.5
Number of rows (---) <1>: 3
Number of columns (||||) <1>: 6
Command: 3DFACE First point:        Repeatedly scrolls by.
```

```
Command: MIRROR                 Mirror it all across the X axis.
Command: ARRAY                  Select them all. Polar around 0,0.
Number of items: 8              And fill 360 degrees. Rotating objects as copied.

Command: PICKVPT                Target 0,0,0 and Eye 4.5,2,4. Use Vpoint -1,-1,0.6 for AutoCAD 2.6.
Command: HIDE                   Take a break!

Command: END
```

  Delete TEST.LSP.

  Rename TEST.LSP to 3DFILL.LSP.

Your screen should look like a 3-D Image Using 3DFILL.

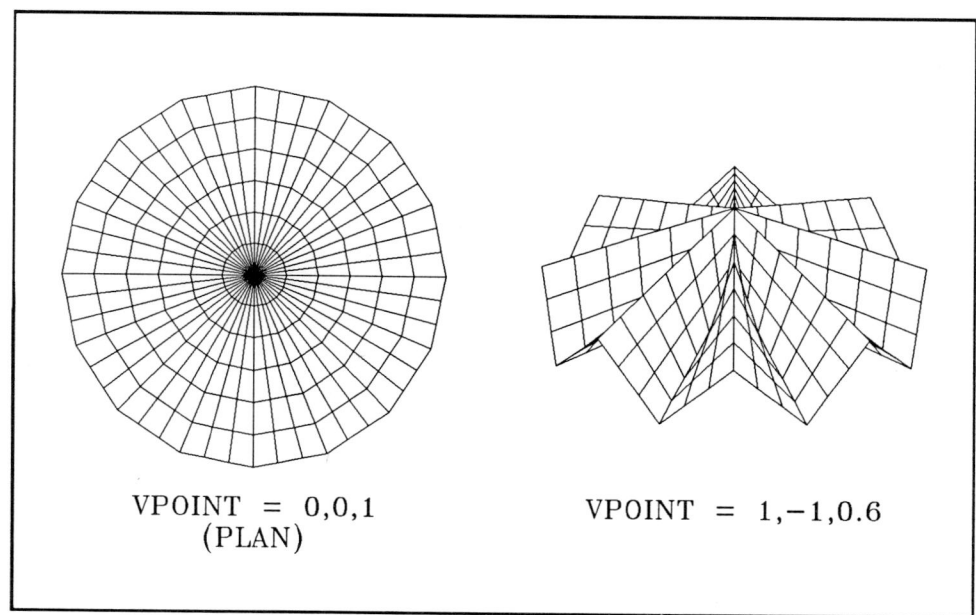

3-D Image Using 3DFILL

If you have the AutoShade program, you can shade this image. An AutoShaded image is shown in Shaded Image. You also can try using 3DFILL to subdivide other images for AutoShade and to see the differences in the renderings.

Shaded Image in AutoShade

Although 3DFILL only works with straight edged sections, you can write more sophisticated programs that work with curved edges. These programs require the use of graph equations.

Converting 2-D Plans to Elevations

Plan views and elevations are generally drawn separately, although they use common coordinate data and entities. Separating plan view and elevation drawings is unnecessary. If you control elevations and thicknesses of the plan entities as they are drawn, AutoLISP gives you access to the database to automatically generate elevations from the plan view.

Generating elevations is simple. You need to:

- Draw your plan with appropriate ELEVations and THicknesses.
- Get a selection set of the plan entities.
- Examine each entity and determine the entity type.
- For each entity type, extract the relevant data needed to show it in elevation.
- Draw the corresponding entity in the elevation.

The MELEV function takes the coordinates, elevation, thickness and layer properties from the entity data and draws the entity in elevation. The function does not include linetype. Linetype is something that you can consider adding to MELEV. The drawn elevation should look like the corresponding 3-D Vpoint of the plan.

To visualize how MELEV works, take a LINE as an example. A LINE has two endpoints, an ELEVation and THickness. It is shown in the elevation view as a THickness high rectangle with its base at ELEV, and its length equal to its projected length at the elevation viewing angle. A CIRCLE projects to look like a rectangle in the elevation view. Actually, everything in the elevation view is drawn as rectangles.

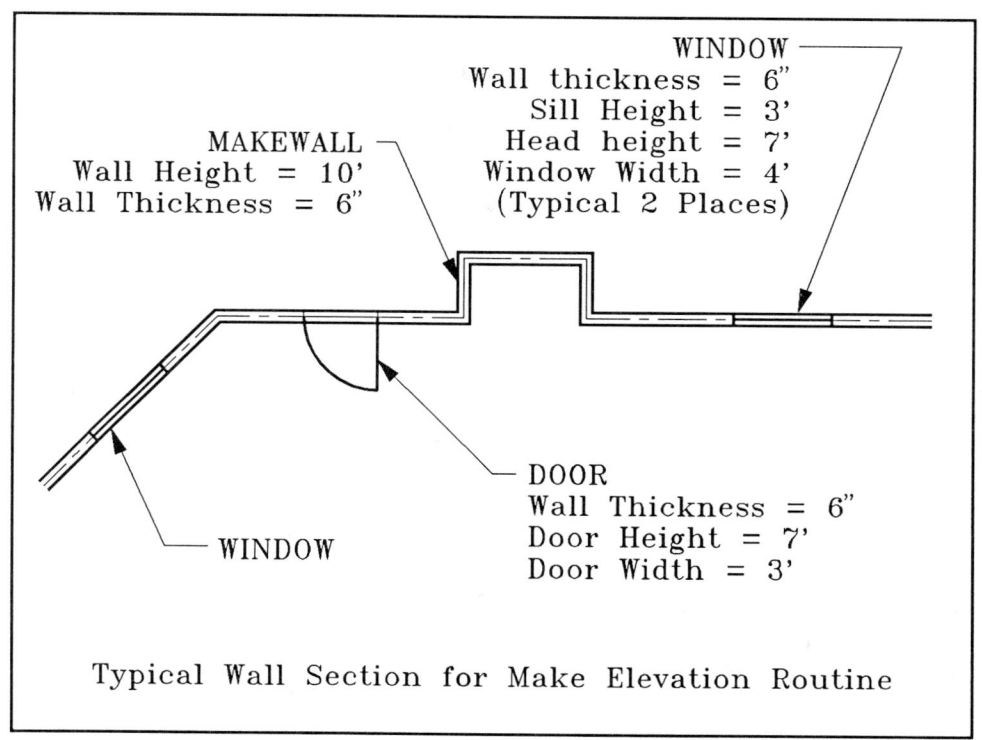

Diagram for the MELEV Program

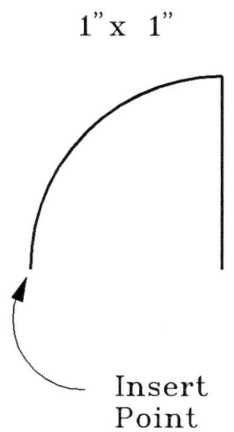

Door Block

If you don't have the CA DISK, you need to create a simple plan view. It must have the appropriate ELEVations and THicknesses set for each entity. The menu items provided in the exercise make it easy for you to draw what you need. What you meed is shown in the Diagram for the MELEV Program. The initial door block that you need is shown in the Door Block illustration.

17—18 Customizing AutoCAD

MELEV A Simple Elevation Program

 If you have the CA DISK, edit the EXISTING drawing MELEV.

 If you don't have the CA DISK, begin a NEW drawing MELEV.

Select **[TEST] [24 x 36] [1/8"] [INITIATE]**   It brings up TEST.MNU.

Create a DOOR Block, then:

Select **[EDIT-MNU]**

Change the [CA00TEST] to [CA17TEST] and add the following page change to the [3D] label on the root page:

```
***SCREEN
**ROOT
[CA17TEST]
[]

[   3D   ]^C^C^C$S=3D $S=FOOTER LAYER S OBJ02;;
```

Change the **EXAMPLES page name to **3D and add these. Count 19 [labels] & []s as usual:

```
**3D
[   3D   ]
[]
[CL-WALL ]^C^C^C(setvar "ELEVATION" 0);+
(setvar "THICKNESS" (setq #wh (udist 1 "" "Enter wall height" #wh nil)));\PLINE;
[MAKEWALL]^C^C^C+
(setq #wthk (udist 1 "" "Enter wall thickness" #wthk nil));+
(setq en (car (setq ent (entsel "^MSelect pline to make wall: "))));\+
(setq pt1 (dxf 10 (entget (setq en (entnext en)))));+
(setq pt2 (dxf 10 (entget (entnext en)))) OFFSET (/ #wthk 2.0);+
!ent (polar pt1 (+ (angle pt1 pt2) 1.57) 1) ;EXPLODE L OFFSET (/ #wthk 2.0) !ent;+
(polar pt1 (- (angle pt1 pt2) 1.57) 1) ;EXPLODE L;+
CHANGE !ent ;P E 0 T 0 LA CEN31 ;SETVAR THICKNESS 0
[]
[WINDOW  ]^C^C^C(setq #wthk (udist 1 "" "Enter wall thickness" #wthk nil));\+
(setq #wsill (udist 1 "" "Enter window sill height" #wsill nil));\+
(setq #whead (udist 1 "" "Enter window head height" #whead nil));\+
(setq #wsize (udist 1 "" "Enter window width" #wsize nil));\+
LINE (setq pt1 (getpoint "^MPick insert point at CL wall")) \(setq pt2 (polar pt1;+
(setq #wangl (uangle 1 "" "Enter window angle" #wangl pt1)) #wsize));\;+
(setq ent (entsel)) @ CHANGE L ;P E !#wsill T (- #whead #wsill) ;OFFSET;+
(/ #wthk 2.0) !ent (polar pt1 (+ (angle pt1 pt2) 1.57) 1));;+
CHANGE L ;P E 0 T !#wsill ;COPY L ;0,0 ;CHANGE L ;P E !#whead;+
T (- #wh #whead);;(setq pt3 (dxf 10 (entget (entlast))));+
OFFSET (/ #wthk 2.0) !ent (polar pt1 (- (angle pt1 pt2) 1.57) 1) ;CHANGE;+
L ;P E 0 T !#wsill ;COPY L ;0,0 ;CHANGE L ;P E !#whead T (- #wh #whead);;+
LINE (dxf 10 (entget (entlast))) !pt3 ;CHANGE L ;P E !#wsill;+
T (- #whead #wsill) ;COPY L ;!pt1 !pt2
[]
```

```
[DOOR     ]^C^C^C(setq #wthk (udist 1 "" "Enter wall thickness" #wthk nil));\+
(setq #dhead (udist 1 "" "Enter door head height" #dhead nil));\+
(setq #dsize (udist 1 "" "Enter door width" #dsize nil));\+
LINE (setq pt1 (getpoint "^MPick insert point at CL wall")) \(setq pt2 (polar pt1;+
(setq #dangl (uangle 1 "" "Enter door angle" #dangl pt1)) #dsize));\;+
(setq ent (entsel)) @ OFFSET (/ #wthk 2.0) !ent;+
(polar pt1 (+ (angle pt1 pt2) 1.57) 1) ;CHANGE L ;P E !#dhead;+
T (- #wh #dhead) ;(setq pt3 (dxf 10 (entget (entlast))));+
OFFSET (/ #wthk 2.0) !ent (polar pt1 (- (angle pt1 pt2) 1.57) 1) ;CHANGE;+
L ;P E !#dhead T (- #wh #dhead) ;LINE (dxf 10 (entget (entlast))) !pt3;;+
CHANGE L ;P E 0 T !#dhead ;COPY L ;!pt1 !pt2 CHANGE !ent ;P E 0 T 0 LA CEN31;;+
INSERT DOOR S !#dsize;\\CHANGE L ;P E 0 T 0;;
[]
[]
[MAKE ELV]^C^C^C(caload "/CA-ACAD/" "MELEV") MELEV
[]
```

For AutoCAD 2.6, change the "INSERT DOOR S ..." line to read:
INSERT DOOR \!#dsize ;\CHANGE L ;P E 0 T 0;;

Save and Exit to AutoCAD. TEST.MNU reloads.

Select [3D]   Test these and draw a plan.
                                      You also can use LINEs and CIRCLEs, with different ELEV and
                                      THickness settings.

[CL-WALL] is a Pline. The other menu items aren't as complex as they look. They get and save variables, draw lines, copy, offset, explode, and change elevation and thickness. For example, [MAKEWALL] takes a pline, turns it into a center line and creates a two sided extruded 3-D wall. It explodes the polylines because MELEV only takes lines and circles as entities. You should break the walls for the openings to avoid obscuring openings in 3-D.

[WINDOW] parametrically builds a window. [DOOR] builds an opening and inserts a unit-sized DOOR block. [DOOR] uses Release 9's block Insert preset scale. For AutoCAD 2.6, change the last line as shown in the exercise.

The [MAKE ELV] item will load and execute MELEV.LSP after you create and rename it.

  Copy MELEV.LSP to TEST.LSP.

  Create a new LISP file called TEST.LSP.

Select [EDIT-LSP]   Edit TEST.LSP, and enter:

```
;* C:MELEV (Make ELEVation) prompts for selection of plan entities
;* and draws an elevation. It considers only LINE and CIRCLE entities.
;* Plan entities must be drawn with appropriate ELEVation and THickness.
;* It uses the following CALCELV, PROJ, and DRWELV subroutines.

(defun C:MELEV( / ss1 bpt pang spt base opt pang90 count emax)
   (prompt "\nSelect plan components for elevation...")
   (setq ss1 (ssget)                                              ;plan components
         bpt (upoint 1 "" "Starting point on plan" nil nil)       ;elev base pt
         pang (uangle 1 "" "Projection plane" pang bpt)           ;projection angle
         spt (upoint 1 "" "Starting point for elevation" nil nil) ;elev start pt
         base (udist 1 "" "Finished floor elevation" (getvar "ELEVATION") nil)
   );setq

   ;* simplify program by readjusting angle if over 180 degree
   (if (> pang pi)
      (progn
         (setq pang (- pang pi))                    ;reduces angle to less than 180 degrees
         (prompt (strcat "\nProjection angle adjusted to " (angtos pang)))
      );progn
   );if

   (setvar "ELEVATION" 0)       ;ensure elevations drawn with no
   (setvar "THICKNESS" 0)       ;evalation and thickness
   (terpri)                     ;newline.

   ;* set some general constant values for the program.
   (setq opt (polar bpt pang 1.0) ;an offset point to describe the line
         pang90 (+ pang #pi90)    ;perpendicular angle to projection angle
         count 0                  ;initialize counter
         emax (sslength ss1)      ;max limit of entities
   );setq

   (while (< count emax)               ;process each entity in selection set
      (calcelv (ssname ss1 count) ;pass data to the calculating function
               bpt spt opt base pang pang90 ;these are constants and various base pts.
      );calcelv function
      (setq count (1+ count))    ;increment to the next entity of set.
   );while
   (grtext)
   (prompt "\nDone")
   (princ)
);defun C:MELEV
;*
```

CALCELV, the "entity manager" function, is the next part of the program. CALCELV determines the entity type, filtering with a COND. It ships the entity data off to the projection function PROJ, and calls DRWELV to draw the elevation entities. It supports LINE and CIRCLE entities. We kept it simple, but you can add support for other entities.

Save what you've done and continue adding to TEST.LSP:

```
;* CALCELV determines entity type (LINE or CIRCLE) sends entity data
;* to PROJ, then calls DRWELV to draw elevation entity.
(defun calcelv (en bpt spt opt base pang pang90 / ed et eel etk elay
                pta ptb pvals cpt rad)
   (setq ed (entget en)                         ;get entity data
         et (dxf 0 ed)                          ;determine it's type
         eel (dxf 38 ed)                        ;entity elevation
         etk (dxf 39 ed)                        ;entity thickness
         elay (dxf 8 ed)                        ;entity layer
   );setq

   (cond                                        ;filter for entity type

      ;* entity is a LINE...
      ((= et "LINE")                            ;entity type is line
         (setq pta (dxf 10 ed)                  ;one end point
               ptb (dxf 11 ed)                  ;second end point
               pvals (proj et pta ptb pang pang90 bpt opt) ;calc projected form.
         );setq

         (if pvals                              ;if object shows up in projected form
            (drwelv (car pvals) (cadr pvals)
                    spt pang eel etk base elay  ;draw it
            );drwelv function
         );if
      );line entity

      ;* entity is a CIRCLE...
      ((= et "CIRCLE")                          ;its a circle
         (setq cpt (dxf 10 ed)                  ;center point
               rad (dxf 40 ed)                  ;radius
               pta (polar cpt 0.0 rad)          ;point on circumference
               ptb (polar cpt pi rad)           ;and another point...
               pvals (proj et pta ptb pang pang90 bpt opt) ;calc projected length
         );setq
         (if pvals                              ;if projection exists
            (drwelv (car pvals) (cadr pvals)    ;draw it
                    spt pang eel etk base elay
            );draw elevation
         );if
      );circle

      ;* if it gets here the user selected an unsupported entity. Tell them.
      (T (prompt (strcat "\n\t\nIgnoring unsupported " et " entity.\n\t")))
   );cond
);defun CALCELV
;*
```

The program lets the user specify the angle of the projection plane. Think of the projection plane as a wall that the entities of the plan are flattened against. PROJ takes the projection plane that MELEV gets, and determines the entity's "True Length" from the plan view entity data fed to PROJ by CALCELV. PROJ then calculates the distorted length for the elevation.

Save what you've done and continue adding to TEST.LSP:

```
;* PROJ calcs projected length of a "line" based on true length & proj'n angle.
(defun proj (et pta ptb pang pang90 bpt opt / tleng lang pleng ppta pptb poff pleng2)
    (grtext -1 (strcat "Processing " et))
    (setq tleng (distance pta ptb)                              ;the line's true length
          ppta (inters bpt opt pta (polar pta pang90 1.0) nil)  ;perp pt along projection
          pptb (inters bpt opt ptb (polar ptb pang90 1.0) nil)  ;perp pt for 2nd pt
    );setq
    (if (> (setq pleng (distance ppta pptb)) 0.000001)          ;length too small-don't report it.
        (list (min (distance bpt ppta) (distance bpt pptb)) pleng) ;return list of offset and
        nil                                                     ;projected dist
    );if
);defun PROJ
```

The next function, DRWELV, takes the entity data and the projection's results and generates a rectangle in elevation.

Save what you've done and continue adding to TEST.LSP:

```
;*
;* DRWELV takes the object information and draws it in elevation view.
(defun drwelv (poff pleng spt pang eel etk base elay / cpt1 cpt2 cpt3 cpt4)
    (if eel                                                     ;if entity has a Z elev
        (setq cpt1 (polar (polar spt pang poff) #pi90 (+ base eel))) ;calc point on entity elev
        (setq cpt1 (polar (polar spt pang poff) #pi90 base))    ;no Z elev
    );if
    (setq cpt2 (polar cpt1 pang pleng))
    (if etk                                                     ;if entity has Z thickness
        (setq cpt3 (polar cpt2 (+ pang #pi90) etk)               ;calc vertical height in elevation
              cpt4 (polar cpt1 (+ pang #pi90) etk)
        );setq
        (setq cpt3 cpt2 cpt4 cpt1)                              ;else it's only a line in elev view
    );if
    (command "layer" "s" elay "" "line" cpt1 cpt2 cpt3 cpt4 "c") ;change layer, draw elev.
);defun DRWELV
;*

(princ)
;*end of MELEV.LSP
;*
```

Save, Exit and reenter AutoCAD. TEST.LSP should load.

One hint before you test MELEV. It can't distinguish which entities should be hidden, so it draws everything selected. You'll have to do a little editing to get rid of entities on the backside of the wall. You also can avoid selecting many of the backside entities by using Vpoint to get a view close to that of your elevation. Make sure the viewpoint has a tiny, but non-zero Z elevation, or you can't select entities. Then, do a Hide and select the entities. It will exclude most backsiders!

```
Command: VPOINT                           Set a 3-D view. Try 0,6,.0000001
Command: HIDE
Command: SELECT                           Preselect them.
Command: ZOOM                             Previous, to the plan view.

Command: MELEV                            Test it.

Select plan components for elevation...
Select objects: P                         Previous. What you just selected.
Starting point on plan:                   Pick point at Lower Left corner of plan entities.
Projection plane:                         Pick point off to the right.
Starting point for elevation:             Pick point above plan.
Finished floor elevation <0">:            Take the default.
Command: LAYER                            Many Layer and Line commands scroll by as it draws.
Done
Command: END                              You'll use it below.
```

 Delete TEST.LSP.

 Rename TEST.LSP to MELEV.LSP.

The generated elevation is shown in the Generated Elevation Drawing.

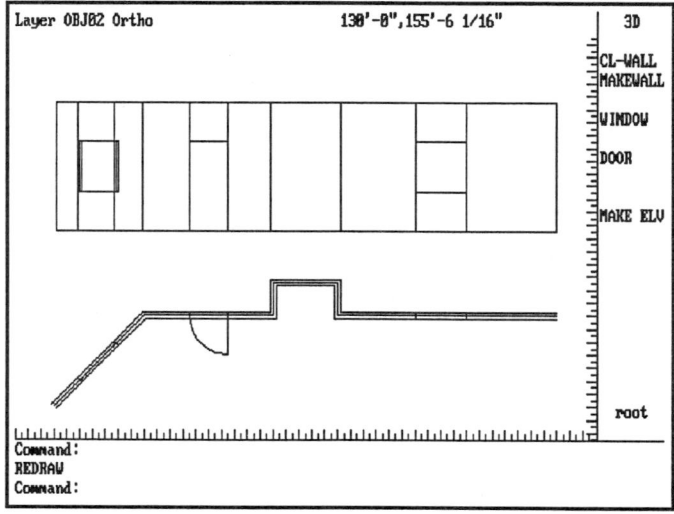

The Generated Elevation

Each entity type that you add to MELEV is a separate case. Adding Arcs and Plines isn't too hard. Use the book's curve formulas and take the projection of the bulge into effect. Plines also have to compensate for width. Blocks require using table access to look at their components, and then calling your modified CALCELV recursively. To handle nested blocks, you also need to make your block routine recursive!

Integration From 3-D to 2-D

You'll be happy to know there is an easier way to generate a 2-D drawing of any 3-D view or elevation, but with a tradeoff in accuracy. AutoCAD can plot any view in a drawing to a DXB format file. This file describes primitive entities in a compact binary format. The process of creating a 2-D image from a 3-D view is as follows:

- Reconfigure AutoCAD to plot ADI to an AutoCAD DXB file.
- Load your drawing, get your desired Vpoint.
- Plot it. It goes to a file with extension .DXB.
- Do a DXBIN to import the file. It comes in as a 2-D drawing of a 3-D view composed of lots of little line segments.
- Edit, Block, Scale and annotate the drawing.

The main limitation is scale and accuracy. Plotting to DXB is like plotting to an electronic sheet of paper. You specify the size of the sheet and the resolution. Resolution is your "plotter step size," the smallest "pen" move. Your SIZE x RESOLUTION cannot exceed 32767 steps. An accuracy of 1000th inches has a maximum size of 32+ inches. You figure your best compromise for what you need.

Using DXB to Create a 2-D Drawing From a 3-D View

```
Enter selection <0>: 5                              From the configuration menu, select 5.  Configure plotter:
Do you want to select a different one? <N> Y

Select device number or ? to repeat list <7>: 2                Pick the ADI driver number.

Select output format:                       Select 2.  AutoCAD DXB file:
Output format, 0 to 3 <0>: 2

Maximum horizontal (X) plot size in drawing units <0'-11">:
Plotter steps per drawing unit <83'-4">: 1000
Maximum vertical (Y) plot size in drawing units <0'-8 1/2">:
Do you want to change pens while plotting? <N>
```

`Remove hidden lines? <N> Y`	The regular plot configuration questions follow, including: Yes. Then save configuration and return to drawing MELEV.
`Command: VPOINT`	Select a 3-D view.
`Command: PLOT` `Enter file name for plot <MELEV>:`	Go through the usual plot dialog. Plot to Fit. Then: Take default file name. It plots.
`Command: VPOINT` `Command: ZOOM`	Return to plan view. Left, corner 0,0, height 9.

`Command: DXBIN`
`DXB file: MELEV`
`Command: END`
	DXB's always come in at 0,0 at their plot scale. If that was your plot file name.

 Delete TEST.MNU.

 Rename TEST.MNU to MY17.MNU.

`Command: QUIT`

The DXB comes in as a lot of entities. You can use the book's MARK and CATCH to collect them if they get entangled with other drawing entities. Plots render curves with vertical lines. You can control their density by adjusting your plot scale, size and resolution, and compensate scale after DXBIN using the Scale command.

Integrating the 3D Menu

If you don't have the CA DISK and want to add the 3D menu to CA-MENU:

- Append the **3D page of your MY17.MNU to the end of the CA-MENU.MNU file.

- Change the [3D] label on the **ROOT page of the CA-MENU to:

  [3D   ]^C^C^C$S=3D $S=FOOTER LAYER S OBJ02;;

Here is this chapter's menu.

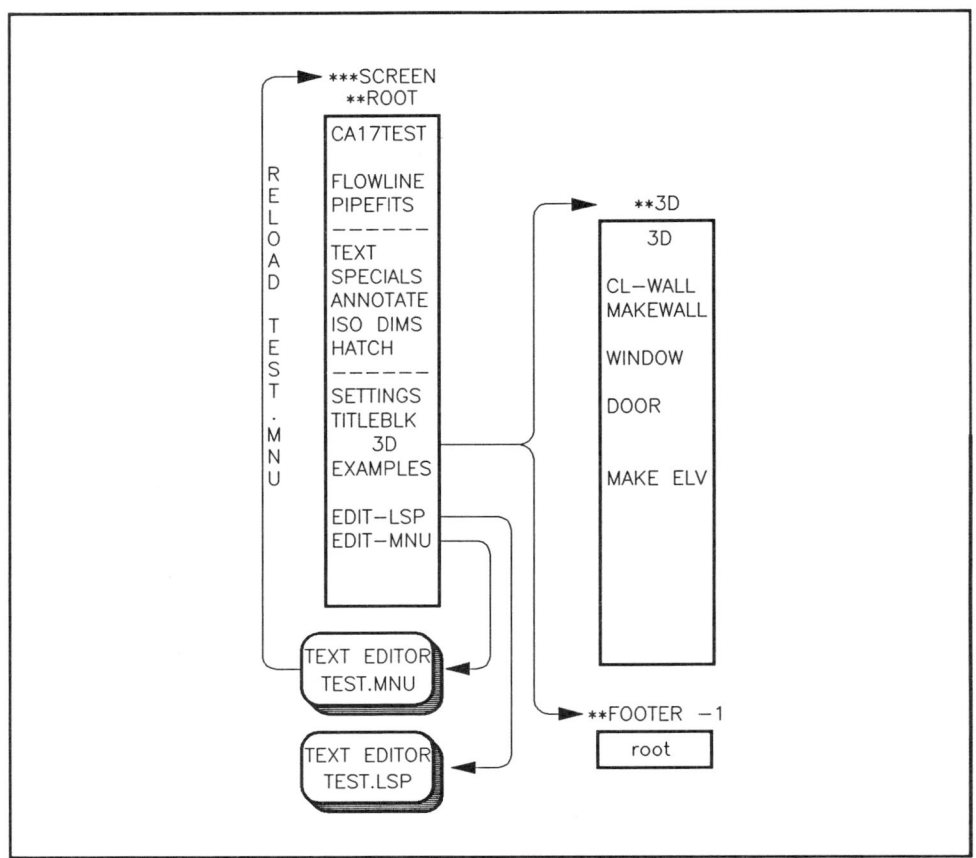

The Chapter 17 Menu

Tips and Techniques.

Your 3-D tools will become more valuable, not obsolete, in true 3-D AutoCAD. By developing general graphing functions you can simplify complex geometric drawings. As AutoCAD progresses towards true 3-D, the functions that you develop will be able to draw on more entity types. Here are some summary tips and techniques:

- ❑ To help your users understand 3-D, make simple programs that aid in drawing or viewing 3-D objects.
- ❑ It makes it easier for your user to draw in 3-D, if you do the point calculations in AutoLISP.
- ❑ Remember AutoCAD needs to be told to return 3-D points.
- ❑ Use 'SETVAR ELEVATION to change current elevations transparently.
- ❑ Use AutoLISP's MAPCAR and APPLY functions when adding, subtracting, multiplying or dividing 3-D points.
- ❑ Generate meshes to help depict 3-D surfaces. They transfer well to Autoshade. Even flat surfaces benefit from meshes.

- ❏ Use 3-DFACES when generating surfaces. AutoShade shades them better.
- ❏ Remember elevation and thickness don't support all entities.
- ❏ You can use the Change command and ENTMOD to apply elevation and thickness to Text and Shapes.
- ❏ For 3-D drafting, develop a Block Library with matching pairs of 2-D and 3-D blocks. Use the 2-D blocks to produce your drawing. Create an AutoLISP routine to swap in your 3-D blocks for 3-D viewing.
- ❏ Develop supporting screen menus for your plan drawings, giving correct elevations and thicknesses.

You have a basic set of tools for working in AutoCAD 3-D. Let's turn to see what you can do with AutoLISP to make a working iso-dimensioning system out of AutoCAD's dimensioning features.

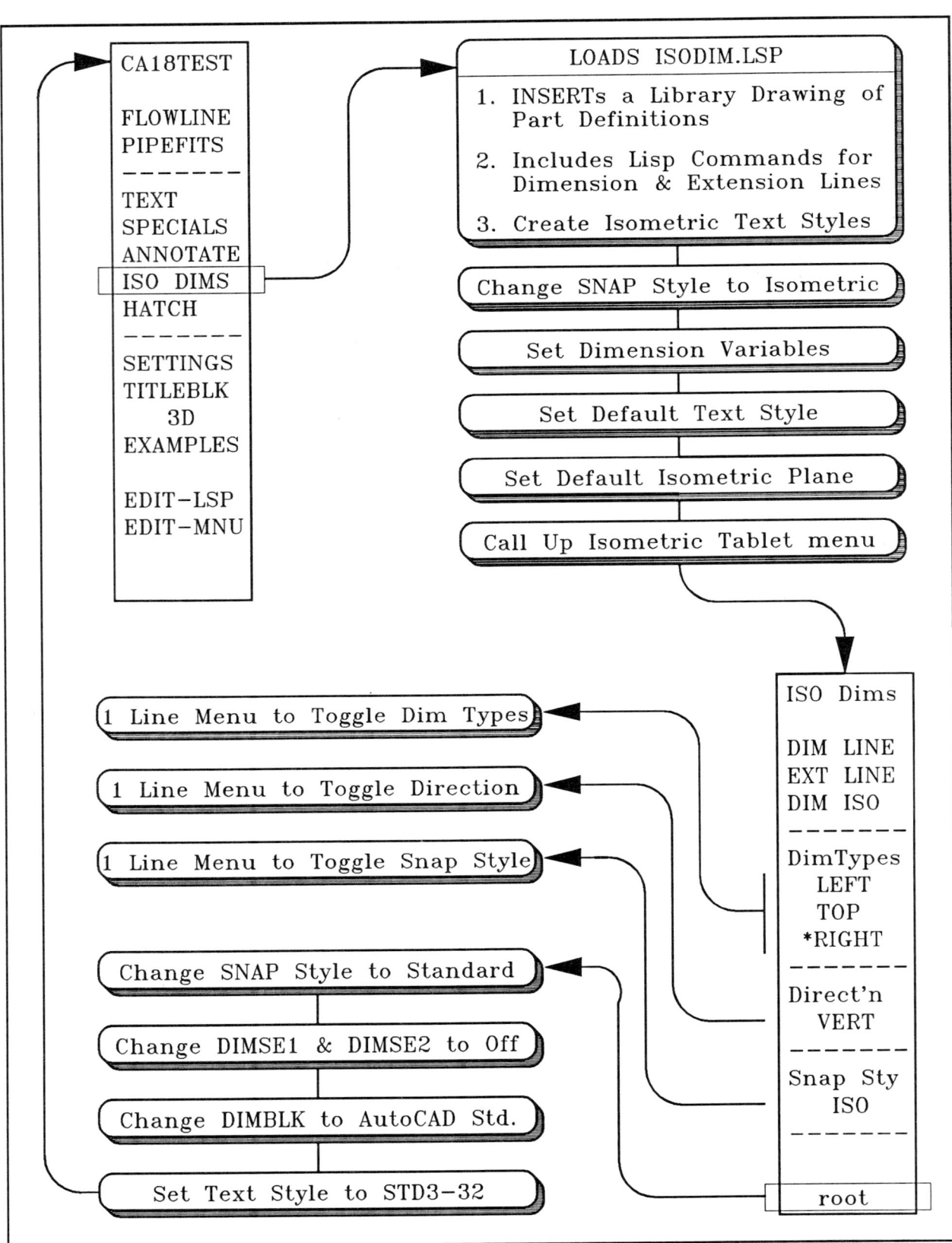

The ISO Dimensioning System

CHAPTER 18

An ISO Dimensioning System

This chapter will show you how to build a complete ISO-Dimensioning System using AutoLISP.

The Benefits of Building an Iso-dimensioning System

AutoCAD's dimensioning system leaves a lot to be desired. Many users just give up and don't attempt to do iso-dimensioning. You don't have to accept AutoCAD's "standard" dimensioning system. You can create your own. The benefits to creating your own iso-system are:

- You can make iso-dimensioning as easy for your users as standard dimensioning.

- By automating your iso-dimensioning, you can indicate the iso-dim type so that your users don't get lost. By showing the iso-dim type, you make iso-drawings easier and more useful.

- By customizing your iso-screen menus, you can give status information to your users to help them navigate your system.

How To Skills Checklist

This chapter will show you how to:

- ❏ Use blocks as dimensioning tics instead of arrows.
- ❏ Make isometric dim arrows.
- ❏ Work in iso-planes.
- ❏ Create symbols and text styles that appear to lie in iso-planes.
- ❏ Interactively manipulate dim variables, including changing text Styles and dim Blocks.
- ❏ Overlay screen menu pages and use them to give an interactive program mode status to your user.
- ❏ Customize the TABLET1 menu area to use iso-dimensioning.
- ❏ Use point of menu entry environment control in application programs.
- ❏ Use an AutoLISP file load to set up and initialize an application environment and tools.

Menus, AutoLISP Tools and Programs

MENUS
**ISO is the main isometric dimensioning page of the isometric menu system. It uses an integrated set of screen menu toggle pages. The ISODIM.LSP functions are required for this menu page.
**TISO is an isometric Tablet area 1 menu that is integrated with the screen menu. Selections made from the tablet are displayed on the screen menu page. The ISODIM.LSP functions are required for this tablet menu.

AutoLISP TOOLS
C:DIMLINE is an AutoLISP command function that draws an isometric dimension line, arrows and text.
C:EXTLINE is an AutoLISP command function that draws isometric extension lines.

PROGRAMS
ISODIM.LSP is a file containing two isometric dimensioning AutoLISP commands and initialization expressions. ISODIM is a program that integrates the isometric menu page system and the ISODIM.LSP file. ISODIM.LSP contains the C:DIMLINE and the C:EXTLINE command functions.

An Overview of Iso-Dimensioning

Isometric dimensioning is simple, if you give it a push from AutoLISP. The Iso-system that you will develop is compact, yet powerful, because it uses AutoLISP to enhance AutoCAD's dimensioning features.

The keys to the Iso-dimensioning system are:

- Altering AutoCAD's standard Dim settings to give the "look" needed for isometric drawings.

- Using the Snap command to set isometric Snap and Grid modes.

- Creating and toggling special text styles and dimension arrow Blocks to the current isoplane.

- Using AutoLISP routines to help draw extension lines, dimension lines and text.

- Placing controls, settings and modes on a dynamic screen menu.

The iso-screen menu is a good example of how to use menu label toggling in an application. The screen menu utilizes several small screen pages to show the user's current iso-dimension type, dimensioning direction and snap mode.

Isometric Dimension Text Styles and Symbols

Isometric text and symbols are skewed to the isometric planes. AutoCAD works in TOP, LEFT and RIGHT Isoplanes. You need to develop symbols to align in these planes. However, you don't need a separate style and symbol for each plane. You only need to create symbols for the LEFT and RIGHT planes. The TOP plane uses the LEFT or the RIGHT symbol and text, depending on which side of the face you are dimensioning.

You need two special text styles, slanted to the LEFT and RIGHT. You distort the text by applying an obliquing angle to the text font with the Style command. You automatically define these styles with a command expression in the ISODIM.LSP file. The command expression will set them with a fixed height, adjusted to the #DWGSC, when it loads. Here are the text styles:

```
NAME     FONT       ANGLE         HEIGHT

DIM-R    SIMPLEX    -30           (* 0.125 #dwgsc)
DIM-L    SIMPLEX     30           (* 0.125 #dwgsc)
```

Let's look at the arrow dimension blocks that you need.

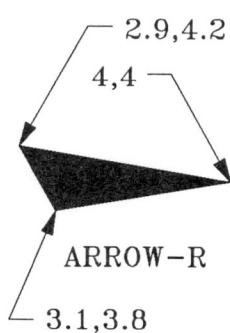

Isometric Arrows Diagram

If you have the CA DISK, you have an ISO-INIT.DWG with arrow blocks. If not, you need to start a blank drawing with an "=." The arrows that you need are shown in the Arrows Diagram. Use a Solid to draw the Iso arrow. Wblock the first arrow. Oops it back. Then, Mirror it to make the second arrow.

Isometric Symbols

 Skip to Making The Iso Screen Page.

 Begin a NEW drawing named ISO-INIT=

```
Command: ZOOM                Center 4,4 with height 4.
Command: SNAP                Set to 0.1.
Command: SOLID               Use the 3 points shown and <RETURN>.

Command: BLOCK               Base point at the tip, 4,4. Name ARROW-R.
Command: OOPS                Bring it back.
Command: MIRROR              Mirror across a horizontal line. Delete the old one.
Command: BLOCK               To ARROW-L. Don't Oops it. The drawing should look empty.
Command: END
```

Making an Iso Screen Page

To simplify writing the screen menu, the exercise provides a chart of the isoplanes, text styles, and dimension blocks for each of the six dimension types. Here is the chart:

ISO-DIMENSIONING DIMTYPES

DIMTYPE-DIRECTION	DIMBLK	TEXT STYLE	AUTOCAD ISOPLANE
RIGHT-VERT	ARROW-R	DIM-R	RIGHT
RIGHT-HORIZ	ARROW-R	DIM-R	TOP
LEFT-VERT	ARROW-L	DIM-L	LEFT
LEFT-HORIZ	ARROW-L	DIM-L	TOP
TOP-LEFT	ARROW-L	DIM-L	RIGHT
TOP-RIGHT	ARROW-R	DIM-R	LEFT

Please note that the exercise's DimType right/left/top do not correspond to AutoCAD's standard Isoplanes. Dimtype-Direction is our convention to relate the six types of iso-dimensions to their positions on a standard cube. Following the DIMBLK and Styles is easy, all Dimtype-Directions with LEFT get ARROW-L and DIM-L, and all the RIGHTs get -Rs. This is shown in the Isometric Reference Drawing.

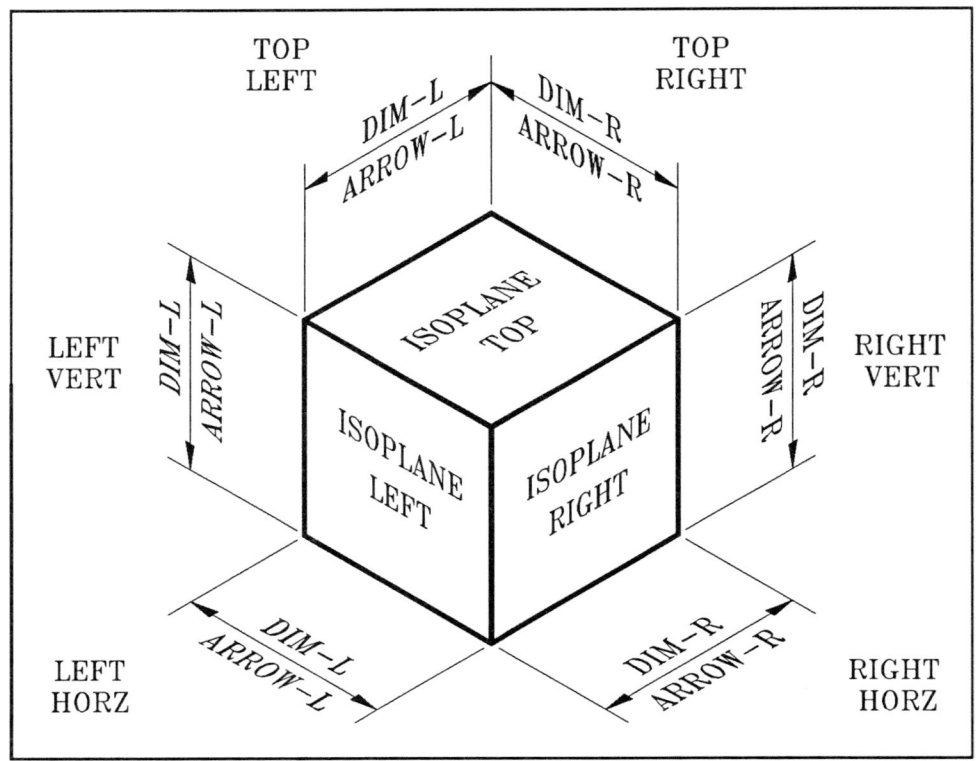

Isometric Reference Drawing

**ISO is the main screen page of the isometric menu. It is called by the [ISO DIMS] item on the root screen menu. [ISO DIMS] establishes the dimensioning environment as it calls the **ISO page. The **ISO menu formats the isometric dimensioning screen, calls toggle menus to switch isoplanes, changes text styles and dim blocks for the six dim-types. The menu also adds an **ISOFOOTER to redefine the normal [root] page switch. This **ISOFOOTER resets the normal environment when you switch back to the root menu. The menu is shown in the Isometric Menu Drawing.

Making an Iso Screen Page 18—5

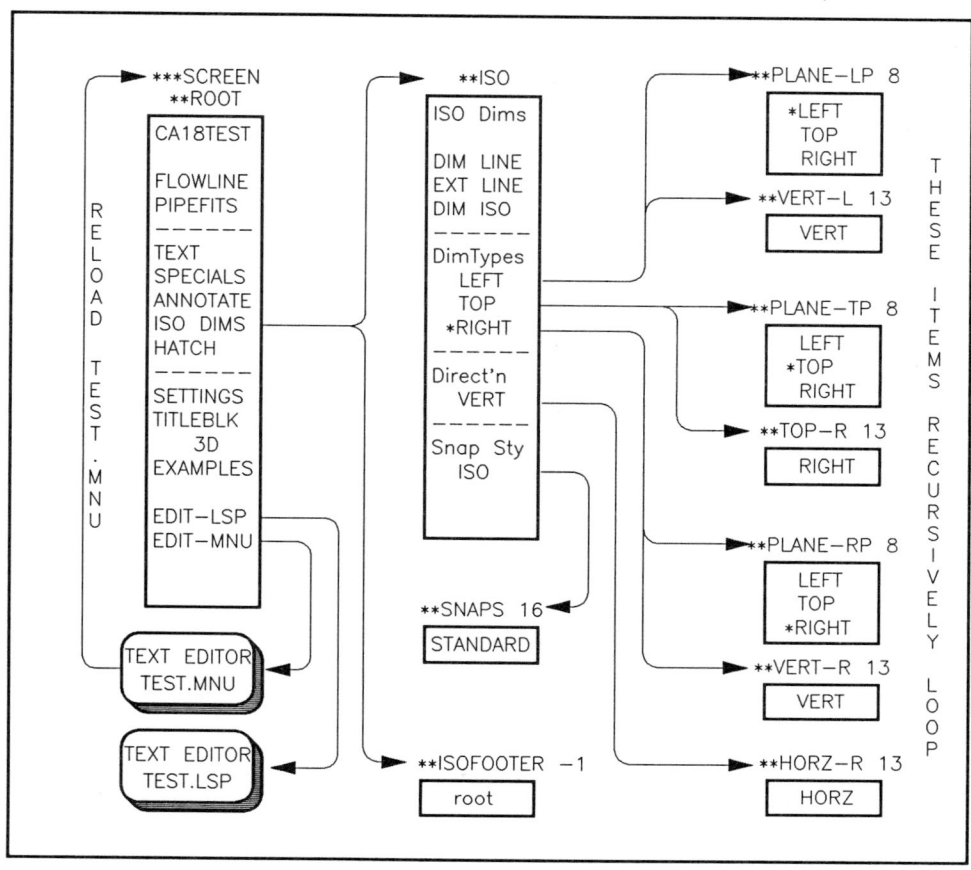

The Isometric Menu

Making the Isometric Dimensioning Screen Menu

Enter selection: Begin a NEW drawing named ISODIM.

  Copy CA18.MNU to TEST.MNU.

  Copy MAIN.MNU to TEST.MNU.

Select **[TEST] [24 x 36] [FULL] [INITIATE]**   It brings up TEST.MNU.
Command: **ERASE**   Erase the border.
Select **[EDIT-MNU]**   Change [CA00TEST] to [CA18TEST]:

***SCREEN
**ROOT
[CA18TEST]

Add the following macro to the [ISO DIMS] label on the root page:

[ISO DIMS]^C^C^C(caload "/CA-ACAD/" "ISODIM") SNAP S I;;+
DIM DIMSE1 ON DIMSE2 ON DIMBLK ARROW-R STYLE DIM-R EXIT +
ISOPLANE RIGHT $S=ISOFOOTER $S=ISO $T1=TISO

Add a short **ISOFOOTER page, change the **EXAMPLES page label to **ISO, and add these:

```
**ISOFOOTER -1
[  root   ]^C^C^CSNAP S S;;DIM DIMSE1 OFF DIMSE2 OFF DIMBLK . STYLE STD3-32 EXIT;+
$S=SCREEN $T1=TABLET1
**ISO
[ISO Dims]
[]
[DIM LINE]^C^C^CDIMLINE
[EXT LINE]^C^C^CEXTLINE
[DIM ISO ]^C^C^CEXTLINE \\EXTLINE \\DIMLINE
[--------]
[DimTypes]
[  LEFT   ]^C^C^CISOPLANE LEFT DIM DIMBLK ARROW-L STYLE DIM-L EXIT +
$S=DIM-LEFT $S=VERT-L
[  TOP    ]^C^C^CISOPLANE LEFT DIM DIMBLK ARROW-R STYLE DIM-R EXIT +
$S=DIM-TOP $S=TOP-R
[ *RIGHT  ]^C^C^CISOPLANE RIGHT DIM DIMBLK ARROW-R STYLE DIM-R EXIT +
$S=DIM-RIGHT $S=VERT-R
[--------]
[Direct'n]
[  VERT   ]^C^C^CISOPLANE TOP DIM DIMBLK ARROW-R STYLE DIM-R EXIT $S=HORZ-R
[--------]
[Snap Sty]
[  ISO    ]^C^CSNAP S S;;$S=SNAPS
[--------]
[]
[]
```

Make sure the [LEFT] label is on line 8, the [VERT] label on 13, and the [ISO] label on 16. The menu system will use these locations in menu toggles.

[ISO DIMS] initializes the necessary settings and styles, loads the AutoLISP file, and calls the **ISO screen menu.

To start, let's look at the [DimTypes] part of the **ISO screen menu. Each menu macro uses the AutoCAD Isoplane command to set the correct plane. Then the Dim command is called to set the DIMBLK block name and text style. The macro's last task is to place a "partial" screen page over the [DimTypes] boxes. This puts an asterisk in the label to show the user the current DimType, like [*LEFT]. Each time a new DimType is selected, this section is toggled.

After the menu and AutoLISP files are completed, [DIM LINE], [EXT LINE] and [DIM ISO] call AutoLISP commands to draw the iso-dimensions.

The menu overlay system depends on AutoCAD's ability to load screen menus starting at a specific box number. Make a small three-item screen page for each overlay. The only difference between each of the pages is the location of the "*" asterisk, indicating the current settings status. Call the three pages **DIM-LEFT, **DIM-TOP and **DIM-RIGHT, for Left, Top and Right DimTypes. Include the "8"s to start the menus at box number 8.

Save what you've done and continue adding just below the **ISO page.

```
**DIM-LEFT 8
[ *LEFT  ]^C^C^CISOPLANE LEFT DIM DIMBLK ARROW-L STYLE DIM-L EXIT +
$S=DIM-LEFT $S=VERT-L
[ TOP    ]^C^C^CISOPLANE LEFT DIM DIMBLK ARROW-R STYLE DIM-R EXIT +
$S=DIM-TOP $S=TOP-R
[ RIGHT  ]^C^C^CISOPLANE RIGHT DIM DIMBLK ARROW-R STYLE DIM-R EXIT +
$S=DIM-RIGHT $S=VERT-R
**DIM-TOP 8
[ LEFT   ]^C^C^CISOPLANE LEFT DIM DIMBLK ARROW-L STYLE DIM-L EXIT +
$S=DIM-LEFT $S=VERT-L
[ *TOP   ]^C^C^CISOPLANE LEFT DIM DIMBLK ARROW-R STYLE DIM-R EXIT +
$S=DIM-TOP $S=TOP-R
[ RIGHT  ]^C^C^CISOPLANE RIGHT DIM DIMBLK ARROW-R STYLE DIM-R EXIT +
$S=DIM-RIGHT $S=VERT-R
**DIM-RIGHT 8
[ LEFT   ]^C^C^CISOPLANE LEFT DIM DIMBLK ARROW-L STYLE DIM-L EXIT +
$S=DIM-LEFT $S=VERT-L
[ TOP    ]^C^C^CISOPLANE LEFT DIM DIMBLK ARROW-R STYLE DIM-R EXIT +
$S=DIM-TOP $S=TOP-R
[ *RIGHT ]^C^C^CISOPLANE RIGHT DIM DIMBLK ARROW-R STYLE DIM-R EXIT +
$S=DIM-RIGHT $S=VERT-R
```

Next, make the direction screen pages. The important thing to remember here is that if your current DimType is LEFT or RIGHT, then the dimension's direction is VERTical or HORIZontal. If the DimType is TOP, then dimensions are drawn on either the LEFT or RIGHT side. Each time a LEFT or RIGHT DimType is chosen, you must make the direction VERTical. VERTical is the initial default. If TOP is selected, you need to make the direction RIGHT (the default) or LEFT. Create single line pages for the directions, making sure that you include the "13"s with the page name.

Save what you've done and add these six pages:

```
**VERT-L 13
[ VERT ]^C^C^CISOPLANE TOP DIM DIMBLK ARROW-L STYLE DIM-L EXIT $S=HORZ-L
**HORZ-L 13
[ HORZ ]^C^C^CISOPLANE LEFT DIM DIMBLK ARROW-L STYLE DIM-L EXIT $S=VERT-L
**VERT-R 13
[ VERT ]^C^C^CISOPLANE TOP DIM DIMBLK ARROW-R STYLE DIM-R EXIT $S=HORZ-R
**HORZ-R 13
[ HORZ ]^C^C^CISOPLANE RIGHT DIM DIMBLK ARROW-R STYLE DIM-R EXIT $S=VERT-R
**TOP-R 13
[ RIGHT ]^C^C^CISOPLANE RIGHT DIM DIMBLK ARROW-L STYLE DIM-L EXIT $S=TOP-L
**TOP-L 13
[ LEFT ]^C^C^CISOPLANE LEFT DIM DIMBLK ARROW-R STYLE DIM-R EXIT $S=TOP-R
```

For convenience, put a Snap Style toggle on the screen menu. It often is necessary to switch back and forth between snap styles to reach points based on the non-rotated grid. Assign the single line pages to box 16. The SNAP pages complete the screen menu. You still need the ISODIM.LSP program file to use it. Add the SNAPs and test the menu page.

Save what you've done and add:

```
**SNAPS 16
[STANDARD]^C^C^CSNAP S I;;$S=SNAPI
**SNAPI 16
[  ISO   ]^C^C^CSNAP S S;;$S=SNAPS
+
```

Save and Exit to AutoCAD. TEST.MNU reloads.

Command: **INSERT**	Insert ISO-INIT to import the arrows for testing.
Command: **STYLE**	Define DIM-R with SIMPLEX for testing. Set:
Height <0'-0">: **(* 0.125 #dwgsc)**	And default the width, then:
Obliquing angle <0.00: **-30**	Default the remaining settings.
Select **[ISO DIMS]**	It sets Snap to Iso. In DIM mode it sets DIMSE1 and DIMSE2 ON to suppress extension lines, sets DIMBLK to ARROW-R and STYLE to DIM-R. Finally, it toggles ISOPLANE to RIGHT and changes to the **ISO screen page:
Current Isometric plane is: Right	[*Right] displays and Direct'n shows [VERT]
Select DimTypes **[TOP]**	Shows [*TOP] and Direct'n [RIGHT]. Try the [Direct'n] toggles:
Select Direct'n **[RIGHT]**	It toggles to Direct'n [LEFT].
Select Direct'n **[LEFT]**	It toggles to Direct'n [RIGHT].
Select DimTypes **[LEFT]**	[*LEFT] displays.
Select Direct'n **[VERT]**	Toggles to [HORZ].
Select **[ISO]**	Toggle Snap Style [STANDARD] and [ISO]. Watch the grid/axis flip normal/iso.
Select **[root]**	Resets normal settings and loads the root screen menu.

Did the page toggles work? OK, let's write the supporting AutoLISP functions next.

The ISODIM Functions

ISODIM.LSP uses the AutoCAD Dim commands to generate dimension lines. However, ISODIM restricts dimension input to picking points to control alignment. AutoCAD does not automatically align dimensions correctly for the isometric planes. AutoCAD's dimensioning doesn't understand snap style or rotation. Otherwise, the AutoLISP function works similar to AutoCAD's Aligned dimension command.

If DIMASO is ON, ISODIM allows associative dimensioning. But use associative dimensioning with caution. Associative editing commands can redefine existing dimensions.

ISODIM contains the DIMLINE command. DIMLINE gets two points from the user, determines the distance between them and uses the distance in the default prompt. After the dimension is verified, the data is passed to AutoCAD's ALIgned dimension command to draw the line, arrows and text. The function's distance prompt uses your USTR function, not a UDIST. It lets the user type in a string, as in normal AutoCAD dimensioning.

ISODIM.LSP

 Copy ISODIM.LSP to TEST.LSP.

 Create a new LISP function file called TEST.LSP.

```
;* ISODIM.LSP is a set of functions to draw isometric dimensions in coordination
;* with an isometric screen (and optional tablet) menu system.
;*
;* C:DIMLINE draws an isometric dimension line between two points.
;* It uses AutoCAD's dimensioning functions so that all DIM Variables will
;* affect the isometric dimension the same way.
(defun C:DIMLINE ( / sp ep distxt)
  (setq sp (upoint 1 "" "Pick first dim. point" nil nil)    ;get first point
        ep (upoint 1 "" "Pick second dim. point" nil sp)    ;get second point
        distxt (rtos (distance sp ep))                      ;get dist & convert to string
        distxt (ustr "Dimension text" distxt T)             ;verify text string
  );setq
  (command "DIM1" "ALIGNED" "NON" sp "NON" ep "NON" ep distxt)  ;draw dimension
);defun
;*
```

The COMMAND function includes "NON" before each point to suppress any running Osnap modes. You do not want a running Osnap to incorrectly snap the calculated points.

The next function in ISODIM is the EXTLINE command. EXTLINE goes hand-in-hand with DIMLINE. Since AutoCAD's dimensioning does not recognize your Iso-mode, you must suppress AutoCAD's extension lines and draw extension line with the correct orientation. The dim variables DIMSE1 and DIMSE2 are turned on (suppressed), when the user selects the **ISO page from the root menu.

The EXTLINE function draws the extension line separately from the DIMLINE. The [DIM ISO] menu item uses both commands to make iso-dimensioning seem similar to normal dimensioning. EXTLINE accesses several dimensioning system variables to determine the extension line length and offset defaults. Enter the EXTLINE function.

Save what you've done and continue adding:

```
;* C:EXTLINE draws a dimension extension line similar to
;* the one AutoCAD draws using DIMEXO, DIMEXE, and DIMSCALE variables
(defun C:EXTLINE ( / sp ep)
   (setq sp (upoint 1 "" "Starting point of Ext. line" nil nil)    ;start of ext. line
         ep (upoint 1 "" "Ending point" nil sp)                    ;end extension line
   );setq
   (command "LINE"                                                 ;start line command
      "NON" (polar sp (angle sp ep) (* (getvar "DIMEXO") (getvar "DIMSCALE")));line offset
      "NON" (polar ep (angle sp ep) (* (getvar "DIMEXE") (getvar "DIMSCALE")));line extension
   "")   ;terminate line
);defun
;*
```

NOTE. If you are using AutoCAD Release 9, you can modify EXTLINE to auto copy the extension line. Use PAUSE in the COMMAND pipeline to obtain the second point of displacement during a Copy command.

You have one more statement for the AutoLISP file, insuring that you insert the necessary block definitions in the drawing. ISO-INIT is automatically evaluated during the function load to Insert ISO-INIT. ISO-INIT has two nested arrow block definitions, but no real entities. The COMMAND function also sets and scales the text Styles.

Save what you've done and add:

```
;* This automatically inserts an "empty" block carrying the dimension arrow definitions.
;* The insert command is cancelled just after the definitions are added to the
;* block symbol table. An "insert" entity is not created. Text Styles are defined and scaled.
;* The command sequence is executed upon the initial function load.
(command "INSERT" "ISO-INIT" nil
         "STYLE" "DIM-R" "SIMPLEX" (* 0.125 #dwgsc) "" "-30" "" "" ""
         "STYLE" "DIM-L" "SIMPLEX" (* 0.125 #dwgsc) "" "30" "" "" ""
)
;*
(princ)
;*end of ISODIM.LSP
;*
```

Save, Exit and reenter AutoCAD. TEST.LSP should load, inserting ISO-INIT and setting the styles.

Next, you want to test the system.

Testing the ISODIM System

 You have the CA18PIPE.DWG to test with.

```
Command: QUIT
Enter selection:                            Begin a NEW drawing named ISODIM=CA18PIPE.
Command (load"ISODIM")
Command: ZOOM                               Window 16',100' to 125',180'
Command: OSNAP                              INT,NEA
```

 If you don't have the disk, stay in this drawing and create an isometric cube.

The cube is shown in the Isometric Reference Drawing at the start of the chapter.
Use the ISO menu to toggle isoplanes as you draw.
Refer to the Isometric Reference Drawing to help you keep DimTypes and directions straight as you test.

```
Select [ISO DIMS]                           Settings and page change should occur.

Select [VERT]                               [HORZ] shows. Match the 50'-0" dimension at the upper left.
Command: EXTLINE                            Draw the left extension line.
Starting point of Ext. line:
Ending point:

Command: EXTLINE                            Repeat for the right extension line.

Command: DIMLINE                            Match to 50' dim line.
Pick first dim. point:                      Pick the left arrow.
Pick second dim. point:                     Pick the right.
Dimension text <50'>:                       <RETURN> for default.
DIM1                                        Uses AutoCAD's ALIGNED dim, draws the line, arrows, and:
Dimension text <50'-0">: 50'-0"             Puts in the text.

Select [TOP]                                Keep testing, trying all six DimTypes.
```

A sample drawing produced with the system is shown in the following Isometric Piping Drawing.

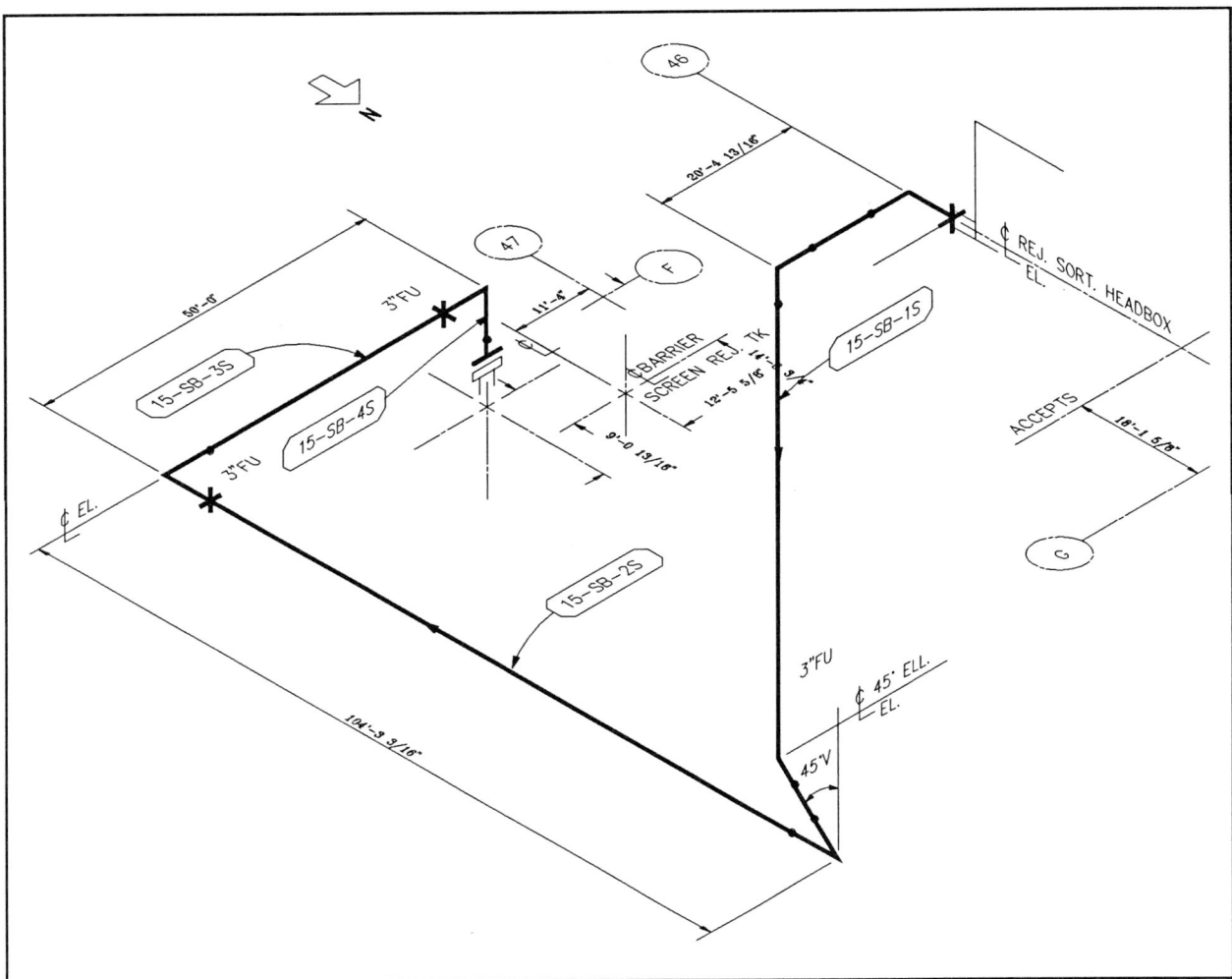

Isometric Piping Drawing

Putting IsoDims on Your Tablet

The various iso-DimTypes, and their relationships to AutoCAD's Isoplanes, are often confusing to first time or infrequent iso users. Putting iso-DimTypes on your tablet menu provides helpful visual cues.

The iso tablet menu consists of six rows and six columns. These are shown in the illustrated Isometric TABLET1 Template. Photocopy the template and tape it to your tablet in the TABLET 1 area that you configured in chapter five. If you have the CA DISK, you have this template as ISO-TAB.DWG.

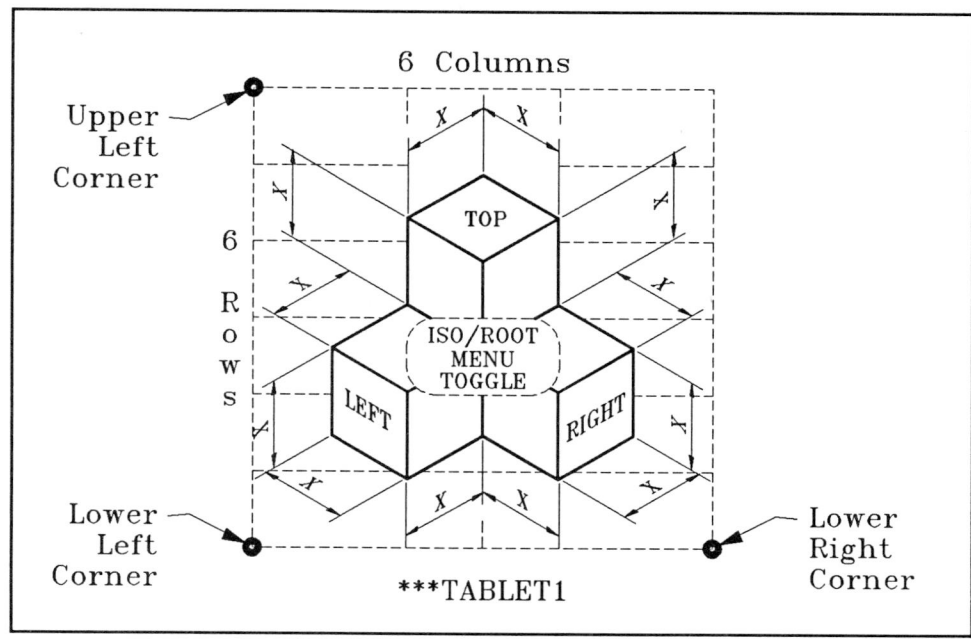

Isometric TABLET1 Template

The tablet menu consist of two pages. The default ***TABLET1 page contains a pair of items that call the **TISO page and flip the screen menu to the **ISO menu. The **TISO tablet menu fills the tablet one area with the **TISO page of macros. The **TISO macros execute the dimensioning commands, and display status by toggling the screen menu. Create the tablet menu.

Making the ISODIM Tablet Menu

Select [EDIT-MNU]   Edit TEST.MNU. Add the following:

***TABLET1
[T1-1]
[T1-2]
[T1-3]
[T1-4]
[T1-5]
[T1-6]
[T1-7]
[T1-8]
[T1-9]
[T1-10]
[T1-11]
[T1-12]
[T1-13]
[T1-14]
[T1-15]
[T1-16]
[T1-17]

[T1-18]
[T1-19]
[T1-20]
[T1-21]^C^C^C(caload "/CA-ACAD/" "ISODIM") SNAP S I;;+
DIM DIMSE1 ON DIMSE2 ON DIMBLK ARROW-R STYLE DIM-R EXIT +
ISOPLANE RIGHT $S=ISOFOOTER $S=ISO $T1=TISO
[T1-22]^C^C^C(caload "/CA-ACAD/" "ISODIM") SNAP S I;;+
DIM DIMSE1 ON DIMSE2 ON DIMBLK ARROW-R STYLE DIM-R EXIT +
ISOPLANE RIGHT $S=ISOFOOTER $S=ISO $T1=TISO
[T1-23]
[T1-24]
[T1-25]
[T1-26]
[T1-27]
[T1-28]
[T1-29]
[T1-30]
[T1-31]
[T1-32]
[T1-33]
[T1-34]
[T1-35]
[T1-36]
+
**TISO
[T1-1]
[T1-2]
[T1-3]^C^C^CISOPLANE RIGHT DIM;DIMBLK;ARROW-L;STYLE;DIM-L;EXIT;$S=TOP-L $S=DIM-LEFT
[T1-4]^C^C^CISOPLANE LEFT DIM;DIMBLK;ARROW-R;STYLE;DIM-R;EXIT;$S=TOP-R $S=DIM-LEFT
[T1-5]
[T1-6]
+
[T1-7]^C^C^CISOPLANE LEFT DIM;DIMBLK;ARROW-L;STYLE;DIM-L;EXIT;$S=VERT-L $S=DIM-LEFT
[T1-8]^C^C^CISOPLANE LEFT DIM;DIMBLK;ARROW-L;STYLE;DIM-L;EXIT;$S=VERT-L $S=DIM-LEFT
[T1-9]^C^C^CISOPLANE RIGHT DIM;DIMBLK;ARROW-L;STYLE;DIM-L;EXIT;$S=TOP-L $S=DIM-LEFT
[T1-10]^C^C^CISOPLANE LEFT DIM;DIMBLK;ARROW-R;STYLE;DIM-R;EXIT;$S=TOP-R $S=DIM-RIGHT
[T1-11]^C^C^CISOPLANE RIGHT DIM;DIMBLK;ARROW-R;STYLE;DIM-R;EXIT;$S=VERT-R $S=DIM-RIGHT
[T1-12]^C^C^CISOPLANE RIGHT DIM;DIMBLK;ARROW-R;STYLE;DIM-R;EXIT;$S=VERT-R $S=DIM-RIGHT
+
[T1-13]^C^C^CISOPLANE TOP DIM;DIMBLK;ARROW-R;STYLE;DIM-R;EXIT;$S=HORZ-R $S=DIM-RIGHT
[T1-14]^C^C^CISOPLANE TOP DIM;DIMBLK;ARROW-R;STYLE;DIM-R;EXIT;$S=HORZ-R $S=DIM-RIGHT
[T1-15]
[T1-16]
[T1-17]^C^C^CISOPLANE TOP DIM;DIMBLK;ARROW-L;STYLE;DIM-L;EXIT;$S=HORZ-L $S=DIM-LEFT
[T1-18]^C^C^CISOPLANE TOP DIM;DIMBLK;ARROW-L;STYLE;DIM-L;EXIT;$S=HORZ-L $S=DIM-LEFT
+
[T1-19]
[T1-20]
[T1-21 root]^C^C^CSNAP S S;;DIM DIMSE1 OFF DIMSE2 OFF DIMBLK . STYLE STD1-8 EXIT;+
$S=SCREEN $T1=TABLET1
[T1-22 root]^C^C^CSNAP S S;;DIM DIMSE1 OFF DIMSE2 OFF DIMBLK . STYLE STD1-8 EXIT;+
$S=SCREEN $T1=TABLET1
[T1-23]
[T1-24]
+
[T1-25]^C^C^CISOPLANE RIGHT DIM;DIMBLK;ARROW-R;STYLE;DIM-R;EXIT;$S=VERT-R $S=DIM-RIGHT
[T1-26]
[T1-27]

```
[T1-28]
[T1-29]
[T1-30]^C^C^CISOPLANE LEFT DIM;DIMBLK;ARROW-L;STYLE;DIM-L;EXIT;$S=VERT-L $S=DIM-LEFT
+
[T1-31]^C^C^CISOPLANE TOP DIM;DIMBLK;ARROW-L;STYLE;DIM-L;EXIT;$S=HORZ-L $S=DIM-LEFT
[T1-32]^C^C^CISOPLANE TOP DIM;DIMBLK;ARROW-L;STYLE;DIM-L;EXIT;$S=HORZ-L $S=DIM-LEFT
[T1-33]^C^C^CISOPLANE RIGHT DIM;DIMBLK;ARROW-L;STYLE;DIM-L;EXIT;$S=TOP-L $S=DIM-LEFT
[T1-34]^C^C^CISOPLANE LEFT DIM;DIMBLK;ARROW-R;STYLE;DIM-R;EXIT;$S=TOP-R $S=DIM-RIGHT
[T1-35]^C^C^CISOPLANE TOP DIM;DIMBLK;ARROW-R;STYLE;DIM-R;EXIT;$S=HORZ-R $S=DIM-RIGHT
[T1-36]^C^C^CISOPLANE TOP DIM;DIMBLK;ARROW-R;STYLE;DIM-R;EXIT;$S=HORZ-R $S=DIM-RIGHT
+
```

Save and Exit to AutoCAD. TEST.MNU reloads. Test it as you did the screen menu.

 Delete TEST.MNU.

 Rename TEST.MNU to MY18.MNU.

Command: **QUIT**

Integrating the Iso Menu

You can integrate the isometric tablet menu into the CA-ACAD.MNU TABLET1 page. Although the screen and tablet menus work together, either can be used alone. If you don't have the CA DISK and want to add the screen menu to CA-MENU:

- Append the **ISO and **ISOFOOTER page of your MY18.MNU to the end of the CA-MENU.MNU file.

- Change the [ISO DIMS] label on the **ROOT page of the CA-MENU to:
  ```
  [ISO DIMS]^C^C^C(caload "/CA-ACAD/" "ISODIM") SNAP S I;;+
  DIM DIMSE1 ON DIMSE2 ON DIMBLK ARROW-R STYLE DIM-R EXIT +
  ISOPLANE RIGHT $S=ISOFOOTER $S=ISO $T1=TISO
  ```

If you want to integrate the isometric tablet menu:

- Merge the ***TABLET1 page items [T1-19] through [T1-22] with the CA-MENU's existing ***TABLET1. Make sure you do not duplicate any numbered labels. You can omit [T1-23] through [T1-36], or fill them with your own macros.

- Append the entire **TISO page.

The chapter's menu is shown in the following diagram.

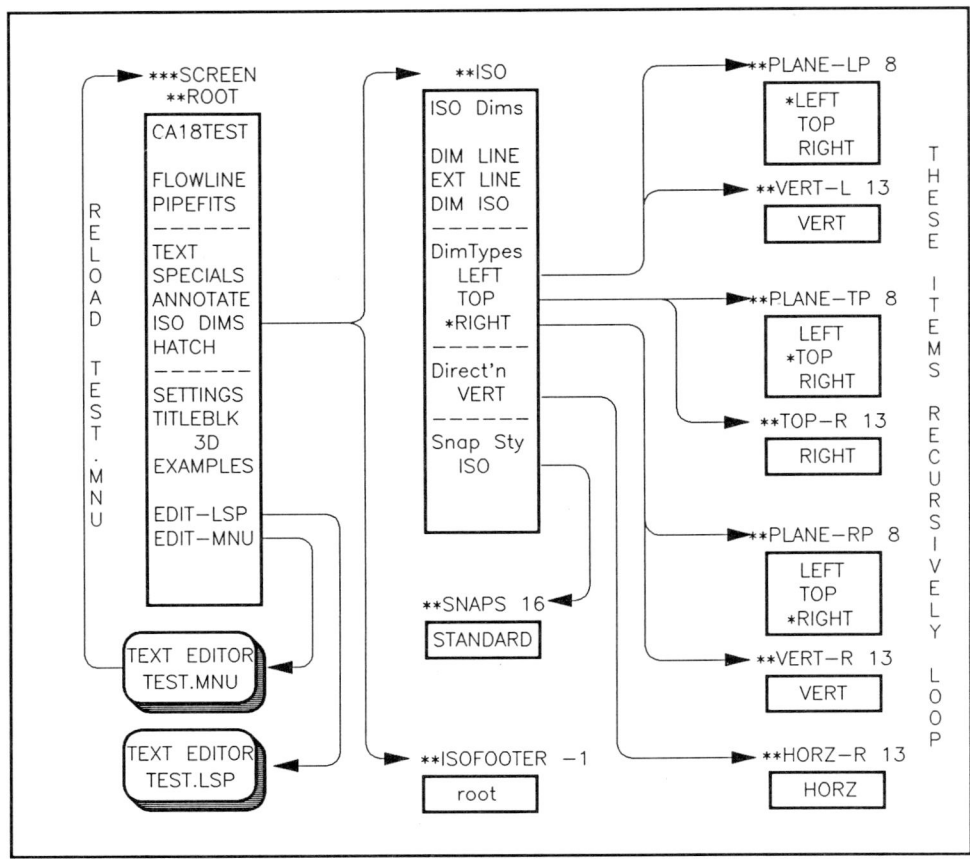

The Isometric Menu

Using Associative Dimensioning with Iso-Dimensioning

If you use associative dimensioning with ISO-DIM, you should redefine (disable) Stretch and all associative dimensioning editing commands. Associative commands update all selected dims, changing your existing arrows and text to the current settings.

Summary Tips and Techniques

Here are some tips and tricks for working with iso-systems.

- ❑ To coax AutoCAD into fitting more text between dimension points, use an "oversize" DIMBLK block, like a 12-inch DIMBLK. Set DIMASZ to a correspondingly "small" value, say 0.78125 or 5/64ths.

- ❑ If you don't need to break an ellipse in your iso-drawing, draw ellipses with a unit block containing a circle. Insert the distorted circle instead of using an actual Ellipse. The distorted circle regenerates faster.

- ❑ When drawing and switching isoplanes, remember to reset your snap basepoint.

- ❑ Remember the current Style applies to both Text and Dimensioning. Use DIM1 STYLE to reset your text style.

If you want to extend the iso-system, try developing an iso-leader routine. Try developing routines to dimension isometric angles.

You can use the iso-dimensioning system developed in this chapter as a model to help you develop more complete applications. Here are some general techniques used in the chapter that we recommend you consider using in your applications:

- ❑ Use screen menu status to help your users. Plan out box numbers for your screens. Use toggling screen pages.
- ❑ Use the insertion of library .DWG files to carry your application block definitions, and layers.
- ❑ Use LOADing of your .LSP files to initialize your environment.
- ❑ Use point of entry screen page environment controls in your application. For example, reset text and dim arrows when changing iso modes. Overkill is better than nothing in point of entry controls. It keeps your users in a controlled environment.
- ❑ You need good templates when doing screen menu overlay pages. Plan ahead.

19—0 Customizing AutoCAD

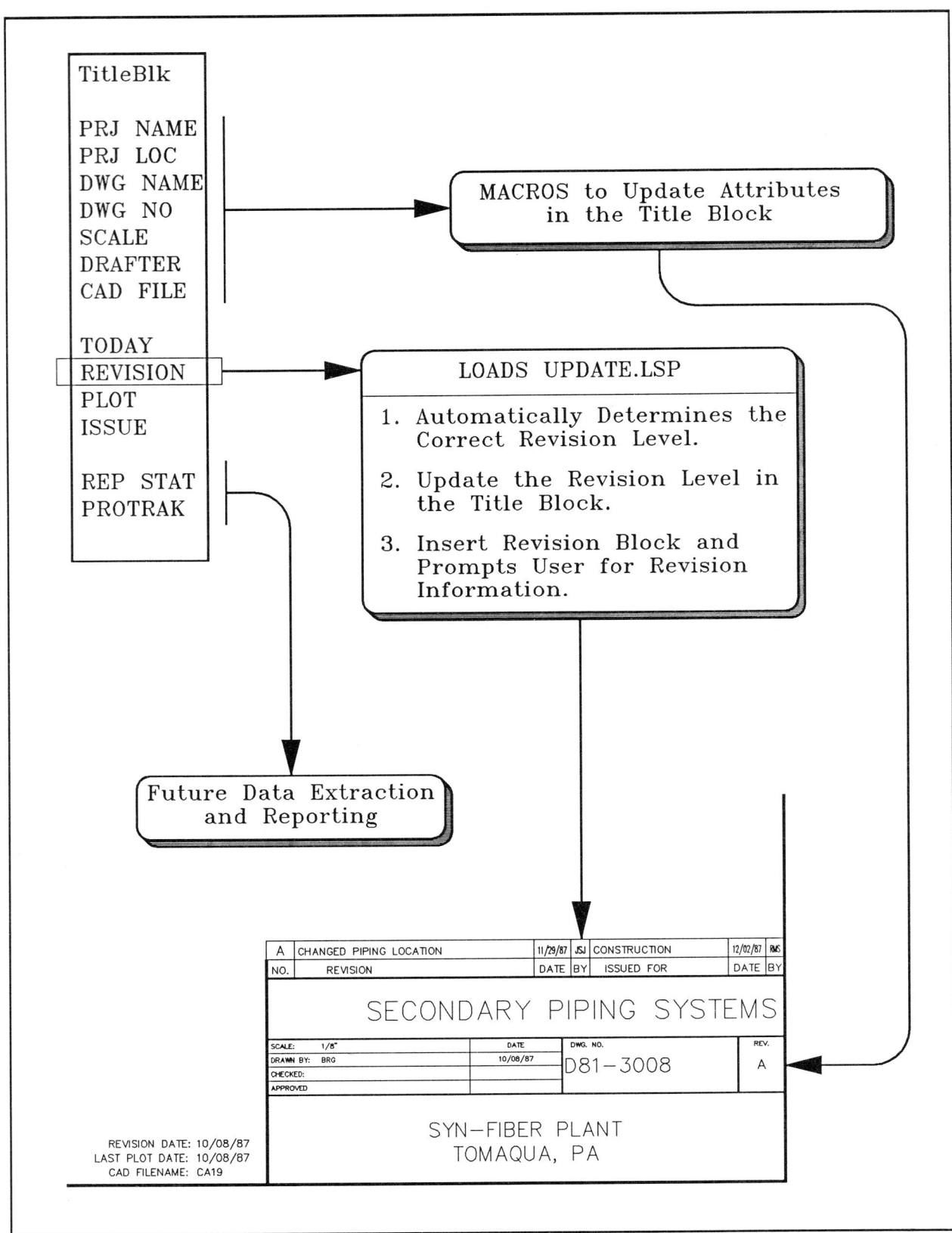

A Custom Title Block Program

CHAPTER 19

Attributes as Data Tools

This chapter covers creating and using attributes, and manipulating them using AutoLISP and macros.

Attributes are intelligent text with controlled style, alignment, size, position, and default values. Attributes are grouped in Blocks. You can define attributes as visible/invisible, and as constant/variable. Attributes can have preset values. You can update their values individually or globally. You can use attributes to store AutoLISP variables, or data to control drawing actions.

This chapter provides three separate applications using attributes. These are:

- A TitleBlock System. This system modifies your setup menu to insert intelligent border sheets with title block information, like Last Plot Date, AutoCAD Drawing Name and Plan Scale. An associated menu helps manage and update this information.

- A Drawing Revision System. The system uses attributes to automatically record and update drawing revision histories, including time and date stamping.

- AutoBreaking Blocks. AutoBreaking blocks are blocks with special scalar attributes to automate their insertion.

The Benefits of Using Attributes

Using attributes in your application provides five main benefits. Using attributes, you can:

- Control text with default values attached to visible or invisible block inserts.

- Assist drawing text updating by grouping information in blocks.

- Move and manipulate an attribute based-schedule as a single entity, updating a number of attributes in one operation. This update is especially useful for drawing schedules.

- Extract attributes to external files for report processing and analysis.

- Tag and count drawing parts and materials for applications, such as inventory or space planning.

How to Skills Checklist

This chapter will show you how to:

- ❑ Use the ATTDEF options. Your attribute definitions determine how AutoCAD stores and manages an attribute.
- ❑ Control the block insertion attribute prompt order. The prompt order is determined by the object selection order of the attributes in the block definition.
- ❑ Integrate attribute input in menu macros.
- ❑ Use the TEXTEVAL system variable to control the interpretation of AutoLISP string input to attribute prompts.
- ❑ Format calendar date, time and year, and feed them to attribute input.
- ❑ Develop a drawing revision bar Block, automatically track revision history and then add new revision levels.
- ❑ Create macros to automatically update and edit attributes.

Macros, AutoLISP Tools and Programs

MACROS
**TITLEBLK is a page of macros to update title block information, like Last Plot Date, CAD Drawing Name, Plan Scale and more. This menu page automates updating title block information.

AutoLISP TOOLS
TIMELIB.LSP is a time and date library of functions.
This file contains time and date library functions.
TODAY returns the current date as mm/dd/yy.
TIME returns the current time as hh:mm:ss.
YEAR returns the current year in form 1988.
GETBK and LINEBLK are functions used by the autobreaking block commands BBLOCK and BLINE.

PROGRAMS
UPDATE.LSP is a program that finds and increments drawing revision levels. It adds a revision bar and prompts for revision notes and drawing reissue data.

C:UPDATE is the AutoLISP command to initiate the program.

AUTOBLK.LSP provides functions for drawing lines with blocks in them. The function automatically inserts a block on the line, and breaks the line at the block based on data stored in the block's attribute. This sequence also is able to place a block on existing lines, breaking the line for the block.
GETBLK gets block name, then searches and returns block and breakdim data.
LINEBLK inserts the autobreaking block and breaks the line or pline.
C:BLINE draws a line with an autobreaking block.
C:BBLOCK inserts an autobreaking block on an existing line or pline.

The Title Block System.

This section shows you how to revise the border sheet from the menu chapters, adding attribute definitions to carry information about the name of a drawing, date of issue and revision history. We recommend using attributes instead of Text for this application because attributes control text style and position. The Title Block System covers defining attributes, managing them in menu macros, updating them with ATTEDIT, and automating the updating with menus.

If you have the CA DISK, you already have the enhanced titleblock and need not redraw it. If you don't have the disk, you need to add to the titleblock drawing you made in the menu chapters so that you can add attributes. The enhanced titleblock is shown in the Border with Title Block drawing.

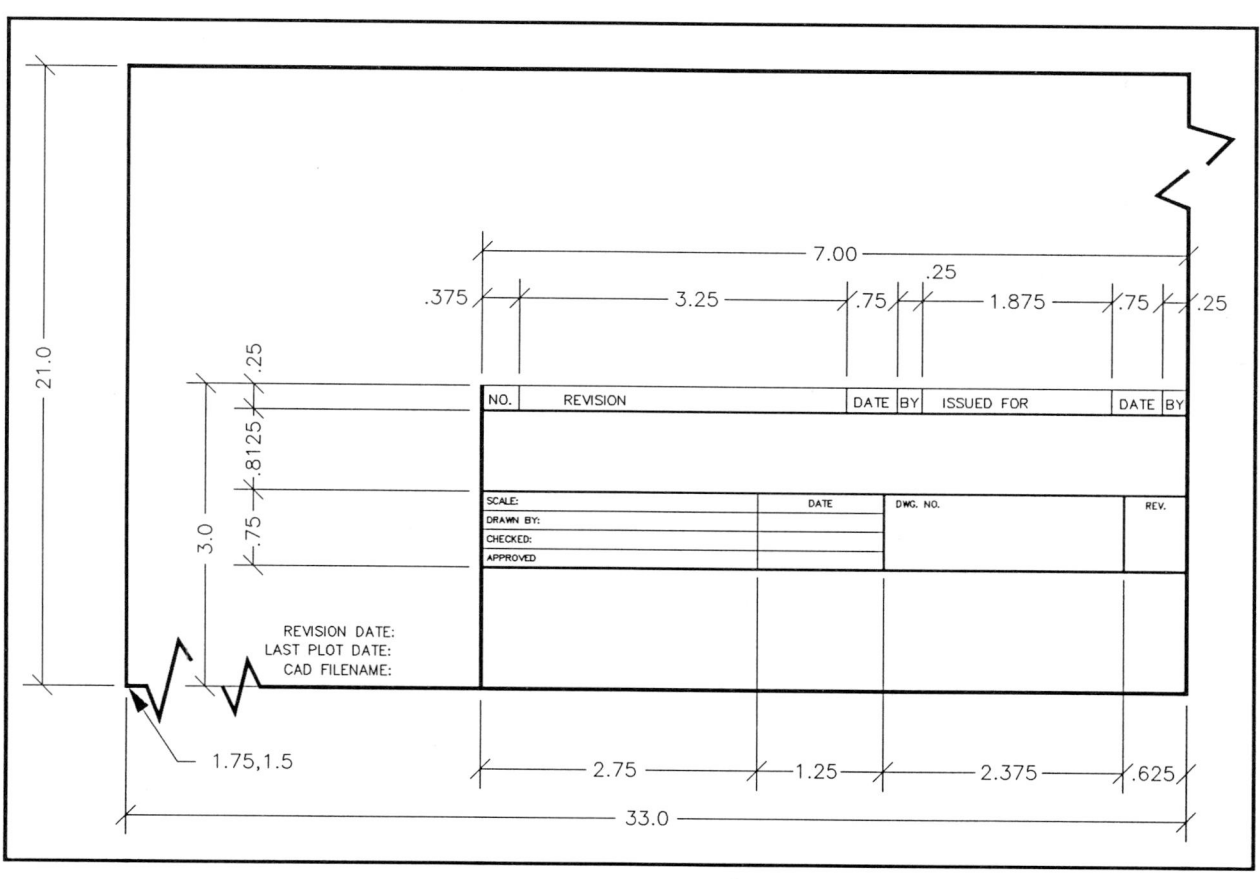

Border with Title Block

Embellishing the Title Block with Attributes

 Copy CA19.MNU to TEST.MNU.

 Copy MAIN.MNU to TEST.MNU.

```
Enter selection:                          Begin a NEW drawing named CA19

Select: [TEST] [24 x 36] [FULL] [INITIATE]
Command: EXPLODE                          Explode Last to edit the border.
Command: ZOOM                             In on the title block area.
Command: LAYER                            Set BORDER current.
```

 Skip the rest of this sequence.

```
Command: PLINE                            Draw the wide lines shown with width 1/64.
Command: LINE                             Draw the narrow lines shown.
Command: DTEXT                            Enter the text shown, using styles STD1-16 and STD3-32.
```

 SAVE it to name ATTSHT-D.

Let's begin the work of defining the attributes for the border sheets.

You need to define your attributes as Variable attributes so that you can change them when you insert them. Their order is important. The attributes that you want are:

ORDER	ATTRIBUTE TAG	PROMPT	DEFAULT	TEXT STYLE
1.	PRJ-NAME	Project name	None	STD3-16
2.	PRJ-LOC	Project location	None	STD3-16
3.	DWG-NAME	Drawing name	None	STD1-4
4.	DWG-NO	Drawing number	0	STD3-16
5.	DWG-SCALE	Drawing scale	None	STD1-16
6.	DWG-FILE	CAD filename	None	STD3-32
7.	DWG-REV	Revision level		STD1-8
8.	DATE-REV	Revision date	None	STD3-32
9.	DATE-PLOT	Last plot date	None	STD3-32
10.	DATE-ISSUE	Issue date	None	STD1-16
11.	DRAFTER	Enter your initials	None	STD1-16

The following Title Block with Attributes drawing shows how the attributes will appear on the Title Block.

The Title Block System. 19—5

[Title block diagram showing a drawing title block with fields for NO., REVISION, DATE, BY, ISSUED FOR, DATE, BY, DWG-NAME, SCALE: DWG-SCALE, DRAWN BY: DRAFTER, CHECKED:, APPROVED, DATE, DATE-ISSUE, DWG. NO., DWG-NO, REV., DWG-REV, PRJ-NAME, PRJ-LOC, REVISION DATE: DATE-REV, LAST PLOT DATE: DATE-PLOT, CAD FILENAME: DWG-FILE]

Title Block with Attributes

To get you started, the next exercise sequence shows the ATTDEF command for the first few attributes in the group. If you don't have the CA DISK, continue defining the rest of the attributes. Use the sample drawing and the table as your guides. Position the attributes in the spots shown in the Title Block with Attributes drawing.

```
Command: SNAP                                     Set to 1/16.

Command: ATTDEF
Attribute modes -- Invisible:N Constant:N Verify:N
Enter (ICV) to change, RETURN when done:          Verify the modes above and <RETURN>.
Attribute tag: PRJ-NAME
Attribute prompt: Project name
Default attribute value: None
Start point or Align/Center/Fit/Middle/Right/Style: S
Style name (or ?) <STD1-8>: STD3-16
Start point or Align/Center/Fit/Middle/Right/Style: C
Center point:                                     Pick location shown.
Rotation angle <0>:

Command: ATTDEF
Attribute modes -- Invisible:N Constant:N Verify:N
Enter (ICV) to change, RETURN when done:
Attribute tag: PRJ-LOC
Attribute prompt: Project location
Default attribute value: 0
Start point or Align/Center/Fit/Middle/Right/Style:   <RETURN> puts below previous tag.
```

 QUIT, and read on.

 Define the others in their table order. The last one looks like:

```
Command: ATTDEF
Attribute modes -- Invisible:N  Constant:N  Verify:N
Enter (ICV) to change, RETURN when done:
Attribute tag: DRAFTER
Attribute prompt: Enter your initials
Default attribute value: BRG
Start point or Align/Center/Fit/Middle/Right/Style:
Rotation angle <0>:
```

If you don't have the CA DISK, you MUST define all attributes in the table in order to do the rest of the exercise. You will need to use the attributes in a later chapter's extraction of the attribute data.

When you are finished, Wblock the attributes and border sheet to the same file, called ATTSHT-D. You need to select the entities for Wblock in the exact order shown in the table. DO NOT window them. If you window, they will reverse their creation order. Your first entity select must be the PRJ-NAME attribute, then the PRJ-LOC, etc. After selecting the attributes, you can Window or use a Crossing to select rest of the border sheet. Duplicate entities are ignored when reselected and will not shuffle your order.

If you have the CA DISK, read on, but skip the next part.

 Do this:

```
Command: ZOOM                              All.
Command: WBLOCK
File name: ATTSHT-D
Block name:
Insertion base point: 0,0
Select objects:                            Remember to select everything in order.

Command: QUIT                              The Wblock saved it.
```

Attributes in Menu Macros

You need to change the [INITIATE] macro in the CA-SETUP menu to fill in the attribute entries. Some of the input data, like the AutoCAD drawing file name, can be filled in automatically. Other information, like the Project name, requires user input and pauses with backslashes. The [INITIATE] macro uses AutoLISP to format the drawing scale, the dates for the drawing and the CAD drawing name. The menu macro toggles the TEXTEVAL system variable, making AutoCAD evaluate AutoLISP expressions.

 These changes are in the CA DISK file named CA19SET.MNU. If you have the CA DISK, you need to copy the revised setup menu named CA19SET.MNU to replace your old CA-SETUP.MNU.

Attributes in Menu Macros 19—7

If you don't have the disk, you will have to modify the CA-SETUP.MNU file.

Let's update the [INITIATE] and drawing size macros.

Revising CA-SETUP.MNU and Inserting ATTSHT-D

Enter selection: Begin a NEW drawing named CA19.DWG.

 Copy CA19SET.MNU to CA-SETUP.MNU.

 Edit your existing CA-SETUP.MNU file as follows.

Command: **ED**   Edit the CA-SETUP.MNU.
Change the [INITIATE] macro as shown in bold:

```
[INITIATE]^c^c^cGRID !#DWGSC ^GSNAP (/ #dwgsc 8) AXIS (/ #dwgsc 2) AXIS OFF;+
SETVAR DIMSCALE !#dwgsc LIMITS 0,0 (list (* x #dwgsc) (* y #dwgsc));+
REGENAUTO ON ZOOM W 0,0 (getvar "LIMMAX") ZOOM .75X REGENAUTO OFF;+
STYLE STANDARD SIMPLEX (* 0.125 #dwgsc) ;;;;;+
STYLE STD1-16 SIMPLEX (* 0.0625 #dwgsc) ;;;;;+
STYLE STD3-32 SIMPLEX (* 0.09375 #dwgsc) ;;;;;+
STYLE STD3-16 SIMPLEX (* 0.1875 #dwgsc) ;;;;;+
STYLE STD1-4 SIMPLEX (* 0.25 #dwgsc) ;;;;;+
STYLE STD1-8 SIMPLEX (* 0.125 #dwgsc) ;;;;;+
VIEW S A LTSCALE (* 0.375 #dwgsc) +
LAYER M BORDER C CYAN ;;SETVAR TEXTEVAL 1 INSERT !STR1 0,0 !#DWGSC;;\\\\+
(cond ((= (setq lu (getvar "LUNITS")) 4)+
(strcat (rtos (/ 12.0 #dwgsc) lu 8) "=" "1'-0" (chr 34)))+
((= lu 3) (strcat "1" (chr 34) "=" (rtos #dwgsc lu 8)))+
(T (strcat "1=" (rtos #dwgsc lu 2))));(getvar "DWGNAME");;;;;\+
LAYER S 0 ;SETVAR TEXTEVAL 0 MENU !#MENU^G AXIS ON
```

Change the names of the border sheets on the **DWG page to the names shown in bold:

```
**DWG
[ Select ]
[ Drawing]
[  Size  ]
[========]
[]
[11 x 17 ](setq x 17.0) (setq y 11.0) (setq str1 "ATTSHT-B");$S=SCALE
[]
[24 x 36 ](setq x 36.0) (setq y 24.0) (setq str1 "ATTSHT-D");$S=SCALE
```

Save and exit to AutoCAD.

Command: **QUIT**   Start a new CA19 drawing to test the revised menu.

Select **[TEST]**   Start the screen menu changes.
Select **[24 x 36]**   For the sheet size.
Select **[1/8"]**   For the scale.
Select **[INITIATE]**   Executes settings & inserts attribute laden block ATTSHT-D.

19—8 Customizing AutoCAD

```
Project name <None>: SYN-FIBER PLANT      Enter all the attribute input as shown for later use.
Project location <None>: TOMAQUA, PA
Drawing name <None>: SECONDARY PIPING SYSTEM
Drawing number <0>: D81-3008
Drawing scale <None>: (cond ((= (setq lu (getvar "LUNITS")) 4)(strcat (rtos (/ 12.0 #dwgsc)
lu 8) "=" "1'-0" (chr 34)))((= lu 3) (strcat "1" (chr 34) "=" (rtos #dwgsc lu 8)))(T (strcat
"1=" (rtos #dwgsc lu 2))))
CAD file name <None>: (getvar "DWGNAME")
Revision number:
Revision date <None>:
Last plot date <None>:
Issue date <None>:
Enter your initials <BRG>: JSJ              Give your initials.

Command: ZOOM                                Zoom in to the title block and check your work.
```

 Do NOT Save.

 Save the drawing now in its current form for later use.

Command: **SAVE**   Enter the name SYNPLANT.

After testing, your title block should look like the Title Block with Intelligent Data drawing.

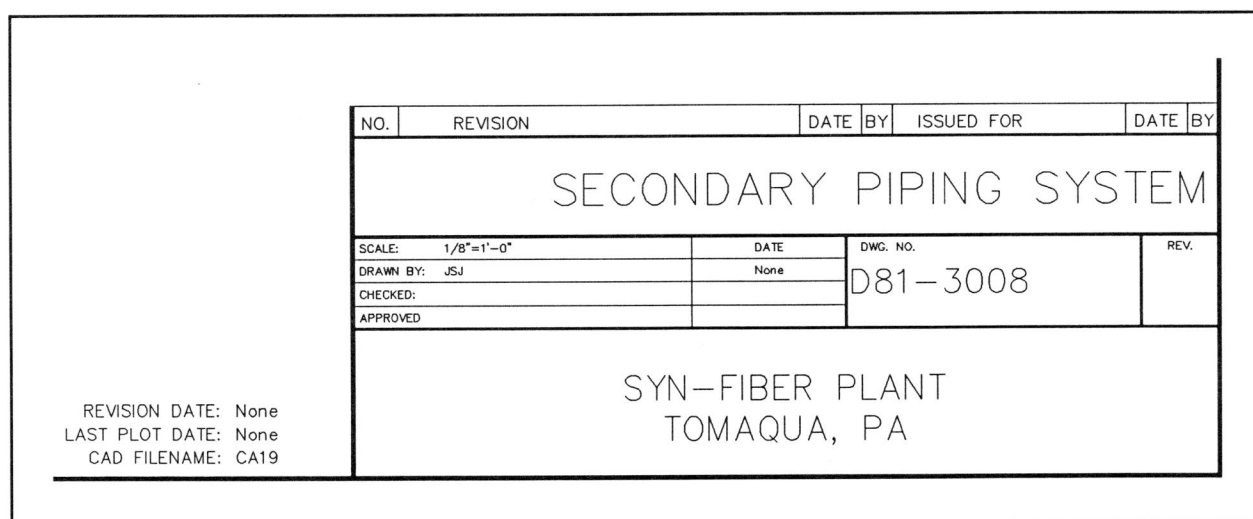

Title Block with Intelligent Data

Here are some comments on the code added to the menu. The [INITIATE] macro's titleblock sheet INSERT is:

LAYER M BORDER C CYAN ;;SETVAR TEXTEVAL 1
This change insures insertion on the correct layer. It makes AutoCAD evaluate AutoLISP expressions at the text prompt.

`INSERT !STR1 0,0 !#DWGSC;;;`
This Insert itself is unchanged, but now it asks for the following attributes:

`\\\\`
A series of four pauses for the Project Name, Project No., DWG Name and DWG No.

```
(cond ((= (setq lu (getvar "LUNITS")) 4)+
(strcat (rtos (/ 12.0 #dwgsc) lu 8) "=" "1'-0" (chr 34)))+
((= lu 3) (strcat "1" (chr 34) "=" (rtos #dwgsc lu 8)))+
(T (strcat "1=" (rtos #dwgsc lu 2))));
```
These four items add more attributes. For Drawing Scale, you first test LUNITS (linear units setting). If LU is 4 architectural, divide scale by 12 to convert for n"=n'-n" format. If LU is 3 engineering, it is formatted n"=n'. Otherwise, it is formatted n=n. All unit types are converted to strings by RTOS with the appropriate precision. If architectural wasn't forced to 8 places precision, 1/16-inch would be rounded to display as 1/8-inch. The (chr 34)s put inch " marks in the menu without using a \". A \" would pause the macro. The terminating semicolon is required to terminate the text value returned from AutoLISP. Without it, AutoCAD would include the following part of the macro in the string.

`(getvar "DWGNAME");`
Enters the AutoCAD system variable DWGNAME as a text string value. Again, you need a semicolon to terminate the text.

`;;;;\+`
Since this is a new drawing, you don't want to assign any values to the Revision number, Revision date, Last plot date and Issue date. The attribute definition defaults are entered by the four semicolons. The final \ pauses for the drafter's name.

LAYER S 0 ;SETVAR TEXTEVAL 0
Resets Layer to 0. Resets TEXTEVAL off to avoid trouble should a user enter something like (El. 145'-6) at a Text: prompt.

OK! Let's talk about updating the title block attributes.

Automatic Title Sheet Maintenance

You can update attribute values using either AutoCAD or AutoLISP. Let's look at how AutoCAD's ATTEDIT command updates attributes. Then, write a group of title block maintenance macros using ATTEDIT.

For a review of the ATTEDIT command, try the following at the keyboard:

```
Command: ATTEDIT
Edit attributes one at a time? <Y>          Default to one at a time.
Block name specification <*>:               Default to all with *.
Attribute tag specification <*>: DATE-PLOT
Attribute value specification <*>:          Default to any value with *.
Select Attributes: C
First corner:                               Select the entire title block.
1 attributes selected.                      An "X" highlights the LAST PLOT DATE.
Value/Position/Height/Angle/Style/Layer/Color/Next <N>: V

Change or Replace? <R>:                     You will replace the value.
New attribute value: 10/12/87               Enter new date.
Value/Position/Height/Angle/Style/Layer/Color/Next <N>:   <RETURN> to finish.
```

Automating ATTEDIT

The Attedit command lets you explicitly search for the specific attribute tag to edit. You can automate Attedit easily with a macro that updates the project name. The title block menu is a combination of macros and functions to update individual attributes.

The [TITLEBLK] page change macro on the root menu turns on TEXTEVAL. It will load the TIMELIB library when you write TIMELIB and when you enable the CALOAD function. The [TITLEBLK] macro changes to the Title block management screen. You need a **TITLEFOOTER to reset TEXTEVAL when returning to the root page.

[PRJ NAME] is the first maintenance macro. The macro uses the system variables EXTMIN and EXTMAX to window the drawing extents. It then finds the project name attribute by its tag.

The following drawing shows the Title Block Menu.

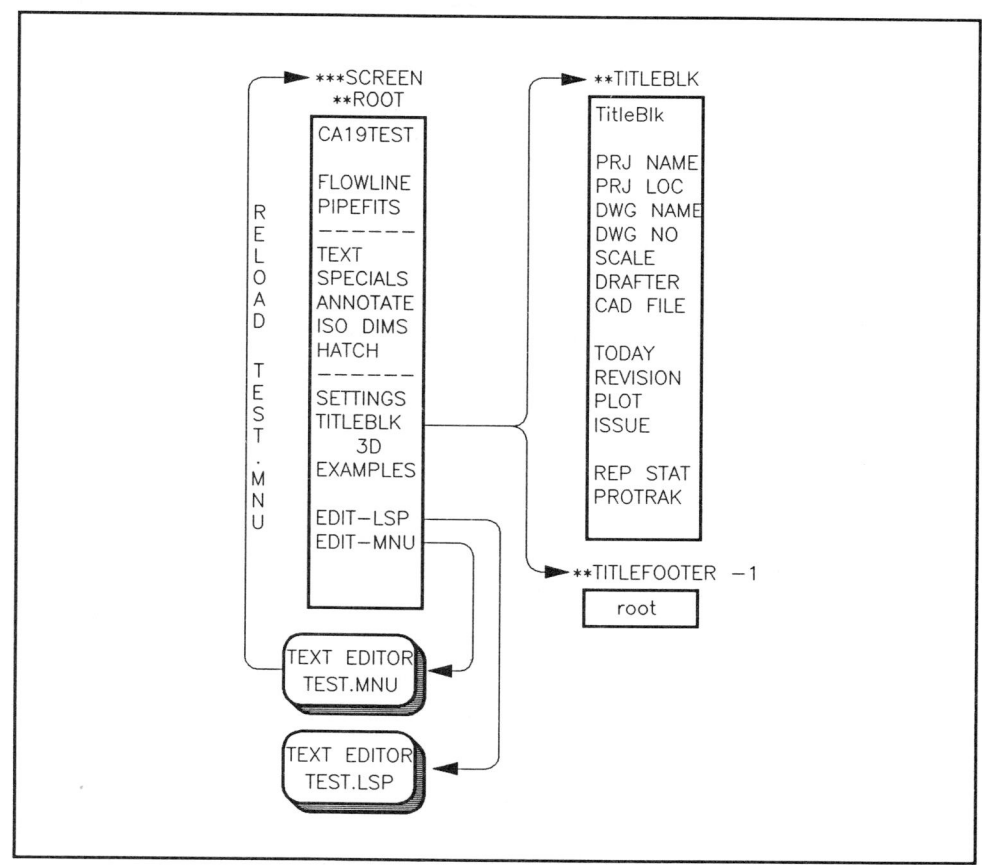

The Title Block Menu

Automating ATTEDIT

Select **[EDIT-MNU]**
On the **ROOT page change [CA00TEST] to [CA19TEST] and add to the [TITLEBLK] label:

```
**ROOT
[CA19TEST]

[TITLEBLK]^C^C^CSETVAR TEXTEVAL 1 LAYER S BORDER;;+
(caload "/CA-ACAD/" "TIMELIB") $S=TITLEFOOTER $S=TITLEBLK
```

Rename the **FOOTER page to **TITLEFOOTER and revise as shown.
Rename **EXAMPLES to **TITLEBLK.
Add labels and the [PRJ NAME] macro:

```
**TITLEFOOTER -1
[   root   ]$S=SCREEN LAYER S 0 ;SNAP ON GRID ON ORTHO ON SETVAR TEXTEVAL 0
+
**TITLEBLK
[TitleBlk]
[]
[PRJ NAME]^C^C^CATTEDIT Y *;PRJ-NAME *;C (getvar "EXTMIN") +
(getvar "EXTMAX") V R \N
```

```
Select [TITLEBLK]                    Test the macro.
Select [PRJ NAME]                    It "X" highlights the PRJ-NAME.
New attribute value:                 Enter a new value.
```

Since the attribute Value string can have spaces within it, the macro can't respond to the attribute <*>: prompt with <SPACE> or *<SPACE>. You need a semicolon to force a <RETURN>. We recommend always using "*;" when accepting the default <*> in Attedit macros.

Next, make more macros to change the other attributes in the border sheet. Some items were left blank. Now, you can include their screen labels. The [SCALE] and [CAD NAME] macros automatically update the current drawing scale and drawing name. The [SCALE] macro duplicates the setup menu's AutoLISP routine, based on the current value of #DWGSC. The drawing name for [CAD NAME] is supplied by the AutoCAD DWGNAME system variable.

Select [EDIT-MNU]
Add the new macros to make the page read:

```
**TITLEBLK
[TitleBlk]
[]
[PRJ NAME]^C^C^CATTEDIT Y *;PRJ-NAME *;C (getvar "EXTMIN") +
(getvar "EXTMAX") V R \N
[PRJ LOC ]^C^C^CATTEDIT Y *;PRJ-LOC *;C (getvar "EXTMIN") +
(getvar "EXTMAX") V R \N
[DWG NAME]^C^C^CATTEDIT Y *;DWG-NAME *;C (getvar "EXTMIN") +
(getvar "EXTMAX") V R \N
[DWG NO  ]^C^C^CATTEDIT Y *;DWG-NO *;C (getvar "EXTMIN") +
(getvar "EXTMAX") V R \N
[SCALE   ]^C^C^CATTEDIT Y *;DWG-SCALE *;C (getvar "EXTMIN") +
(getvar "EXTMAX") V R (cond ((= (setq lu (getvar "LUNITS")) 4)+
(strcat (rtos (/ 12.0 #dwgsc) lu 8) "=" "1'-0" (chr 34)))+
((= lu 3) (strcat "1" (chr 34) "=" (rtos #dwgsc lu 8)))+
(T (strcat "1=" (rtos #dwgsc lu 2))));N
[DRAFTER ]^C^C^CATTEDIT Y *;DRAFTER *;C (getvar "EXTMIN") +
(getvar "EXTMAX") V R \N
[CAD FILE]^C^C^CATTEDIT Y *;DWG-FILE *;C (getvar "EXTMIN") +
(getvar "EXTMAX") V R (getvar "DWGNAME");N
[]
[TODAY   ] (today);
[REVISION]^C^C^C(caload "/CA-ACAD/" "UPDATE") UPDATE
[PLOT    ]^C^C^CATTEDIT Y *;DATE-PLOT *;C (getvar "EXTMIN") +
(getvar "EXTMAX") V R \N
[ISSUE   ]^C^C^CATTEDIT Y *;DATE-ISSUE *;C (getvar "EXTMIN") +
(getvar "EXTMAX") V R \N
[]
[REP STAT]
[PROTRAK ]
[]
[]
```

Save and Exit to AutoCAD. TEST.MNU reloads.

```
Command: (setq #dwgsc 192.0)        Change it to test.
Lisp returns: 192.000000

Select [TITLEBLK]
Select [SCALE]                      Did the scale change to 1/16"=1'-0"?
Select [CAD NAME]                   You'll see it blink and update, but it is the same name.
Select [PLOT]                       Enter the date 10/01/87 and then try the other macros.
Command: (setq #dwgsc 96.0)         Set it back.
```

[TODAY] and [REVISION] aren't active yet, unless you have the CA DISK. The [REP STAT] and [PROTRAK] are presented in the dBASE reporting chapter later in the book.

Let's turn to look at how you time and date stamp drawings, accessing attributes in the drawing database.

The Drawing Revision System

Keeping track of revisions is important in any drawing management scheme. The drawing revision system that follows uses attribute-laden revision blocks to record and update drawing revision histories, including time/date stamping. Time and date are handy items to put on macros. You can use them in automatic functions that enter the date for the user.

First, you want to look at time and date inquiry and formatting, then make a revision block containing several attributes. After looking at how AutoCAD handles the storage and retrieval of attribute data in the BLOCK table and INSERT entities, you will write a function to update attributes in the titleblock, as well as Insert a revision block, REVBLOCK, with current data.

AutoCAD returns the date in the form of YYYYMMDD.hhmmss:

```
YYYY = Year
MM   = Months
DD   = Day
hh   = Hours
mm   = Minutes
ss   = Seconds
```

It is a simple task to use AutoLISP's SUBSTR to extract the date and time. You simply parse the string and STRCAT it back together in the form that you need.

AutoCAD keeps a system variable, called CDATE. CDATE includes the time. Take a look at date and time with CDATE.

Time and Date Stamping

Command: **(rtos (getvar "CDATE") 2 6)**   Convert to a string.
Lisp returns: "19871008.152009"   Yours will vary.

Let's create a small library of date and time functions to use in your later programs.

   Copy TIMELIB.LSP to TEST.LSP

   Begin a new TEST.LSP file.

Write these date and time functions.

```
;* TIMELIB.LSP contains commonly used date and time conversions based on
;* the Julian calender and clock format.

;* TODAY returns the current date as mm/day/yr
(defun today ( / d yr mo day)
   (setq d (rtos (getvar "CDATE") 2 6)       ;gets the Julian date.
         yr (substr d 3 2)                   ;parses out year as '87
         mo (substr d 5 2)                   ;takes the month number
         day (substr d 7 2)                  ;and the day number
   );setq
   (strcat mo "/" day "/" yr)                ;shuffles order and puts it together
);defun

;* TIME returns current time as hh:mm:ss
(defun time ( / d yr m s)
   (setq d (rtos (getvar "CDATE") 2 6)       ;gets Julian date.
         hr (substr d 10 2)                  ;hours
         m  (substr d 12 2)                  ;minutes
         s  (substr d 14 2)                  ;seconds
   );setq
   (strcat hr ":" m ":" s)                   ;mends together
);defun

;* YEAR returns year in form 1987
(defun year ()
    (substr (rtos (getvar "CDATE") 2 6) 1 4)  ;full year as in 1987 is returned
);defun
(princ)
;* end of TIMELIB.LSP file
;*
```

Save, Exit and reenter AutoCAD. TEST.LSP should load. Try them. Your time and date will vary.

Command: **(today)**   Returns: "10/08/87".
Command: **(time)**   Returns: "22:22:05".
Command: **(year)**   Returns: "1987".

Command: **QUIT**   Quit, you saved it earlier as SYNPLANT.

 Delete TEST.LSP.

 Rename TEST.LSP to TIMELIB.LSP.

Tracking Drawing Revisions

The next section shows you how you can automate the common task of tracking drawing revisions by adding and manipulating revision blocks in the drawing. You are going to make a revision block, called REVBLOCK. REVBLOCK is designed to align at the position you want when it is inserted in the drawing.

If you have the CA DISK, you already have REVBLOCK.DWG and can skip making it. If you don't have the disk, you must make it so that you can add it the titleblock drawing. Draw the revision block outline shown in the illustration called Revision Block with Attributes.

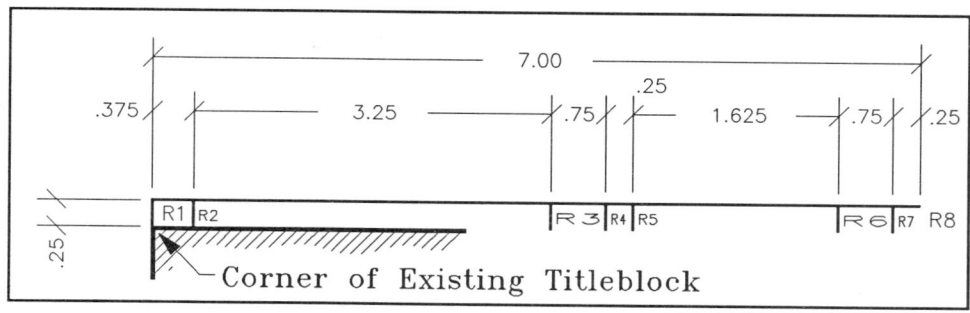

Revision Block with Attributes

After you make the drawing, add the following eight attribute definitions in exact order. Be sure to make the last attribute for revision time invisible.

ORDER	ATTRIBUTE TAG	PROMPT	DEFAULT	TEXT STYLE	ALIGNMENT
1.	R1	Revision number:	(blank)	STD3-32	Center
2.	R2	Description:	""	""	Start
3.	R3	Revision date:	""	""	Fit
4.	R4	Revised by:	""	""	Fit
5.	R5	Issued for:	""	""	Start
6.	R6	Revised issue date:	""	""	Fit
7.	R7	Issued approved by:	""	""	Fit
8.	R8	Revision time:	""	""	Start

(R8 is invisible)

Making a Revision Box with Attributes REVBLOCK

  Make the revision block.

Enter selection:	Begin a NEW drawing named CA19=ATTSHT-D.
Command: **ZOOM**	In on the title block area.
Command: **LAYER**	Set BORDER current.
Command: **LINE**	Draw the revision block lines.
Command: **DTEXT**	Enter the table text, using style STD3-32.
Command: **ATTDEF**	
Attribute modes -- Invisible:N Constant:N Verify:N	
Enter (ICV) to change, RETURN when done:	Check the modes shown.
Attribute tag: **R1**	
Attribute prompt: **Revision number**	
Default attribute value:	
Start point or Align/Center/Fit/Middle/Right/Style: **C**	
Center point:	Select the point.

Continue defining the rest of the attributes in the table above.

You need to key the REVBLOCK insertion to the title block. The title block inserts at 0,0 so you need to define REVBLOCK to use the 0,0 base point. If you have the CA DISK, you can skip Wblocking the REVBLOCK, but insert REVBLOCK to see it.

If you don't have the disk, Wblock REVBLOCK. Again, select the attributes in the exact order shown in the table.

  Skip the WBLOCK.

Command: **WBLOCK**	Give the name REVBLOCK. Use 0,0 as insertion base point.
Select objects:	Select the attributes in the right order.
Select objects:	Select the lines of the revision block.
Command: **INSERT**	Test the order of the attributes.
Block name (or ?): **REVBLOCK**	
Insertion point: **0,0**	Default the scale and rotation.
Enter attribute values	Enter appropriate values:
Revision number:	
Description:	
Revision date:	
Revised by:	
Issued for:	
Revised issue date:	
Issued approved by:	
Command: **QUIT**	Don't update the drawing, Wblock saved REVBLOCK.

AutoCAD's Treatment of Attribute Data

AutoCAD stores all constant attribute data, including values and prompts in the block definition. They can only be changed through block redefinition.

Variable or Preset attribute values vary, so they are stored with each INSERT entity, not in the block definition. However, their default values and prompts are stored in the block definition as well as the INSERT entity.

NOTE. Release 9 introduced Preset attributes and ATTREQ, the ATTribute REQuest system variable. ATTREQ supresses attribute prompting at insertion time. However, Preset values are still assigned to the INSERT entity regardless of the ATTREQ setting.

Let's look more closely at the data of the attributes you defined and entered in ATTSHT-D block:

Accessing Attribute Data

Enter selection: Edit the EXISTING drawing SYNPLANT.

Command: **<F1>**   Flip to the text screen.

Command: **(setq edata (entget (setq en (entlast))))**   An INSERT entity, the border block:

Lisp returns: ((-1 . <Entity name: 60000028>) (0 . "INSERT") (8 . "BORDER") (66 . 1) (2 . "ATTSHT-D") (10 0.000000 0.000000) (41 . 96.000000) (42 . 96.000000) (50 . 0.000000) (43 . 96.000000) (70 . 0) (71 . 0) (44 . 0.000000) (45 . 0.000000))

Command: **(setq edata (entget (entnext (dxf -1 edata))))**   Next, the INSERT's 1st ATTRIB entity:

Lisp returns: ((-1 . <Entity name: 6000003C>) (0 . "ATTRIB") (8 . "BORDER") (10 2947.714000 107.477900) (40 . 18.000000) (1 . "SYN-FIBER PLANT") (2 . "PRJ-NAME") (70 . 0) (73 . 0) (50 . 0.000000) (41 . 1.000000) (51 . 0.000000) (7 . "STD3-16") (71 . 0) (72 . 1) (11 3072.000000 107.477900))

Command: **(repeat 11 (princ (setq edata (entget (entnext (dxf -1 edata)))))) (princ "\n\n")**

Look at the rest, and you'll get 10 more ATTRIBs, then a SEQEND:

((-1 . <Entity name: 60000118>) (0 . "SEQEND") (8 . "BORDER") (-2 . <Entity name: 60000028>))

An ATTRIB is an entity that stores the data of Variable attributes in the drawing data immediately following the block INSERT entity. You only can access ATTRIBs by getting the INSERT, then following the trail with ENTNEXT. The end of the ATTRIBs is marked by a SEQEND entity, so you know when to make your programs stop. A DXF group 66 code of 1 indicates that an INSERT has attributes. You can see the group 66 code in the previous "BORDER" INSERT example. Some other DXF groups you will encounter are:

DXF CODE	ATTRIB MEANING
1	The attribute value.
2	The TAG name.
11	The attribute text location

The Constant, default and prompt attribute data are stored differently. They are stored as an ATTDEF entity accessible only through the BLOCK table.

Command: `(setq tbdata (tblsearch "block" "attsht-d"))`
Lisp returns: `((0 . "BLOCK") (2 . "ATTSHT-D") (70 . 2) (10 0.000000 0.000000 0.000000)`
`(-2 . <Entity name: 400006C8>))`   The DXF -2 group points to first subentity.

Command: `(setq atdata (entget (dxf -2 tbdata)))`   Get 1st subentity data (-2 group).

Lisp returns: `((-1 . <Entity name: 400006C8>) (0 . "ATTDEF") (8 . "BORDER")`
`(10 31.263390 1.119561) (40 . 0.187500) (1 . "None") (3 . "Project name")`
`(2 . "PRJ-NAME") (70 . 0) (73 . 0) (50 . 0.000000) (41 . 1.000000)`
`(51 . 0.000000) (7 . "STD3-16") (71 . 0) (72 . 1) (11 32.000000 1.119561))`

Command: `(repeat 11 (princ (setq atdata (entget (entnext (dxf -1 atdata)))))) (princ "\n\n")`

You will get 10 more ATTDEFs, then:

`((-1 . <Entity name: 400025AA>) (0 . POLYLINE) (8 . BORDER) (66 . 1) (70 . 1) (40 . 0.031250) (41 . 0.031250))`

You'll notice that many of the entity data fields are the same for both the ATTDEF in the block definition and the ATTRIB in the inserted entity. At first glance it appears wasteful, but it lets you edit attributes after their insertion, using the ATTEDIT command.

Also notice that the PRJ-NAME attribute is first in the block definition. It is first because you selected it first while Wblocking. The 2 value of the DXF group 70 in the BLOCK "ATTSHT-D" indicates the BLOCK contains ATTDEFs. But there is no SEQEND to tell you when you are through all the ATTDEFs. Any SEQENDs you encounter in a BLOCK definition belong to POLYLINES or nested INSERTs with their own ATTRIBs. ENTNEXT walks through all the entities of the BLOCK definition, so you saw a POLYLINE in the example. ENTNEXT will eventually return NIL at the end of all the BLOCKs entities. When searching a BLOCK, test what ENTNEXT returns to trap the NIL, and test the 0 DXF group to see if it is "ATTDEF."

Some other DXF group codes you will encounter in the ATTDEF are:

DXF CODE	ATTDEF MEANING
1	The DEFAULT value.
2	The TAG name.
3	The PROMPT string.

Next let's look at making an update function, called UPDATE, to manipulate the drawing revision data.

Drawing Revision Updates Function

The C:UPDATE function automatically updates related dates. It checks and updates revision levels, inserts a revision bar in the drawing and prompts for the revision data. It is a good function to automatically call when ENDing the drawing. If you are using Release 9, you want to put in a Redefined END command.

The revision updates are shown in the sample drawing called Drawing Revision Blocks.

E	ARCHITECT'S WHIM	12/15/87	RBG	BID	12/15/87	JSJ
D	REDRAWN AND REVISED	12/11/87	DMP	CONSTRUCTION	12/13/87	WMP
C	CHANGED PUMP 1 & 2	12/07/87	TBB	CONSTRUCTION	12/09/87	PTM
B	ADDED CONDUIT	12/06/87	PJS	CONSTRUCTION	12/06/87	DBS
A	CHANGED PIPING LOCATION	11/29/87	JSJ	CONSTRUCTION	12/02/87	RMS
NO.	REVISION	DATE	BY	ISSUED FOR	DATE	BY

SECONDARY PIPING SYSTEM

SCALE: 1/8"=1'-0" DATE: None DWG. NO. D81-3008 REV. E
DRAWN BY: JSJ
CHECKED:
APPROVED:

SYN-FIBER PLANT
TOMAQUA, PA

REVISION DATE: 10/08/87
LAST PLOT DATE: 10/08/87
CAD FILENAME: CA19

Drawing Revision Blocks

UPDATE first asks the user if the session is a drawing revision. Then, it uses SSGET X to get the title sheet data by looking for an INSERT entity on layer BORDER with attribute flag of 1 (true). One common mistake is to assume an insert resides on the layer of its definition. Recall that inserts reside on their layer of insertion. To control the system and drawing integrity, the [INITIATE] macro forces the insert of ATTSHT-D on layer BORDER.

UPDATE searches the SS1 selection set to find an entity with an "ATTSHT" substring. The intelligent border names all begin with "ATTSHT," permitting substring comparison. If a block INSERT starting with "ATTSHT" is found, UPDATE feeds its entity name to the REVALL function. REVALL modifies the titleblock attribute entity data lists with ENTMOD, then UPDATE updates them in the drawing with ENTUPD.

Work on the main UPDATE function first.

Creating UPDATE

 Copy UPDATE.LSP to TEST.LSP

 Begin a new TEST.LSP file.

Select [EDIT-LSP]   Edit TEST.LSP, and enter:

```
;* UPDATE.LSP revises information in drawing titleblock "ATTSHTxx"
;* It changes dates, records revision information and insert a block containing
;* the revision data. It uses DXF and UKWORD and the TIMELIB.LSP functions.
;* If you have AutoCAD 2.6, substitute:
;*              (prompt "\nSelect ATTSHTxx title block: ") (setq ss1 (ssget))
;* for the     (setq ss1 (ssget "X"..... expression.

;* C:UPDATE is the main revision function, which calls REVALL to update.
(defun C:UPDATE( / ss1 count emax en ed blkn found)
   (if (= "Y" (ukword 1 "Y N" "Is this a revision" "Y"))   ;permission to update
     (progn                                           ;Then run
       (setq ss1 (ssget "X" '((0 . "INSERT")          ;Gather all inserts
                              (8 . "BORDER")          ;on border layer
                              (66 . 1)))              ;with attributes following.
     );setq ssgetx
     (if ss1                                          ;If found possible
        (progn
           (setq count 0                              ;initialize count
                 emax (sslength ss1)                  ;max entities in SS1
           );setq
           (setvar "TEXTEVAL" 1)                      ;force eval on
           (while (< count emax)
              (setq en (ssname ss1 count)             ;entity name
                    ed (entget en)                    ;entity data
                    blkn (dxf 2 ed)                   ;block name
              );setq
              (if (= "ATTSHT" (substr blkn 1 6))      ;test insert's block name substring
                 (setq count emax found T)            ;it's the border sheet - set flag T
                 (setq count (1+ count))              ;it's some other block - keep looking
              );if
           );while                                    ;last "en" value is our border
           (if found                                  ;if we have the titleblock
              (progn
;*(princ en) ;prints the block insert ename, delete ;* to test, delete this line when done
                 (revall en)                          ;call revision function
                 (entupd en)                          ;update the title block
              );progn
              (prompt "\nError: No border sheet found")   ;else tell them
           );if found
           (setvar "TEXTEVAL" 0)                      ;force text evaluation off
        );progn
     );if ss1
    );progn then run
  );if
  (princ)
);defun C:UPDATE
;*
```

Even if have the CA DISK, remove the ";*" from (princ en) and add a leading ";*" each to (revall en) and (entupd en) to disable them to test the function:

```
(princ en)      ;prints the block insert ename, delete ;* to test, delete this line when done
;*         (revall en)                    ;call revision function
;*         (entupd en)                    ;update the title block
     );progn
```

Save, Exit and reenter AutoCAD. TEST.LSP should load.

Command: **UPDATE**	Test it.
Is this a revision <Y>:	Answer Y for yes.
Lisp returns: <Entity name: 400006A2>	The entity name shows it found the block.

REVALL searches all ATTRIButes of the block INSERT and updates using a COND structure. It loops until the SEQEND entity indicates that there are no more subentities. For each of the supported atibute tags, like DATE-REV and DATE-PLOT, the attribute entity name and new value are passed to a small function, UPD. UPD performs the entity modifications. This saves having to repeat the entmod code.

DWG-REV is the attribute for the current drawing revision level. It is initially blank in the title block. The REVALL function assigns an initial revision level of "A," and automatically increments the value found. The ASCII and CHR functions are used to increment the level. The new revision level is stored in a variable for later insertion of the REVBLOCK.

The REVBLOCK revision bar insertion point is calculated relative to 0,0 and offset by 0.25 times #DWGSC times the number of revision levels of the drawing (DWG-REV A=1, B=2, etc). During insertion, attributes are requested and either filled in automatically, or entered by the user at the pauses in the COMMAND function. The TODAY function is used inside of the command pipeline for "Revision date:." Unfortunately, the user cannot enter it at the keyboard for "Issue date:" because AutoLISP does not allow reentry. Any value input at the pause in the COMMAND function goes to AutoCAD, but does NOT get passed back to AutoLISP.

Enter the REVALL and UPD functions.

Select **[EDIT-LSP]**   Edit TEST.LSP
Remove the disabling ;* from the (revall en) and (entupd en) lines. Delete the (princ en) line.

```
     (progn
        (revall en)             ;call revision function
        (entupd en)             ;update the title block
     );progn
```

Now add REVALL and UPD:

```lisp
;* REVALL is called by C:UPDATE to search and update attributes.
(defun revall(en / date ed ctime etime atag lstrev ipt)   ;en is border sheet with
  (setq date (today)                          ;save the date
        en (entnext en)                       ;get the first subentity of title block
        ed (entget en)
        ctime (getvar "DATE")                 ;get the current Julian time
  );setq
  (while (/= "SEQEND" (dxf 0 ed))             ;while not at the end of insert entity
    (if (= "ATTRIB" (dxf 0 ed))               ;if it's an attribute
        (progn
          (setq atag (dxf 2 ed))              ;then get the tag id
          (cond                               ;test id in a cond loop
            ((= atag "DATE-REV")              ;it's a revision date
              (upd ed en date)                ;call the updater function
            );last revision date
            ((= atag "DATE-PLOT")             ;it's the plot date
              (if (= "Y" (ukword 1 "Y N" "Update last plot date?" "Y"))  ;ask them
                (upd ed en date)              ;if ok - call the updater
              );if
            );last plot date
            ((= atag "DATE-ISSUE")            ;it's the drawing issue date
              (if (= "" (dxf 1 ed))           ;if its not filled in
                (if (= "Y" (ukword 1 "Y N" "Issue the drawing?" "N"))   ;ask them
                  (upd ed en date)            ;and if they say yes - update it
                ) );if
            );issue date
            ((= atag "DWG-REV")               ;it's the revision level letter
              (if (= "" (dxf 1 ed))           ;if its not filled in
                  (progn
                    (upd ed en (setq lstrev "A"))   ;make it the A rev
                    (setvar "USERR1" ctime)         ;save the original creation time
                  );progn
                  (upd ed en                  ;increment the letter
                    (setq lstrev (chr (1+ (ascii (dxf 1 ed))))))
              );if                            ;that is in there
            );revision level
          );cond - the rest of attributes can be skipped
        );progn
    );if = ATTRIB
    (setq en (entnext en)
          ed (entget en)
    );setq
  );while /= SEQEND

  ; Now that the main title block is updated, insert a new rev block
  ; at an insert point based on the revision level letter - A comes to 0,0
  ; but B is at 0,0.25 (adjusted for the dwg scale) and so on...
;* If you have AutoCAD 2.6, add GET functions to get user input before the
;* COMMAND function, and feed the GET input as variables to the COMMAND
;* replacing the pauses.

  (setq ipt (polar '(0 0) (* pi 0.5) (* 0.25 #dwgsc (- (ascii lstrev) 65))))
  (command "insert" "revblock" ipt #dwgsc #dwgsc 0.0       ;insert a revblock
           lstrev pause (today) pause pause pause pause    ;pause for entry from user
           (revtime ctime)                                 ;time stamp for revision period
  );command
);defun REVALL
;*
```

The REVTIME function calculates the amount of time allotted to the current revision. It stores the current time and date at the moment the revision command is started. It uses the user system variables from AutoCAD. The variables USERR1 through USERR5 store real numbers. USERI1 through USERI5 store integers. The function returns the time difference between the current date/time and the last revision date/time. It uses minutes as the base unit of time. Add the REVTIME function.

```
;* USERR1 is the time period of original creation for the drawing.
;* USERR2 is the time period from last update.
(defun revtime(ctime / period)
  (setq period (- ctime (getvar "USERR2")))    ;rev period = current time - last rev time
  (setvar "USERR2" ctime)                       ;update rev time to start next rev period
  (rtos (* period 1440) 2 0)                    ;time in minutes as a string, no places.
);defun

;* UPD is a simple ENTMOD function called by REVALL to reduce repetitive code.
(defun upd (ed en nval / el)
  (setq el (subst (cons 1 nval) (assoc 1 ed) ed)) ;swaps the entity data
  (entmod el)                                     ;sends updated list to AutoCAD
);defun
;*
(princ)
;*end of UPDATE.LSP
;*
```

Save, Exit and reenter AutoCAD. TEST.LSP should load.

Updating SYNPLANT with REVBLOCKs

☞ Even if you have the CA DISK, you need to update and save your SYNPLANT.DWG. You need this updated file for exercises in later chapters.

If you are not in the existing SYNPLANT drawing, edit it again.

Command: **(load "TIMELIB")**	If not loaded, it must be loaded.
Command: **UPDATE**	Test it.
Is this a revision <Y>:	Take default Yes.
Update last plot date? <Y>:	Yes. It inserts REVBLOCK, then:
Enter attribute values	
Revision number: A Description:	Enter a description. Rev. no. filled itself in.
Revision date: 11/28/87 Revised by:	Enter initials.
Issue the drawing? <N>:	Enter Y for Yes. This only prompts if blank.
Issued for <CONSTRUCTION>:	
Issue date:	Enter the date.
Issued by:	Enter initials.
Revision time:	Enter the number of minutes for the revision.
Command: **UPDATE**	Make a few more revisions using UPDATE, entering the information for each as illustrated.
Command: **SAVE**	Enter SYNPLANT to replace the previous drawing for use in a later chapter.

Now some finishing touches. Add the TODAY and C:UPDATE functions to the menu system.

Select **[EDIT-MNU]** Edit TEST.MNU
Add the bolded code to the last part:

```
[TODAY   ] (today);
[REVISION]^C^C^C(caload "/CA-ACAD/" "UPDATE") UPDATE
[PLOT    ]^C^C^CATTEDIT Y *;DATE-PLOT *;C (getvar "EXTMIN") +
(getvar "EXTMAX") V R \N
```

Save and Exit to AutoCAD. TEST.MNU reloads. Test them.

Command: **QUIT**

 Delete TEST.LSP and TEST.MNU.

 Rename TEST.MNU to MY19.MNU. Rename TEST.LSP to UPDATE.LSP.

AutoBreaking Blocks

An often overlooked use of attributes is storing parametric data with blocks. The concept is simple. You can design blocks to store information as Constant attributes within the block definitions. A function that controls the insertion gets the attribute value from the BLOCK definition table, processes it and does something with it. What you do with it can be extremely complex, even extending to controlling entire parametric drawings.

The example that we use to show parametric attribute use is an "autobreaking block." The autobreaking block breaks out a line segment as it inserts. The way the block is defined is important. To avoid extensive BLOCK table subentity searches, your attribute must be tagged BREAKDIM and it must be the first entity in the block definition.

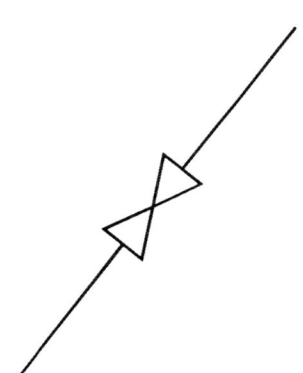

The BLINE command

If you have the CA DISK, you have the block as VALVE-B.DWG, so you can just do the setup below. Otherwise you must make and Wblock a valve as shown in the drawing, VALVE-B Block. Design the valve so the breakout distance is a horizontal dimension centered on the insertion base point.

Making the AutoBreak Block

```
Enter selection:                        Begin a NEW drawing again named CA19.

Select [TEST]
Select [24 x 36]
Select [FULL]
Select [INITIATE]                       Then enter 5 <RETURN>s to default the attributes.

Command: ZOOM                           Center with 4" height.
Command: SNAP                           Set to 1/16.
```

 Skip to the next sequence.

```
Command: LINE                           Draw the valve as dimensioned.

Command: ATTDEF                         Define the attribute Invisible and Constant.
Attribute modes -- Invisible:Y Constant:Y Verify:N     Set these modes.
Attribute tag: BREAKDIM                 The special tag name.
Attribute value: 0.25                   Break out distance.
                                        The attribute can be any text style, alignment or location you wish.

Command: WBLOCK                         Only if you don't have the CA DISK.
File name: VALVE-B
Block name:                             <RETURN> To select.
Insertion base point:                   Pick a point at center.
Select objects:                         You must select the attribute first, then the lines.
```

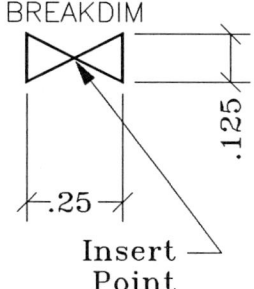

AutoBreaking VALVE-B Block

Now you need to develop a set of routines to help draw lines while inserting autobreaking blocks. The system is modular so that you can insert individual blocks, breaking existing Lines and Plines. Let's look at the core routine that gets the block's break information.

The GETBK routine takes the block name and break dimension as arguments. It uses them as default prompts, if they are non-NIL. GETBK prompts the user for a block name, looks up the break data and returns the block name and the data.

The AUTOBLK.LSP Functions

 Copy AUTOBLK.LSP to TEST.LSP

 Begin a new TEST.LSP file.

```
Select [EDIT-LSP]                       Enter the GETBK function:
```

```
;* AUTOBLK.LSP contains routines that allow the placement of a BLOCK on a
;* PLINE or LINE and automatically break the line out of the Block area.
;* The block must be built with its first entity a constant invisible attribute
;* tagged BREAKDIM, containing a value equal to the length of break required.
;* The insertion point must be in the center of the horizontal BREAKDIM location.
;* The DXF and USTR functions are required.

;* GETBK prompts for block name, searches & returns block name & breakdim data.
;* If BLNAME argument is non-nil it serves as the prompt default.
;* It must be supplied a valid BKDIST argument if supplied BLNAME is non-NIL.
(defun getbk (blname bkdist / ok tbdata bldata tmp)
   (while (not ok)
      (while (not (setq tmp (ustr "Block name" blname nil))))   ;Get BLOCK name
      (if (/= tmp blname)                         ;If new block name given
         (progn                                   ;Then
            (while
               (not
                  (if (setq tbdata (tblsearch "BLOCK" tmp))  ;Search table for BLOCK name
                     (setq bldata (entget (dxf -2 tbdata)))  ;Then get first subentity
                     (command "INSERT" tmp nil)   ;Else insert BLOCK in database, and loop
   ) ) ) ;while not if
            (setq attag (dxf 2 bldata))
            (if                        ;If it's an attribute and has the proper format
               (and (= (dxf 0 bldata) "ATTDEF") (= (dxf 2 bldata) "BREAKDIM"))
               (setq bkdist (atof (dxf 1 bldata)) ;Then get breakdim data
                     blname tmp                   ;and save bkname
                     ok T                         ;exit flag
               )
               (prompt "\nError: Block must have BREAKDIM attribute as 1st entity.")
            ) ;if an attrib...
         ) ;progn then
         (setq ok T)                              ;Else used same block name
      ) ;if new
   ) ;while
   (list blname bkdist)                           ;return the block name and break distance
) ;defun GETBK
;*
```

Save, Exit and reenter AutoCAD. TEST.LSP should load.

```
Command: (getbk "test" 99)
Block name <test>: VALVE-B
INSERT Block name (or ?): VALVE-B        It inserts and cancels if not already in dwg database.
Lisp returns:  ("VALVE-B" 0.250000)
```

GETBK is only the engine to get the data. The following C:BLINE command function calls the GETBK function, gets the points from the user and draws the line. BLINE then calls another function, LINEBLK, to insert and break. Add the C:BLINE and LINEBLK functions.

Select [EDIT-LSP]   Edit TEST.LSP, and add:

```lisp
;* C:BLINE gets a block name, gets its data, draws a line and calls LINEBLK to
;* insert the block and break the line.
(defun C:BLINE ( / linent blkprmt spt ept)    ;bbname & bbdata float for defaults
  (setq bkdata (getbk bbname bbdist)          ;Get block name and break data
        bbname (car bkdata)
        bbdist (cadr bkdata)
  )
  (setq spt (getpoint "\nStart point of line: "))    ;Get first point of line
  (if spt
    (while (setq ept (getpoint "\nPick end point of line: " spt)) ;Get 2nd point of line
      (command "LINE" spt ept "")             ;Draw LINE
      (setq linent (entlast))                 ;Get insert point.
      (setq pt1 (osnap ept "MIDP,QUI"))       ;Determine midpoint of LINE
      (lineblk bbname bbdist pt1 linent)      ;Perform Block insert routine
      (setq spt ept)                          ;Establish a new start point
    );while
  );if
  (setvar "LASTPOINT" spt)                    ;Resets properly to end of line
  (princ)
);defun C:BLINE
;*

;* LINEBLK inserts the autobreaking block and breaks the line or pline
(defun lineblk (blname bkdist pt1 linent / blname brkdim blkname blkent validatt atttag ang)
  (setq pt2 (osnap pt1 "ENDP,QUI"))           ;Get end point for rotation.
  (setq ang (angle pt1 pt2))                  ;Get angle of line
  (command "INSERT" blname pt1 #dwgsc "" pt2) ;Insert BLOCK
  (setq pt1 (polar pt1 ang (* (/ bkdist 2) #dwgsc)))  ;Calculate 1st break point
  (setq pt2 (polar pt1 (+ ang pi) (* bkdist #dwgsc))) ;Calculate 2nd break point
  (command "BREAK" linent pt1 pt2)            ;Break LINE or PLINE
);defun LINEBLK
;*
```

Save, Exit and reenter AutoCAD. TEST.LSP should load.

Command: **BLINE**	Test it.
Block name: **VALVE-B**	Your special block.
Start point of line:	Pick.
Pick end point of line:	Pick.
Command: LINE	It draws the line,
Command: INSERT Block name (or ?) <VALVE-B>: VALVE-B	Inserts the valve,
Command: BREAK Select object:	Breaks the line,
Pick end point of line:	Prompts for the next point.

The last function, BBLOCK uses the previous routines to break an existing entity, then places the autobreak block at the selected point.

Select [EDIT-LSP]   Edit TEST.LSP, and add:

```
;* C:BBLOCK uses GETBK and LINEBLK to insert an autobreaking block on existing entities.
(defun C:BBLOCK ( / linent bkdata)            ;bbname & bbdata float for defaults
  (setq linent (entsel "\nPick BLOCK insert point: ")) ;Get entity and point
  (if linent
    (progn
      (setq bkdata (getbk bbname bbdist)       ;Get the block data
            bbname (car bkdata)                ;store the block name
            bbdist (cadr bkdata)               ;and its break distance
      )
      (setq pt1 (osnap (cadr linent) "NEA"))   ;Get nearest point on line
      (lineblk bbname bbdist pt1 (car linent));Perform BLOCK insert & break
    );progn
    (prompt "\nMust select Line or Pline. ")
  );if
  (princ)
);defun C:BBLOCK
;*

(princ)
;* end of AUTOBLK.LSP
```

Save, Exit and reenter AutoCAD. TEST.LSP should load.

Command: **BBLOCK**   Try it:
Pick BLOCK insert point: Pick point on the line.
Block name: **VALVE-B**   It breaks it.

Command: **QUIT**

  Delete TEST.LSP.

  Rename TEST.LSP to AUTOBLK.LSP.

A Word about Block Redefinition and Lost Attributes

When blocks are redefined several management compromises occur with attribute data.

- Constant block attributes are lost, or updated if they are just redefined.

- Constant block attributes always reflect the current definition of the block.

- Old Variable attributes remain with their block inserts, but new insertions observe the new definitions of block attributes. It is possible to have two identically named block inserts, but with different variable attributes if the block has been redefined.

Integrating the TitleBlk Menu

If you don't have the CA DISK, and and want to add the Titleblk menu to CA-MENU:

- Append the **TITLEBLK and **TITLEFOOTER page of your MY19.MNU to the end of the CA-MENU.MNU file.

- Change the [TITLEBLK] label on the **ROOT page of the CA-MENU to:
```
[TITLEBLK]^C^C^CSETVAR TEXTEVAL 1 LAYER S BORDER;;+
(caload "/CA-ACAD/" "TIMELIB") $S=TITLEFOOTER $S=TITLEBLK
```

Tips and Techniques

Here are some summary tips and techniques for using attributes.

- ❏ Code visible attributes with a different color than the drawing's Text to reflect what is permanent text and what are values are entered or updated. Use a dark color for attributes, like DWG NO., DRAWN BY:, or LAST PLOT DATE:. Use a lighter color for attributes that get input.

- ❏ You can use attributes to invisibly store user variables. The title block example used in this chapter is a visible use of attributes as "variables."

- ❏ Don't over use attributes. Use them for system level information, not individual drawings. Use them for more stable types of data.

- ❏ Use preset attribute definitions for material specifications, like chair color, supplier, weight. Leave cost and title block data as variable attributes.

- ❏ Write menu macros to update commonly changed attributes.

- ❏ If you need to control attribute location independently of block insert point, use two blocks, or incorporate the ATTEDIT command in your macro.

- ❏ To "erase" an attribute from a block, simply change it to an empty value.

- ❏ You may want to make the UPDATE function part of the end sequence in your application system. If you are using Release 9 or a later version, redefine the END command to include the UPDATE capability.

- ❏ Attributes are difficult to edit manually. Consider writing a function that edits text in the same way whether it is in an attribute, or Text string.

20—0 Customizing AutoCAD

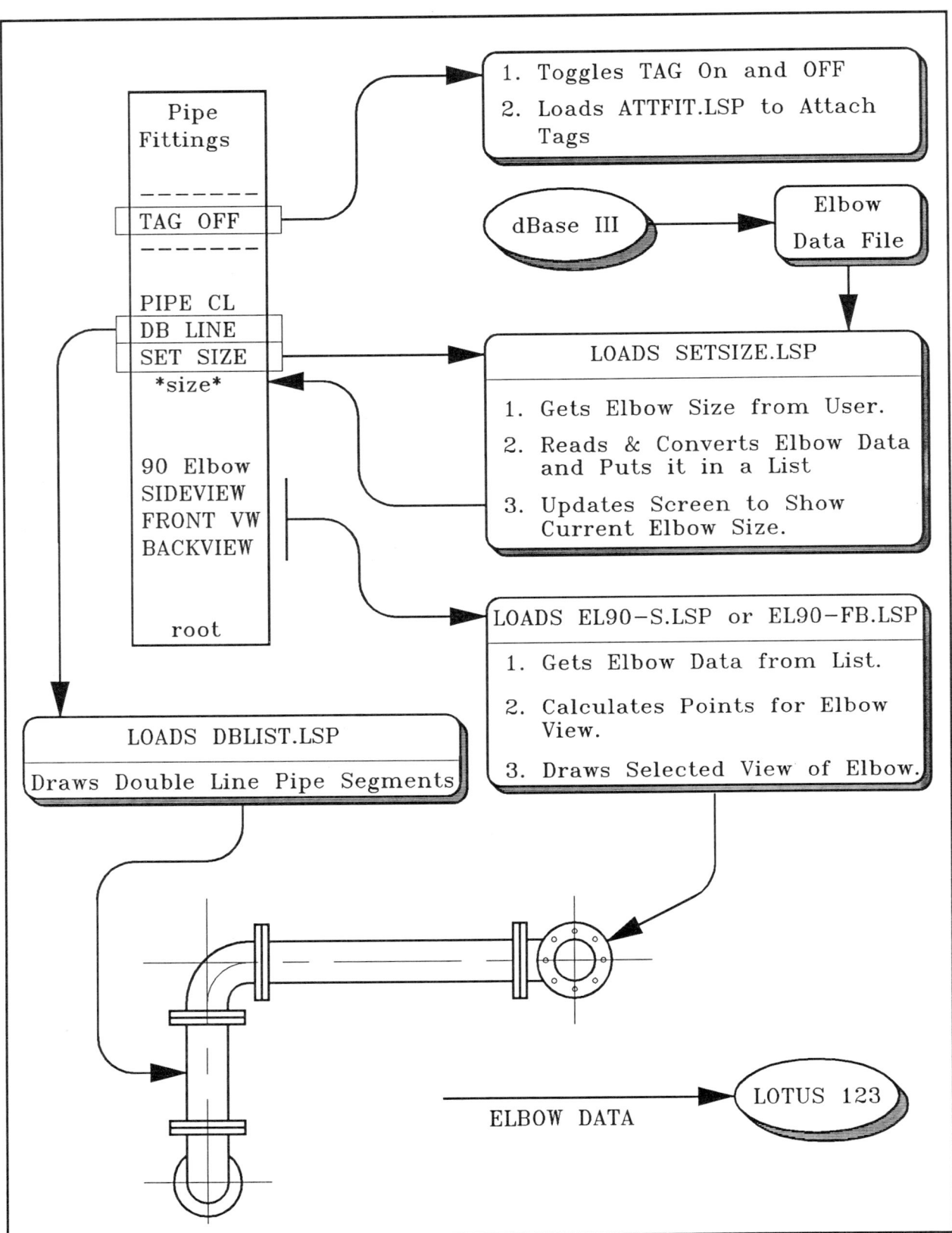

Program Flow of a Parametric System

CHAPTER 20

Parametrics and Material Tagging

An Overview of Parametric Systems

AutoCAD parametric programs are AutoLISP-based programs that get parameter data from the user or from external files, and generate an image in an AutoCAD drawing. One program is used to draw many different sizes of a common shape. The AutoCAD Insert command is parametrics at its simplest.

The parts of a more elaborate parametric component system include:

- External software to make data files.
- An external file containing the component data.
- An AutoLISP function to read the data file.
- A function to calculate and draw an image.
- A menu screen with a selection of parametric viewing functions.
- A help support screen with a list of components.
- Design properties.
- Material tags.

Spreadsheet and database programs can create the data files, but not necessarily in the form that you need. However, you can create data files with any ASCII text editor. This chapter shows you how to create a parametrics program that does pipe fittings with materials tagging. While the chapter's example is based on pipe fittings, the basic ideas and many of the routines are applicable to any parametrics system.

The Benefits of Parametrics

Parametrics offer flexibility and efficiencies in drawing design and the drawing processing. Here are some key benefits in using parametric systems:

- Parametrics provide drawing flexibility, generating all views of a component type without the overhead of duplicate data storage.
- You can compile data from other sources outside of AutoCAD. You can use Lotus 123, dBASE and other application programs to make files suitable for parametric processing.

- You can program AutoLISP based parametrics to make intelligent decisions, alert users to potential component problems and suggest alternative solutions.

- Parametrics provide faster updates and additions to the component library. Programs can automatically tag components with bill of material data. There is no need for user interaction, reducing the chances of error.

How To Skills Checklist

This chapter will show you how to:

- ❏ Design a parametric screen menu.
- ❏ Make and format data files for use with **parametrics**.
- ❏ Handle component query, telling the user what components are available.
- ❏ Parse a line of component data to make a numerical list of the component's characteristics, like size and design properties.
- ❏ Decide the drawing requirements you need to help position the parametric image.
- ❏ Assign large numbers of variables.
- ❏ Take advantage of geometric similarities in the parametric image.
- ❏ Use AutoLISP to calculate the geometry and to drive AutoCAD to draw the component.
- ❏ Supplement the parametrics with parallel line tools.
- ❏ Expand the system's scope by automating bill of material data entry.

Macros, AutoLISP Tools and Programs

MACROS
**PIPEFITS is a screen menu for the parametric application. It includes:
[TAG ON]/OFF toggles component material tagging on and off.
[DB LINE] draws double line pipes and flanges between fittings with dimensions based on current pipe data.
[SET SIZE] loads the requested pipe data.
[SIDEVIEW], [FRONT VW] and [BACKVIEW] draw the parametric pipe fittings.

AutoLISP TOOLS
GETSIZE parses strings read from a data file. It separates each item of the data record.
ATTFIT and ATTLINE are tools that automatically attach material tags to the elbows and pipe segments.

PROGRAMS
SETSIZE.LSP is the program that manages the data retrieval process. It uses GETSIZE and informs the user of the current component size.
EL90-S.LSP is a parametric program. It makes a 90 degree pipe elbow in side view.
DBLINE.LSP draws double line pipes between pipe fittings.
The EL90-FB.LSP program draws elbows in front and back views.

The Parametric Screen Menu

Let's start by designing and writing a screen menu for the user interface of the parametric system. Then, add the AutoLISP programs, menus and text data files to support the menu. The program flow for the parametric system is shown in the illustration at the beginning of the chapter.

When the user selects the [PIPEFITS] option from the **ROOT screen page, it changes the layer, sets a flag to turn off component tagging, and changes to the **PIPEFITS and **FOOTER screen menus. It then uses GRTEXT to display a size on the screen menu if it has been set. The menu is shown in the following PIPEFITS Menu drawing.

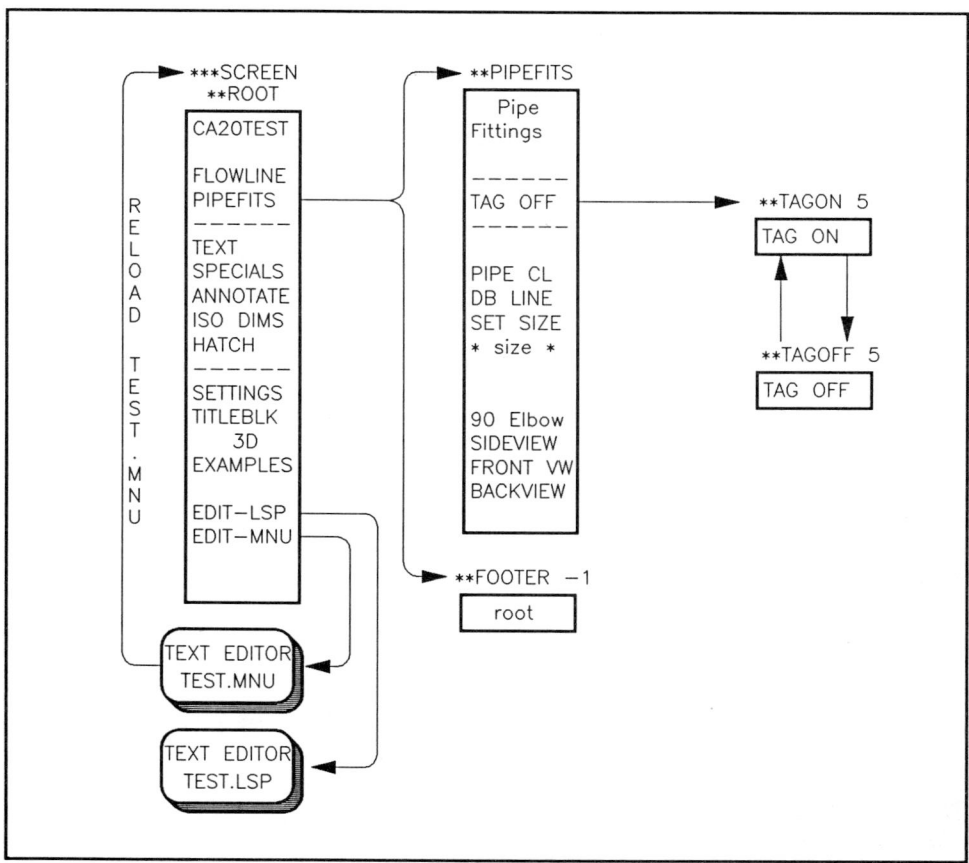

The PIPEFITS Menu

Making the PIPEFITS Menu

Enter selection: Begin a NEW drawing named CA20

   Copy CA20.MNU to TEST.MNU.

   Copy MAIN.MNU to TEST.MNU.

Select **[TEST]**
Select **[24 x 36]**
Select **[FULL]**
Select **[INITIATE]**   Then enter 5 <RETURN>s to default the attributes.

Command: **ERASE**   Erase the titleblock for a clear screen.

Select **[EDIT-MNU]**

Revise the top of the **ROOT page to be:

```
***SCREEN
**ROOT
[CA20TEST]
[]
[FLOWLINE]
[PIPEFITS]^C^C^CLAYER S OBJ01;;(setq #tag nil) $S=FOOTER $S=PIPEFITS +
(if #arglist (grtext 10 (strcat "* " (rtos (car #arglist) 4 2))))
```

Next, create the **PIPEFITS screen page of parametric sizes and options. It contains tools to help create the base drawing and parametric shapes. It uses two techniques to help the user. First, it uses menu label toggles to display the ON or OFF status of the pipe component labeling feature. Second, GRTEXT dynamically displays the pipe diameter data.

Save what you've done and rename the **EXAMPLES page name to **PIPEFITS. Add the following macros, labels and **TAG pages:

```
**PIPEFITS
[ Pipe  ]
[Fittings]
[]
[--------]
[TAG OFF ]^C^C^C(caload "/CA-ACAD/" "ATTFITS") (setq #tag T) $S=TAGON
[--------]
[]
[PIPE CL ]^C^C^CLAYER S CEN31 ;LINE
[DB LINE ]^C^C^CLAYER S OBJ01 ;(caload "/CA-ACAD/" "DBLINE") (dbline #tag #arglist)
[SET SIZE]^C^C^C(caload "/CA-ACAD/" "SETSIZE");+
(setq #arglist (setsize "/CA-ACAD/PIPEFIT.DAT" "Nominal pipe size (or ?)" ));
[* size *]
[]
[]
[90 Elbow]
```

```
[SIDEVIEW]^C^C^CLAYER S OBJ01 ;(caload "/CA-ACAD/" "EL90-S") (el90-s #tag #arglist)
[FRONT VW]^C^C^CLAYER S OBJ01 ;(caload "/CA-ACAD/" "EL90-FB") (el90-fb #tag #arglist T)
[BACKVIEW]^C^C^CLAYER S OBJ01 ;(caload "/CA-ACAD/" "EL90-FB") (el90-fb #tag #arglist nil)
[]
[]
**TAGOFF 5
[TAG OFF ](setq #tag T) $S=TAGON
**TAGON 5
[TAG ON  ](setq #tag nil) $S=TAGOFF
```

Save and Exit to AutoCAD. TEST.MNU reloads.

Select **[PIPEFITS]**	Then try [root] to test the page changes.
Select **[TAG OFF]**	Its (setq #tag T) returns T and toggles the label [TAG ON].
Select **[TAG ON]**	It toggles it off and returns NIL.

You need to write some AutoLISP functions and create data files before you can test the menu further. Let's take a look at the basic file format for the parametric data.

The External File Format

When you write parametric systems, standardize your data file formats with your entire system in mind. Then, you can write one type of function to process the standard format. The format should have a KEY element to mark the component name, and a DATA element holding the component data. The data can be anything, STRing, REAL, INTeger, LIST, etc. The file format must allow you to distinguish between the KEY and the DATA elements of the file. Recall in the discussion of file I/O that there are several file formats that you can use. Decide which one works best for your type of data and application. The example parametrics system uses a comma delimited file with an asterisk tag keying the material name. A data record in this format appears in the data file as:

```
*PIPE-1
12.5,3000,1800,0,134.0
```

You can put all the DATA for each record on one line, or separate it on several lines. The example uses the first line for the material ID, and the second line for the data.

For the example, you need a file whose KEY elements are the pipe diameters and whose DATA elements describe the pipe turning radius, bend diameters and flange dimensions. You might find this sort of data in charts printed in your typical reference book. Convert this data into a comma delimited format. If you have the CA DISK, the file PIPEFIT.DAT has the parametric data in it already. Otherwise, you need to make a new file and enter the data.

20—6 Customizing AutoCAD

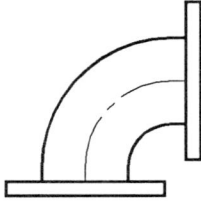
SIDE VIEW

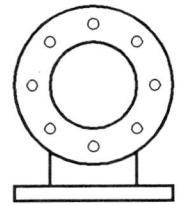
FRONT VIEW

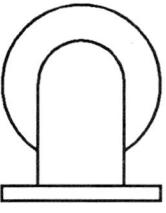
BACK VIEW

Parametric Elbow Fittings

A Pipe Fitting Data Library

 Examine the PIPEFIT.DAT file.

 If you don't have the CA DISK, do this:

Command: **ED**

```
*1
1.0,3.5,4.25,0.41375,4.0,0.5
*1-1/4
1.25,3.75,4.625,0.5,4.0,0.5
*1-1/2
1.5,4.0,5.0,0.5625,4.0,0.5
*2
2.0,4.5,6.0,0.625,4.0,0.625
*2-1/2
2.5,5.0,7.0,0.6875,4.0,0.625
*6
6.0,8.0,11.0,1.0,8.0,0.75
```

Start a new file PIPEFIT.DAT and enter:

Save and exit to AutoCAD.

Even though a chart may give some dimensions in INTeger form, we put the dimensions in the data file as REALs by adding a .0 to the end each one. The example uses dimensions as reals not integers. Once the sizes are in the file, you need a way to tell the user which sizes are available. One way to do this is to use AutoCAD's help facility.

A Little Help

The AutoCAD Help command looks up and displays information stored in a file called ACAD.HLP. It is stored in an ASCII format. You can modify and add to it. For fast access, AutoCAD uses a file called ACAD.HDX (Help inDeX). If it can't find the .HDX file, AutoCAD creates a new index.

By adding to the ACAD.HLP file, you can display help in the parametric command by typing a ? to get a list of components. AutoCAD's help facility handles this task better than reading your entire parametric data file, formatting it, and displaying the data.

The changes to ACAD.HLP are in a CA DISK file named PIPEFIT.HLP. Let's modify a copy of the ACAD.HLP file. Remember to use your AutoCAD directory name in the following COPY and REName steps.

The PIPEFIT Help Screen

Command: **DUP**   Copy your original ACAD.HLP to \CA-ACAD\ACAD.HLP.
Command: **ED**   Edit \CA-ACAD\ACAD.HLP.

   Append the file PIPEFIT.HLP to the file.

   Add the following to the end of the file:

```
\PIPEFIT
        Available pipe diameter fitting sizes

Nominal         Center       Flange      Flange        No.       Diam.
Pipe size       to End       Diam        Thickness     Bolts     Bolts
PDIAM           CE           FDIAM       FTHK          NB        DB
  1             3.5          4.25        0.41375       4         0.5
  1-1/4         3.75         4.625       0.5           4         0.5
  1-1/2         4            5           0.5625        4         0.5
  2             4.5          6           0.625         4         0.625
  2-1/2         5            7           0.6875        4         0.625
  6             8            11          1             8         0.75

Example of use:

Select [SET SIZE] from screen menu.

Nominal pipe size (or ?): 1-1/4          Enter as 1-1/4 (not 1.25 or 1-1/4").

Select [SIDEVIEW]                        Or another component draw command.
Select two lines...
First pipe center line:
Second pipe center line:

Save and exit to AutoCAD.
```

```
Command: SH
DOS Command: RENAME \ACAD\ACAD.HLP ACADHLP.HLP

Command: ?                                    Test the new help.
"C:\CA-ACAD\ACAD.hdx":  Can't open file
HELP index being created.
Command name (RETURN for list): PIPEFIT       And the help screen displays.
```

The "\PIPEFIT" is the name AutoCAD will use to find the help screen. ACAD.HLP uses the "\" to identify each help screen name. If the help is more than one screenful, divide it into pages by inserting "blank" lines containing only a leading "\" backslash. A page can also have multiple labels, like:

```
\PIPEFIT
\PIPESIZE
         Available pipe diameter fitting sizes

Nominal      Center    Flange    Flange    No.       Diam.
```

You must delete the file called ACAD.HDX, if it exists in the new ACAD.HLP file's directory. Rename your original ACAD.HLP. AutoCAD only recognizes one help file at a time, and gives preference to any found in its program directory. DON'T FORGET to rename your original file back when you finish this chapter, or when you leave the system for others to use!

Retrieving External Data

Data retrieval is handled by two functions. SETSIZE handles the input and prompting. It uses a second function called GETSIZE to get the data. The data is returned as a LIST.

The data file name and a prompt are passed into the SETSIZE function as arguments. The SETSIZE function determines if the data file exists, then prompts for the component name to draw, and searches the file for the component data. If the component is not found, it prompts for another one. Once found, the data is read, parsed and returned in the form of an AutoLISP argument LIST, or NIL if it could not find the data.

The argument list returned by SETSIZE is saved by the calling macro as a global #ARGLIST variable. The idea behind parametrics is to use a current set of data for as many purposes as needed. The SETSIZE function uses GRTEXT to print the current size on the screen menu page. Continue in the CA20 drawing and write the SETSIZE function.

Creating SETSIZE.LSP

 Copy SETSIZE.LSP to TEST.LSP

 Begin a new TEST.LSP file.

Select **[EDIT-LSP]**   Edit TEST.LSP, and enter:

```
;* SETSIZE.LSP prompts for and retrieves requests of parametric data files.
;* Data retrieval uses two functions. SETSIZE handles input and prompting.
;* GETSIZE gets the data, which is returned as a LIST.
;*
;* SETSIZE is for the PIPEFIT parametric system. Support for the AutoCAD Help
;* facility is added to the prompting sequence of the program.
;* The user get functions are required. It requires filespec and prompt string
;* args and returns NIL or a data list.
(defun setsize (datfile prmpt / fp found size arglist)
  (if (setq fp (open datfile "r"))            ;test for file
  (progn
     (while (not found)  ;flag to terminate loop, local "found" initially nil
       (while (or
                 (not (setq size (ustr prmpt size nil)))   ;get the pipe size
                 (= size "?")                    ;allow a ? to list'em
              );or
         (if (= size "?")                       ;use the acad.hlp to display pipe sizes
            (command "'?" "PIPEFIT")            ;call acad for help.
         );if
       );while or
       (if (setq arglist (getsize fp (strcat "*" size))) ;gets size, puts * in front of name
         (setq found T)                         ;stop size prompting
         (progn
            (prompt "\nSorry, that size not found! ")  ;prompt for size again
            (close fp)                          ;close the file
            (setq fp (open datfile "r"))        ;reopen to rewind to top
         );progn
       );if
     );while not found
     (close fp)                                 ;close the file
     (grtext 10 (strcat "* " size))             ;print size on screen label
  );prong
  );if
  arglist                                       ;return data list
);defun SETSIZE
;*
```

GETSIZE is the core routine. It reads through an asterisk "*" tagged comma delimited file looking for the shape name. When it finds the name, it parses the string looking for commas. Each piece of data is converted to a real value and put on a list. The function returns the data list at the end of process, or it returns NIL.

Creating GETSIZE

Save what you've done and continue adding:

```
;* GETSIZE is a general purpose data search and retrieval function for
;* asterisk tagged "*name" comma delimited data format. Its FP arg is the file
;* handle and SIZE is the "*name" to find data of. It returns NIL or a data list.
(defun getsize (fp size / item data dline dlist maxs count chrct numb)
  (setq item (read-line fp))              ;first line is a label for file
  (while item                             ;process each line of file
    (if (= item size)
      (setq data (read-line fp)           ;read a line
            item nil                      ;stop searching for item
      );setq
      (setq item (read-line fp))          ;keep searching for item
    );if
  );while
  (if data                                ;if the size has been found
    (progn
      (setq maxs (strlen data)            ;establish length of input
            count 1   chrct 1             ;initialize count and char position
      );setq
      (while (< count maxs)               ;process string one chr at a time
        (if (/= "," (substr data count 1)) ;look for the commas
          (setq chrct (1+ chrct))         ;increment to next pos
          (setq numb (atof (substr data (1+ (- count chrct)) chrct))  ;convert to real
                dlist (append dlist (list numb))    ;add it to the list
                chrct 1                   ;resets field ct
          );setq
        );if
        (setq count (1+ count))           ;increment the counter
      );while
      (setq numb (atof (substr data (1+ (- count chrct))))) ;convert to real
            dlist (append dlist (list numb))      ;add it to the list
      );setq
    );progn
  );if data
  dlist                                   ;may be nil or a list of reals
);defun GETSIZE
;*
(princ)
;* end of SETSIZE.LSP
;*
```

Save, Exit and reenter AutoCAD. TEST.LSP should load.

Select **[PIPEFITS] [SET SIZE]**   Test it with the screen menu.
Nominal pipe size (or ?): **?**   Test the help facility with a ?.
                                          The help screen displays.

```
Nominal pipe size (or ?): 6                Select nominal size 6, & the data is returned.
Lisp returns: (6.000000 8.000000 11.000000 1.000000 8.000000 0.750000)

Command: !#ARGLIST                          The current data is now available:
Lisp returns: (6.000000 8.000000 11.000000 1.000000 8.000000 0.750000)
```

 Delete TEST.LSP.

 Rename TEST.LSP to SETSIZE.LSP.

The current pipe diameter is now shown in the screen menu area as [* 6]. Now you are ready to generate an image.

Generating the Image

When you plan out your system, sketch each component view on paper. Label the variables, relationships and critical points on the image. Show linetypes and identify common graphic constructions. For instance, elbow fittings all have a least one flange drawn as a rectangle. That's one routine you can centralize.

Plan how the user will locate and orient the images in the drawing. Use centerlines, line endpoints, midpoints, or other intersections of objects. Figure out which points you will need to have the user select, and those you need to calculate. Label the points you'll pass back to AutoCAD when drawing the image. The variables and points for the fittings are shown in the Variables and Fittings drawing.

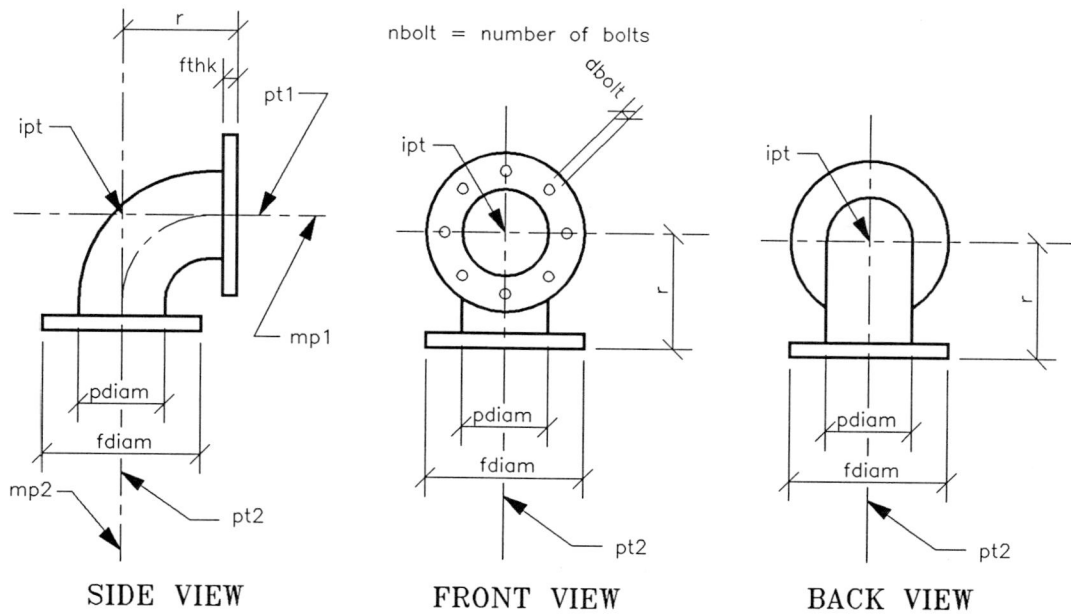

Variables and Fittings

Drawing a Side View of a 90 degree Elbow

The EL90-S function draws a 90 degree elbow in side view. It takes two arguments, a flag to toggle component tagging and the list of pipe fitting data. The component tagging toggle flag is off for now. The user selects two orthogonally intersecting centerlines (lines or plines), and the fitting is drawn around the centerlines.

Parametric functions must carefully control Osnap and object selection to avoid wrong points and entities, especially when the drawing is zoomed way out. This program sets NEArest for object selection. The selection points are OSNAPed END with a small aperture to get the other two points to feed to INTERS to obtain the intersection point.

Let's enter this much of the program now.

Creating the EL90-S.LSP Function

  Copy EL90-S.LSP to TEST.LSP.

  Begin a new TEST.LSP file.

Select **[EDIT-LSP]**   Edit TEST.LSP, and enter:

```
;* EL90-S.LSP generates parametric 90 degree elbow fittings in side view.
;* It requires selecting two intersecting orthogonal centerlines about
;* which the fitting is drawn at the preselected fitting size.
;* Its TAG arg is an attrib tagging flag T or NIL. ARGLIST is the pipe data,
;* generally global variable #ARGLIST. The user get functions are required.
(defun el90-s (tag arglist / $op $ap pt1 pt2 mp1 mp2 ipt pdiam r fdiam fthk nbolt dbolt
                             alist h-fdiam r-fthk tp1 tp2 tp3 tp4 tp5 ent)
  (if arglist                                     ;if user selected a size
    (progn
      (setq $ap (getvar "APERTURE"))              ;store current aperture
      (setq $op (getvar "OSMODE"))
      (setvar "OSMODE" 512)                       ;set Osnap NEA for obj selection
      (setvar "APERTURE" 5)                       ;set it to 5
      (prompt "\nSelect two lines...")            ;info prompt
                                                  ;get 2 points, be sure entity is selected
      (while (not (setq pt1 (upoint 1 "" "First pipe center line" nil nil))))
      (while (not (setq pt2 (upoint 1 "" "Second pipe center line" nil nil))))
      (setvar "APERTURE" 1)                       ;set a smaller aperture
      (setvar "OSMODE" 0)                         ;NONe
      (setq mp1 (osnap pt1 "end")                 ;get the endpoint of first entity
            mp2 (osnap pt2 "end")                 ;end of second entity
            ipt (inters mp1 pt1 mp2 pt2 nil)      ;intersection of the points
      );setq
```

The next part of the program starts by testing that an intersection point was found. If an intersection point is not found, EL90-S aborts.

The program sets all data in the arglist data to a variable for the pipe fitting dimension calculations. It employs an extremely efficient MAPCAR technique for setting a group of variables to their corresponding values. An alternative would be to have six program statements, one for each variable. The following lines compare both ways:

(setq d (nth 0 arglist))   The slow way.

(mapcar 'set '(pdiam r fdiam fthk nbolt dbolt) arglist)   Faster.

The program maps the SET function to each element of the first list and its corresponding element of the second list. SET is used instead of SETQ to evaluate each quoted element of the first list. With the currently set #ARGLIST, it sets PDIAM to 6.000000, R to 8.000000, FDIAM to 11.000000, and so on.

The ANGLE function calculates angles for the two objects, and the angles are combined in a list. Making a list of angles cuts part of the program code in half. Certain constant angles like 90 are already defined in the ACAD.LSP file. These defined angles, like #PI90, are called here. Continue adding to the function.

Save what you've done and continue adding:

```
(if ipt                                          ;if an intersection exists
   (progn
      (mapcar 'set '(pdiam r fdiam fthk nbolt dbolt) arglist) ;sets all local vars
      (setq e1ang (angle ipt pt1)                ;angle of first entity
            e2ang (angle ipt pt2)                ;angle along second entity
            alist (list e1ang e2ang)             ;make a list of the angles
            h-fdiam (/ fdiam 2.0)                ;set a var to half the flange width
            r-fthk (- r fthk)                    ;radius minus the flange thickness
      );setq
```

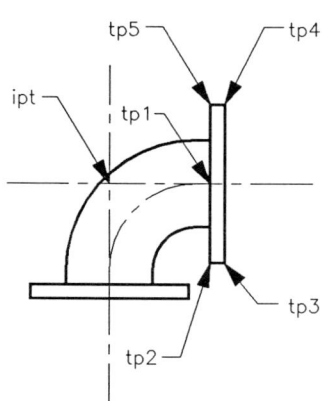

Calculated Points for Fitting Sideview

Next, draw the flanges for the fitting. An elbow in side view has two such flanges. This is where the list of angles works well. The orientation of each flange is governed by the angles in the list. Each time through the FOREACH loop the angle is different.

With four flanges, you could use a list of four angles. At some point though, an ARRAY, or a COPY then ROTATE command combination will outperform the FOREACH loop. It depends on the number and complexity of objects that you need to draw. All the point calculations are done outside of the COMMAND pipeline for ease of readability of the program. Before the loop starts again, you use the Pline command to draw the flange. The calculated points are shown in the Calculated Points drawing.

Save what you've done and continue adding:

```
    (foreach ang alist                          ;process each angle of the list
        (setq tp1 (polar ipt ang r-fthk)        ;calculate the temporary points
              tp2 (polar tp1 (+ ang #pi270) h-fdiam) ;for each of the flanges
              tp3 (polar tp2 ang fthk)
              tp4 (polar tp3 (+ ang #pi90) fdiam)
              tp5 (polar tp4 (+ ang pi) fthk)
        );setq
        (command "pline" tp2 tp3 tp4 tp5 "c")   ;draw the flange
    );foreach of the flanges
```

Now the hard part! Or the easy part, it depends on how you use your AutoCAD tools. Why calculate curves when AutoCAD can draw the center line of the fitting as straight Pline segments? You can Fillet to create the curve, and Offset to make the inner and outer pipe curves. The offset distance is half of the pipe diameter. The center of the pipe curve line is given a CENTER linetype.

Offset requires selecting entities by a point. If you are zoomed way out, the pickbox may find the wrong entity. Control your point selection by building an ENTSEL style list with ENTLAST and TP1, and using it to select the entity to offset.

That's it. Reset the system variables and close out the function, adding error prompts as the "else" of the two IF tests. Finish the function and test it.

Save what you've done and continue adding:

```
        ;* add the arcs
        (command "pline" (polar ipt e1ang r-fthk) ipt ;Pline arc along center of fitting
            (polar ipt e2ang r-fthk) ""
            "fillet" "r" pdiam "fillet" "p" "l"        ;fillet polyarc with fitting diameter
            "offset" (/ pdiam 2.0)                     ;offset center arc to each side
            (setq ent (list (entlast) tp1)) tp5 ent tp2 ""  ;using ENTSEL style list
            "change" tp1 "" "p" "la" "cen31" ""        ;change center arc to a center line
        );command
        (if tag (attfit "ELBOW" arglist ipt))          ;attach material tag if tagging is ON
        (setvar "APERTURE" $ap)                        ;reset the aperture
        (setvar "OSMODE" $op)
      );progn
      (prompt "\nError: No intersection point found. ")  ;abort if no intersection found
    );if
  );progn
  (prompt "\nSelect a size first. ")                  ;abort if no size selected
);if
);defun EL90-S
;*
(princ)
;*end of EL90-S.LSP
;*
```

Save, Exit and reenter AutoCAD. TEST.LSP should load.

```
Command: ZOOM                              Zoom Center, height 6 feet for more space.
Command: PLINE                             Draw a series of 12"+ lines at 90 degrees to each other.

Select [SIDEVIEW]                          Try it. Size should show [* 6 ].
Select two lines...
First pipe center line:                    Pick first line.
Second pipe center line:                   Pick second. It draws the fitting and fillets the centerline.
```

 Delete TEST.LSP.

 Rename TEST.LSP to EL90-S.LSP.

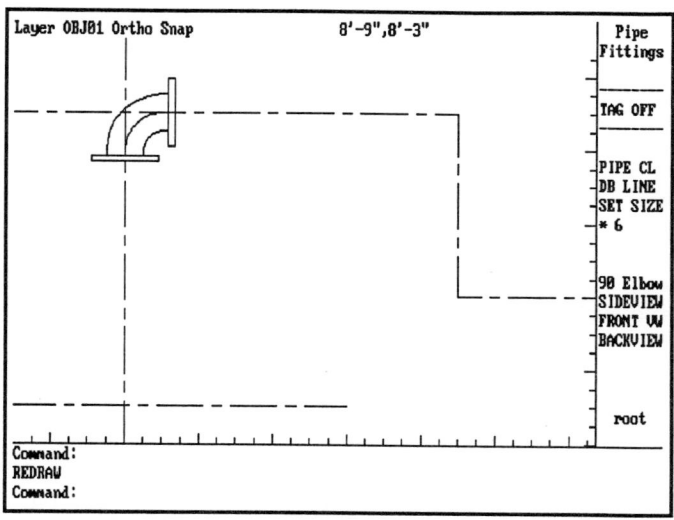

Parametric Elbow   Side View

Now that one is drawn, creating the others is easy.

Multiple Views: The Beauty of Parametrics

The beauty of parametrics is that you get more views with less overhead! You can make a FRONT and BACK view of the same elbow. Most of the code is the same, and all the data is present. We considered putting all three views in one function. The trade off depends on how often all three views are used during the same drawing session. One giant function means slower loading time and more memory to swap. Individual functions mean faster response, but more duplicated code if all the functions are loaded. We will compromise, combining the front and back views in one more function.

Make the next function EL90-FB (Front and Back view). Use a flag in the argument list. Call it FRONT. Use the flag to choose which view should be drawn. Set the flag in the menu macro that calls the function. Input is the same as the previous function except the center of the fitting and a second flange direction is requested. You can copy portions of the code from EL90-S.LSP.

Creating EL90-FB for Front and Back Elbows

 Copy EL90-FB.LSP to TEST.LSP

 Begin a new TEST.LSP file.

Select **[EDIT-LSP]**   Edit TEST.LSP, and enter:

```
;* EL90-FB.LSP parametrically draws a Front/Back view 90 degree elbow fitting.
;* The fitting is drawn at the preselected fitting size.
;* Its TAG arg is an attrib tagging flag T or NIL. ARGLIST is the pipe data,
;* generally global variable #ARGLIST. If the FRONT arg is T, it draws the
;* front, otherwise the back is drawn. The user get functions are required.

(defun el90-fb (tag arglist front / e1 e2 $op pt2 ipt pdiam h-pdiam
                r fdiam fthk nbolt dbolt elang h-fdiam r-fthk h-inca
                tp1 tp2 tp3 tp4 tp5 tp6 tp7 tp8 tp9)
  (if arglist                                    ;if user selected a fitting size
    (progn
      (setq $op (getvar "OSMODE"))
      (setvar "OSMODE" 0)
                                                 ;get intersection point and point on entity
      (while (not (setq ipt (upoint 1 "" "Centerpoint of 1st flange" nil nil))))
      (while (not (setq pt2 (upoint 1 "" "2nd Flange direction" nil ipt))))
```

These views require several more TPxx temporary points, like the points on the lines that intersect the curved surfaces of the pipe diameter. The program uses INCANG from the 3DTOOLS.LSP library, but instead of loading the entire library, only INCANG is included in the parametric file. INCANG finds the points along the curve. As an alternative, you could use the AutoCAD's TRIM command to find the points. The calculated points are shown in the drawing of Front and Back Views.

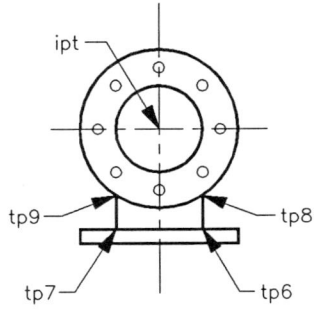

Front and Back Views

Save what you've done and continue adding:

```
    (mapcar 'set '(pdiam r fdiam fthk nbolt dbolt) arglist) ;sets all local vars
    (setq nbolt (fix nbolt)                ;convert from a real for ARRAY command
          elang (angle ipt pt2)            ;angle of selected line
          h-fdiam (/ fdiam 2.0)            ;half the flange width
          r-fthk (- r fthk)                ;radius less the flange thick
          h-pdiam (/ pdiam 2.0)            ;half the pipe diameter
          h-inca (/ (incang pdiam h-fdiam) 2.0) ;half the included angle
          tp1 (polar ipt elang r-fthk)     ;calculate the temp points
          tp2 (polar tp1 (+ elang #pi270) h-fdiam)
          tp3 (polar tp2 elang fthk)
          tp4 (polar tp3 (+ elang #pi90) fdiam)
          tp5 (polar tp4 (+ elang pi) fthk)
          tp6 (polar tp1 (+ elang #pi90) h-pdiam)
          tp7 (polar tp6 (- elang #pi90) pdiam)
          tp8 (polar ipt (+ elang h-inca) h-fdiam)
          tp9 (polar tp8 (- elang #pi90) pdiam)
          tp10 (polar ipt (+ elang #pi90) h-pdiam)
          tp11 (polar ipt (- elang #pi90) h-pdiam)
    );setq
    (command "pline" tp2 tp3 tp4 tp5 "c")          ;draw the flange
```

Here is where you add the difference in views. The FRONT flag is used to direct the program to use one set of drawing instructions versus another. If the front view is drawn, full circles and ARRAYed bolt holes show in the view. Otherwise, a back view is constructed using the ARC command, and the corresponding lines are drawn. Finish the function.

Save what you've done and continue adding:

```
      (if front                                    ;check which view
        (command "circle" ipt "d" pdiam            ;draw the front view
                 "circle" ipt "d" fdiam
                 "line" tp6 tp8 "" "line" tp7 tp9 ""
                 "circle" (polar ipt elang (+ h-pdiam (/ (- fdiam pdiam) 4.0)))
                 "d" dbolt "array" "l" "" "p" ipt nbolt 360 "n"   ;array the bolt holes
        );command
        (command "line" tp6 tp10 "" "line" tp7 tp11 ""  ;else draw back view
                 "arc" "C" ipt tp10 tp11
                 "arc" "C" ipt tp8 tp9
        );command
      );if direction
      (setvar "OSMODE" $op)
      (if tag (attfit "ELBOW" arglist ipt))   ;chk for mat'l tagging, insert attblock
    );progn
    (prompt "\nSelect a size first. ")              ;no size was set
  );if data
  (princ)
);defun EL90-FB
;*
```

```
;* We took a copy of INCANG from 3Dtools.lsp.
;* INCANG calculates the included angle given a chord distance and radius
(defun incang (chord radius)
  (* 2.0 (atan                                    ;twice the arc tangent
     (/ chord 2.0)                                ;of opposite over adjacent
       (sqrt (- (expt radius 2) (expt (/ chord 2.0) 2)))  ;calcs adjacent leg of triangle
     ));divide & atan
  );
);defun INCANG
;*
(princ)
;*end of EL90-FB.LSP
;*
```

Save, Exit and reenter AutoCAD. TEST.LSP should load.

Select **[PIPEFITS]**	Try out the new views.
Select **[FRONT VW]**	Draw one.
Centerpoint of 1st flange:	Pick point.
2nd Flange direction:	Pick, and it scrolls and draws.
Select **[BACKVIEW]**	Test this one also.

   Delete TEST.LSP.

Rename TEST.LSP to EL90-FB.LSP.

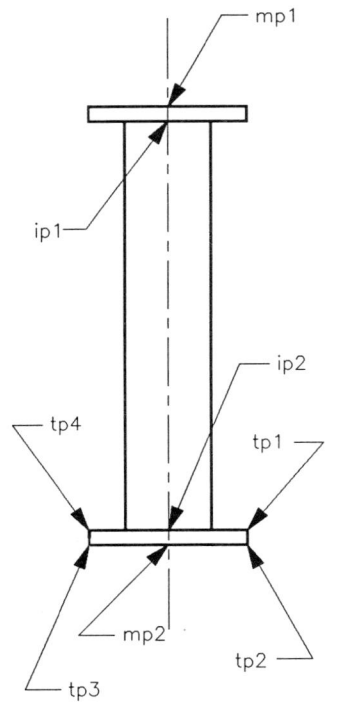

Fittings in Three Views

PIPE SEGMENT

Double Line Pipe
Points

A Double Line Pipe Command

To finish the drafting task for this application, you give the user a command to connect all these fittings. The DBLINE function creates double pipe lines and draws in the additional flange fittings at each pipe end.

DBLINE's beginning is similar to the elbow functions. But this time you externally set Osnap MIDP through SETVAR to get the two points. The UPOINT with its INITGET feature ensures that the user enters the second point. To draw the two flanges, you use a FOREACH for the two points. The points are shown in Double Line Pipe Points.

Creating DBLINE.LSP Pipes

 Copy DBLINE.LSP to TEST.LSP

 Begin a new TEST.LSP file.

Select [EDIT-LSP]   Edit TEST.LSP, and enter:

```
;* DBLINE constructs double line pipes at the preselected fitting size by
;* selecting points on two flanges. Its TAG arg is an attrib tagging flag T or
;* NIL, controlling the option of calculation and recording of pipe segment
;* lengths and parametric properties. The ARGLIST argument is the pipe data,
;* generally global variable #ARGLIST. The user get functions are required.

(defun dbline (tag arglist / pdiam r fdiam fthk nbolt dbolt h-fdiam h-pdiam r-fthk
                             mp1 mp2 tmpang ip1 ip2 tp1 tp2 tp3 tp4)
   (if arglist                                          ;if the size data exist
     (progn                                             ;then run
       (mapcar 'set '(pdiam r fdiam fthk nbolt dbolt) arglist) ;sets all local vars
       (setq h-fdiam (/ fdiam 2.0)                      ;half the flange diameter
             h-pdiam (/ pdiam 2.0)                      ;half the pipe diameter
             r-fthk (- r fthk)                          ;radius minus the flange thickness
       );setq
       (setvar "OSMODE" 2)                              ;turn on MIDP Osnap mode
       (while (setq mp1 (getpoint "\nFrom point: ")) ;start WHILE loop, get the first point
         (setq mp2 (upoint 1 "" "To point" nil mp1))    ;second point
         (setvar "OSMODE" 0)                            ;turn Osnap off
         (setq ang (angle mp1 mp2)                      ;calc the angle
               tmpang ang                               ;store the angle in a temporary variable
               ip1 (polar mp1 ang fthk)                 ;point on first flange
               ip2 (polar mp2 (+ ang pi) fthk)          ;point on second flange
         );setq
         (foreach pt (list ip1 ip2)                     ;draw pipe flange at each fitting flange
           (setq tp1 (polar pt (+ tmpang #pi90) h-fdiam) ;calc the flange points
                 tp2 (polar tp1 (+ tmpang pi) fthk)
                 tp3 (polar tp2 (+ tmpang #pi270) fdiam)
                 tp4 (polar tp3 tmpang fthk)
           );setq
           (command "pline" tp1 tp2 tp3 tp4 "c")        ;draw it with a Pline
           (setq tmpang (+ tmpang pi))                  ;swap direction of angle for next loop
         );foreach of the flanges
```

The pipe is easy. Just draw lines and loop back around for more points.

Save what you've done and continue adding:

```
        (command "line" (polar ip1 (+ ang #pi90) h-pdiam)    ;draw parallel lines
                        (polar ip2 (+ ang #pi90) h-pdiam) ""  ;for the double line
                "line"  (polar ip1 (+ ang #pi270) h-pdiam)   ;pipe diagram
                        (polar ip2 (+ ang #pi270) h-pdiam) ""
        ) ;command
        (if tag (attpipe arglist ip1 ip2))   ;if material tagging ON, insert attribute block
        (setvar "OSMODE" 2)                  ;turn on MIDP again for next pipe segment
      ) ;while
      (setvar "OSMODE" 0)                    ;turn it off after all pipes are drawn
    ) ;progn
    (prompt "\nSelect a size first. ")       ;on size selected
  ) ;if
) ;defun DBLINE
;*
(princ)
;* end of DBLINE.LSP
;*
```

Save, Exit and reenter AutoCAD. TEST.LSP should load.

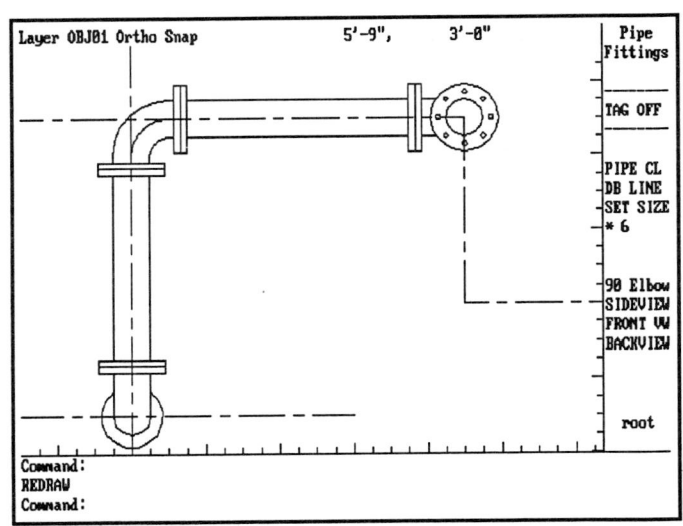

Pipe Segments Between Flanges

Select **[DBLINE]**   Draw between two opposite fittings.
From point: Pick one flange point.
To point: Pick other flange. It draws pipe and flanges.

 Delete TEST.LSP.

 Rename TEST.LSP to DBLINE.LSP.

Since you already have all the data, why not tag the elbows and pipe with attributes and write a report of the materials?

Labeling Components with Material Tags

First, you have to set up the drawing for the reporting. By using invisible attributes to hold information, you can access them using AutoLISP and then generate reports of materials in the drawing.

You need some attributes to carry the information about the pipe fittings. Make two pipe material blocks: ATTFIT and ATTPIPE containing only Invisible, Variable attributes. Note that the ATTFIT tags are in the same order as the #ARGLIST data. If you have the CA DISK, you have ATTPIPE.DWG and ATTFIT.DWG. Otherwise, create the tags with Attdef and Wblock them as shown.

```
For Block Name:  ATTFIT                    ATTPIPE
Assign Tags:  1.  F_TYPE              1.  P_DIAMETER
              2.  F_DIAMETER          2.  P_LENGTH
              3.  F_RADIUS            3.  P_FLANGE_DIAM
              4.  F_FLANGE_DIAM       4.  P_FLANGE_THICK
              5.  F_FLANGE_THICK      5.  P_NUMBER_BOLTS
              6.  F_NUMBER_BOLTS      6.  P_BOLT_DIAM
              7.  F_BOLT_DIAM
```

You can define the blocks with any default prompts or values you wish, or none. They will be filled in automatically.

Material Tagging.

 You have the ATTPIPE.DWG and ATTFIT.DWG. Skip the sequence.

 Do the following.

```
Command: ERASE                                    Clean up the drawing.
Command: ZOOM                                     Center with height 8.
Command: LAYER                                    Set 0 current.

Command: ATTDEF                                   Define the ATTFIT set 1st.
Enter (ICVP) to change, RETURN when done: I       Set Invisible:Y  Constant:N
Attribute tag: F_TYPE                             The first tag.
Attribute prompt:                                 None.
Default attribute value:                          None.
Start point or Align/Center/Fit/Middle/Right/Style:   Pick and default rotation.

Command: ATTDEF                                   Repeat for the rest of the ATTFIT set.

Command: WBLOCK                                   ONLY if you don't have the CA DISK
File name: ATTFIT
Block name (or ?):                                <RETURN> to select entities.
Select objects:                                   Select the attributes in exact table order.
```

```
Command: ATTDEF                              Repeat for the ATTPIPE set.
Command: WBLOCK                              If you don't have the CA DISK, Wblock to ATTPIPE.

Command: INSERT                              Test them.
Block name (or ?): ATTFIT

Enter attribute values                       <RETURN> to default the values.
Fitting type:                                Your prompt appears, or tag F_TYPE if you set no prompt.
Pipe diameter:
Fitting curve radius:
Flange diameter:
Flange Thickness:
Number of bolts:
Bolt diameter:                               And you see nothing. Invisible!
```

You need two functions. ATTFIT will place the fitting attributes. ATTPIPE will record the pipe length and flange data of a pipe section. These are called automatically from the parametric elbow and DBLINE functions when the global #TAG variable is ON (non-NIL). They are loaded when the user toggles material tagging ON from the **PIPEFITS screen page. Let's write the functions.

ATTFIT takes three arguments, the fitting description, the fitting data, and the point at which the attribute block gets inserted. It is simple since our data order matches our attribute order.

 Copy ATTFITS.LSP to TEST.LSP

 Begin a new TEST.LSP file.

```
;* ATTFITS.LSP inserts attribute data with the elbows and pipes drawn by
;* the EL90-S, EL90-FB and DBLINE functions, if the #TAG variable is non-NIL.

;* ATTFIT assigns fitting data to an attribute block and inserts it
;* at the fitting intersection
(defun attfit (type arglist pt)
   (if (and type arglist pt)                    ;tests if all defined
      (progn
         (command "insert" "attfit" pt 1 1 0 type)  ;inserts attribute data tag
         (foreach item arglist                      ;send each one to attribute prompts
            (command (rtos item 2 6)))              ;rounds output to 6 places
      );foreach
   );progn
 );if
);defun ATTFIT
;*
```

ATTPIPE is a little more complex. It calculates the length of pipe, and inserts the attribute block. It assigns the ARGLIST variables with MAPCAR so it can use part of the list.

Save what you've done and continue adding:

```
;* ATTPIPE inserts attribute block for pipe length segments
(defun attpipe (arglist pt1 pt2 / leng pdiam r fdiam fthk nbolt dbolt mpt)
   (if (and arglist pt1 pt2)                                 ;if everything is defined
     (progn
        (mapcar 'set '(pdiam r fdiam fthk nbolt dbolt) arglist) ;sets all local vars
        (setq leng (- (distance pt1 pt2) fthk fthk)            ;compute the pipe length
              mpt (polar pt1 (angle pt1 pt2) (* leng 0.5))     ;calc the midpoint
        );setq
        (command "insert" "attpipe" mpt 1 1 0)                 ;insert the attribute block
        (foreach item (list pdiam leng fdiam fthk nbolt dbolt) ;builds a new list for output
           (command (rtos item 2 6))                           ;convert to decimal string
        );foreach
     );progn
   );if
);defun ATTPIPE
;*
(princ)
;* end of ATTFITS.LSP
;*
```

Save, Exit and reenter AutoCAD. TEST.LSP should load.

Command: **QUIT**   To get a clean start for testing.

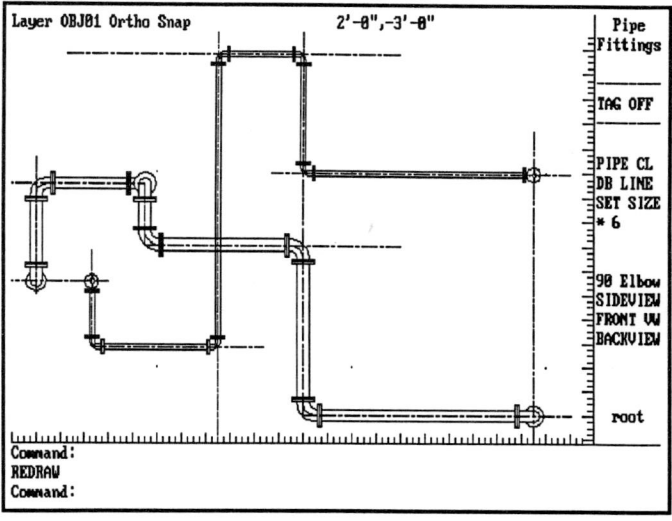

Populated Drawing

Customizing AutoCAD

Reporting the component tag information is covered in the next chapter. You need to generate a set of tagged materials for reporting. Draw about twelve fittings and pipe lengths, like those shown in the screen shot. Use [SET SIZE] to change sizes. Make sure you keep the [TAG ON]. Your drawing should look like our screen shot called Populated Drawing

Making Some Fittings

```
Enter selection:                          Begin a NEW drawing, again named CA20.
Loading Customizing AutoCAD LISP Tools...done.

Select [TEST]
Select [24 x 36]
Select [1/2"]
Select [INITIATE]                         Then enter 5 <RETURN>s to default the attributes.

Command: (load "EL90-S")                  They must be loaded.
Command: (load "EL90-FB")                 (CALOAD will automate these at the book's end.)
Command: (load "DBLINE")
Command: (load "ATTFITS")
Command: (load "SETSIZE")

Command: ATTDISP                          Set ON so you can see the Invisible attributes.
Select [PIPEFITS]

Select [SET SIZE]
Nominal pipe size (or ?): 2-1/2
Lisp returns: (2.500000 5.000000 7.000000 0.687500 4.000000 0.625000)

Select [TAG OFF]                          Toggle TAG ON.

Select [PIPE CL]                          Draw some centerlines.
Select [SIDEVIEW]                         Draw one, see how it puts in data.
Enter attribute values
Fitting type: ELBOW Pipe diameter: 2.500000 Fitting curve radius: 5.000000 Flange diameter:
7.000000 Flange Thickness: 0.687500 Number of bolts: 4.000000 Bolt diameter: 0.625000
                                          And the ATTFIT block with values appears.

Command: ZOOM                             Zoom in to see the attributes. Then Zoom P.

Select [FRONT VW]                         Try it too.
Select [BACKVIEW]                         Try it. And some more.
Select DBLINE                             Draw some pipe lines.
```

```
Command: SH                                  Restore your ACAD.HLP file.
DOS Command: RENAME \ACAD\ACADHLP.HLP ACAD.HLP
```

 Delete TEST.LSP and TEST.MNU.

 Rename TEST.MNU to MY20.MNU. Rename TEST.LSP to ATTFITS.LSP.

```
Command: SAVE                         ONLY if you don't have the CA DISK.
File name <CA20>: PIPEMATL            It is used in the next chapter.
Command: QUIT
```

Your screen should show the tags.

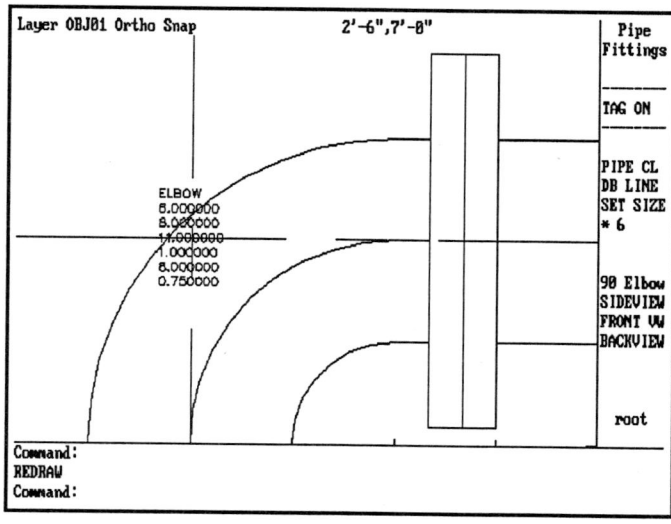

Elbow with Tag

Integrating the Pipefits Menu

If you don't have the CA DISK and want to add the menu to CA-MENU:

- Append the **PIPEFITS, **TAGOFF and **TAGON pages of your MY20.MNU to the end of the CA-MENU.MNU file.

- Change the [PIPEFITS] label on the **ROOT page of the CA-MENU to:
  ```
  [PIPEFITS]^C^CLAYER  S  OBJ01;;(setq #tag nil) $S=FOOTER
  $S=PIPEFITS +
  (is #arglist (grtext 10 (strcat "* " (rtos (car #arglist) 4 2))))
  ```

Here is the chapter's menu.

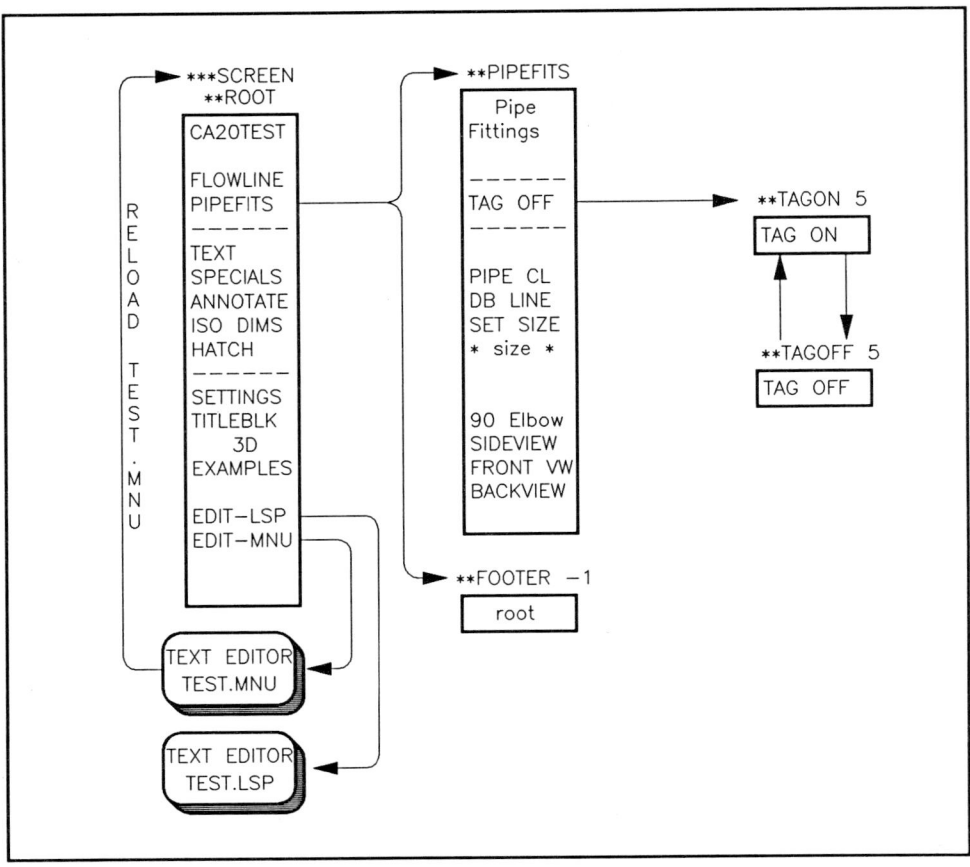

The Chapter 20 Menu

Tip and Techniques

Here are some summary tips and techniques for making parametric systems.

- ❏ Your system's speed influences how complex you can make your parametric system. If a program seems slow, you'll have bored users, even if it draws 10x faster than they could.
- ❏ Balance data file size with disk storage. Don't make one giant file or a million tiny files.
- ❏ Single component data files with preformatted AutoLISP expressions, or lists, are faster than any string parsing function when your application requires frequent changes of component data in memory.
- ❏ If you have to use one large file over 10K in size, consider writing an external program in BASIC, FORTRAN, C or some other language. It can extract the data via SHELL, and output data to a temporary file that is read by the AutoLISP program.
- ❏ Use medium sized component library files if you only need one of several components from the library during a drawing session. Group components in families, where the family is depth or curve radius, diameter, etc.

- Use the AutoCAD help command to supply the name of components and an example of command dialogue for libraries with stable component names. Make a custom help entry in the ACAD.HLP file. Don't forget to delete the old ACAD.HDX index file.
- Use the MAPCAR function to assign a list of data to a list of variable names.
- Make a mini-tool box of subroutines to generate images piece-by-piece. Centralize the common routines and let the specific parametric viewing commands use them.
- Don't stop with the parametric image. Use the component data to drive any other command that might supplement the system. Make programs that automatically tag materials with attributes or draw other parts of a system.

21—0 Customizing AutoCAD

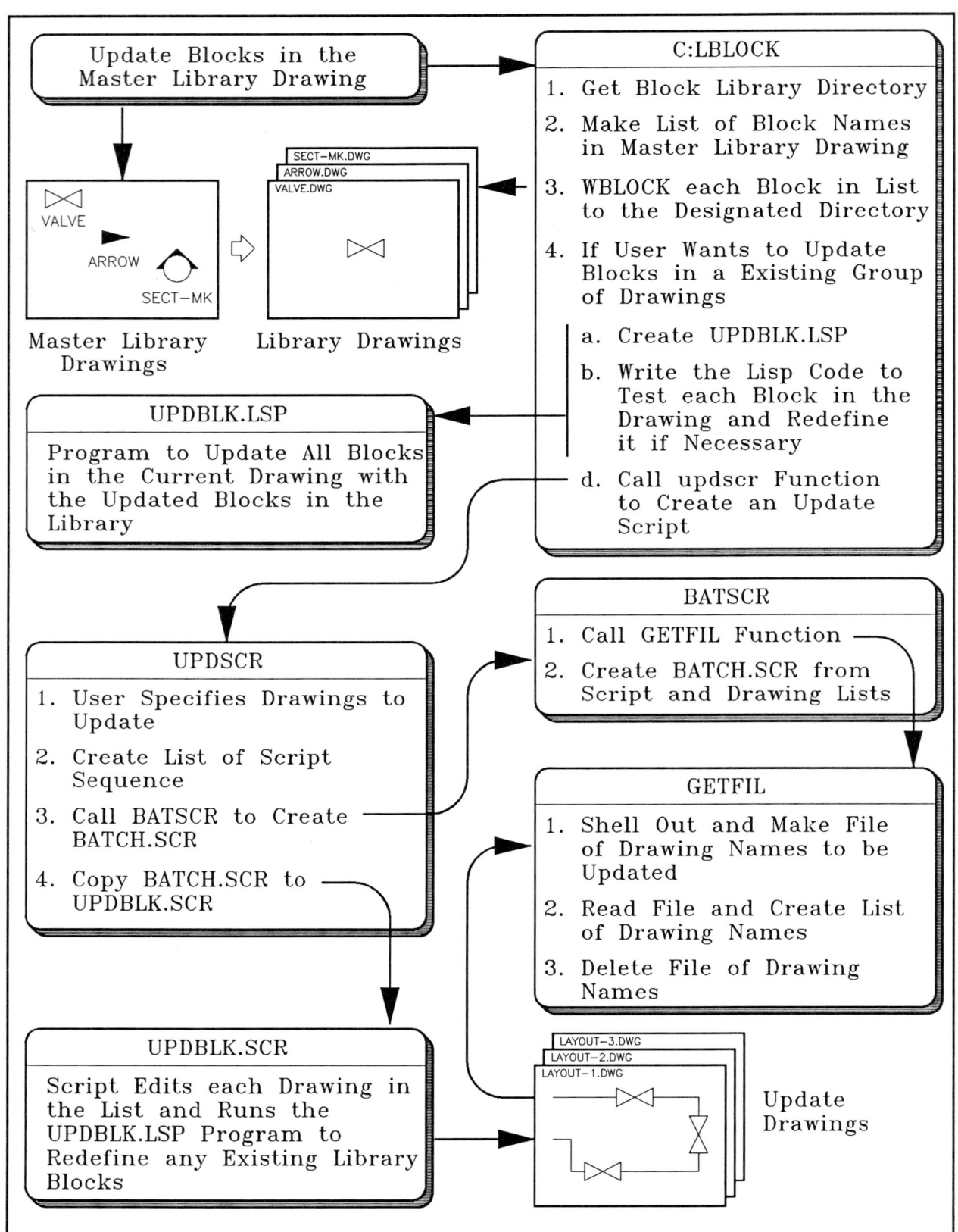

LBLOCK A Custom Block Library Update Program

CHAPTER 21

AutoLISP, Script and DXF Batch Processing

AutoCAD's flexibility usually means there are several ways to do what you want to do, including batch processing. Many times you will find that using a combination of AutoLISP, menus, scripts and external programs is more efficient than using a single method of batch processing.

This chapter will discuss the strengths and weaknesses of scripts, menus, AutoLISP, and DXF for batch processing, giving you some guidelines and hands-on examples to help you choose the best methods for your application.

The Benefits of Batch Processing

There are benefits to using scripts, AutoLISP and DXF files for batch processing. Here are some key benefits to look for:

Scripts and AutoLISP Processing

- Using AutoLISP to automate building repetitive scripts can reduce script errors. You can use automated scripts to assist in plotting, make a series of Slides, show Slides, extract attribute information from multiple files, load multiple .LSP files, update "block" libraries, and even redefine Blocks in a series of drawings.

DXF Processing

- You can convert drawings back to earlier versions of AutoCAD.

- With some limitations, you can translate drawings from other CAD programs into AutoCAD's format.

- In some cases, you can update drawing information faster through DXF processing than through the drawing editor.

- You can write programs to read the DXF file and use the information for tasks, like material reporting and design analysis. Or you can pass the DXF file to other programs that accept DXF files, like Ventura Publisher, to merge drawings into text documents.

How To Skills Checklist

In this chapter, you will gain skills in how to:

- ❏ Read in external data through Scripts, DXF and AutoLISP.
- ❏ Automate script building to reduce error.

- Automate the input slide listing for AutoCAD's SLIDELIB program.
- Write scripts that call AutoLISP functions, pause for user input, stop and resume automatically.
- Use an AutoLISP function to create an AutoLISP file.
- Maintain a master Block library in a single file and automatically update a library directory of Wblocked "block" .DWG files.
- Automatically update a group of drawing files with a group of redefined blocks.
- Interpret, modify and import a DXF file.
- Convert between AutoCAD versions by altering the DXF file.
- Evaluate transferring DXF files to and from other CAD software programs.

Macros, AutoLISP Tools and Programs

AutoLISP TOOLS
GETFIL is an AutoLISP routine that builds a list of file names from a wildcard search of user specified directories.
GETSCR builds a list of AutoCAD commands in a script file format. It has special symbols that are replaced by file names from the GETFIL directory listing.
BATSCR is an engine to build scripts from a directory listings. It is used by several functions.
FSLASH verifies the existence and converts backslashes in the path name to forward slashes for use with AutoLISP and Scripts.
BSLASH verifies the existence and converts forward slashes in the path name to backslashes for use with DOS.
VFPATH obtains a path name from the user and verifies its existence.
P_EXIT is an error handling tool. It puts traps in the programs you write to handle the errors returned from AutoLISP. The function also implements a clean exit from within AutoLISP programs.

PROGRAMS
C:FTEXT reads an external text file, formats the text and places it in the drawing at a user given text width.
C:MSCRIPT is a general purpose program for BATSCR, handling the user interface and formatting.
C:CSLIDE and C:SSLIDE are examples of specialized script building programs. C:CSLIDE uses BATSCR to make a Script that produces .SLD slide files from a directory listing. C:SSLIDE, for ShowSLIDEs, makes a script to display each .SLD slide in a directory.
LBLOCK automatically updates a library directory of Wblocked "block" files from a single Block library file. Optionally, it uses UPDSCR to automatically update a group of drawing files with a group of redefined blocks.
CONVDXF is an AutoLISP based program to convert a Release 9 drawing to AutoCAD version 2.6. It shows how to organize a program to read the DXF format.

OTHER PROGRAMS

CONVDXF.EXE is a much faster C language version of CONVDXF. It is available on the optional CA BONUS DISK.

Script vs. Menu vs. AutoLISP vs. External Software

Let's start looking at the different methods for batch processing by comparing how each method imports a group of text notes. We will look at the differences, and discuss the advantages and disadvantages of each method. The example text is:

```
A SCRIPT CAN READ IN EXTERNAL TEXT BUT IT
MUST BE FULLY PREFORMATTED.

AN EXTERNAL PROGRAM CAN PREFORMAT TEXT AND
PUT IT IN THE FORM OF A .DXF FILE OR A SCRIPT
FOR IMPORTATION.

MENUS CAN INCLUDE STANDARD TEXT NOTES IN
PREFORMATTED FORM, BUT ARE INFLEXIBLE.

AUTOLISP OFFERS FLEXIBILITY AND DYNAMIC
FORMATTING, BUT REQUIRES MORE PROGRAMMING.
```

Of course, if you wanted to fit the text into a given area on the drawing, you could type it in with DTEXT. However, typing text in doesn't offer an automated solution. Here, we are looking for automated solutions. Start with scripts.

Scripts

To import the example text with a script, you have to format the text file. The format (only the top few lines are shown) is shown below:

```
TEXT

A SCRIPT CAN READ IN EXTERNAL TEXT

BUT IT MUST BE FULLY PREFORMATTED.

AN EXTERNAL PROGRAM CAN PREFORMAT

TEXT AND PUT IT IN THE FORM OF A .DXF
```

You can generate this script manually with a text editor, or automatically using another external program. However, scripts cannot stop gracefully for input. You must preset the line lengths, text height, rotation, Style and start point so that the script can use defaults. To use a script, for example, you would first preset a start point with the ID command so that the script could start the text at "@." You would use a menu item like the following:

```
[IN-TEXT ]^C^C^CID \SCRIPT
```

Using Long Menu Items

Some users make very long multi-line menu items to do batch processes, like drawing setup. In general, it is better to handle drawing setups with scripts, prototype drawing settings, or AutoLISP files. Obviously you could make menu items to input the example text notes. Menu items offer flexibility for settings, but you need to hard code the text itself and the line length. You would have to edit the menu when you wanted to change the line length. We recommend restricting menus to invoking and handling input and settings for scripts and AutoLISP batch processes.

DXF Importation

An external program can generate drawing data to import into AutoCAD in the form of a .DXF file. AutoWORD, the popular text formatting utility for AutoCAD, uses DXF. The following is a ready to load DXF file, containing a single text entity.

Importing a DXF Text File

Enter selection: Begin a NEW drawing named CA21.

   Copy CA21.MNU to TEST.MNU.

   Copy MAIN.MNU to TEST.MNU.

Select **[TEST]**
Select **[24 x 36]**
Select **[FULL]**
Select **[INITIATE]**   Then enter 5 <RETURN>s to default the attributes.

Command: **ED**   Create this TEST.DXF text file, even if you have the CA DISK.

```
  0
SECTION
  2
ENTITIES
  0
TEXT
  8
0
 10
6.5
 20
7.5
 40
0.25
  1
AN EXTERNAL PROGRAM CAN PREFORMAT
  0
ENDSEC
  0
EOF
```
                                                   Be sure to <RETURN> at the end.

Save and exit to AutoCAD.

Command: **DXFIN**   Test it. It inserts the text.

Using the .DXF file to import the text file has several drawbacks besides being hard to read and edit. It is inflexible, requiring that you hard code the Layer, Color, start point, height, rotation, and line length. However, Style can assume the current setting. The DXF offers the advantage of absolute control in spite of the current environment of the drawing.

Reading in Data with AutoLISP

The most flexible approach is to format your data in an external file and import it with AutoLISP running AutoCAD commands.

The following example file, SAMP.TXT, consists of four long lines, separated into "paragraphs" by blank lines. The example can be created in word processing mode, including formatting codes, and then printed to an ASCII text file. The file has no print or control characters, it has no effect on AutoLISP.

If you have the CA DISK, you have the text file SAMPTEXT.TXT. If you don't, you need to create the file so that you can work with it.

AutoLISP Text Importation

   Skip this sequence.

   Do this:

Command: **ED**   Create the SAMPTEXT.TXT file:

```
A SCRIPT CAN READ IN EXTERNAL TEXT BUT IT
MUST BE FULLY PREFORMATTED.

AN EXTERNAL PROGRAM CAN PREFORMAT TEXT AND
PUT IT IN THE FORM OF A .DXF FILE OR A SCRIPT
FOR IMPORTATION.

MENUS CAN INCLUDE STANDARD TEXT NOTES IN
PREFORMATTED FORM, BUT ARE INFLEXIBLE.

AUTOLISP OFFERS FLEXIBILITY AND DYNAMIC
FORMATTING, BUT REQUIRES MORE PROGRAMMING.
```

Save, exit and return to AutoCAD.

You can read this file into AutoCAD using an AutoLISP program. You can put as much flexibility as you wish into the AutoLISP program. If you use text import often, or wish to preset defaults, consider incorporating the AutoLISP routine into a menu item. The following program is simple, with a few limitations, but you can enhance it.

 Copy FTEXT.LSP to TEST.LSP.

 Begin a new TEST.LSP file.

```
;* C:FTEXT imports an ASCII file into AutoCAD as Text. Default style
;* must be fixed height. It optionally wordwraps, or imports unformatted as a
;* table. If a table, current style should be MONOTXT font.
;* If wrap, it prompts for the maximum characters per line.
;* The variables MAXC and ROT float for defaults.
(defun C:FTEXT ( / mtable file pt lftovr instr str len pos c)
  (if (setq file (open (getstring "\nEnter text source filename.ext: ") "r"))
    (progn                                              ;then do it
      (setq mtable (ukword 1 "MT WW" "MonoTable or Word Wrap? MT/WW" "WW")
            pt (upoint 1 "" "Start point" nil nil)      ;text start point
            lftovr ""                                   ;initialize leftover string
      );setq
      (if (= mtable "MT")                               ;If mono table
        (progn                                          ;then warn and
          (prompt "\nLines over 132 chars will truncate. Current font should be MONOTXT.\n")
          (setq rot 0)                                  ;hardcode rotation
        )
        (progn                                          ;else
          (setq mtable nil                              ;nil flag and get format
                maxc (uint 1 "" "Maximum characters per line" maxc)
                rot (uangle 1 "" "Rotation angle" rot pt)
      ) ) );setq, progn & if mono table
      (command "TEXT" pt (angtos rot))                  ;start Text command
                           ;WHILE LOOPS FOR EACH LINE OF TEXT
      (while (setq instr (read-line file))              ;loop for each line of file
        (cond                                           ;filter for conditions
          (mtable (command instr "TEXT" "" ))           ;enter txt & start new line
          ((and (= lftovr "") (= instr ""))             ;nothing left, and blank line
            (command "" "TEXT" "")                      ;start a new paragraph
          )
          ((and (/= lftovr "") (= instr ""))            ;something left, but new line
            (command lftovr "TEXT" "" "" "TEXT" "")     ;finish string, then new paragraph
            (setq lftovr "")                            ;reset left over to nothing
          )
          (T                                            ;all other cases
            (if (= lftovr "")                           ;if nothing left
              (setq str instr)                          ;only the input string
              (setq str (strcat lftovr " " instr))      ;else add left over to new input
            );if
            (setq len (strlen str))                     ;determine string length
            (while (> len maxc)                         ;loop while string is too long
              (setq pos maxc)                           ;set or reset position
              (while (not (member
                    (setq c (substr str pos 1)) '("" "-" " "))) ;loop in search of characters
                (setq pos (1- pos))                     ;decrement position
              );while
              (command (substr str 1                    ;pass to waiting Text command
```

```
                (if (= c "-") pos (1- pos))) "TEXT" ""    ;then restart Text command
          );command
            (setq str (substr str (1+ pos))                ;chop old part of string
                  len (strlen str)                         ;get length of (remaining) string
            );setq
          );while > len
          (setq lftovr str)                                ;save remaining as left over
        );T
      );cond
    );while setq instr
    (command lftovr)                                       ;finish Text, might be a left over
    (close file)                                           ;close file
  );progn
  (prompt "\nSource text file not found.\n")               ;file was never found
);if
(princ)
);defun FTEXT
;*
(princ)
;*end of FTEXT.LSP
;*
```

Save, exit and return to AutoCAD.

Select **[EDIT-MNU]**
Relabel TEST.MNU to [CA21TEST] and add an [FTEXT] item:

```
***SCREEN
**ROOT
[CA21TEST]
[]

**EXAMPLES
[Examples]
[]
[FTEXT]^C^C^C(caload "/CA-ACAD/" "FTEXT") LAYER S TXT01 ;FTEXT
```

Select **[EXAMPLES] [FTEXT]**	Test it.
Command: FTEXT	
Enter text source filename.ext: **SAMPTEXT.TXT**	
MonoTable or Word Wrap? MT/WW <WW>:	<RETURN> for word wrap.
Start point:	Pick it.
Maximum characters per line: **30**	If you repeat FTEXT, it'll default.
Rotation angle: **0**	The next time it'll show a default.
Command: **ZOOM**	Zoom in to see it.
Select **[FTEXT]**	Test Monotable format on the same file.
	It should come in identical to .TXT file format

 Delete TEST.LSP.

 Rename TEST.LSP to FTEXT.LSP.

Your screen should look like Formatted Text from FTEXT.LSP. It isn't in mono-tabular form because the current style is still STD1-8.

```
┌─────────────────────────────────────────────────────────────┐
│ Layer 0 Ortho Snap            0'-11 1/8",1'-2 7/8"  │CA21TEST│
│                                                     ├────────┤
│                                                     │FLOWLINE│
│       A SCRIPT CAN READ IN EXTERNAL                 │PIPEFITS│
│       TEXT BUT IT MUST BE FULLY                     ├────────┤
│       PREFORMATTED.                                 │TEXT    │
│                                                     │SPECIALS│
│       AN EXTERNAL PROGRAM CAN                       │ANNOTATE│
│       PREFORMAT TEXT AND PUT IT IN                  │ISO DIMS│
│       THE FORM OF A .DXF FILE OR A                  │HATCH   │
│       SCRIPT FOR IMPORTATION.                       ├────────┤
│                                                     │SETTINGS│
│       MENUS CAN INCLUDE STANDARD                    │TITLEBLK│
│       TEXT NOTES IN PREFORMATTED                    │   3D   │
│       FORM, BUT ARE INFLEXIBLE.                     │EXAMPLES│
│                                                     ├────────┤
│       AUTOLISP OFFERS FLEXIBILITY                   │EDIT-LSP│
│       AND DYNAMIC FORMATTING, BUT                   │EDIT-MNU│
│       REQUIRES MORE PROGRAMMING.                    │        │
│                                                     │        │
│                                                     │        │
├─────────────────────────────────────────────────────┴────────┤
│ Command:                                                     │
│ REDRAW                                                       │
│ Command:                                                     │
└──────────────────────────────────────────────────────────────┘
```

Formatted Text from FTEXT.LSP

The FTEXT program uses the defaults for Style and height, but lets you specify rotation, start point and line length. It will work with any ASCII input text file so long as the line length is less than the current AutoLISP READ-LINE string limit of 132 characters.

What, When and Where for Scripts, AutoLISP and DXF?

You have dealt with menus and AutoLISP, but when do scripts or DXF processes make sense?

We recommend using scripts only for repetitive batch processes where you need to transcend the AutoCAD drawing editor. You can use scripts to automate the Main Menu, Configuration Menu and Plot Menu. Some examples include batch plotting, making a series of Slides, showing Slides, extracting attribute information, loading multiple .LSP files, updating "block" libraries, and even updating redefined Blocks in a series of drawings.

The DXF is designed to allow drawing data transfer between AutoCAD versions and other CAD programs. It's a comprehensive standard for drawing data, intended for intensive applications. We will show you how to use it for AutoCAD version conversion, and for globally updating drawing data that is difficult to get at with AutoLISP. Most DXF applications involve importing complex graphic data generated by external programs.

Script Batch Builder

Like menus, scripts are a pain to write. It would be nice to automate the building process. Since scripts often deal with groups of filenames, it is logical to think of writing a function that could get a directory listing of a group of files and perform a batch of operations on each file.

Here are the functions that you need to develop a Script Batch System:

- GETFIL returns a list of filespecs.
- GETSCR builds a list of script commands, with a variable for filename.
- BATSCR is the core function. It builds a script.
- C:MSCRIPT is a user interface command.

The functions also use the file utility functions from the File I/O chapter to make your task here easier.

First, develop the GETFIL routine to generate a list of files from a directory and the filespec (file specification) supplied by the user.

GETFIL Getting the File List

 Copy BATCHSCR.LSP to TEST.LSP.

 Begin a new TEST.LSP file.

Select **[EDIT-LSP]**   Edit TEST.LSP, and enter:

```
;* BATCHSCR.LSP contains 6 Script building functions.

;* GETFIL builds the listing of specified files used by the batch builder.
;* GETSCR gets the script commands from the user and returns them in a list.
;* BATSCR builds a script from a GETFIL directory and a script command listing.

;* C:MSCRIPT is a general purpose calling program for BATSCR.
;* C:CSLIDE uses BATSCR to make a Script to produce slides of a dir's dwgs.
;* C:SSLIDE - ShowSLIDEs makes a script to display each slide in a dir.
;*
;* GETFIL returns a listing of files matching the path and file spec provided
;* by its arguments. Wildcards are optional for filename but not the extension.
;* The dir listing "\path\CA*.DWG" is OK, but *.* is illegal.
;* PATH must be in the form "\path\" -- Trailing \ req'd, (& ..\ dots illegal).
(defun getfil (fspec path / files fp dirlin v inc flist)
  (if path
     (setq files (strcat path fspec))       ;test for other than current path
     (setq files fspec path "")             ;string together path and file
  );if                                      ;leave alone, make path null string
  (command "SH" (strcat "DIR " files " > DIR.$"))  ;make temp filename file
  (setq fp (open "DIR.$" "r"))              ;open temp file to read filenames
  (repeat 4 (read-line fp))                 ;skip 1st 4 lines
  (setq dirlin (read-line fp))              ;5th line, are there any filenames
  (if dirlin                                ;if there are files in dir
    (progn
      (prompt "\nMaking file listing.")
      (while dirlin                         ;loop for each file name
        (prompt ".")
        (setq v (substr dirlin 1 1))
        (if (not (member v '("." " ")))     ;illegal characters
```

21—10 Customizing AutoCAD

```
          (progn
            (setq inc 9)                              ;max file name + 1
            (while (= " " (substr dirlin (setq inc (1- inc)) 1))) ;strip trailing spaces
            (setq flist (append flist                 ;put file name on list
                (list (strcat path (substr dirlin 1 inc)))))
            );setq
          );progn
        );if
        (setq dirlin (read-line fp))                  ;get next filename
      );while
      (close fp)                                      ;close temp dir file
      (command "FILES" "3" "DIR.$" "" "")             ;delete temp dir file
    );progn
    (prompt (strcat "\nNo " files " found\n"))        ;no files were found
  );if
  flist
);defun GETFIL
;*
```

Save, Exit and reenter AutoCAD. TEST.LSP should load.

```
Command: (getfil "*.DWG" "\\ACAD\\")     Test with any dwg directory. Your listing will differ from ours.
Making file listing........              The FILES menu scrolls by, then:
Lisp returns: ("\\ACAD\\ACAD" "\\ACAD\\BORDER" "\\ACAD\\CHROMA" "\\ACAD\\NOZZLE"
"\\ACAD\\POINTS" "\\ACAD\\TABLET")
```

Now, you need a routine to create the script command listing. Since the script will run one loop for each filename, the function must allow the user to indicate where the file name goes in the command sequence. The key FNAME is input when the file name is desired. The function treats this FNAME input as an AutoLISP SYMbol. FNAME is processed later by the BATSCR function, where the FNAME symbol is replaced by each actual file name from the file listing for each script loop. Let's create the GETSCR function.

GETSCR Getting the Script's Commands

Select **[EDIT-LSP]**   Edit TEST.LSP, and add:

```
;* GETSCR assists the user in building the script file of commands.
;* It prompts for command input and returns a list in the form:
;*    ("string1" "str2" FNAME "srt3" "str4" ...)  where each string becomes
;* 1 script line, & FNAME is a symbol to be replaced by each file name in sequence.
(defun getscr ( / script input item)
  (setq script '()                                    ;initialize script list
        input T                                       ;set a flag for input control
  );setq
  (while input                                        ;get script input
    (setq item (ustr "Enter commands, FNAME or . to exit" nil T))
    (if (= item ".")                                  ;they're exiting
      (setq input nil)                                ;set flag to nil
      (if (= (strcase item) "FNAME")                  ;test for file name
        (setq script (append script (list 'FNAME)))   ;append atom to list
        (setq script (append script (list item)))     ;otherwise, append command string
      );if
    );if
```

```
    );;while
    script
);;defun GETSCR
;*
```

Save, Exit and reenter AutoCAD. TEST.LSP should load.

```
Command: (getscr)                       Test the program.
Enter commands, FNAME or . to exit: 2
Enter commands, FNAME or . to exit: FNAME
Enter commands, FNAME or . to exit: ZOOM ALL MSLIDE
Enter commands, FNAME or . to exit: FNAME
Enter commands, FNAME or . to exit: END
Enter commands, FNAME or . to exit: .
Lisp returns: ("2" FNAME "ZOOM ALL MSLIDE" FNAME "END")     This is the COMMAND list.
```

You have the most important functions in place. Next, you need to create a script file to process the commands and the file name listing. The BATSCR function creates the script file by processing each list of data (script command and file name) and writing the combined string out to a file.

BATSCR merges a list of filespecs, like GETFIL's, with a list of commands, like GETSCR's. BATSCR writes a script file named BATCH.SCR. BATCH.SCR is executed like any other script. It writes one loop of the script for each filespec. Each loop consists of the full list of commands, with each occurrence of the FNAME symbol replaced by each filespec in sequence. The script file is terminated by a stop character string. Stopping with "RSCRIPT" reruns the script, "0" exits to DOS from the main menu, and "STOP" stops it cold by forcing an intentional error.

BATSCR Writing the Script

Select **[EDIT-LSP]**   Edit TEST.LSP, and add:

```
;* BATSCR process a GETFIL list and script command list into a script file named
;* BATCH.SCR, which is executed in the normal manner for scripts.
;* Its SCRIPT arg is the script list, if nil it calls GETSCR.
;* Its file/path args are same as GETFIL's. STOP is the script termination char.
;* Each string becomes 1 script line. If STOP is nil, BATSCR prompts for it.
(defun batscr (fspec path script stop / files fp bfp item)
  (setq files (getfil fspec path))
  (if files
    (progn
      (if (not script) (setq script (getscr)))    ;if no script, call subr to get it
      (setq bfp (open "BATCH.SCR" "w"))           ;open output script file
      (prompt "\nCreating script file.")
      (foreach name files                         ;process each file name
        (prompt ".")
        (setq FNAME name)                         ;assign value to FNAME atom
        (foreach item script                      ;process each script list
          (write-line (eval item) bfp)            ;eval & output script commands
        );foreach
      );foreach
      (if (not stop)
```

```
          (progn
            (prompt "\nR=rerun script, 0=exit to DOS, S=Stop")
            (setq stop (ukword 1 "0 Rscript Stop" "Script terminator" "R"))
          );progn
        );if
        (write-line stop bfp)                        ;write script terminator
        (close bfp)                                  ;close script file
      );progn
    );if
    (if bfp T nil)                                   ;return T if script file was opened
);defun BATSCR
;*

Command: (batscr "*.DWG" "\\ACAD\\" nil nil)
Making file listing.......
Enter commands, FNAME or . to exit: 2
Enter commands, FNAME or . to exit: FNAME
Enter commands, FNAME or . to exit: ZOOM A MSLIDE
Enter commands, FNAME or . to exit: FNAME
Enter commands, FNAME or . to exit: END
Enter commands, FNAME or . to exit: .
Creating script file..........
R=rerun script, 0=exit to DOS, S=Stop
Script terminator <R>: 0
Lisp returns: T
```

Look at the contents of your BATCH.SCR file. Make sure the contents are correct. A portion of the batch file from the above series looks like:

```
2
\ACAD\ACAD
ZOOM A MSLIDE
\ACAD\ACAD
END
2
\ACAD\BORDER
ZOOM A MSLIDE
\ACAD\BORDER
END
0
```

and so on.

You also can feed BATSCR the STOP string and a script command list as arguments. This is useful if you have standard scripts you use on different file sets. A predefined list avoids input errors.

It is a nuisance to call BATSCR from the command line. A command function makes it easier for the user. C:MSCRIPT, the command function, gets the input to feed BATSCR and GETFIL. The program flow for C:MSCRIPT is shown in the following diagram.

Script Batch Builder 21—13

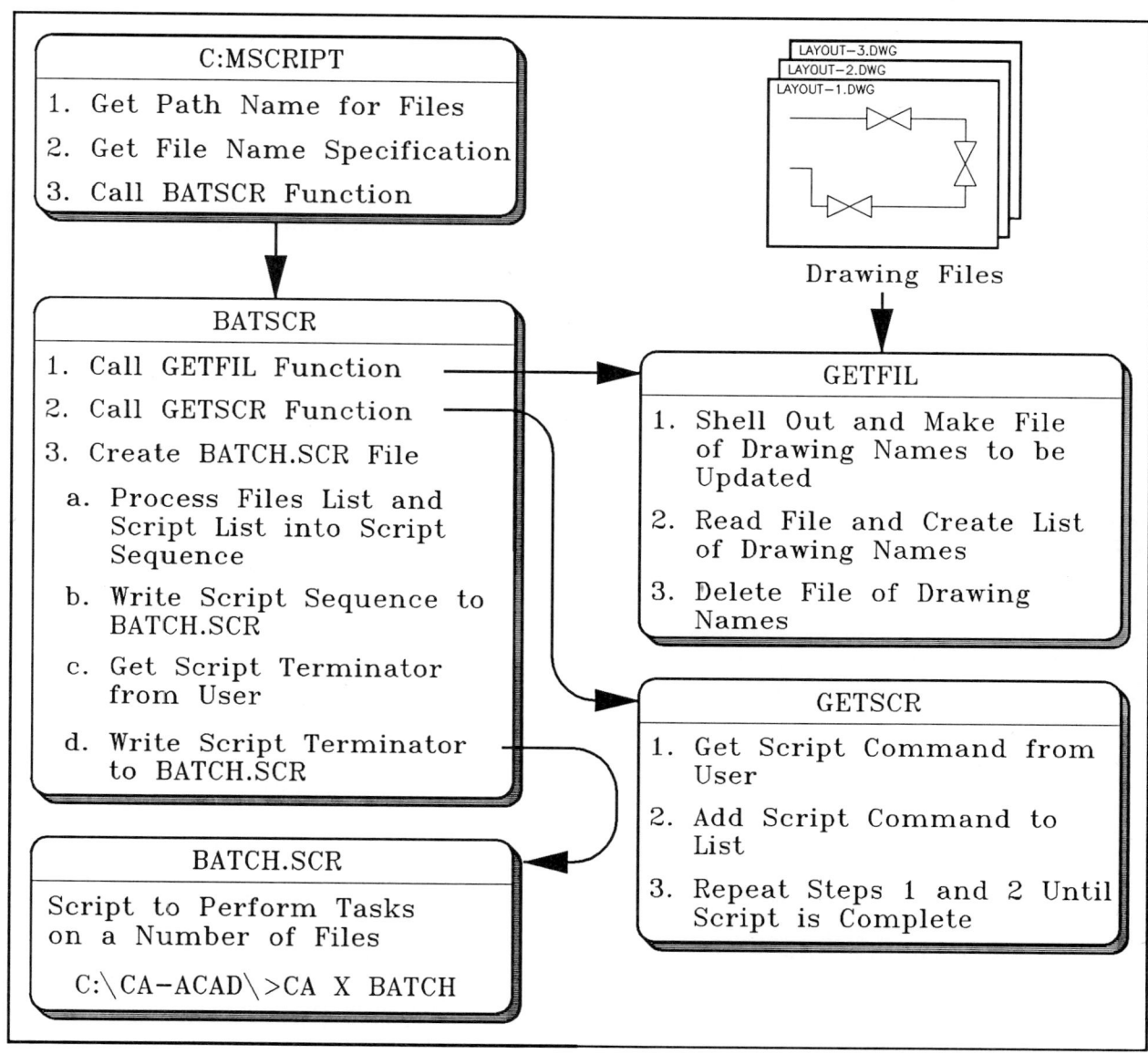

Diagram of C:MSCRIPT

Putting the Pieces Together with C:MSCRIPT

Select **[EDIT-LSP]**   Edit TEST.LSP, and add:

```
;* C:MSCRIPT is a general purpose calling program for the above BATSCR.
(defun C:MSCRIPT ( / script path fspec)
   (setq path (ustr "Path name (or RETURN for none)" nil nil)   ;get path name
         fspec (ustr "Enter files to list" "*.DWG" nil)          ;get file spec
   );setq
   (if (not path) (setq path ""))
   (batscr fspec path nil nil)                        ;call batscr, no script or terminator
   (princ)
);defun
;*
```

Save, Exit and reenter AutoCAD. TEST.LSP should load.

Command: **MSCRIPT**
Path name (or RETURN for none): **\ACAD**
Enter files to list <*.DWG>:

Test it.

Take the default.

You have seen the rest of the program, try some new script commands.

For some purposes, like generating and showing Slides, it makes sense to have separate command functions. C:CSLIDE creates a group of slides. C:SSLIDE makes a script and shows any set of slides, not necessarily the same set created by CSLIDE. Both functions preformat the script lines, so all the user has to input is the filespec and path. You can easily modify CSLIDE to add cleanup functions to the SCRIPT string list.

Creating C:CSLIDE

Select **[EDIT-LSP]**   Edit TEST.LSP, and finish it up:

```
;* C:CSLIDE - CreateSLIDEs uses BATSCR to make a Script to
;* produce .SLD slides of all drawings in a given directory.
(defun C:CSLIDE ( / script path)
  (setq script '("2" FNAME "MSLIDE " "QUIT Y")           ;make script string
        path (ustr "Path name (or RETURN for none)" nil nil)  ;get path to search
  );setq
  (if (not path) (setq path ""))
  (if (batscr "*.DWG" path script "0")                    ;If script builds OK
    (progn                                                ;Then copy script
      (command "FILES" "5" "BATCH.SCR" (strcat path "CSLIDE.SCR")
       "" "" "FILES" "3" "BATCH.SCR" "" "" "GRAPHSCR")    ;delete original file
      );command
      (prompt  (strcat "\nTo make slides, QUIT or END and restart ACAD with: \n"
               "\t\\path\\ACAD X " path "CSLIDE"))
      );prompt
    );progn
  );if
  (princ)
);defun
;*

;* C:SSLIDE - ShowSLIDEs makes a script and displays each .SLD slide in a given
;* directory with a delay between each.
(defun C:SSLIDE ( / script path delay)
  (setq delay (* 1000 (uint 1 "" "Delay between slides in seconds" 5))  ;get delay time
        script (list "VSLIDE" 'FNAME "DELAY" (itoa delay))               ;script list
        path (ustr "Path name (or RETURN for none)" nil nil)             ;get path
  );setq
  (if (not path) (setq path ""))
  (if (batscr "*.SLD" path script "RSCRIPT")                             ;if script OK
    (command
      "FILES" "5" "BATCH.SCR" (strcat path "SSLIDE.SCR") "" ""           ;copy script file
      "FILES" "3" "BATCH.SCR" "" "" "GRAPHSCR"                           ;delete original
      "SCRIPT" (strcat path "SSLIDE"))                                   ;start show
    );command
```

```
  );if
   (princ)
);defun
;*
(princ)
;*end of BATCHSCR.LSP
;*
```

Save, Exit and reenter AutoCAD. TEST.LSP should load.

`Command: `**`CSLIDE`**	Test it.
`Path name (or RETURN for none):`	It only needs a path name.
`SH`	
`DOS Command: DIR *.DWG  DIR.$`	
`Making file listing...`	And the FILES menu scrolls by, then:
`To make slides, QUIT or END and restart ACAD with:`	
`\path\ACAD X CSLIDE`	

The X in the prompt "\ACAD X CSLIDE" is a place keeper for the default filename. It could be any name. The "\path\ACAD" is a reminder to start AutoCAD in your normal manner, adding "X CSLIDE." If you want to test SSLIDE, you will have to make some slides first. QUIT and run ACAD X CSLIDE at the DOS prompt, or use the MSLIDE command to make a few Slide files for testing.

`Command: `**`QUIT`**	Exit to DOS.
`C:\>`**`CA X CSLIDE`**	Starts ACAD with CA.BAT. Runs the script. Makes slides,...
...or...	
`Command: `**`MSLIDE`**	Make a few slides. Then:
`Command: `**`(load "test")`**	Reload the TEST.LSP file.
`Command: `**`SSLIDE`**	
`Delay between slides in seconds <5>: `**`3`**	
`Path name (or RETURN for none):`	
`SH`	
`DOS Command: DIR *.SLD  DIR.$`	
`Command:`	
`Making file listing....`	And FILES scrolls, then the slide show starts.
	Hit <^C> to stop it.
	Stay in the current drawing.

 Delete TEST.LSP.

 Rename TEST.LSP to BATCHSCR.LSP.

Slide Libraries

AutoCAD's Slide Library files are groups of slides in single .SLB files. You need to feed the SLIDELIB command, which is executed at the DOS prompt, a clean list of .SLD file names. You can use the MSCRIPT command function to create a list. Respond to MSCRIPT with a path, filespec.SLD as filespec, and FNAME as the only script input. You can supply the resulting BATCH.SCR to SLIDELIB as its required slide list. If you do this often, modify a copy of MSCRIPT to automate it further. You can even execute SLIDELIB automatically via Shell.

Scripts and Plotting

Batch plotting is one batch process where you can take advantage of a script's unique ability to control all parts of AutoCAD. Batch pen plotting requires a roll or automatic sheet feed plotter and reliable pens. Laser, dot matrix and other raster output devices really benefit from batch plotting. You can use the BATSCR program as the engine for a program to create Plot or Prplot batch Scripts. You also can create one shot Plot scripts to handle the plot menu input in a friendlier manner, so that you only have to respond to the settings that vary in your application. Prompt for input in an AutoLISP routine, write a Script file, then call the Script to run the plotter.

Starting, Stopping and Resuming Scripts

One of the limitations of scripts is that they supposedly cannot stop for input. If you try to put an AutoLISP function in your script to pause or get input, the script won't work. The script will try to fill the pause or provide input and cause an error, stopping the script.

If you need to stop a script, you can cause a harmless error, stopping the script. This is how the BATSCR program scripts stop. Then, you can restart the script with a RESUME command, used transparently as 'RESUME.

You also can use the AutoLISP COMMAND function to protect your scripts from errors in execution, when the script may encounter unpredictable conditions. It also may pay to use COMMAND when a script runs a lengthy standard sequence on multiple drawing files. The AutoLISP expression can resume the script, in this form:

```
SCRIPT STUFF....
(command nil (if something then else) "RESUME")
MORE SCRIPT STUFF
```

The nil cancels the script and AutoLISP takes over until the RESUME restarts the script. You cannot make RESUME transparent within AutoLISP. You can use AutoLISP conditionals to branch to other scripts.

Before going on to work with another program, you need to add three quick functions to your file handling library FILELIB.LSP file. If you have the CA DISK, they are already in the file. Let's do those functions now.

More File Handling Functions

You can skip this.

Do this:

Command: **ED**   Edit FILELIB.LSP.
Add the following three functions:

```
;* FSLASH (Forward SLASH) forces "\\"s or "/"s in path strings to "/"s
;* & forces a trailing "/".
(defun fslash (path / inc wpath char)
  (setq inc 1 wpath "")                                    ;initialize variables
  (while (/= "" (setq char (substr path inc 1)))           ;test each char
    (setq wpath (strcat wpath (if (= "\\" char) "/" char)));append proper char back
    (setq inc (1+ inc))                                    ;increment counter
  );while
  (if (and (/= wpath "") (/= (substr wpath (strlen wpath) 1) "/" ))
    (setq wpath (strcat wpath "/"))
  );if
  wpath
);defun FSLASH
;*

;* BSLASH (Back SLASH) forces "\\"s or "/"s to "\\"s in path strings
;* & forces a trailing "\\" for sending strings to DOS.
(defun bslash (path / inc wpath char)
  (setq inc 1 wpath "")                                    ;initialize variables
  (while (/= "" (setq char (substr path inc 1)))           ;test each char
    (setq wpath (strcat wpath (if (= "/" char) "\\" char)));append proper char back
    (setq inc (1+ inc))                                    ;increment counter
  );while
  (if (and (/= wpath "") (/= (substr wpath (strlen wpath) 1) "\\" ))
    (setq wpath (strcat wpath "\\"))
  );if
  wpath
);defun BSLASH
;*

;* VFPATH gets and verifies a PATH. If the test file name TESTFN is nil, "$." is
;* used. The test filename will be left behind in the successful directory, as a
;* 0 byte file. Adding (command "FILES" "3" (strcat path testfn) "" "") will delete it.
;* Otherwise, supply a name which is sure to exist or to be later overwritten.
(defun vfpath (testfn / path fp)
  (or testfn (setq testfn "$"))                            ;assigns name if nil
  (while (not path)                                        ;loops until valid path
    (setq path (getstring "\nEnter path name <current>: ") );Gets path from user
    (if (/= "" path)                                       ;user want current path
      (progn
        (setq path (bslash path))                          ;check for valid path syntax
        (if (setq fp (open (strcat path testfn) "a"))      ;test users path entry
          (close fp)                                       ;path found, close temp file
          (setq path (prompt "\nInvalid path, try again.")));path could not be accessed
      );if
    );progn
  );if
  );while
  path                                                     ;return the path
);defun VFPATH
;*
(princ)
;* end of FILELIB.LSP
```

```
;*
```

Save, exit and return to AutoCAD.

```
Command: (load "FILELIB")

Command: (fslash "\\abc/def")          Test slash conversion.
Lisp returns: "/abc/def/"              OK!

Command: (bslash "/brg/roo")
Lisp returns: "\\brg\\roo\\"

Command: (vfpath "temp")               Try a path.
Enter path name <current>:             Take current default.
Lisp returns: ""                       Empty string. OK.
Command: (vfpath "temp")               Try again.
Enter path name <current>: \abc        Give a nonexistent name.
Invalid path, try again.
Enter path name <current>: \ca-acad    Try a good name.
Lisp returns: "\\ca-acad\\"            Proper slashes and trailing slashes.
```

We will assume the file functions are OK, and are now in your FILELIB.LSP file.

Coordinated Scripts and AutoLISP

The next program C:LBLOCK is an example of coordinating scripts and AutoLISP. C:LBLOCK updates a Library of BLOCKS. It is executed from within a master "block library" drawing. LBLOCK Wblocks out all defined Blocks to update a designated "block library" drawing file directory.

The exercise C:LBLOCK also has an option to create an UPDBLK.LSP file and an UPDBLK.SCR file, which combine to perform a batch insertion of all blocks into a designated list of drawing files. Defined blocks within each drawing file are updated. However, no additional library "blocks" are added to any drawing where they do not exit. Block names can't exceed 8 characters.

UPDBLK.LSP and UPDBLK.SCR are saved in the block library directory, along with the Wblocked drawings. You need the CA LISP libraries: ACAD.LSP, FILELIB.LSP and BATCHSCR.LSP. Use this brief discussion to write your own header for C:LBLOCK.

Making a Block Library Update Program LBLOCK

 Copy LBLOCK.LSP to TEST.LSP

 Begin a new TEST.LSP file.

Write your header explanation as a ;* header description.
Enter the following:

```
(defun C:LBLOCK ( / flag blkdef blist path wpath expert cmd file fun)
  (if (setq blkdef (tblnext "BLOCK" T))              ;test for blocks in drawing
    (progn
      (textscr)                                       ;flip to text screen
      (prompt "\e[2J")                                ;clears the screen
      (prompt (strcat "\n** NOTE: Existing .DWG files in designated directory **"
                      "\n   ** will be overwritten if Blocknames match. **\n")
      );prompt
      (prompt "\nSpecify block library directory...");instruct user what type directory
      (setq path (vfpath "updblk.lsp"))               ;assign verified path name
      (setq blist (list (dxf 2 blkdef)))              ;start block list

      (while (setq blkdef (tblnext "BLOCK"))          ;loop and make list of block defs
        (setq blist (cons (dxf 2 blkdef) blist))      ;append to block list
      );while

      (setq expert (getvar "EXPERT")                  ;store users current setting
            cmd (getvar "CMDECHO")                    ;save users echo state
      );setq
      (setvar "CMDECHO" 0)                            ;turn off echo
      (setvar "EXPERT" 3)                             ;suppress file overwrite, expert mode
      (foreach blknam blist
        (command "WBLOCK" (strcat path blknam) blknam)    ;wblock each out
        (prompt (strcat "\r                                      "
                        "\rWblocking " blknam))       ;prompt which block
      );foreach
      (prompt "\r                                      ");wipe prompt line clean

; IF YOU DON'T HAVE A NEED for such updating of existing blocks within
; a group of drawings, you can omit/delete the entire following IF expression:
      (if (= "Y" (ukword 1 "Y N" (strcat
                 "You may create a program to update blocks in other drawings."
                 "\nDo you want to do so?") "Y"))
        (progn                                        ;Yes create an UPDBLK.LSP file
          (setq file (open (strcat path "updblk.lsp") "w"))
          (setq wpath (fslash path))                  ;convert back to forward slashes
          (write-line "(setq blist '(" file)          ;write list of blocks
          (foreach blkdef blist                       ;loop for each block name
            (print blkdef file)                       ;print each, quoted 1 per line
          );foreach
          (write-line (strcat ") path \"" wpath "\" )" ) file)  ;write out path name
          (setq fun '((setq regen (getvar "REGENMODE"))   ;make a list for our output file
                      (setvar "REGENMODE" 0)              ;set regen auto off
                      (foreach blkdef blist               ;loop through each block
                        (if (tblsearch "BLOCK" blkdef)    ;check in block table
                          (command "INSERT" (strcat blkdef "=" path blkdef)  ;insert file
                            (not (setq flag T))           ;set save flag to true
                          );command
                          (setq flag nil)                 ;block not in drawing, set flag
                        );if
                      );foreach
                      (setvar "REGENMODE" regen)          ;reset regenauto
                      (if flag (command "SAVE" ""))       ;if flag T save drawing
                      (command "RESUME")                  ;resume the script
                     );list of expressions to write out
          );setq
          (foreach exp fun                            ;write ea above expr to file
            (print exp file)
```

```
            );foreach
            (close file)                                          ;done writing updblk.lsp
            (updscr wpath)                                        ;make the update script UPDBLK.SCR
            (setvar "EXPERT" expert)                              ;reset expert mode
            (setvar "CMDECHO" cmd)                                ;reset echo
        );progn
      );if create program -- end of optional omit/delete

    );progn
    (prompt "\nNo Blocks found in current drawing.\n")            ;Else of if blist
  );if blocks
  (princ)
);defun C:LBLOCK
;*
```

The optional UPDSCR generates a list of files to update, using the block library. It gets a path and drawing specification from the user and creates a batch script using the batch builder. The resulting BATCH.SCR file is copied to the name UPDBLK.SCR.

In C:LBLOCK, if you omitted or deleted the IF to create a program to update existing blocks within a group of drawings, you can omit or delete the entire following function.

Otherwise, add UPDSCR.

Save what you've done. Add a ;* header description.

Add the following:

```
(defun updscr (wpath / fspec dpath)
  (prompt "\nSpecify drawings to update...")                      ;tell user type of file spec.
  (setq dpath (vfpath "batch.scr")                                ;get the verified path
        fspec (ffname (ustr "Drawing names" "*" nil) "DWG")       ;get filespec w/ extension
        scrlst (list "2" 'FNAME                                   ;the script line, enter drawing
                (strcat "(load \"" wpath "UPDBLK\")")             ;load "updblk.lsp" function
                "QUIT" "Y"
              );list
  );setq
  (batscr fspec dpath scrlst "0")                                 ;call the batch builder
  (command "FILES" "5" (strcat wpath "BATCH.SCR")                 ;copy the file
                       (strcat dpath "UPDBLK.SCR") "" "")
  );command
  (prompt (strcat "\nStart AutoCAD from DOS by"                   ;instruct on use
                  "\n\\path\\ACAD X " dpath "UPDBLK"))
  );prompt
  (princ)
);defun UPDSCR
;*
(princ)
;* end of LBLOCK.LSP
;*
```

Save, Exit and reenter AutoCAD. TEST.LSP should load.

Command: **(load "FILELIB")**   Make sure FILELIB.LSP

```
Command: (load "BATCHSCR")            and BATCHSCR.LSP are loaded.

Command: (setq inc 0)                 This will make 6 test Blocks named T1, T2, T3, T4, T5 & T6.
Command: (repeat 6 (setq inc (1+ inc))
1> (command "CIRCLE" "@" inc "BLOCK" (strcat "T" (itoa inc)) "@" "L" ""))

Command: BLOCK                        Type a ? to check that you have Defined blocks:
                                      T1, T2, T3, T4, T5 & T6

Command: SAVE                         Save to name TX.

Command: LBLOCK                       It prints on text screen:
** NOTE: Existing .DWG files in designated directory **
** will be overwritten if Blocknames match. **

Specify block library directory...
Enter path name <current>: \CA-ACAD   Or return.
Wblocking T5                          Block names print here.

You may create a program to update blocks in other drawings.
Do you want to do so? <Y>:            <RETURN> for yes if you wrote and kept UPDSCR.

Specify drawings to update...
Enter path name <current>: \CA-ACAD
Drawing names <*>: T?
Making file listing......
Start AutoCAD from DOS by             This is the final message.
\path\ACAD X \CA-ACAD\UPDBLK

                                      Before you run the UPDBLK combo, check the UPDBLK.LSP and
                                      UPDBLK.SCR files.

Command: SHOW                         Another PGP command.
File to list: UPDBLK.LSP              It should be:

(setq blist '(
"T6"
"T5"
"T4"
"T3"
"T2"
"T1"
) path "/CA-ACAD/" )
(SETQ REGEN (GETVAR "REGENMODE"))
(SETVAR "REGENMODE" 0)
(FOREACH BLKDEF BLIST (IF (TBLSEARCH "BLOCK" BLKDEF) (COMMAND "INSERT" (STRCAT BLKDEF "="
PATH BLKDEF) (NOT (SETQ FLAG T))) (SETQ FLAG nil)))
(SETVAR "REGENMODE" REGEN)
(IF FLAG (COMMAND "SAVE" ""))
(COMMAND "RESUME")

Command: SHOW
File to list: UPDBLK.SCR              The last part should be:
2
\ca\ca-acad\T2
(load "/ca-acad/UPDBLK")
QUIT Y
-- More --                            Not part of file. <RETURN> to continue.
2
```

```
\ca\ca-acad\T1
(load "/ca-acad/UPDBLK")
QUIT Y
0
```
UPDBLK saves the drawing if it updates any blocks.

Command: **QUIT**

To test the UPDBLK combo. Quit your drawing to DOS.

C:\>**CA X UPDBLK**

Start AutoCAD with CA.BAT. Run the script from the DOS command line. It loads TX.DWG, and:

```
Command: (load "UPDBLK")
INSERT Block name (or ?): T6=T6 Block T6 redefined
Insertion point:
```
This repeats for T5 thru T1, then:
```
Command: INSERT Block name (or ?): ATTSHT-D=ATTSHT-D Block ATTSHT-D redefined
Insertion point:
Command: SAVE File name <TX>:
Command: RESUME
Command: QUIT Y
```

It cycles thru the rest of the T1 - T6 dwgs, but doesn't change them since they have no defined Blocks within. It leaves you in DOS.

 Delete TEST.LSP.

 Rename TEST.LSP to LBLOCK.LSP.

When you do scripted batch drawing file loads, you may want to temporarily rename your ACAD.LSP to avoid the time it takes to load in each drawing. Unless, of course, your process needs ACAD.LSP!

XINSERT External Block Extraction

There is another powerful command function on the CA DISK that processes Block libraries. XINSERT.LSP contains C:XINSERT (eXtracts and INSERT). XINSERT extracts a Block from another drawing file, without leaving the current drawing. It doesn't leave anything else of the other drawing file in the current drawing's database. We don't have room to show it all here, but you will find it useful. Several readers have asked for the program.

The CA DISK file XINSERT.TXT contains an example showing how it works. Examine XINSERT.TXT and XINSERT.LSP with your text editor to see it.

AutoCAD Drawing File Formats

You can store AutoCAD drawing files in two formats, DWG and DXF. A DWG drawing file is a protected AutoCAD file, its contents are hidden to the user. However, the AutoCAD program can create drawing files in a form readable by other software programs. This format is called the DXF file.

A DXF file is an ASCII file, designed for general drawing data transfers. Its rigid, well defined manner of describing a "CAD" drawing has helped it gain wide acceptance as a common exchange form for drawing data. You already created a small DXF file at the beginning of the chapter.

When to Use the DXF

Working with the DXF file is not complicated, just tedious. DXF programs have a reputation for being complicated only because DXF processing is often used for sophisticated applications, like engineering design and analysis. If you are evaluating DXF file processing, some factors to consider are:

- Does the application require you to scan the entire drawing file?
- Will the program depend on any design or analysis data?
- Do you have to determine relationships between components within the drawing?
- Will the drawing be modified by the results of the program's calculations?
- Do you need to make global changes to the drawing that AutoLISP can't accomplish, like Table Data changes?
- Do you need to import complex graphic data generated by an external program?
- Do you need to convert drawings to an older AutoCAD version, or between AutoCAD and another CAD program?

If you get "Yes" to several of these questions, then you are looking at candidate applications for DXF files. Sample applications are calculating the air circulation requirements for an office building, the rentable storage area for real estate lease management, or optimizing the spacing of seats in an arena or outdoor stadium. These are not casual applications. We recommend approaching DXF file processing applications with caution.

The AutoLISP Input/Output chapter introduced you to file processing. The techniques for opening, reading, writing and closing a file are common to all files, regardless of type. The only problem with reading and writing DXF files, which often are large, is the relatively slow pace of AutoLISP file processing. Most DXF applications use more efficient external program languages, like C.

One useful applications is to use DXF file processing for AutoCAD version conversions, converting from AutoCAD Release 9 to 2.6, for example. In fact, this is the example!

21—24 Customizing AutoCAD

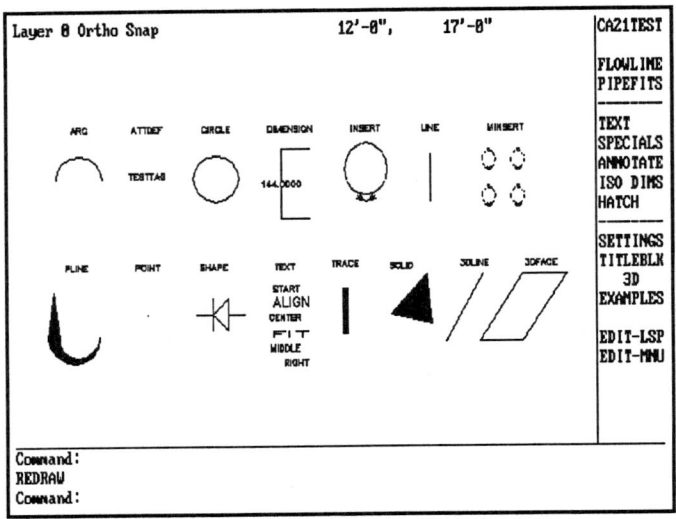

Sample Entities to Examine

First, let's look at the DXF file format. AutoCAD creates a DXF file with the DXFOUT command. You need to create a DXF file, using the ENTITY.DWG. The entity drawing is shown in the screen shot of Sample Entities to Examine.

Examining the DXF File Format

Enter selection: Begin a NEW drawing named DXFTEST=ENTITY.

 You have the ENTITY.DWG.

 You made ENTITY.DWG in Chapter 12.

```
Command: DXFOUT                                            Make a DXF file.
File name <DXFTEST>:                                       Take default name and 6 decimals.
Enter decimal places of accuracy (0 to 16) (or Entities) <6>:
                                                           It creates DXFTEST.DXF.
Command: ED                                                Edit DXFTEST.DXF to see it.
```

Inspect the DXF file as you read the next several sections. Try finding the items in the DXF file as they are described.

DXF Group Codes and Data Elements

Always think in terms of DXF groups, a DXF code followed by the data entry. AutoCAD precedes each piece of drawing data with a DXF code to indicate what type of data follows. Application programs reading the DXF file process the data according to the data type. The major types of data are:

GROUP CODE	DATA TYPE
0 - 9	Strings
10 - 59	Reals
60 - 79	Integers

Chapter twelve (and Appendix C) contain a complete DXF CODE chart. Refer to the DXF CODE chart for the meaning of specific group codes. Group codes are not absolute in meaning. For example, the CIRCLE entity uses the 40 code for the radius value, while TEXT uses 40 for text height. A sampling of the DXF file shows how these groups appear.

Look at the beginning of the DXFTEST.DXF file:

```
0           String follows, start of a new record.
SECTION     Identifies this as the start of a section.
2           String value follows, section name follows.
HEADER      This marks the start of the header information.
9           String value follows, next item is system setting.
$ACADVER    AutoCAD version system setting.
1           Next value is a string.
AC1004      The version number of DXF format file.
9           String value follows, start of next header record.
$INSBASE    Insertion base system variable.
10          X coordinate follows (in this case).
0.0         The data value 0.0.
20          Real value follows, Y coordinate.
0.0         The Y data value 0.0.
9           Next header variable name, etc...
```

DXF File Format, Data Types and Codes

A DXF file contains four basic groups of information about the drawing. Each type of group has a named section in the file:

```
HEADER    groups the drawing environment settings.
TABLES    lists reference symbols: layers, linetypes, views and
          styles.
BLOCKS    defines all block references.
ENTITIES  lists the entities.
```

To further distinguish items in the DXF file, AutoCAD uses a set of key words that indicate the start or end of groups, header, tables, blocks or entities. The key start and end words are:

```
SECTION   marks the beginning of a section.
ENDSEC    marks the end of a section.
TABLE     marks the start of a table entry.
ENDTAB    ends a table entry.
BLOCK     starts a block definition.
ENDBLK    ends the block definition.
SEQEND    ends an entry for complex entities
          (INSERTs with attributes and POLYLINEs).
EOF       end of the file marker.
```

Header Information

The header information contains system and drawing setup data. Information like the current menu name, the snap rotation angle and the current drawing view is stored in header variables. These header variables are similar to the

AutoCAD SETVAR variables, yet they are not always named the same. Here is a partial listing of some of the header variables:

NAME	SAMPLE INFORMATION	PURPOSE
$MENU	CA-MENU	Current menu file.
$SNAPUNIT	0.25,0.25	X,Y snap unit values.
$SPLFRAME	0	Show Spline Curve Frame.
$TDCREATE	2447143.9502840280	Time and date of creation.

Reference Table Information

The AutoCAD program uses tables to store lists of Layers, Linetypes, Views and Styles. As you would expect, the DXF file groups each entry under the appropriate table name. The table names are LTYPE, LAYER, STYLE and VIEW.

Find and examine these DXF groups:

```
0                     Start of a record.
SECTION               New section.
2                     Section name follows.
TABLES                Tables section.

0                     Start of new record.
TABLE                 A table entry.
2                     Table name.
LTYPE                 This is the linetype table beginning.
70                    Flag, number of table entries.
    1                 Only one linetype in our drawing.

0                     Start of a record.
LTYPE                 Linetype record.
2                     Linetype name follows.
CONTINUOUS            Specific linetype name.
70                    Flag for maximum number of dash dot entries.
    64                Maximum number.
3                     String follows.
Solid line            Verbal description of linetype.
72                    Alignment code.
    65                A type alignment.
73                    Number of dash dot entries.
     0                None.
40                    Real value follows.
0.0                   Pattern length.

0                     New record.
ENDTAB                End of Table.
```

Other Table entries are similarly listed here...

```
0                     
ENDSEC                End of section.
```

Block References

Although AutoLISP entity access treats Blocks similar to the Tables, DXF keeps block references in a separate BLOCKS section. This section contains information, like the block name, insertion base point, layer of creation, and a list of entities included in the block definition. Find the records for the first block reference in the DXF file.

0	Again, the start of a record.
SECTION	The record is a new section.
2	Section name.
BLOCKS	Block reference section.
0	Start of new record.
BLOCK	Record type is a BLOCK definition (only type so far).
8	Layer of creation.
0	Layer name of 0.
2	Block reference name.
LADYBUG	Our name.
70	Information flag.
66	Flag number, (attributes defined in this case).
10	X coordinate follows.
6.78125	X value of insertion base point.
20	Y coordinate follows.
1.125	Y value of insertion base point.
0	Start of new record.
ATTDEF	Entity name. Entity data follows.

The list of entities continues until the end of the block definition.

0	
ENDBLK	End of block definition.

The Entities Section

This section contains the primitive entities in the drawing file. It is usually the largest section of the file. AutoCAD Version 2.6 and Release 9 support fifteen types of entities, two are subentities of the complex INSERT and POLYLINE entity types. Here are the entity types:

ARC	INSERT	TEXT
ATTDEF	LINE	TRACE
ATTRIB (subentity)	POINT	VERTEX (subentity)
CIRCLE	POLYLINE	3DLINE
DIMENSION	SHAPE	3DFACE

Each entity type has different requirements for its data. There are five properties common to all entities: layer, linetype, color, elevation and thickness. Elevation and thickness are not relevant to 3DLine and 3DFace. All properties, with the exception of Layer, are optional. Optional properties are only output to the DXF file if they are assigned non-default values. The default settings are: linetype BYLAYER, color BYLAYER, elevation 0 and thickness 0.

Look for the entry of a typical LINE:

`0`	Start of a record.
`LINE`	Record is a LINE entity.
`8`	Layer name code.
`WHISKERS`	Name of layer.
`10`	X coordinate, first point.
`6.869023`	X value
`20`	Y coordinate, first point.
`1.15309`	Y value.
`11`	X coordinate, second point.
`6.970822`	X value.
`21`	Y coordinate, second point.
`1.233508`	Y value.
`0`	End of record.

`Exit back to AutoCAD.`

This Line did not have any optional properties assigned. If properties were assigned, you would find the following DXF groups used:

`6`	Linetype group code.
`DASHED`	Linetype.
`62`	Color group code.
`3`	Color green.
`38`	Elevation group code.
`12.000000`	1'-0" elevation.
`39`	Thickness.
`144.000000`	12'-0" thickness value.

As AutoCAD has evolved, so have its data file formats. The DWG drawing file format often undergoes changes which aren't upwardly compatible. However, the DXF file format is documented and was designed to be upwardly compatible. It remains reasonably stable with new releases of AutoCAD. The few DXF files changes are usually additions due to new entity types and drawing environment variables. The changes may be relatively simple, but the task of finding all occurrences of these items is a tedious error prone task. Although Autodesk does not provide a drawing conversion utility, you can use the DXF file format to convert AutoCAD drawings to previous versions.

A Version Conversion Reversion Utility

CONVDXF.LSP demonstrates the processing of the DXF file and converts Release 9 drawings to AutoCAD 2.6. It is included on the CA DISK. As an AutoLISP based program, it reads each line of a DXF file, determines the contents of the line and writes the line out to a new DXF file if the information was supported under the previous AutoCAD version. The program is **very slow**. We do not recommend it in any real world application. It is written in AutoLISP so that you can study the program and gain some understanding of DXF file processing. CONVDXF.LSP is too long to print here. Use your text editor to examine it.

For real world DXF file processing, we recommend faster computer languages with more flexible file handling. You can rewrite it in the compiled language of your choice. The CA-BONUS DISK includes an improved version of the program written in the C language. The C version is much faster and a useful utility. If you are faced with incompatible AutoCAD drawings, this program will help solve the problem.

If you have the CA DISK, and AutoCAD Release 9 or a Release 9 DXF file, try the CONVDXF program. Use a small DXF file that contains Spline-fit polylines to run the program.

A DXF Look Backwards

```
Command: CONVDXF
Release 9 DXF file name <dxftest>:        Use your .DXF file's name.

Converting file to AutoCAD version 2.62
Processing HEADER.                        It keeps you informed.
Processing TABLES.
Processing BLOCKS.
Editing a POLYLINE                        And goes through all the entities.
```

The CONVDXF program creates an output file with the same filename as the input file, but with a .NEW extension. Rename it with a .DXF extension to import it. You must use a "new" drawing to import it. Begin a new drawing with an "=." If you have the Customizing AutoCAD BONUS DISK, try the CONVDXF.EXE version and compare its speed!

Translating from Other CAD Programs

Although other CAD programs can read and write DXF files, most translations lose information along the way. Most often the other CAD program does not support a feature offered in AutoCAD. Sometimes AutoCAD does not provide a feature the other program offers. If you are investing in translation software, make sure the translation is bidirectional.

Effective CAD translations are possible. The key to a successful translation is testing the effects of placing one drawing database in the environment of another. The only dependable translation is one which has been tested. Develop a test drawing like the Translator Exerciser shown below. Include all entity types, text formatting options, attributes and layers. Test the translation and make provisions to accommodate problems.

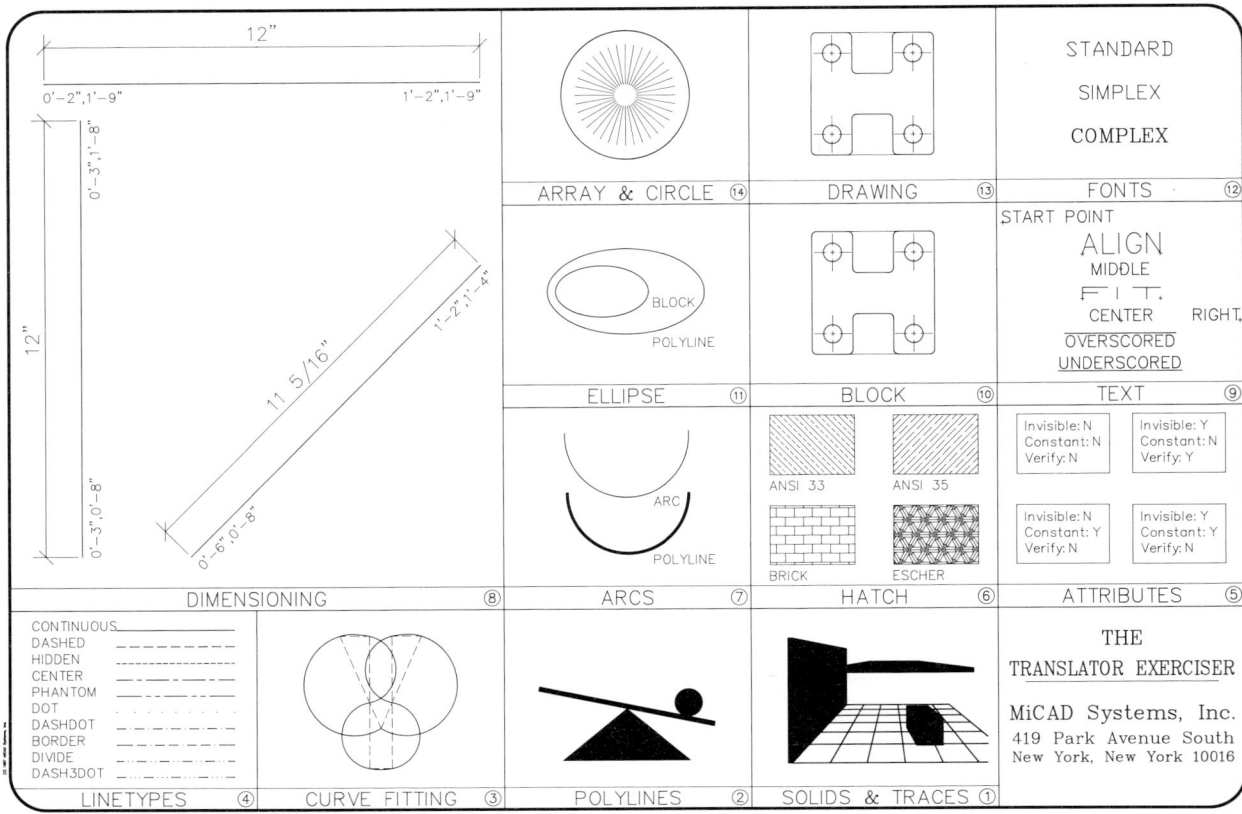

The Translator Exerciser, Courtesy of MiCAD Systems, Inc.

The most common areas of translation problems occur in text formatting, block definitions and Pline entity widths. By testing the conversion, the problem areas are isolated. Drafting and office procedures can restrict the use of problem features of a CAD package. You can convert some problem entities to an acceptable entity by writing a DXF processing program. You can handle others, like DIMENSION entities, inside AutoCAD. You can Explode dimensions before DXFing to programs that can't handle them.

Integrating the Example Menu

You do not need to integrate this chapter's menu into the CA-MENU. If any of these macros are useful, add them to your own menus. Here is the chapter's menu.

```
                    ►***SCREEN          ►**EXAMPLES
                      **ROOT
                    ┌──────────┐        ┌──────────┐
                    │ CA21TEST │        │  FTEXT   │
                    │          │        │          │
  R                 │ FLOWLINE │        │          │
  E                 │ PIPEFITS │        │          │
  L                 │ ──────── │        │          │
  O                 │   TEXT   │        │          │
  A                 │ SPECIALS │        │          │
  D                 │ ANNOTATE │        │          │
                    │ ISO DIMS │        │          │
  T                 │  HATCH   │        │          │
  E                 │ ──────── │        │          │
  S                 │ SETTINGS │        │          │
  T                 │ TITLEBLK │        │          │
  .                 │    3D    │        │          │
  M                 │ EXAMPLES │────┐   └──────────┘
  N                 │          │    │
  U                 │ EDIT-LSP │    │   ►**FOOTER -1
                    │ EDIT-MNU │──┐ │   ┌──────────┐
                    │          │  │ └──►│   root   │
                    └──────────┘  │     └──────────┘
                    ┌──────────┐  │
                    │TEXT EDITOR◄─┘
                    │ TEST.MNU │
                    └──────────┘
                    ┌──────────┐
                    │TEXT EDITOR│◄──
                    │ TEST.LSP │
                    └──────────┘
```

The Chapter 21 Menu

Summary Tips and Techniques

Here are some tips and tricks:

- ❏ Look for alternatives before you resort to a script or DXF process.

- ❏ You cannot load multiple .LSP files cleanly with AutoLISP. But you can call BATSCR easily with an AutoLISP program to write a script to LOAD a directory full of .LSP files. The program can call the script to execute the multiple loads. The end of the script can even call another AutoLISP function.

- ❏ You can run drawings with heavy amounts of text (via DXF) through a customized ASCII spelling checker. Build a special dictionary with all the "words" that AutoCAD uses, and include your abbreviations and terms.

- ❏ You may be able to save corrupted DWG files by using DXF. When a DXFIN fails, AutoCAD reports the problem line in the file, so you may be able to fix it.

Let's turn to look at file processing with two popular programs, Lotus 123 and dBASE III.

22—0 Customizing AutoCAD

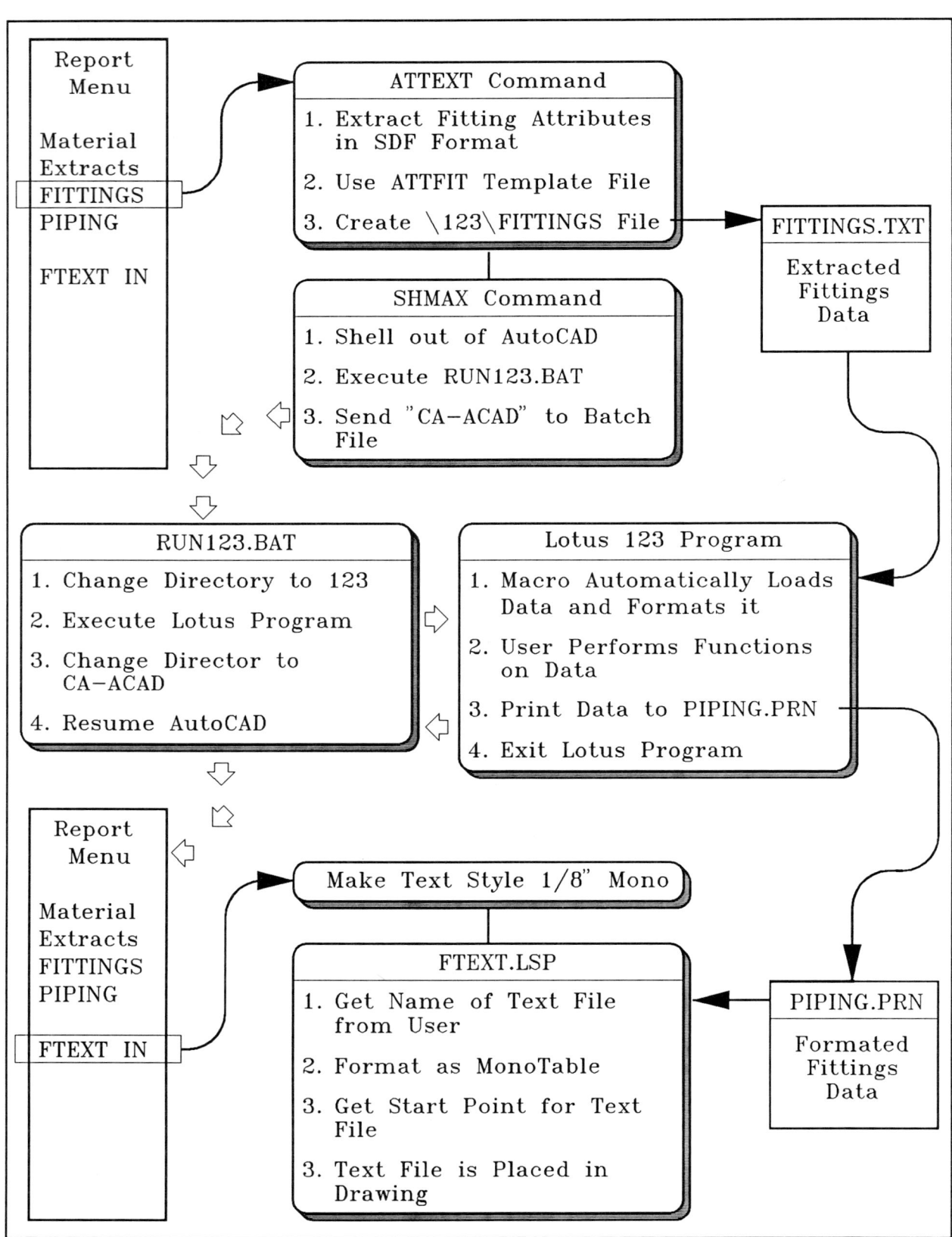

Program Flow Diagram Passing Data to Lotus 123

CHAPTER 22

LOTUS and dBASE

IMPORTING AND EXPORTING DATA

The Reporting Scenario

You can make an effective system to extract design information from AutoCAD drawings with very little effort. This chapter shows you how to put together systems for two types of reporting: material estimating reporting, and project drawing schedule reporting. The chapter provides a simple, complete reporting system, making a seamless transition between AutoCAD and Lotus 123; and AutoCAD and dBASE III. If you wish to run the Lotus 123 exercise, the chapter assumes that you have the Lotus 123 program setup and running in a directory called C:\123. If you have the CA DISK, the dBASE III program is not required. The CA DISK contains an executable file of CLIPPER-compiled dBASE code. Here is the general scenario for the chapter:

- Create formatted ASCII files using either AutoCAD or AutoLISP.

- Import those files into Lotus and dBASE via their import facilities.

- Massage the data into an appropriate form, present it to the user and then output it to a file.

- Take the output file and bring it back into AutoCAD.

The Benefits of Importing and Exporting AutoCAD Data

There are two key benefits to importing and exporting AutoCAD data.

- It launches your "CAD" system into another dimension of material reporting, providing up to date counts of parts and component descriptions. You know how much material the design requires. Using Lotus 123, you can add costs and total the amounts for different drawings.

- Tracking project drawings keeps you informed with the latest revisions and drafting time for a project. Project schedules assist in office coordination of CAD drawings and in preparing financial billing records.

How to Skills Checklist

In this chapter, you will learn how to:

- ❑ Make template files that control drawing data extraction, the output data format and field spacing.
- ❑ Extract component data from an AutoCAD drawing.
- ❑ Integrate the Lotus 123 spreadsheet program with AutoCAD.
- ❑ Set up a Lotus 123 spreadsheet to accept data from AutoCAD, import, format and tally the data in Lotus 123.
- ❑ Write a Lotus macro to automate the data import process.
- ❑ Send the spreadsheet file back to AutoCAD and use AutoLISP to import it into the drawing.
- ❑ Write AutoCAD macros to automate the entire report process.
- ❑ Report the drawing status to a master project status file.
- ❑ Output reference drawing data to a dBASE III program, process the data in dBASE, and report and print the data to a reference drawing schedule file.

Macros, AutoLISP Tools and Programs

MACROS
**REPORT menu page
[PIPING] and [FITTINGS] are macros that extract pipe length and fitting attributes from a drawing and call the Lotus 123 program.
[FTEXT IN] imports ASCII print file output from Lotus using the FTEXT AutoLISP program developed in the previous chapter.
[REP STAT] and [PROTRAK] call the REPSTAT.LSP and PRO_TRAK.EXE programs.

AutoLISP PROGRAMS
REPSTAT.LSP generates a project tracking record for the current drawing and adds to the master PROJECT.DAT file.

EXTERNAL PROGRAMS
PRO_TRAK.EXE is a CLIPPER compiled dBASE III program that tracks drawing information, like project name and location, drawing number and revision level, dates of issue, plotting, last revision, and management information about each revision.

The Template File

Let's start with the Lotus interface and report the pipe data generated in the earlier chapter on Attributes. The program flow is shown in the Lotus Program Diagram shown at the beginning of the chapter. The first step is preparing an AutoCAD Attribute Extract TEMPLATE file.

A special ASCII template file is used to tell AutoCAD which materials to report and how to format the report during the attribute extraction process. The template file controls attributes included in the report. It controls Numeric and Character data, and the width of each output field. The template also can include data about block insert scaling, coordinates and nesting levels.

The purpose of the template file is to match attribute TAG names in the file with those in the drawing and to extract the attribute VALUE attached to each block insert. Attribute tags not listed in the template file are ignored. Tags listed in the template report as blank, if they are not found in the drawing.

Let's make two template files and then look at the format of the files. You need one file for pipe fittings and one for pipe length segments. If you have the CA DISK, you have ATTFIT.TXT and ATTPIPE.TXT. Otherwise, create the files. If you have the CA DISK, you have PIPEMATL.DWG. Otherwise, you need the PIPEMATL.DWG that you created in the Parametrics chapter.

Templates and Attribute Extraction

DO NOT use tabs or trailing spaces. Be SURE to <RETURN> after the last line.

Enter selection: Edit the EXISTING drawing PIPEMATL.

 Read, but skip doing the rest of this sequence.

 Do this:

Command: **ED**   Create ATTFIT.TXT first, with:

```
F_TYPE              C010000
F_DIAMETER          N009006
F_RADIUS            N009006
F_FLANGE_DIAM       N010006
F_FLANGE_THICK      N009006
F_NUMBER_BOLTS      N003000
F_BOLT_DIAM         N009006
```

Save ATTFIT.TXT, and create ATTPIPE.TXT with:

```
P_DIAMETER          N009006
P_LENGTH            N012006
P_FLANGE_DIAM       N010006
P_FLANGE_THICK      N009006
P_NUMBER_BOLTS      N003000
P_BOLT_DIAM         N009006
```

Save and exit the file.

Attribute extract files MUST have a .TXT file extension. You need a unique attribute tag for each type of component taken off. You need at least one attribute tag in a template file. You need at least one space between the attribute TAG name and the format string.

Format strings, like C010000 and N009006, control the form of the output data. "C" or "N" indicates Character or Numeric output. The first three digits (like 010 and 009) tell AutoCAD the total output field character width (010 allots a 10 character wide field). The last three digits specify the number of decimal places to output (always 000 for Char fields).

Integers may be output by using 000 decimal precision. For example, N009006 has 6 decimal digits following the decimal point (which counts as a char). This leaves (9-6-1=2) 2 leading digits for the whole part of the number. It outputs in the form: 12.123456. The distinction between Char and Number is strictly for formatting. The data is written as ASCII characters in the output file.

Several problems may arise from incorrect field length numbers. AutoCAD issues a "field overflow" warning and truncates the output if the extracted attribute values don't fit in the template's specified field width. The output record may run together as one string, making processing of the output with Lotus difficult. This run-on is not reported as an error. To avoid a run-on string, increase the field width by one character. AutoCAD fills any extra spaces. For example:

```
ELBOW    6.0000008.00000011.0000001.00000080.750000    Won't work.
ELBOW    6.000000 8.000000 11.000000 1.000000 8 0.750000   OK!
```

Include Block insert "tags" in the template, using these standardized keys:

KEY NAME	FORMAT	DESCRIPTION
BL:LEVEL	Nwww000	(block nesting level)
BL:NAME	Cwww000	(block name)
BL:X	Nwwwddd	(X coordinate of insertion point)
BL:Y	Nwwwddd	(Y coordinate of insertion point)
BL:Z	Nwwwddd	(Z elevation of insertion point)
BL:LAYER	Cwww000	(layer name inserted on)
BL:ORIENT	Nwwwddd	(insert rotation angle)
BL:XSCALE	Nwwwddd	(X scale factor)
BL:YSCALE	Nwwwddd	(Y scale factor)
BL:ZSCALE	Nwwwddd	(Z scale factor)

AutoCAD outputs attribute fields in the order listed in the template file. There is no automatic control for the order of records in the output file. AutoCAD outputs the records in the entity order. To control the order, you must select entities, using Attext's Entity select option to get the order that you want. For most applications, record order does not matter.

Try the attribute extract command ATTEXT, and make a Standard Data Format (SDF) output file. You should be in the PIPEMATL drawing:

```
Command: ATTEXT
CDF, SDF or DXF Attribute extract (or Entities)? <C>: S
Template file: ATTFIT
Extract file name <PIPEMATL>: \123\FITTINGS
15 records in extract file.                          Yours may vary.

Command: TYPE                                        Look at the output.
File to list: \123\FITTINGS.TXT

ELBOW      6.000000  8.000000  11.000000  1.000000   8  0.750000
ELBOW      6.000000  8.000000  11.000000  1.000000   8  0.750000
ELBOW      6.000000  8.000000  11.000000  1.000000   8  0.750000
ELBOW      6.000000  8.000000  11.000000  1.000000   8  0.750000
ELBOW      6.000000  8.000000  11.000000  1.000000   8  0.750000
ELBOW      6.000000  8.000000  11.000000  1.000000   8  0.750000
ELBOW      6.000000  8.000000  11.000000  1.000000   8  0.750000
ELBOW      6.000000  8.000000  11.000000  1.000000   8  0.750000
ELBOW      2.500000  5.000000   7.000000  0.687500   4  0.625000
ELBOW      2.500000  5.000000   7.000000  0.687500   4  0.625000
ELBOW      2.500000  5.000000   7.000000  0.687500   4  0.625000
ELBOW      2.500000  5.000000   7.000000  0.687500   4  0.625000
ELBOW      2.500000  5.000000   7.000000  0.687500   4  0.625000
ELBOW      2.500000  5.000000   7.000000  0.687500   4  0.625000
ELBOW      2.500000  5.000000   7.000000  0.687500   4  0.625000
```

The default output file name is the current file name, with a .TXT file extension. It is unfortunate that the output file and template file both use a .TXT extension. Many unwary users have overwritten their template file. DO NOT use the same name for both template and output files.

Next, let's see how to use the data with Lotus 123.

Importing Data Into Lotus

The reporting example will only show a few features of Lotus 123. Lotus is a popular program for making engineering and financial calculations with AutoCAD imported data. Even if you don't have Lotus, we encourage you to read along to gain some understanding of integrating a spreadsheet, like Lotus, with AutoCAD. The exercise can help you decide if it is a worthwhile extension to your AutoCAD application.

To start, the exercise shows you how to get the data into Lotus and shows you a little about automating the transfer steps. Even if you have the CA DISK, create a new spreadsheet in Lotus. Shell out of AutoCAD, change over to the 123 directory and start 123.

Manually Importing to 123

```
Command: SHMAX                    Use the book's max memory PGP command.
DOS Command:                      <RETURN> to go out to DOS.
C:\CA-ACAD>>CD \123               Change to Lotus 123 directory.
C:\123>>123                       Start 123. You'll be in a new file.
```

In the Lotus screens illustrated in this sequence, the Lotus commands to choose are shown "highlighted" at the top of the screen, like the word "Worksheet," shown in the illustration the Worksheet Display.

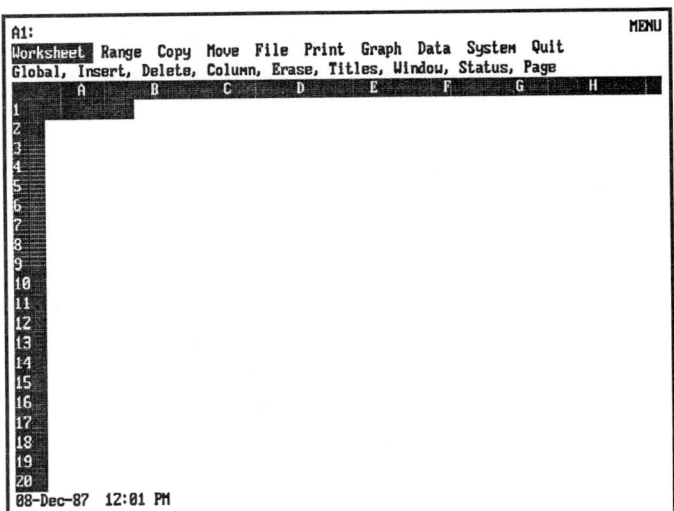

The Worksheet Display

Use Lotus commands to bring the data into Lotus. First, position the cursor at the top left cell where you want the data. Bring up the command menu with the slash character "/." This is Lotus's activate key. Enter a file name, and tell Lotus to interpret the file as Text information.

Let's import the data.

```
Enter  /                          To activate the Lotus menu.
Choose FILE                       This gives the file utility menu.
Choose IMPORT                     You want to import data.
Choose TEXT                       Choose the text form.
```

To clear out the default name given by Lotus hit <ESC> twice.

Then enter name of attribute extract file. It's in the 123 directory:
Enter name of file to import: **\123\FITTINGS.TXT**

The screen sequence is shown in the Importing Procedure screen shots.

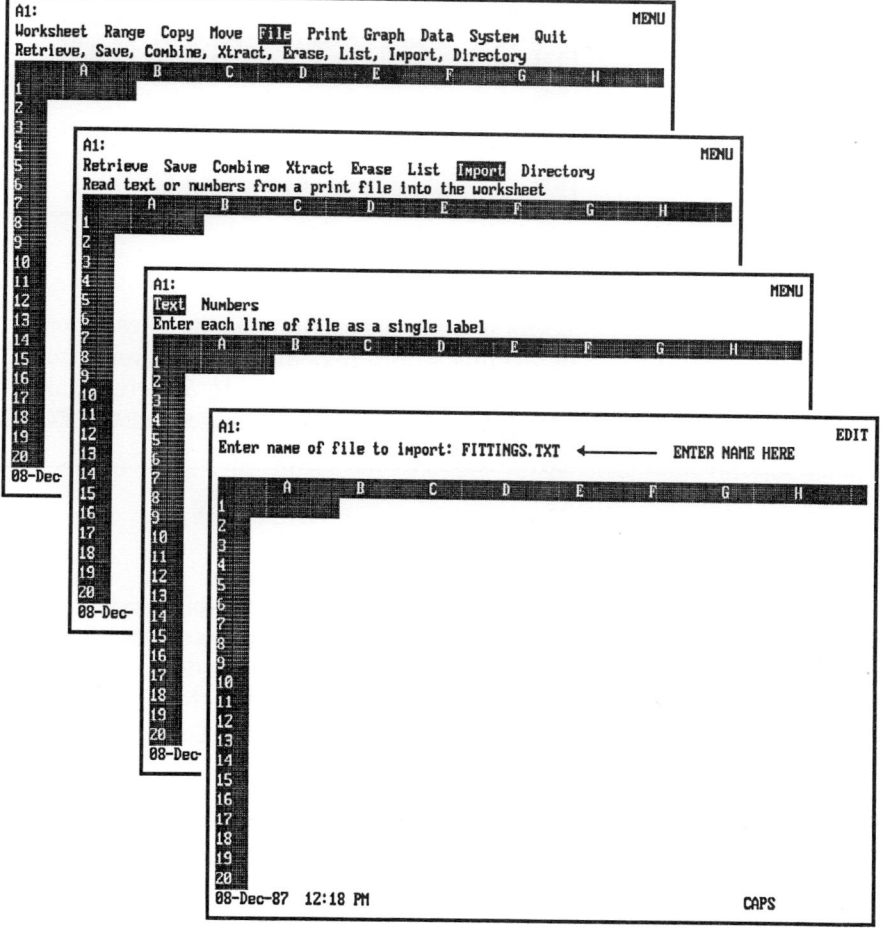

Importing Procedure

Lotus will make a long single-cell entry for each line of the imported file. The numeric values are still ASCII text characters. They are not yet numbers in Lotus. Converting the lines of text into individual numeric fields is a two part process using the PARSE command from the DATA menu.

Your screen will shown the imported file.

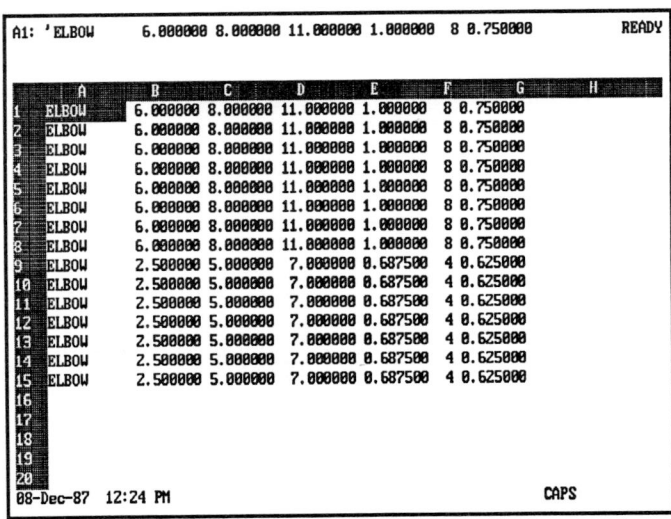

The Imported File

Lotus is a great assistant in the parsing operation. Parsing requires that you determine where text lines are broken into fields, and whether to convert the data into string or numeric data. Lotus makes a "Best guess" at the data types and the positions for each field when the FORMAT-LINE option is selected.

Move the cursor to the upper left cell of the imported data, then:

Choose **/**	Call the menu.
Choose **DATA**	Get the data commands.
Choose **PARSE**	Presents the Parse menu.
Choose **FORMAT-LINE**	Gives some options.
Choose **CREATE**	Use this to make a format line.

The Format-Line command looks at the spaces and decimal points in the text line and guesses at a format. The format line is placed above the imported data. It uses special characters to reveal its guess to you:

CHAR	DESCRIPTION
L	Beginning of a text label.
V	Starts a numeric value.
*	Places holder for future characters of preceding cell.
>	Occupied character positions in the cell.

The parsing screen sequence is shown in Parsing Input Data.

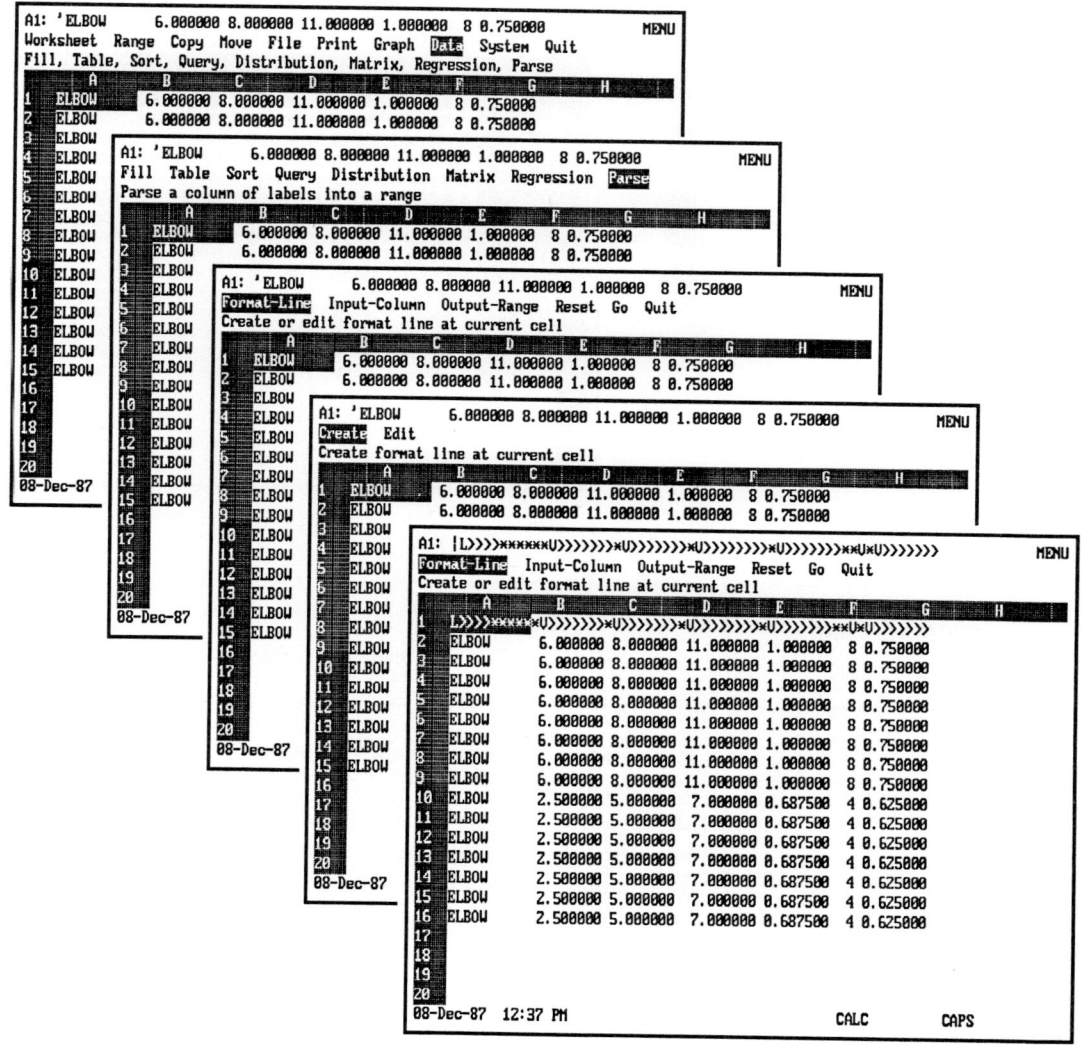

Parsing Input Data

INPUT-COLUMN tells Lotus which rows of text strings to convert. Select the command and highlight the range of cells the imported data occupies. Notice that only the first column is marked. Each line is still a single text string. You can see the entire entry in the cell edit buffer at the top of the screen.

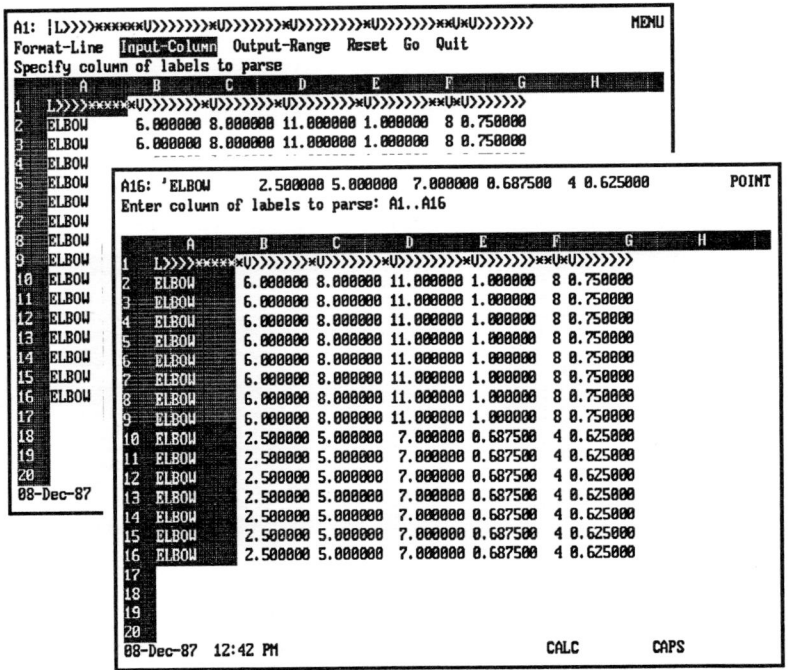

Selecting the Input Columns

Choose **INPUT-COLUMN**
Choose **A1..A16**   For the columns. Yours may differ.

You can place converted data in any cell range of the spreadsheet. Lotus needs to know where to put the data. To keep from jumping around the spreadsheet, make Lotus overwrite the imported text strings. Use the OUTPUT-RANGE command to highlight all the data lines including the blank, but apparently filled, cells. DO NOT include the format string in the output range. When both ranges are set, select the GO command. Go performs the actual parsing.

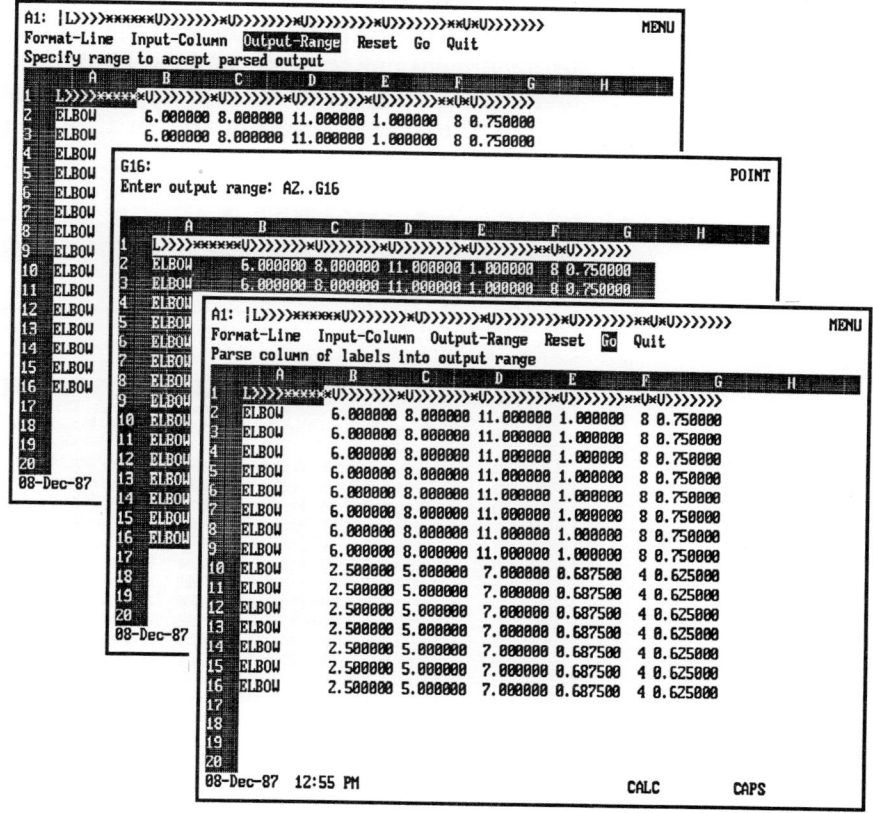

The Output Range

Choose **OUTPUT-RANGE**
Choose **A2..G16**   For the book's data. Yours may differ.
Choose **GO**   Start the parser.

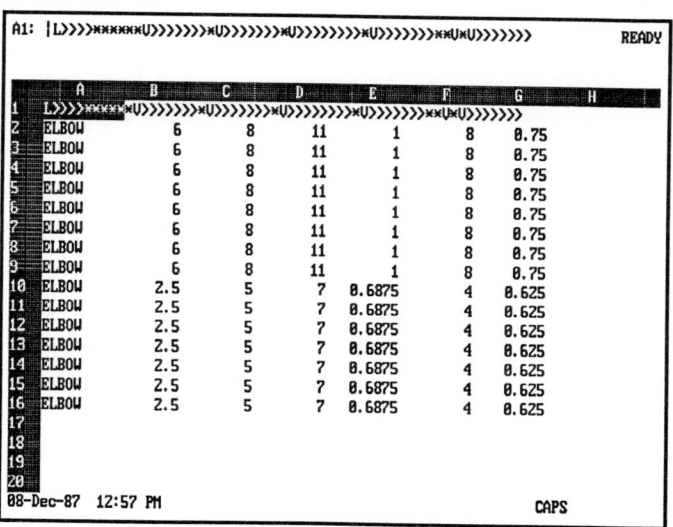

Raw Labels and Numeric Data in Lotus

Lotus converts the data to labels for the "ELBOW" column, and numbers for the fitting dimensions and components. Trailing 0s and decimal points are omitted and the cells are formatted to the maximum precision of the original number. This is shown in the screen shot of Raw Labels. The translation process is complete.

You can add additional information, headers and totals to embellish the spreadsheet file. This is shown in the Pipe Fitting Spreadsheet screen shot.

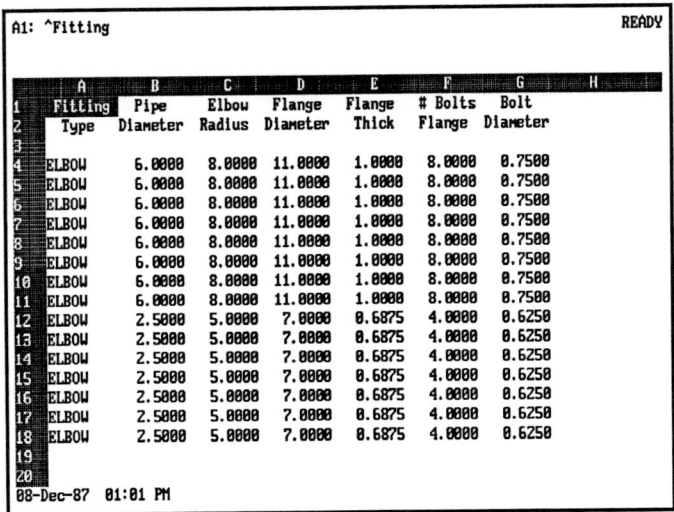

The Pipe Fitting Material Spreadsheet

Save and exit the spreadsheet and go back to AutoCAD.

Although the book does not require it, you may want to save the spreadsheet for your own purposes.

```
Choose /
Choose FILE
Choose SAVE
Enter save file name: \123\FITTINGS      It writes FITTINGS.WK1.
Choose /
Choose QUIT
Choose YES

C:\123>>CD \CA-ACAD                      Go back to your directory.
C:\CA-ACAD>>EXIT                         Exit Shell to AutoCAD.
```

Importing Data Into Lotus 22—13

You can automate the link between AutoCAD and Lotus with several menu macros. [FITTINGS] will extract the fittings, and [PIPING] will extract the pipe lengths. Both call Lotus. [FTEXT IN] imports the printed Lotus file back into AutoCAD.

If you have the CA DISK, you have the completed CA22.MNU file. If not, you need to add to the Chapter 19 TITLEBLK menu. You should be back in the PIPEMATL drawing.

Automating the Lotus Link

 Copy CA22.MNU to TEST.MNU.

 Copy MY19.MNU to TEST.MNU.

Select **[EDIT-MNU]**
Change the root page label to [CA22TEST].
Change the [EXAMPLES] page call to [REPORT] as shown.
Add a **REPFOOTER:

```
***SCREEN
**ROOT
[CA22TEST]
[]
[FLOWLINE]
[PIPEFITS]
[--------]
[TEXT    ]
[SPECIALS]
[ANNOTATE]
[ISO DIMS]
[HATCH   ]
[--------]
[SETTINGS]
[TITLEBLK]^C^C^CSETVAR TEXTEVAL 1 LAYER S BORDER;;+
(caload "/CA-ACAD/" "TIMELIB") $S=TITLEFOOTER $S=TITLEBLK
[   3D   ]
[EXAMPLES]$S=REPFOOTER $S=REPORT
[]
[EDIT-LSP]LSP (load "TEST")
[EDIT-MNU]MNU MENU TEST
[]
**REPFOOTER -1
[   root   ]$S=SCREEN LAYER S 0 ;SNAP ON GRID ON ORTHO ON DIM1 STYLE STD1-8
```

Fill in the **EXAMPLES page, with 19 [items] and [] blanks:

```
**EXAMPLES
[ Report ]
[  Menu  ]
[]
[Material]
[Extracts]
[FITTINGS]^C^C^CATTEXT S /CA-ACAD/ATTFIT /123/FITTINGS SHMAX RUN123 CA-ACAD;
[PIPING  ]^C^C^CATTEXT S /CA-ACAD/ATTPIPE /123/PIPING SHMAX RUN123 CA-ACAD;
[]
[FTEXT IN]^C^C^CSTYLE MONO1-8 MONOTXT (* 0.125 #dwgsc) ;;;;+
(caload "/CA-ACAD/" "FTEXT") FTEXT \MT
[]
```

Save and Exit to AutoCAD. TEST.MNU reloads.

The new macros execute the ATTEXT command, outputting a SDF file of the attribute extract data, writing it to the Lotus 123 directory. They call a batch file that automates the DOS directory changes. For those of you who have the CA DISK, and are running Lotus 123, make this small batch file called RUN123.BAT. Put it in the CA-ACAD subdirectory:

```
Command: ED
File to edit: RUN123.BAT

CD \123
123
CD \%1
```

Make the batch file.

Remember to use your Lotus directory name, if it is not \123.

Save, exit, and return to AutoCAD.

 Copy the PIPING.WK1 file from \CA-DISK to \123 (or your Lotus directory).

Select **[PIPING]**

Report the pipe segment lengths. [PIPING] makes the output file PIPING.TXT. And 123.BAT takes you right into Lotus 123.

The %1 in the RUN123.BAT batch file is replaced by "CA-ACAD" from the end of the [FITTINGS] and [PIPING] macros to restore the directory, when they are run.

An annoying limitation of the Lotus 123 program is that you cannot have Lotus go right into a worksheet file of your choice. You have to load the spreadsheet manually.

You can automate the steps with Lotus' macro capability. Lotus can import and convert the data without instructions from you. Here are the rules for the game:

- Macro definitions are stored within the worksheet cells.

- A macro is continued in a row-by-row fashion.

- Commands in one row are immediately followed by commands in the next row.

- Macros can call subroutines, like one LISP function calls another.
- Macro names are two characters long.
- All names begin with "\" backslash and then a single letter, A thru Z.
- A special macro called \0 automatically executes each time a worksheet file is retrieved.
- Macros are executed with <ALT> keys. Holding down <ALT> and hitting I (that's <ALT-I>) would execute \I.

Develop a simple macro to import the data and parse it. Document your macros. Keep them to the side or top of your calculation area. Put the name of your macros to the left so that you know where and what they are.

Since there is no telling how many text lines are imported from the attribute output file, they might overwrite your macro space. In this case, the macro is put above the actual calculation area. Plan your sheet to accommodate the six fields of output as six Lotus columns.

Go to an open area of the spread sheet, like cell A1. The macro to import and parse the data is shown in the screen shot called the Import Macro.

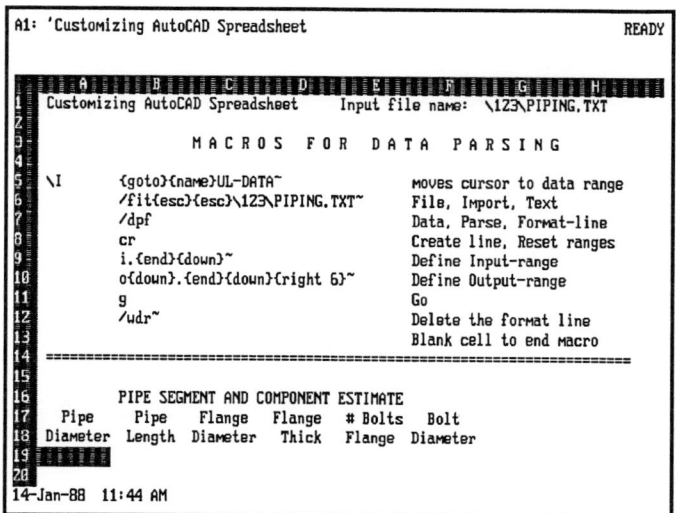

The Import Macro

Select **[PIPING]** if you are not already in 123.

 Retrieve your existing PIPING.WK1 file in Lotus.

 Go to cell A1 and type in the macro, including comments, as shown in the screen shot.

In this macro, you use two range names, UL-DATA and \I, to hold the data and the macro's positions inside the worksheet. If you are unfamiliar with Lotus programming tools for use with macros. Here is a partial list:

MACRO COMMAND	FUNCTION
{GOTO}	Moves cursor to a cell position.
{NAME}	Used to call a range name in macros.
{ESC}	The <ESC> key.
{END}	Jumps to the end of a filled range.
{UP n}	Moves up *n* positions.
{DOWN n}	Moves cursor down *n* cells.
{LEFT n}	Moves left.
{RIGHT n}	Moves right.

The macro \I is used to find the last line of text brought in from the import command. It positions the cursor at the range location UL-DATA, and imports the file. A parse format string is created. The input and output ranges are identified, and the data parsed.

Define the ranges and try the macro. Call the macro using the <ALT> key and the macro letter. Lotus will begin the operation.

 Skip this sequence.

 Do this:

Choose **/**
Choose **RANGE**
Choose **NAME**
Choose **CREATE**
Enter name: **UL-DATA**
Enter range: **A19**   In the book's spreadsheet!

Repeat the same steps to define the macro range.

Enter name: **\I**
Enter range: **B5**   The top cell of the macro.

Reposition the cursor to cell A1, then save:

Choose **/**
Choose **FILE**
Choose **SAVE**   Save to \123\PIPING only if you do not have the CA DISK.

<ALT-I>   Hold down <ALT> and hit I to call the macro.

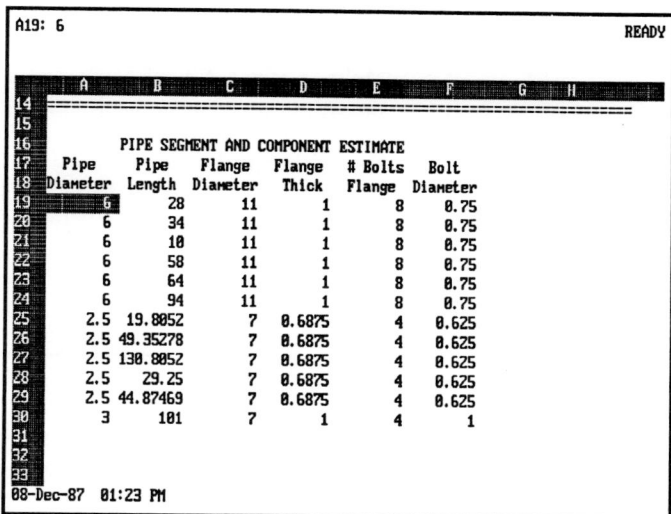

Data After Importing

If you defined the ranges and entered the macro correctly, the data will automatically be imported, parsed and formatted. The format line is erased at the end of the operation. Your screen should look like Data After Importing.

Next, add some descriptive labels and total the fields as shown in the Tally of Pipe Length and Bolt Numbers.

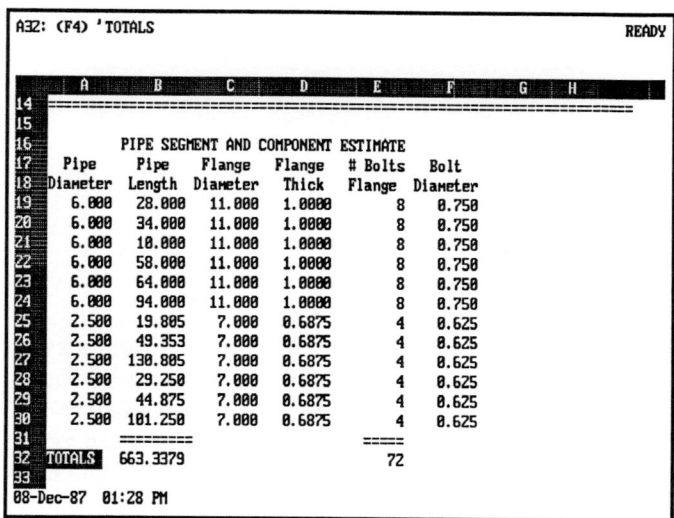

Tally of Pipe Length and Bolt Numbers

```
Choose /                                    Call the menu.
Choose PRINT                                Make a print file of your work.
Choose FILE                                 Later, it is brought back into AutoCAD.
Enter print file name: \CA-ACAD\PIPING      Hit <ESC> twice to clear name.
Choose RANGE                                To establish the print range.
Choose A16..F32                             Give this range. It is the book's spread sheet.
Choose GO                                   Starts the printing to PIPING.PRN file.
Choose QUIT                                 The print menu.
Choose /QUIT                                It was saved in the master form.
Return to AutoCAD.
```

Your screen should look like Print Out Range.

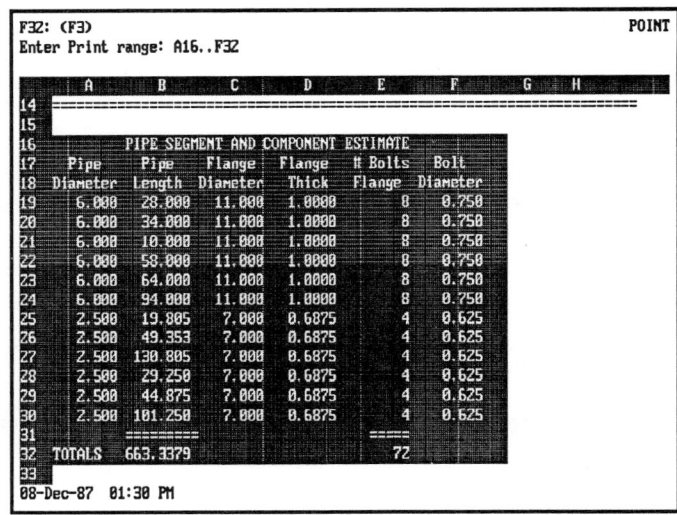

Print Out Range

Bringing It Back In

In reality, you would extract and process a good deal more data. You would have your spreadsheet program produce and print a summary report, then you would import the report to AutoCAD as a BOM (Bill Of Materials). The next exercise imports the Lotus reformatted data as an example of importing data. Whatever your application, the import process is the same.

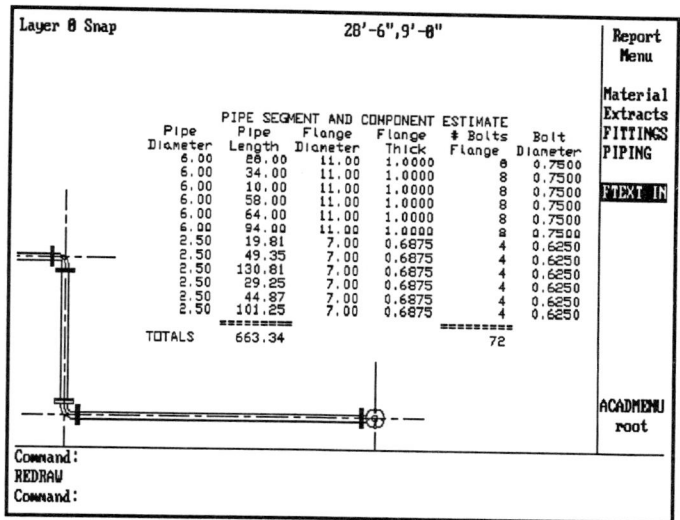

Imported Text in Drawing

There is no direct link to import Lotus spreadsheets to AutoCAD drawings. However, with a little help from AutoLISP, you can use the FTEXT function from the chapter on Batch Processing. You put it on the **REPORT screen menu. This menu uses FTEXT in its monotable mode. Import the Lotus output, PIPING.PRN:

Importing the .PRN File into AutoCAD

Command: **(load "FTEXT")**   Because CALOAD is still disabled.
Select **[REPORT] [FTEXT IN]**
Command: STYLE Text style name (or ?) <MONO1-8>: MONO1-8   Runs Style command and sets MONO1-8.
MONO1-8 is now the current text style.
Command: FTEXT
Enter text source filename.ext: **\CA-ACAD\PIPING.PRN**
MonoTable or Word Wrap? MT/WW <WW>: MT
Start point: Pick, and it puts tabular text in. Zoom to see it all.

Command: **END**   You are finished with the drawing.

Using dBASE with AutoCAD

You can use dBASE III and many other database programs to process data from an AutoCAD extract file. The program outlined here is for tracking AutoCAD project drawings. Many of the steps shown in the exercise are common to any data transfer between AutoCAD and database programs. You can extrapolate steps from the exercise to your own application.

Let's start with a look at the drawing attributes that will go into the reporting system. The chapter on Attributes developed title block and drawing revision routines that create the data that you will use here. You need two more attribute template files for project tracking. If you have the CA DISK, you already have the template files, TITLEBLK.TXT and REVBLOCK.TXT. The files should contain:

FILE NAME:	TITLEBLK.TXT	FILE NAME:	REVBLOCK.TXT
ATT TAG	FORMAT CODE	ATT TAG	FORMAT CODE
PRJ-NAME	C031000	R1	C001000
DWG-NAME	C031000	R2	C031000
DWG-NO	C010000	R3	C008000
DWG-FILE	C031000	R4	C003000
DRAFTER	C003000	R5	C016000
DWG-REV	C001000	R6	C008000
DATE-REV	C008000	R7	C003000
DATE-ISSUE	C008000	R8	C010000

You need the SYNPLANT drawing file for the macros and programs in this section. You need to get the drawing file from the CA DISK, or from your work in the Attributes chapter. Load the drawing with its title block and revisions. Then, make the two template files. Run the ATTEXT command to extract the attributes.

Testing the Attribute Files

```
Enter selection:                              Begin a NEW drawing named CA22=SYNPLANT
Command: ED                                   Create the 2 attribute template files as shown above.

Select [TITLEBLK] [CAD FILE]                  To update the DWG-FILE attribute.

Command: ATTEXT                               This time, use the default CDF extract.
CDF, SDF or DXF Attribute extract (or Entities)? <C>:
Template file: TITLEBLK                       Do the title block information first.
Extract file name <CA22>: $PRJ                Give it a temporary name.
1 records in extract file.                    Should only be one!

Command: TYPE                                 Look at it:
File to list: $PRJ.TXT                        It should be similar to:

'SYN-FIBER PLANT','SECONDARY PIPING SYSTEM','D81-3008','CA22','JSJ','E','12/15/87','None'

Command: ATTEXT
CDF, SDF or DXF Attribute extract (or Entities)? <C>:
Template file <TITLEBLK>: REVBLOCK             Do the revision blocks.
Extract file name <CA22>: $REV                 Use another temporary name.
```

```
Command: TYPE                          Look at the revision file.
File to list: $REV.TXT                 It should be similar to:

'A','CHANGED PIPING LOCATION','11/29/87','JSJ','CONSTRUCTION','12/02/87','RMS','42870'
'B','ADDED CONDUIT','12/06/87','PJS','CONSTRUCTION','12/06/87','DBS','1441'
'C','CHANGED PUMPS 1 & 2','12/07/87','TBB','CONSTRUCTION','12/09/87','PTM','751'
'D','REDRAWN AND REVISED','12/11/87','CMP','CONSTRUCTION','12/13/87','MMP','6063'
'E','ARCHITECTURAL ADJUSTMENTS','12/15/87','RBG','BID','12/15/87','JSJ','6090'
```

The generated records of the $REV.TXT file include all data concerning revision history, including revision time. Yours may be different. Now make the master project data file.

Preparing the Input Record

Normally, you would like to do one attribute extraction, but the Attext command cannot provide the data in the order that you want. A little file manipulation is required. Open three files at one time, read data from the temporary file $PRJ.TXT, splice it together with each record of the $REV.TXT file and append the combined record to the master project file, called PROJECT.DAT. The REPSTAT routine that you need to do this is short and sweet.

REPSTAT A Report Status Program

  Copy REPSTAT.LSP to TEST.LSP

  Create a TEST.LSP file.

Inspect or enter the function.

```
;* The C:REPSTAT command extracts the title sheet and drawing revision information
;* from the current drawing and records the information in the master project
;* file called PROJECT.DAT, which is processed by the PRO_TREK data base program.

(defun C:REPSTAT ( / pf p$ r$ prdat rvdat)       ; MAKE THE ATTRIB EXTRACT FILES
    (command "ATTEXT" "C" "TITLEBLK" "$PRJ")     ;extract title block data
    (command "ATTEXT" "C" "REVBLOCK" "$REV")     ;extract revision block data

                                                 ; OPEN THE EXTERNAL FILES
    (setq pf (open "PROJECT.DAT" "a")            ;open master project data file to append
          p$ (open "$PRJ.TXT" "r")               ;open title block data file to read
          r$ (open "$REV.TXT" "r")               ;open revision block data file to read
    );setq
```

Splice together the original drawing creation time with a dummy revision record. This is done to hold the correct number of fields for later importing to dBASE. The original creation time is placed in the same field as the revision time of a revision record. In effect, original creation time is treated the same way as any revision. If no revision level has been assigned, the value of USERR1 is 0.0 and the programs use the AutoCAD system variable TDINDWG. The program truncates the time to minutes precision. Continue typing.

```
                                                    ; BUILD AND OUTPUT TITLE BLOCK RECORD
  (if (and p$ (setq prdat (read-line p$)))          ;check for file, read title block data
    (progn
      (close p$)                                    ;close the file, we're done
      (if (= 0.0 (getvar "USERR1"))                 ;no revisions have been made yet
        (setq dwgtime (rtos (* 1440 (getvar "TDINDWG")) 2 0)) ;user ACAD sys var time
        (setq dwgtime (rtos (getvar "USERR1") 2 0)) ;use our original dwg creation time
      );if
      (write-line (strcat prdat                     ;merge title block record with
                  ",'','','','','','','','"         ;dummy revision record (7 fields)
                      dwgtime "'") pf)  ;add in 8th field, original creation time
    );write-line
```

Now process each of the revision records, add them to the project title block data, and output a record in the master project data file.

```
                                                    ; PROCESS THE REVISION FILE
    (if r$                                          ;if a revision file exists.
      (progn
        (while (setq rvdat (read-line r$))          ;loop and read each revision entry
          (write-line (strcat prdat "," rvdat) pf)  ;put together & write to master file
        );while
        (close r$)                                  ;close the file, we're done
      );progn
    );if
  );progn
  (prompt "\nNo title block information extracted from drawing.")
  );if
  (close pf)                                        ;close the project file
  (princ)
);defun
```

Save, exit and return to AutoCAD. Try the function.

Next, generate the revision records.

Generating Revision Records

Command: **REPSTAT** This starts the revision status program.

Check the contents of the PROJECT.DAT file with the PGP command SHOW.

Command: **SHOW**
File name to type: **PROJECT.DAT** The master file will be similar to:

```
'SYN-FIBER PLANT','SECONDARY PIPING
SYSTEM','D81-3008','CA22','JSJ','E','12/15/87','None','','','','','','','','','2447130'
'SYN-FIBER PLANT','SECONDARY PIPING
SYSTEM','D81-3008','CA22','JSJ','E','12/15/87','None','A','CHANGED PIPING
LOCATION','11/29/87','JSJ','CONSTRUCTION','12/02/87','RMS','42870'
'SYN-FIBER PLANT','SECONDARY PIPING
SYSTEM','D81-3008','CA22','JSJ','E','12/15/87','None','B','ADDED
CONDUIT','12/06/87','PJS','CONSTRUCTION','12/06/87','DBS','1441'
'SYN-FIBER PLANT','SECONDARY PIPING
SYSTEM','D81-3008','CA22','JSJ','E','12/15/87','None','C','CHANGED PUMPS 1 &
2','12/07/87','TBB','CONSTRUCTION','12/09/87','PTM','751'
'SYN-FIBER PLANT','SECONDARY PIPING
SYSTEM','D81-3008','CA22','JSJ','E','12/15/87','None','D','REDRAWN AND
REVISED','12/11/87','CMP','CONSTRUCTION','12/13/87','MMP','6063'
'SYN-FIBER PLANT','SECONDARY PIPING
SYSTEM','D81-3008','CA22','JSJ','E','12/15/87','None','E','ARCHITECTURAL
ADJUSTMENTS','12/15/87','RBG','BID','12/15/87','JSJ','6090'
```

Generate some additional records by changing the project name, the drawing number, dates, and locations.

Select **[DWG NO]** Then run REPSTAT again.
New attribute value: **D81-4000**

Command: **REPSTAT**

Select **[PRJ NAME]** Enter a new value.

Command: **REPSTAT** Run it again.

Generate several records in this manner.

The rest of the process is done in the database program. But first, let's put the items on the **TITLEBLK screen page developed in the Attribute chapter.

Select **[EDIT-MNU]** Edit the CA22.MNU.

Fill in entries for the [REP STAT] and [PROTRAK] labels:

```
**TITLEBLK
[TitleBlk]

[REP STAT]^C^C^C(caload "/CA-ACAD/" "REPSTAT") REPSTAT
[PROTRAK ]^C^C^CSHMAX PRO_TRAK;
[]
```

Save, exit and reload AutoCAD. Try the [REP STAT] macro.

Tracking CAD Drawings

The PRO_TRAK data base program has been kept simple, but it is still too long to print here. PRO_TRAK.EXE is a Clipper compiled version of the dBASE III program. The source code is available on the Customizing AutoCAD BONUS DISK.

This section shows you how the program was developed, why certain reporting options are used, and gives an outline of the program's steps. PRO_TRAK.EXE is on the CA DISK for your use.

The dBASE III program gives you help in creating database structures. Virtually anyone can effectively use dBASE for their application. Clipper creates easy to use, fast running compiled database programs, with the added advantage of protecting the source code.

The PRO_TRAK program starts with a main menu, giving several database reporting choices to the user. One menu choice exits the program. Another reads the master project data file, and transcribes the records found there into the dBASE environment. The options available are shown in the PRO_TRAK Main Menu.

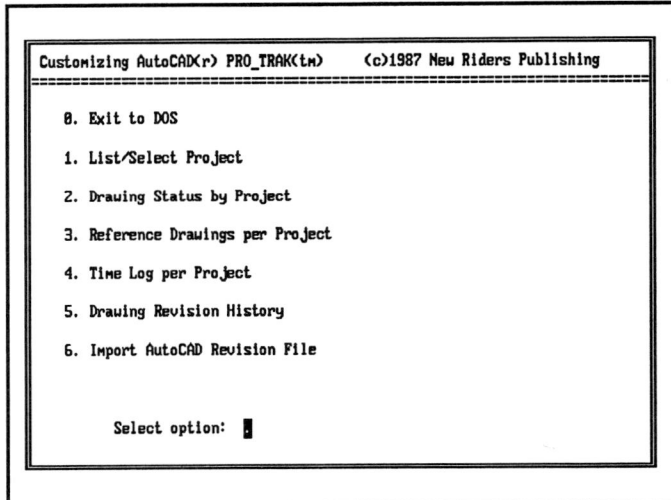

The PRO_TRAK Main Menu

Like AutoLISPs' DEFUN ability to create named subroutines, dBASE has a feature called a PROCEDURE. PROCEDUREs define most of the program's tasks. They are then mended together to form the main program. Using these subroutines lets you concentrate on one small task at any given moment. The procedures keep the program from becoming overwhelmingly complex and permit easier maintainence. In addition to the program's PROCEDUREs, two database structures are defined, CATEMP.DBF and CADATA.DBF. These also are included on the Customizing AutoCAD BONUS DISK.

A PROCEDUREs can be as simple as the header for the program's menu. Take a look at the TITLE PROCEDURE.

```
PROCEDURE TITLE
    CLEAR
    @ 2, 3  SAY "Customizing AutoCAD(r) PRO_TRAK(tm)     (c)1987,1988 New Riders Publishing"
    @ 3, 2  SAY "======================================================================"
RETURN
```

First, the procedure clears the screen with a CLEAR command, then SAY is used to print a message on the screen at "@" location 2,3, for row and column. A RETURN command finishes the procedure to make the program jump back to the calling procedure.

Procedures can be called from within any other procedure, just like defuns in AutoLISP. For instance, TITLE is called by the MAINMENU procedure "DO TITLE." The PRO_TRAK Main Menu is defined in the MAINMENU procedure as:

```
PROCEDURE MAINTITLE
    DO TITLE
    @ 5,  6  SAY "0. Exit to DOS"
    @ 7,  6  SAY "1. List/Select Project"
    @09,  6  SAY "2. Drawing Status by Project"
    @11,  6  SAY "3. Reference Drawings per Project"
    @13,  6  SAY "4. Time Log per Project"
    @15,  6  SAY "5. Drawing Revision History"
    @17,  6  SAY "6. Import AutoCAD Revision File"
    @21,  9  SAY "   Select option: " GET OPTION PICTURE "!"
    @ 1,  1  TO 23, 78    DOUBLE
    READ
RETURN
```

Prompting on the screen is as simple as SAYing something. Two other parts of the program are the database structure, and a program that will massage the data into the form needed.

The PRO_TRAK Database Structure

The database structures for the program were defined using the dBASE program menu CREATE command and the DATABASE file option. A database structure is like a template that formats the database. dBASE presents a screen to help create the files. CADATA.DBF is the filename for the database structure. The next screen shot shows how CADATA is created.

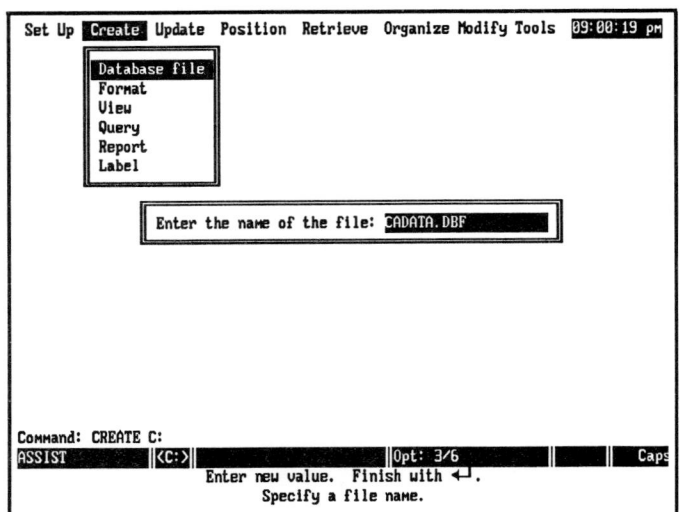

Creating A Database Structure

The program itself is one giant WHILE loop, similar to that used in the DDRAW.LSP function. The DDRAW.LSP function monitored the users key strokes and cursor movements. The dBASE WHILE loop waits for the user to select an option from the program main menu. If the option is one of the supported numbers, the CASE program command filters the input and determines the user's desired task. The CASE program command is similar in function to AutoLISP's COND.

Each CASE option is specific to a task. Let's look at case 6, the most challenging CASE in the program.

Importing the Project Data

CASE 6 is the project data file import option. It prints the screen headers, tells the user which options are available and waits for a selection. Then the program opens two data files for the project reporting, CATEMP.DBF and CADATA.DBF. Since this task is performed by a call to a PROCEDURE, you don't actually see the file names in the example. The call to open a data file is simple: USE CATEMP.

dBASE can open many data files for use at one time. The SELECT command tells dBASE which open file is being addressed. This is similar to file pointers in AutoLISP, but with a friendlier touch. The data files are assigned the SELECT file slot number current at the time the files are first opened. In the program, SELECT 1 is the master data file CADATA. SELECT 2 is the temporary data file CATEMP. The temporary data file is used only while the PROJECT.DAT file is being transferred to dBASE.

CASE 6 flips between the two file slots, 1 and 2, and compares the recently read project name and drawing number with entries found in the master database. If a matching existing record is found in the master database the record is updated. Revisions are added to the record. A maximum of 5 revisions may be recorded for each drawing number.

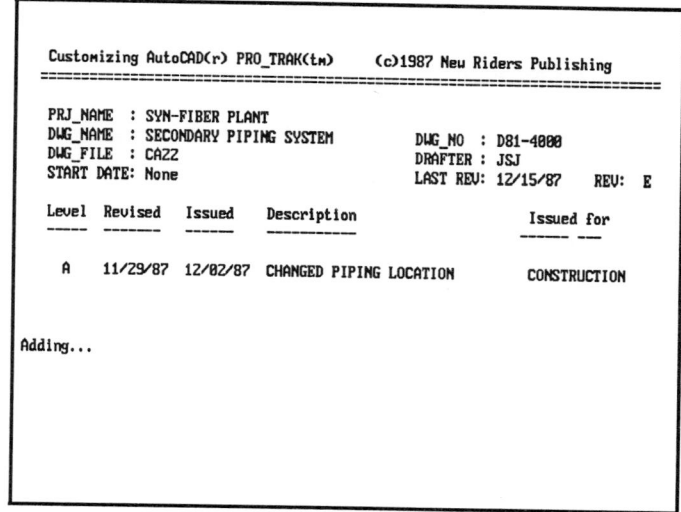

Importing Project Records

The routine branches at this point, one direction to update an existing record, the other direction to add the new record. If the record is being updated, the transfer of data is made from the temporary data file CATEMP to the master data file CADATA. The transfer is done by the TRANSFER procedure. It looks like:

```
PROCEDURE TRANSFER
    SELECT 1
    REPLACE CADATA->PRJ_NAME    WITH CATEMP->PRJ_NAME
    REPLACE CADATA->DWG_NO      WITH CATEMP->DWG_NO
    REPLACE CADATA->DWG_NAME    WITH CATEMP->DWG_NAME
    REPLACE CADATA->DWG_FILE    WITH CATEMP->DWG_FILE
    REPLACE CADATA->DRAFTER     WITH CATEMP->DRAFTER
    REPLACE CADATA->L_DWG_REV   WITH CATEMP->L_DWG_REV
    REPLACE CADATA->L_DATE_REV  WITH CTOD(CATEMP->L_DATE_REV)
    REPLACE CADATA->START_DATE  WITH CTOD(CATEMP->START_DATE)
    REPLACE CADATA->REV_LTR     WITH CATEMP->REV_LTR
    REPLACE CADATA->DESC        WITH CATEMP->DESC
    REPLACE CADATA->DATE_REV    WITH CTOD(CATEMP->DATE_REV)
    REPLACE CADATA->WHO_REV     WITH CATEMP->WHO_REV
    REPLACE CADATA->RES_ISU     WITH CATEMP->RES_ISU
    REPLACE CADATA->DATE_CHK    WITH CTOD(CATEMP->DATE_CHK)
    REPLACE CADATA->WHO_CHK     WITH CATEMP->WHO_CHK
    REPLACE CADATA->REV_TIME    WITH VAL(CATEMP->REV_TIME)
RETURN
```

It behaves the way it looks, slipping and sliding data from one file to the other with the dBASE REPLACE command. It replaces the named data field of one file with the named data field of another file. The field names like "prj_name" don't have to match across the replacement, but the data types do. Strings that are too long to transfer to another field are simply truncated.

New project entries also are transferred, but a blank database record is added to CADATA prior to the transfer. In effect, the program transfers the data into a new record.

At the file import option's end, the program indexes the CADATA file using the project name PRJ_NAME as the key name. After indexing, it ZAPs the temporary file with a procedure. ZAP clears the data records from the .DBF file, but leaves the data structure intact for the next file import.

NOTE. The PROJECT.DAT file is deleted each time the Import Data command is selected. If you want to save the file, make a copy of it to another name prior to running PRO_TRAK.

Reporting Project Data

The PRO_TRAK Main Menu provides selections of report types. You can report project data in a variety of ways depending on your application. You may want to take the dBASE source code and design reports for yourself. The example selection discussed here is for tracking drawing revisions.

```
Customizing AutoCAD(r) PRO_TRAK(tm)     (c)1987 New Riders Publishing
====================================================================

PRJ_NAME  : SYN-FIBER PLANT
DWG_NAME  : SECONDARY PIPING SYSTEM      DWG_NO  : D81-3008
DWG_FILE  : CA22                         DRAFTER : JSJ
START DATE:  / /                         LAST REV: 12/15/87   REV:  E

Level  Revised    Issued    Description              Issued for

         / /       / /

  A    11/29/87  12/02/87  CHANGED PIPING LOCATION   CONSTRUCTION

  B    12/06/87  12/06/87  ADDED CONDUIT             CONSTRUCTION

  C    12/07/87  12/09/87  CHANGED PUMPS 1 & 2       CONSTRUCTION

  D    12/11/87  12/13/87  REDRAWN AND REVISED       CONSTRUCTION

  E    12/15/87  12/15/87  ARCHITECTURAL ADJUSTMENTS BID

Hit [RETURN] when ready.
```

Revision History of Drawing D81-3008

The CASE 5 drawing revision option presents information about individual drawing revision levels, a description of the revisions, date of revision, who did it, purpose of issue, date of reissue and the person who checked the revision. It represents a fair sampling of the type of data recorded for drawing revisions.

```
CASE OPTION = "5"
@15,   3 SAY "Please enter Drawing Number" GET MDWG_NO PICTURE "!!!!!!!!!!"
READ
DO O_FILES
SELECT 1
SET INDEX TO DWG_NO
SEEK MDWG_NO
IF .NOT. EOF()
  DO TITLE_TOP
  DO &TITLE
  DO SET_VAR1
  DO WHILE DWG_NO = MDWG_NO
    DO TITLE_REV
    SKIP
  ENDDO
ELSE
  ? CHR(7)
  @15,   3 SAY "I'm sorry " + TRIM(MDWG_NO) + " NOT  FOUND!!"+SPACE(47)
ENDIF
DO PAUSE
CLOSE DATA
```

CASE 5 begins by opening the project data files and setting the INDEX search name to the drawing number. It looks for the database record of the named drawing using dBASE's SEEK. It formats the screen with the standard menu. The program sets the entries of drawing records to some temporary variables using a different procedure. It then loops through the database comparing subsequent drawing numbers with the original drawing number.

"How does this work?" The program design ordered the database by drawing number. There may be several entries in the database with the same drawing number, but different revision levels. Each subsequent entry with the same drawing number is considered a revision level. The WHILE loop continues displaying the revision data until all revisions are found.

Time Log Reports

You can write a simple routine to add up the drawing creation and revision time. Customizing AutoCAD BONUS DISK owners can inspect the PRO_1.PRG program file to look at CASE 4. It looks at the time field in each revision record and produces a report similar to the one shown in the screen shot Project Log Time.

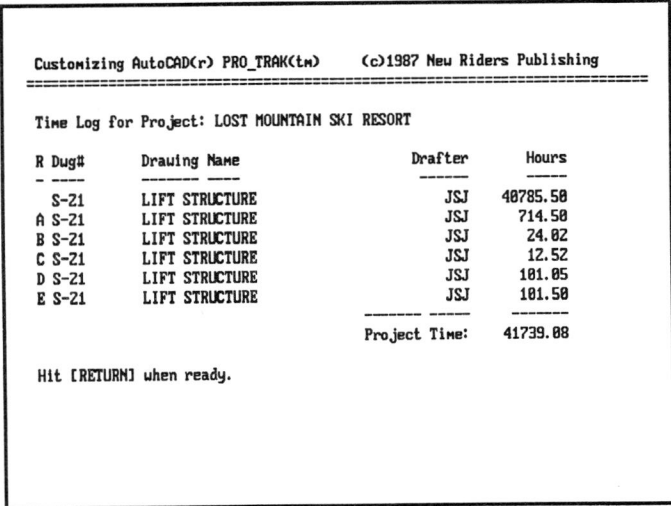

Project Log Time

Instructions for Using PRO_TRAK

Start the program with the PRO_TRAK command from the DOS prompt. It is a single directory program, meaning that all the support, project and data files are expected in the local directory. Run it from the DOS prompt in the C:\CA-ACAD directory by typing PRO_TRAK, or run it from AutoCAD using the [PROTRAK] menu macro.

Using PRO_TRAK

You need the PROJECT.DAT file you created earlier in this chapter in the Generating Revision Records exercise.

Select [PROTRAK]	Starts it.
Select 6	To import the project information.
Select 1	Get a listing of the projects.
Select 2	List the drawing status for a project.
Select 3	Look at a list of reference drawings.
Select 4	Get the time for a project using this options.
Select 5	Look up the revision history for a drawing.
Select 0	Exit to AutoCAD
Command: **QUIT**	And exit to DOS.

 Delete TEST.MNU.

 Rename TEST.MNU to MY22.MNU.

Your screen should show the PRO_TRAK main menu screen when you make your report choices.

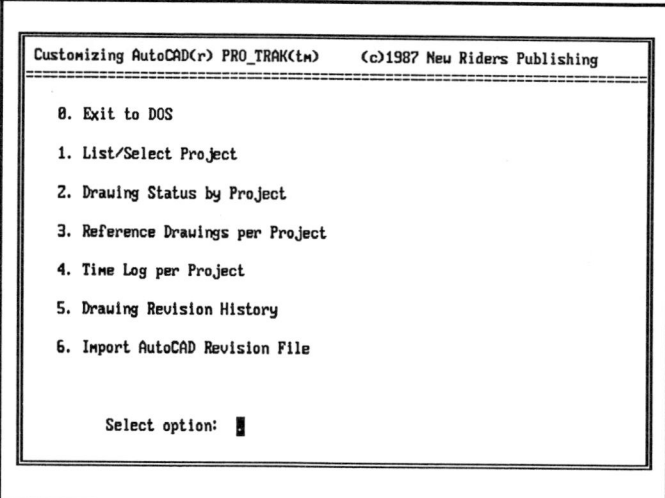

The PRO_TRAK Main Menu

Integrating the Reports Menu

This is your last chapter menu, you no longer need the [EXAMPLES] selection. In the next chapter, we will ask you to integrate the Reports menu and clean up the root page of the CA-MENU.MNU.

Tips and Techniques

Here are some tips and tricks on importing and exporting data.

- ❏ Make sure you keep a backup copy of your template files. The *.TXT extension makes it easy to delete these files.

- ❏ Use the SDF file format for Attext output for Lotus-type data import. Use CDF for dBASE and other general program imports.

- ❏ Combine Attext's entity selection with Release 9's SSGET "X" to filter entities for inclusion in the output file.

- ❏ Make sure you have adequate space between each field in the Lotus import file. The Lotus parser needs at least one space between each field.

- ❏ Make your Lotus macros autoexecute by naming the special 0 (zero) macro in Lotus. Prompt the user immediately for a new file name so the user reassigns the spreadsheet file with the imported data and the initial Lotus spreadsheet file is saved intact.

- ❏ Use a batch file to control directory changes and the execution of your external programs. Use replaceable parameters like %1 in your .BAT file to substitute the current AutoCAD working directory.

- ❏ Plan your dBASE structure first. Create your dBASE program to manage the data files. Test it with a dummy data file. Then write your AutoLISP routine to extract the data from AutoCAD in the form that you need.

23—0 Customizing AutoCAD

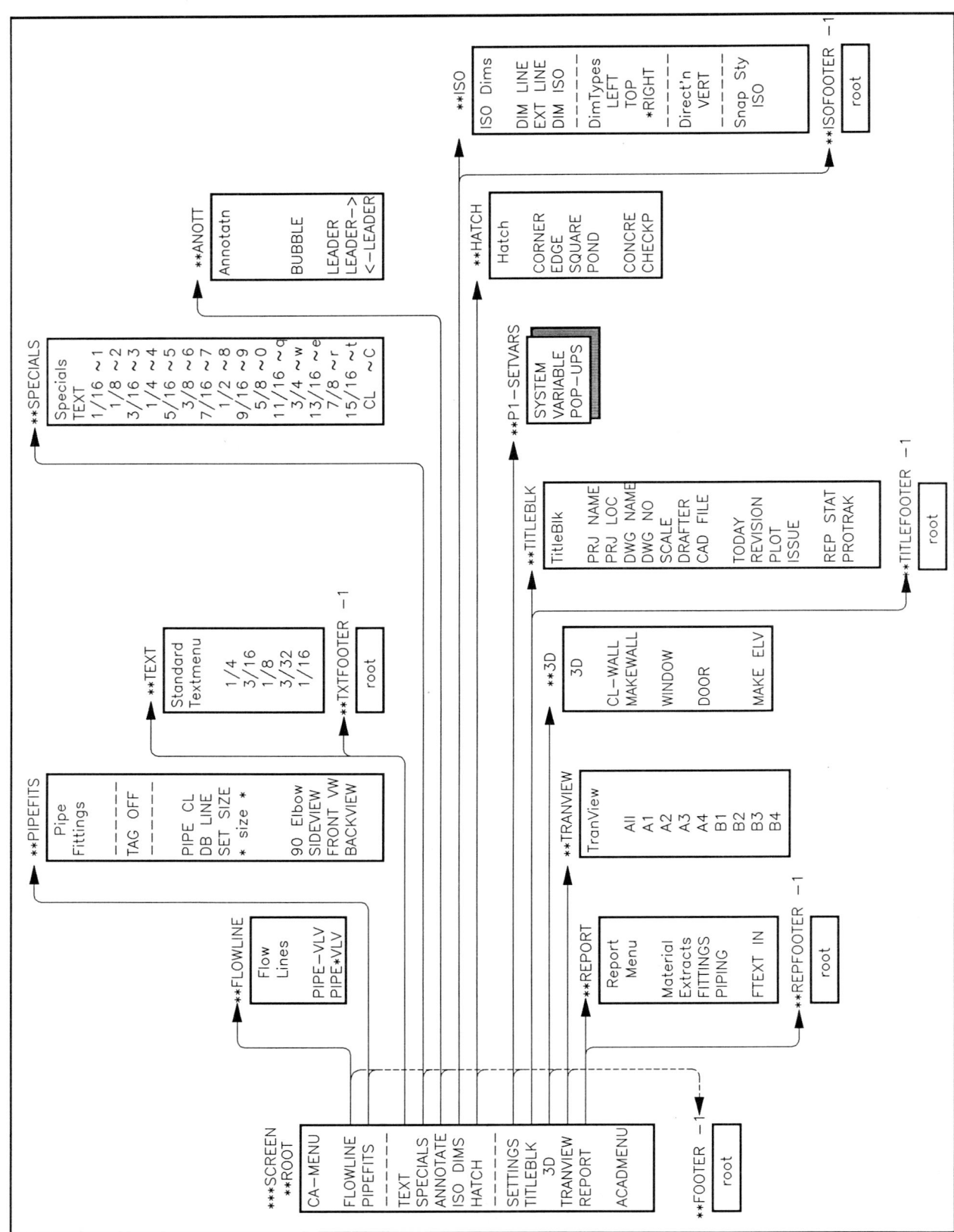

The Complete Customizing AutoCAD Menu

CHAPTER 23

Summing Up with System Controls

The trend in successful CAD installations beyond one or two workstations is to have a system manager customize and maintain users' systems. This chapter discusses system controls and system documentation from a manager's viewpoint.

The Benefits of System Controls

There are several benefits to system controls and system documentation. Four key benefits are:

- System control enhances productivity, giving you consistent output from your applications and a faster startup with new users. The best balance of benefits is gained by giving your users the power to draw efficiently, but restricting their ability to change or alter the system.

- A customized system is valuable property to your company. Trained operators are valuable investments. Simple security measures can help you keep your investments from benefiting your competitors.

- System documentation, program help and application consistency all benefit your company and users by making the system easier and faster to learn.

- Using small predefined AutoLISP commands in the ACAD.LSP file can conserve AutoLISP memory, yet automate transparent loading when the commands are used.

How To Skills Checklist

In this chapter, you will learn how to:

- ❏ Manage and maintain your customized system, including using ACAD.LSP, using setup drawings and resetting your application environment.

- ❏ Redefine AutoCAD commands to restrict use, or add functionality to your application environment.

- ❏ Assign keyboard macros to keys without using macro key definition programs.

- ❏ Make your application independent of DOS directory setups.

- ❏ Create on-the-fly menu items without modifying your menu, using AutoLISP to emulate menu items.

- ❏ Secure your system using software and hardware controls, including using tools available from Autodesk, Inc.
- ❏ Document your application using AutoCAD's Help facility, Slides, external text files, and the book's LSPSTRIP program to provide your users with AutoLISP help about your programs.

Macros, AutoLISP Tools and Programs

AutoLISP TOOLS
C:END is a custom END command.
GPATH is a subroutine to get a path to manage support file calls.
ACAD.KEY is an external text file to redefine keys for use with AutoCAD.
TOGGLE.KEY toggles key definitions off for standard use outside of AutoCAD.
KILL.KEY clears key definitions.

PROGRAMS
C:MFLY is a program to save and repeat a series of commands.
LSPSTRIP.EXE reduces an AutoLISP file's size by stripping comments and white space. It also provides a help display for AutoLISP programming.

MENU and AutoLISP TOOLS available (on request) from Autodesk, Inc.
MNUCRYPT.EXE
PROTECT.EXE
KELV.EXE, also called the Kelvinator

Controlling Your System

Most customized systems incorporate standards for disk directory organization, Layer, Color, Linetype, standard Block, Text, Dimensioning and other organization and graphic settings that are important for maintaining drawing consistency. Approaches for controlling and maintaining these standards range from wishful expectation that users will observe them, to very rigid, fully automated systems where the typical user doesn't know certain commands exist. You will have to judge for yourself how far you want to take control. This chapter adds some techniques to supplement the systems controls already covered in the book. Many of these suggestions assume that your application will pass through a system manager.

Directory and File Control

If you are implementing a management system, all routines, macros, and Wblocked drawings should pass through the system manager BEFORE they are added to the customized menu system, AutoLISP files or Block library. Effective system management starts with simple DOS directory management and extends to implementing directory settings in menus and AutoLISP. For example, the book's CALOAD function accepts a directory name argument to locate AutoLISP functions. For simplicity's sake, the book has kept all the programs, drawings, and configuration and support files in the CA-ACAD directory. We recommend organizing your system into directories for configuration, Block library, AutoLISP library, support, and drawings for individual projects.

If your application is complex or varied, you may have more than one directory type. If this is the case, you will need a way to direct AutoLISP, menu and drawing paths to appropriate directories. You need a way to manage all disk directories whether you just want to adapt your programs to anticipate future disk setups, or to run a Local Area Network, where individual node configurations are not identical.

Create a function, called GPATH. GPATH appends DOS directory names to Block file names, Slide libraries or any other type of supporting file requested by your AutoLISP programs. Add it to your ACAD.LSP. Then, you can SETQ path variables in your ACAD.LSP file to establish the actual path names for individual user's workstation. These simple changes to the ACAD.LSP file eliminate the need to edit all program files to accommodate a new directory structure. If you have the CA DISK, you have the GPATH.LSP file. Edit the ACAD.LSP file, adding GPATH.

GPATH Directory Path Management

`Enter selection:`   Begin a NEW drawing named CA23.

 Merge GPATH.LSP into the ACAD.LSP file and add settings.

 Enter the GPATH function in the ACAD.LSP file.

`Command: ED`   Edit the ACAD.LSP file.

```
;* GPATH takes a \path\ name and a file name. It returns the joined file spec.
(defun gpath (path file)
  (if path
    (strcat path file)         ;if the path is not nil
    file                       ;combine with the file name
  );if                         ;else just return the file name
);defun
```

Add some global directory settings to the top of ACAD.LSP:

```
(setq #dwg "\\CA-ACAD\\")
(setq #lsp "\\CA-ACAD\\")
(setq #blk "\\CA-ACAD\\")
(setq #lsp "\\CA-ACAD\\")
```

Save and exit to AutoCAD.
`Command: (load "acad")`   Reload ACAD.LSP.

`Command: (gpath #dwg "STUFF")`   Test it.
`Lisp returns: "\\CA-ACAD\\STUFF"`

Use the GPATH function with Insert, Wblock, Attext, LOAD, or any other command or function that handles files. For example, use it on the **DWG sheet of the CA-SETUP.MNU file. Use GPATH to locate the border sheet drawings:

```
Command: ED                              Edit the **DWG page of CA-SETUP.MNU:
[11 x 17 ](setq x 17.0) (setq y 11.0) (setq str1 (gpath #dwg "ATTSHT-B"));$S=SCALE
[]
[24 x 36 ](setq x 36.0) (setq y 24.0) (setq str1 (gpath #dwg "ATTSHT-D"));$S=SCALE

Save, exit, and reload the menu.
Select [TEST]
Select [24 x 36]
Select [FULL]
Select [INITIATE]                        Then enter 5 <RETURN>s to default the attributes.
Command: !STR1                           Check it.
Lisp returns: "\\CA-ACAD\\ATTSHT-D"
```

You can keep different project border sheet files in project support directories and set the #dwg variable in ACAD.LSP. Your standard setup menu will apply to any project. You can use a similar approach to toggle between different libraries of support files. Set a user selectable option to switch between libraries of 2-D and 3-D blocks by resetting the #DWG directory name. Use a menu label toggle, like the book's ISODIM menu system, to provide status information.

To keep the book's examples simple, we have not used GPATH in all examples. We recommended GPATH to control directory access for your programs.

DOS SUBST Directory Path Control

An alternative method for controlling paths is available to you in DOS 3.2 (or later DOS versions). The following SUBST command lets you substitute an imaginary (logical) drive name for any drive\path. For example, if drive D: does not exist, you can enter:

```
C:\> SUBST D: C:\CA-ACAD
```

Then, if you preface a file name with D:, like:

```
(setq str1 D:ATTSHT-B")
```

AutoCAD will look in the \CA-ACAD directory of drive C:. You can establish logical drives for any of your support directories, like:

```
D:        for Drawings.
I:        for drawings to Insert.
L:        for AutoLISP.
M:        for Menus.
```

You can set these logical drives to paths in a startup batch file, similar to CA.BAT, for each application or project. If you use logical drives, your menus and AutoLISP programs only need to use the logical drive letters, instead of paths. You may need to include a LASTDRIVE=X: in your CONFIG.SYS file, where X is the highest letter used. See your DOS manual or a DOS reference book for more detail.

Drawing Control

You can use prototype drawings to establish Styles, Layers, Colors, Linetypes, Views, Snap, Grid and other options, and to set system variables, like REGENAUTO, MIRRTEXT, and to set Dim Variables to standard defaults.

You can use menus, like the book's CA-SETUP.MNU, to further enforce and adjust your standard settings. A SET-UP menu makes the user's drawing selection process easy, automatic and consistent. However, when you edit an existing drawing, your system needs to reset a portion of the drawing environment. Here is a sample list of settings that you may have to check or reset.

- Reset global variables like the drawing scale #DWGSC.
- Check path names for blocks #BLK, and programs #LSP.
- Undefined commands.
- Change function key definitions.
- Set layer to correspond to initial screen menu.
- Set AutoLISP constants like angles, etc.
- Load AutoLISP functions.

ACAD.LSP provides a more automatic way of enforcing these and other wake-up default settings. Using AutoLISP routines is similar to using a setup menu's actions, but AutoLISP routines are more automatic and more transparent to the user. The advantage to using ACAD.LSP is that it enforces setup on existing drawings, regardless of the menu loaded. Examples of the settings to include in ACAD.LSP are:

```
(setvar "pdmode" 0 )
(setvar "pdsize" -8)
(setvar "aperture" 6)
(setvar "skpoly" 1)
(setvar "regenmode" 0)
```

When you consider the choice of prototype drawing vs. ACAD.LSP vs. startup menu to control your setup defaults, you need to consider: transparency; how automatic you want the setup; and whether your setup requires input choices. You must use startup menu control for anything requiring actual command execution, since ACAD.LSP can't access the command processor during the initial load. However, you can include AutoLISP GET functions in the ACAD.LSP to prompt for input.

New or Existing Drawing?

You may also need to determine whether a drawing is "new" or an existing drawing that is being re-edited. If the drawing is an existing drawing, you may not want to reset some settings. To flag a new drawing, you can process a system variable in ACAD.LSP. The sequence is:

- Set a system variable to an unrealistic value in the prototype drawing.
- Test the system variable's setting in your ACAD.LSP.
- Adjust whatever settings are needed for a new or existing drawing.
- Reset the system variable to a proper value.

Comparing the TDCREATE and TDUPDAT System Variables may seem like a safer alternative, but it is not safe. An existing drawing created by Wblock shows identical dates for both settings!

Reset Controls

Automatic resets help you keep a consistent environment for macros and functions. Your application menus can control the drawing environment by performing functions prior to loading a new screen menu. This point of entry control has been used in several ways throughout the book. However, users may still find some way to get entangled in your system. Professional software developers spend much of their time testing and making programs boiler plate tough.

A good tool is a RESET command to assist the user when a problem occurs. Put the RESET command in a far-off corner of the tablet, or defun it in the ACAD.LSP file. Your RESET might perform the following functions:

- Reload the menu to get back to the **ROOT page.
- Reset all the system variables: DIMs, Osnap, Regen, etc.
- Set the layer, text style, etc.

You also can incorporate reset controls in redefined AutoCAD commands.

Command Redefinition

A well organized system can control itself through menus and AutoLISP commands. However, since a user can type any C:DEFUN or standard AutoCAD command at the keyboard, as well as call it by menu, the potential for users to disorganize the drawing is difficult to control. If you are using Release 9, your most effective control is to UNDEFINE troublesome commands.

Release 9 includes the UNDEFINE and REDEFINE commands. When a command is undefined, AutoCAD does not recognize it. If a C: AutoLISP command is defined with the same name, it is executed instead. For example, to impede the creation of non-standard styles, you could Undefine the Style command.

Using Undefine and Redefine Commands

Continue in drawing CA23.

Command: **UNDEFINE**
Command name: **STYLE**

Only for Release 9 and later versions.

Command: **STYLE**
Unknown command. Type ? for list of commands.

Now try to use Style, and you get:

Use DEFUN to create a prettier message:

Command: **(defun undef () (prompt"\nCommand reserved for System Manager!"(princ)))**
Lisp issues: UNDEF

Command: **(defun C:STYLE () (undef))**
Lisp issues: C:STYLE

Command: **STYLE**

Try it again and you get:

Command reserved for System Manager!

Defining an UNDEF function and calling it from the C:STYLE function lets any number of Undefined commands issue the standard message. It reduces the overhead of a prompt in each replacement "command."

The following example shows you how to ensure that a drawing is ENDed in the format that you want. By replacing AutoCAD's End command, you can eliminate having your users accidently END the drawing, for example, when they meant to just use an ENDpoint Osnap. They can still type END, or you can display END on your screen and tablet menus.

  Copy UNDEFINE.LSP to TEST.LSP.

  Start a new TEST.LSP file.

Select **[EDIT-LSP]**   Edit TEST.LSP, and enter:

```
;* UNDEFINE.LSP
;* C:END replaces the AutoCAD END command with a customized version that
;* resets the enviroment, and optionally makes a slide of the drawing.

(command "UNDEFINE" "END")                    ;turns off AutoCAD's end command.
(defun C:END ()                               ;define our own.
  (if (= "Y" (ukword 1 "Y N" "Are you sure? (Y/N)" "Y"))
    (progn
      (command "SETVAR" "EXPERT" "3" "LAYER" "ON" "*" ""
        "FILL" "ON" "QTEXT" "OFF" "ATTDISP" "N" "ZOOM"
      );command
      (if (= "Y" (ukword 1 "Y N" "Make a record slide? (Y/N)" "N"))
        (command "ALL" "MSLIDE" "" ".END")    ;use AutoCAD's END
        (progn
          (prompt "\nDON'T wait for the regen! Hit  ^C  to continue END...\n")
```

```
            (command "ALL" ".END")
        );progn
      );if
    );progn
  (princ)
);defun
```

```
Command: END                                   Try our command now!
Are you sure? (Y/N) <Y>:                        Take default.
Make a record slide? (Y/N) <N>: Y               Make the slide.
DON'T wait for the regen! Hit   ^C  to continue END...   Cancel with a <^C>.
```

 Delete TEST.LSP.

 Rename TEST.LSP to UNDEFINE.LSP.

The best place for global system control through command redefinition should be the ACAD.LSP file. But as you have seen that doesn't work because AutoCAD commands cannot execute while the drawing is being loaded. This is likely to change. But for now your best choice is adding command redefinition to other global system controls as part of a forced initial menu pick, similar to the book's [INITIATE] macro.

NOTE. Unfortunately, PLOT and PRPLOT currently wipe out any Undefined commands that you have set. They return to normal after plotting. You should disable plotting, or redefine the PLOT and PRPLOT commands if you want to depend on Undefine control. You could create a new PLOT command that gets and filters the plot input, then runs a script to plot. The last item in the script should load an AutoLISP file to reset your Undefined commands.

Keyboard Control

Some system managers actually forbid the typed entry of commands. Rather than forbid keyboard entry, we recommend making menu alternatives more attractive so that your users prefer them.

You can control the keyboard by redefining keys with ANSI.SYS, or macro programs, such as Prokey, Keyworks or Superkey, or abbreviation programs, such as JOT! or PRD+. These programs can provide quick command entry with predefined keystrokes for your users.

Memory resident abbreviation programs let you assign any sequence of keystrokes to a two to eight character abbreviation. When you hit a punctuation key, space or <RETURN>, the program looks back at the preceding characters. If they match a definition, it replaces them with the keystroke sequence.

Memory resident macro programs let you redefine <CTRL> or <ALT> key combinations, as well as function keys. If you define <ALT-C> to be CIRCLE<RETURN>, then whenever you hit <ALT-C> the program will play back the sequence CIRCLE<RETURN>.

You also can redefine keys using the ANSI.SYS driver, as shown in the following exercise. You can learn more about ANSI.SYS key redefinition in a good DOS book.

ANSI.SYS Key Redefinition

The advantage to using ANSI key redefinition vs. memory resident programs is that ANSI key redefinition does not reduce RAM available to AutoCAD. You can define any key to be any other key or string. You can change key definitions through AutoCAD or DOS. This exercise uses DOS by creating a file of ANSI redefinition sequences and directing the file to the display screen. DOS has to "see" the redefinitions for them to take effect.

Here is an example showing how to use <ALT> key combinations to define commonly used AutoCAD commands.

KEY	EXTENDED ASCII CODE	AutoCAD COMMAND
<ALT-E>	0;18	Erase Auto
<ALT-T>	0;20	Type UNDEF.KEY (toggles defs Off)
<ALT-S>	0;31	Save to default name
<ALT-D>	0;32	Dtext
<ALT-F>	0;33	Fillet
<ALT-K>	0;37	Type KILL.KEY (kills key defs)
<ALT-L>	0;38	Line
<ALT-Z>	0;44	Zoom W
<ALT-X>	0;45	Zoom P
<ALT-C>	0;46	Copy Auto
<ALT-M>	0;50	Move Auto

If you have the CA DISK, you have the files to examine. Otherwise, make files to redefine the keys. If you do not have Release 9, omit the "Auto" from the strings.

NOTE. The "^[" at the start of each line represents the single ASCII character <ESCAPE>. Text editors differ in their procedure for entering <ESCAPE>. Do whatever your text editor requires.

Using ANSI.SYS Key Redefinition

Enter selection: Edit the EXISTING drawing named CA23.

 Examine your ACAD.KEY, KILL.KEY and TOGGLE.KEY files.

 Create the files:

Command: **ED**   Start a new file ACAD.KEY and enter:

```
            *** THIS REDEFINES several ALT KEYS to AutoCAD COMMANDS ***
^[[0;18;3;3;3;"Erase Auto";13p
^[[0;20;0;20p
^[[0;20;3;3;3;"Type TOGGLE.KEY";13p
^[[0;31;3;3;3;"Save ";13p
^[[0;32;3;3;3;"Dtext";13p
^[[0;33;3;3;3;"Fillet";13p
^[[0;37;3;3;3;"Type KILL.KEY";13p
^[[0;38;3;3;3;"Line";13p
^[[0;44;3;3;3;"Zoom W";13p
^[[0;45;3;3;3;"Zoom P";13p
^[[0;46;3;3;3;"Copy Auto";13p
^[[0;50;3;3;3;"Move Auto";13p
            ***      HIT  <ALT-T>  TO TOGGLE DEFINITIONS ON/OFF     ***
            ***         HIT  <ALT-K>  TO KILL DEFINITIONS           ***
```

Save it. Before you test it, make a file to clear the definitions, resetting original key definitions.

Start a new file KILL.KEY and enter:

```
            ***      THIS UNDEFINES AutoCAD COMMAND ALT KEYS        ***
^[[0;18;0;18p
^[[0;20;0;20p
^[[0;31;0;31p
^[[0;32;0;32p
^[[0;33;0;33p
^[[0;37;0;37p
^[[0;38;0;38p
^[[0;44;0;44p
^[[0;45;0;45p
^[[0;46;0;46p
^[[0;50;0;50p
   *** Enter:  TYPE ACAD.KEY   to Redefine ALT KEYS  to AutoCAD COMMANDS ***
```

Save it. Make another file toggle the definitions off.

Start a new file TOGGLE.KEY and enter:

```
            ***    THIS TOGGLES AutoCAD ALT KEY DEFINITIONS OFF     ***
^[[0;18;0;18p
^[[0;20;0;20p
^[[0;20;"Type ACAD.KEY";13p
^[[0;31;0;31p
^[[0;32;0;32p
^[[0;33;0;33p
^[[0;37;0;37p
^[[0;38;0;38p
^[[0;44;0;44p
^[[0;45;0;45p
^[[0;46;0;46p
^[[0;50;0;50p
            ***       HIT  <ALT-T>  TO TOGGLE DEFINITIONS ON        ***
```

Save and exit to AutoCAD.

The ACAD.KEY file redefines the <ALT> keys. For example, in the first:

`^[[0;18;3;3;3;"Erase Auto";13p`

The ^[represents <ESCAPE>. DOS recognizes what follows as an ANSI sequence by the second [after the ^[.

Semicolons are used to separate the following ASCII codes. The 0 indicates that the following code is an extended ASCII code, above ASCII 127. 18 is the extended code for <ALT-E>. 3 is the ASCII code for <^C>. The example uses three 3s for three <^C>s. Strings are put in quotes, like "Erase Auto." 13 is the ASCII code for <RETURN>. The closing p code identifies the sequence as an ANSI key sequence.

The KILL.KEY file redefines each key code. The TOGGLE.KEY file is the same, but resets <ALT-T> to reload ACAD.KEY. In ACAD.KEY, <ALT-T> is defined to load TOGGLE.KEY, so it toggles on/off. The ^[[0;20;0;20p line above each <ALT-T> line in ACAD.KEY clears the definition before redefining it. If a toggle isn't cleared, you may get garbage when toggling.

If you have problems with the definitions at the end of your file, you have exceeded the available ANSI storage space and you must reduce definitions. Use a memory resident macro program if you need many key definitions.

Let's load ACAD.KEY.

You should be in the CA23 drawing.

```
Command: TYPE
File to list: ACAD.KEY            You should see only:

    *** THIS REDEFINES several ALT KEYS to AutoCAD COMMANDS ***

                          And several blank lines, then:

        ***     HIT ALT-T  TO TOGGLE DEFINITIONS ON/OFF     ***
        ***         HIT ALT-K  TO KILL DEFINITIONS          ***
```

If you saw anything else, like part of the ANSI sequence codes themselves, it didn't work. The screen must "see" them, but you shouldn't. Now, test it.

```
Command: <ALT-M>                  Hit <ALT-M> and you get:
Command: Move
Select objects: Auto

Try each of the others, saving <ALT-T> and <ALT-K> for last.

Command: <ALT-T>                  Types TOGGLE.KEY which gives you:

    ***   THIS TOGGLES AutoCAD ALT KEY DEFINITIONS OFF   ***

                          And several blank lines, then:
```

```
                    ***     HIT  ALT-T  TO TOGGLE DEFINITIONS ON         ***
Command: <ALT-T>                            Again, types ACAD.KEY to toggle them on.
Command: <ALT-K>                            Types KILL.KEY which gives you:
           ***      THIS UNDEFINES AutoCAD COMMAND ALT KEYS              ***
                                            And several blank lines, then:
*** Enter:  TYPE ACAD.KEY    to Redefine ALT KEYS   to AutoCAD COMMANDS ***
Command: <ALT-T>                            Check to make sure they're gone.
```

ANSI key redefinition will affect anything you do when you SHELL out. It will affect your system if you use a multitasking program, such as SOFTWARE CAROUSEL. For example we use Norton's Editor, which uses <ALT-K> and others extensively. Having a handy way of toggling the keys off avoids conflict. We recommend adding a "^C^C^CTYPE TOGGLE.KEY" to the start of menu macros that start up the editor. Then, add a "^C^C^CTYPE ACAD.KEY" to the end of the macros to reinstall the definitions to make it automatic.

You can make your setup batch file automatically load the key definitions before AutoCAD, and restore the keys when leaving AutoCAD. Place it just before and after the line that calls AutoCAD.

You also can put key definitions in your ACAD.LSP file. We like to redefine the <ESCAPE> key to a <^C> followed by an <ESCAPE>. If you wish to do this, insert the following lines into your ACAD.LSP file.

```
(textscr)
(write-line (strcat (chr 27) "[27;3;27p"))
```

This <^C> is harmless to virtually all programs, and doesn't bother you at the DOS command line. But you should test it to see if you can use it with the other programs you use.

Limited Power to the User

It pays to give users limited controlled access to customization on their own. If you don't wish to let users create their own menu macros or functions, the following MFLY function is a limited compromise that lets users create a repeating macro command on-the-fly. You can recall the macro any time during the drawing session with the FLY command. MFLY defuns the FLY command. You can use the Release 9 "\" pause in AutoLISP to run the command interactively. This is shown below.

Making a Macro-on-the-Fly Command MFLY

  Copy MFLY.LSP to TEST.LSP.

  Start a new TEST.LSP file.

```
;* C:MFLY prompts for input and creates a repeatable LISP macro on the fly.
;* To run the macro it creates, type FLY.
;* With Release 9 or later, it interactively runs commands as it makes C:FLY.
;*
(defun C:MFLY ()
  (setq inlist '(command))        ;Initialize string, building an expression...
                                  ;...to later evaluate.
  (prompt
"\nSpaces are NOT interpreted as RETURNS! Input one command or option per line."
  )
  (prompt "\nType \"\\\" for pause, \"EXIT\" to terminate.")
  (while (/= "EXIT"
                                                              ;get command input
      (strcase (setq input (getstring T "\nInput, \"\\\", or \"EXIT\": "))))
      (setq inlist (append inlist (list input)))
      (command input)    ;*(Disable for V2.6 / enable for R9)   Run command interactively
  );while
  (defun C:FLY () (eval inlist) (princ)) ;make a command out of it to call on-the-fly
;*  (C:FLY)                ;*(Enable this for V2.6 / disable for R9)  Call the function to run
  (princ)
);defun
;*
```

Save, Exit and reenter AutoCAD. TEST.LSP should load.

```
Command: MFLY                             Test it.
Spaces are NOT interpreted as RETURNS! Input one command or option per line.
Type "\" for pause, "EXIT" to terminate.
Input, "\", or "EXIT": CIRCLE
CIRCLE 3P/2P/TTR/<Center point>:
Input, "\", or "EXIT": \   Tell it to pause for center point.
                                          It pauses. Pick point.
Diameter/<Radius>:
Input, "\", or "EXIT": 5                  Enter a radius.
5                                         It passes radius to command.
Command:
Input, "\", or "EXIT": EXIT

Command: FLY                              Test it.
CIRCLE 3P/2P/TTR/<Center point>:          Pick a point.
Diameter/<Radius>: 5                      It draws another circle.
```

  Delete TEST.LSP.

  Rename TEST.LSP to MFLY.LSP.

Encryption and Security

File security and control are on-going management issues. Here we offer a few recommendations for you to consider in your applications. There are DOS and other file management utilities that you can use to make your files "read only," to hide and unhide files and to keep files in hidden directories. You can control these utilities with your .BAT files, making them transparent to your user if ECHO is off. Look at the DOS ATTRIB command or Norton Utilities FA (File Attribute) if you want to hide files or make them read-only.

You can control where files are stored by SUBSTituting drives for paths, or coding path variables in your functions and macros. Eliminating user input from paths reduces errors and misplaced files. Users can deposit user defined Wblocked symbols,.DWG files and on-the-fly menu macros in assigned user directories for the system manager to examine and consider for inclusion in the system.

Customization Security

You don't want your customization effort to walk out the door and into your competitors' offices. In the days of manual drafting, it was easy to tell one person's work from another's. On disk, everybody's work looks the same. It helps to take precautions to protect your menu and AutoLISP files.

Menus are easy to protect. Keep the .MNU files under lock and key, and distribute only the .MNX files.

You can encrypt AutoLISP by using a utility program called PROTECT.EXE, available upon request from Autodesk, Inc. The PROTECT program takes three command line arguments, the name of the source code file, an encryption key and the output file name. An example is:

```
C:\CA-ACAD>PROTECT source key target
```

Since AutoLISP loads .LSP files by default, rename your master source .LSP files as .LSC (for LISP Source Code) before encryption. Do not confuse the file extensions. There is no way to decrypt files again once they are encrypted. This is one drawback to encryption. The only thing that you can read in an encrypted file is the statement: AutoCAD PROTECTED LISP file.

Unfortunately, software hackers have found a way to break the encryption codes. Another Autodesk utility called the KELVINATOR strips extra whitespace from the code and systematically substitutes random nonsense names for variables. Although not true encryption, it turns code into gibberish that still works. Kelvination followed by encryption offers the best security. The KELV utility program is run from the DOS prompt. It redirects its output to a file name with the DOS > redirection code. Kelvinating the ATEXT.LSP file looks like:

```
C:\>KELV ATEXT.LSC > ATEXT.LSP

The Kelvinator (September 2nd, 1987)
Release 2a:  The Kelvinator Strikes Back
(C) Copyright 1986, 1987  Autodesk, Inc.
(DEFUN C:ATEXT(/ Qj Q@ QQ Ql Q& Ql Q# Q0 Q$ Q0 Ql Q% Q?j Qjj
Q@j) (SETVAR"HIGHLIGHT"0) (SETVAR"CMDECHO"0) (SETQ
```

```
Q@(GETPOINT"\nPick radius point: ")Qj(GETPOINT"\nPick middle
point of text: "Q@)QQ(GETSTRING"\nText: "QQj)Ql(DISTANCE Q@
Qj)Q&(STRLEN QQ)Ql(CDR(ASSOC
41(TBLSEARCH"STYLE"(GETVAR"TEXTSTYLE")))))(IF(=(SETQ Q#(CDR(ASSOC
40(TBLSEARCH"STYLE"(GETVAR"TEXTSTYLE")))))0)(SETQ
Q#(GETDIST"\nText height: ")Q0(quote (COMMAND"TEXT""C"Qjj Q# Q?j
Q@j)))(SETQ Q0(quote (COMMAND"TEXT""C"Qjj Q?j Q@j))))(SETQ
Q%(STRCASE(GETSTRING"\nIs base of text towards radius point <Y>:
")))(IF(OR(= Q%"")(= Q%"Y"))(SETQ Ql(- Ql(/ Q# 2)))(SETQ Ql(+
Ql(/ Q# 2))))(SETQ Q$(* Q& Ql Q#)Ql(/ Q$ Q&)Q0(/ Q$
Ql)Q|(-(+(ANGLE Q@ Qj)(/ Q0 2))(/ Ql Ql 2))Qlj 1)(REPEAT
Q&(IF(OR(= Q%"")(= Q%"Y"))(SETQ Q?j(ANGTOS(- Q|(/ Q&j 2)))Qlj
Ql)(SETQ Q?j(ANGTOS(- Q|(* Q&j 1.5))0)Qlj(-(1+ Q&)Qlj)))(SETQ
Qjj(POLAR Q@ Q| Ql))(SETQ Q@j(SUBSTR QQ Qlj 1))(EVAL Q0)(SETQ
Qlj(1+ Qlj))(SETQ Q|(- Q|(/ Ql
Ql))))(SETVAR"HIGHLIGHT"1)(SETVAR"CMDECHO"1)(PRINC))(PRINC)
```

A bit confusing? You might figure it out by the prompts in the program, but anyone not familiar with your code will have a difficult time.

Although this level of security is not required for many users, Kelvination has a beneficial side effect. It reduces AutoLISP file size resulting in faster loading, smaller functions.

Cleaning .LSP Files with LSPSTRIP

Many users just want to strip white space and comments out of their AutoLISP programs. This chapter's LSPSTRIP program takes out all comments and white space from files. It does not encrypt the files. LSPSTRIP is written in Basic. The source and compiled code is on the CA DISK. LSPSTRIP can provide most of the efficiency advantages of the Kelvinator without the problems associated with cryptic variable and function names. Call LSPSTRIP from the DOS command line, or through SHELL.

Use it on the GPATH.LSP file. Stripped of comments and white space, GPATH reduces from 268 bytes to a mere 68 bytes.

Using LSPSTRIP.EXE

Command: **DUP**	Copy GPATH.LSP to TEST.LSP.
Command: **SHELL** DOS Command: C:\CA-ACAD>>**LSPSTRIP TEST.LSP /S**	Strip it. <RETURN> The /S option strips the file.
C:\CA-ACAD>>**TYPE TEST.LSP**	Views the resulting file.

`(defun gpath(path file)`
`(if path`
`(strcat path file)`
`file`
`)`
`)`
	See, no ;comments or whitespace.

LSPSTRIP creates a backup file with a .LSB extension to protect your original source code. If you try to strip the same file twice and LSPSTRIP finds an existing .LSB file, it displays "File has been stripped" and refuses to strip it to avoid overwriting the source code in the .LSB file.

LSPSTRIP also has a help feature that will display specially noted comments of an AutoLISP file. Throughout the book, we've used a leading ;* to code header comments and other key comments in .LSP files. If you use LSPSTRIP without the /S option, it displays all comment lines starting with the ;* code:

```
C:\CA-ACAD>>LSPSTRIP ANSILIB.LSP

;WARNING: ANSILIB.LSP uses ; semicolons in strings - Do Not run LSPSTRIP/S

ANSILIB.LSP library of functions consist of the most common
tasks required for screen formatting using the ANSI escape calls

C:\CA-ACAD>> EXIT                       Return to AutoCAD.
```

There is one thing LSPSTRIP can't handle. It can't strip .LSP files which have semicolons imbedded in strings, like the ANSILIB.LSP file. LSPSTRIP can't distinguish between these semicolons, and comments. To prevent accidental stripping of such files, LSPSTRIP looks at the top line of the .LSP file. If you begin the first line with a ;*; instead of ;* then LSPSTRIP will refuse to strip the file. It will display the top line, which should identify the file and give a warning. The top line of ANSILIB is:

```
;*;WARNING: ANSILIB.LSP contains ; semicolons in strings. Do Not LSPSTRIP/S
```

Documenting and Presenting an Application

You can always enhance your program's effectiveness by providing easily accessible and clear documentation.

You've seen examples of AutoLISP documentation throughout the book. Here, let's recall a few examples.

```
**DWG-DOC
[This selection establishes the drawing size and saves it as variables.]
[The final limits will be determined by scale factor in the next selection.]
**DWG
[ Select ]
[ Drawing]

**ALT-BUT 2
[PTFILTER alt. def. of 2 button]$S=PTFILTERS $B=BUTTONS
```

**DWG-DOC, which is from the CA-SETUP.MNU, is not a "real" menu page. **DWG-DOC is never called. It is used to store some comments about the following **DWG page. The two lines of comments are "stored" in [brackets]. They are ignored by AutoCAD when it compiles the menu. They do not increase the .MNX menu size.

Although the screen displays only the first 8 characters of a menu [label], you can make the actual label much longer. (We've tested up to 256 characters). Like the [PTFILTER] item above (a button item), you can use long labels for documenting screen menu items. AutoCAD ignores all but the first eight characters, so the compiled menu size is unaffected.

The first rule of good program presentation is not to confuse your users with misleading or hard to remember command names. AutoCAD is much better than most CAD programs in using clear names for command functions. Try to use familiar terms in your customization. Make it as much like your design and drafting application as you can, taking terms from familiar templates, operations and tools.

The second rule is to keep your users informed.

- Use menu labels.
- Give clear concise prompts.
- Display program status using GRTEXT and menu label toggles.
- Provide help screens and external help files.

The third rule is to make your prompts consistent in appearance. Placing similar commands in similar positions on all your applicable screen menus makes a system much easier and friendlier to use.

Provide on screen help to your users. The HELP command gives quick access to ACAD.HLP. As the book has shown, ACAD.HLP is an ASCII text file and you can modify or replace it to give your users help with your customized features. You can even use ANSI format codes in the file to change the color, boldness or appearance of help screens. A simple <ESC> code sequence at the beginning and end of the help page will help format your text, setting it off for clear readablilty.

You can provide a help screen for each screen menu page. Put the call as a macro behind the top label of the screen menu. For example, the **PIPEFITS screen page has the top label [PIPE]. Add the call to help just behind the label:

```
[   Pipe   ]'? PIPEFITS
```

Don't forget to use   Slides for help. Slides can show how a command works. Again, you can use the top menu label for this. For example, you could add a slide to the book's ISODIM commands to show the isoplanes, arrow and text styles, dimension and extension lines. You could add the macro to the top [ISO Dims] label.

```
[ISO Dims]^C^C^CVSLIDE ISODIM
```

If you store your help slides in a Slide Library filename.SLB file, use the form:

```
[ISO Dims]^C^C^CVSLIDE filename(ISODIM)
```

The help slide will look like the following screen shot.

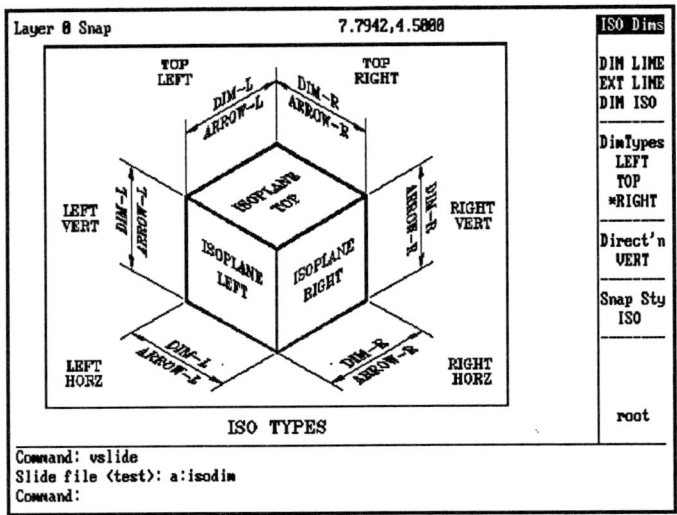

Help Slide

Finalizing ACAD.LSP

Let's look at steps that you want to consider in finalizing your ACAD.LSP file for use. One refinement is command function loading.

The next exercise shows how to use CALOAD to automate the loading of C:functions. The exercise uses C:MFLY as the example command.

You need to edit the ACAD.LSP file, turning the *ERROR* and CALOAD functions on. Recall that you disabled these functions, using a DEBUG flag to preserve the development environment during the course of the book.

Self Loading Functions

Command: **ED** Edit ACAD.LSP.

Make sure the setting (setq #lsp "\\CA-ACAD\\") is in it.

Find and change the (setq DEBUG... line to be:

(setq DEBUG **nil**)   ;allows global control of error and program

Add the following :

```
;********************* SELF-LOADING FUNCTIONS *********************
;*Temporarily defines C:MFLY to self-load the first time it is used.
(defun C:MFLY ()
   (caload #lsp "mfly")   ;feed CALOAD path and filename
   (princ)
)
```

Save ACAD.LSP and exit to AutoCAD.
Command: **ED**   Edit MFLY.LSP and add, at the end of the file:

(C:MFLY)   ;executes command when self-loaded

Save and exit to AutoCAD.

Command: **(load "acad")**   Reload ACAD.LSP.

Command: **MFLY**   Test it.

Initial load. Please wait...   And it loads MFLY.LSP, then executes:
Spaces are NOT interpreted as RETURNS!
Input one command or option per line.
Type "\" for pause, "EXIT" to terminate.
Input, "\", or "EXIT": **EXIT**

Command: **ED**   Edit ACAD.LSP to leave CALOAD and the custom *ERROR* on.

Make sure the (setq DEBUG... line near the top of the file is:

(setq DEBUG nil)   ;allows global control of error and program

Add a leading semicolon to the error line near the end of the file:

;(setq *error* #err) ;*remove leading ; if any to enable the standard *ERROR* function.

Save ACAD.LSP and exit to AutoCAD.

Finalizing Menus

One final step in any application is pulling together your menus. It's time to pull together the final CA-MENU.MNU file. You can integrate the menu with the Standard AutoCAD Template tablet menu.

If you have integrated your menu with the Integrating the Menu section at the end of each chapter, your CA-MENU is complete except for the chapter 22 menu, CA22MENU. The final CA-MENU is shown at the start of this chapter.

Let's integrate the CA22 menu. If you skipped integration earlier, you can skip this section and just read it. If you have the CA DISK, you have the completed menu under the file name CAMASTER.MNU.

Finalizing the CA-MENU

  Copy CAMASTER.MNU to CA-MENU.MNU and examine it as you edit.

  Edit your CA-MENU.

Command: **ED**   Edit CA-MENU.MNU:

Append the **EXAMPLES and **REPFOOTER pages of your MY22.MNU to the end of CA-MENU.

Change the newly appended **EXAMPLES page name to **REPORT:

**REPORT
[Report]
[Menu   ]

Change [EXAMPLES] on the **ROOT page to: [REPORT], with the code shown below.
Replace the [EDIT-LSP] and [EDIT-MNU] items with [] blanks.
The final **ROOT page should be:

```
***SCREEN
**ROOT
[CA-MENU ]
[]
[FLOWLINE]^C^C^CLAYER S OBJ01;;$S=FOOTER $S=FLOWLINE +
(if #dwgsc nil (setq #dwgsc (getvar "DIMSCALE")))
[PIPEFITS]^C^C^CLAYER S OBJ01;;(setq #tag nil) $S=FOOTER $S=PIPEFITS +
(if #arglist (grtext 10 (strcat "* " (rtos (car #arglist) 4 2))))
[--------]
[TEXT    ]$S=FOOTER $S=TEXT
[SPECIALS]$S=FOOTER $S=SPECIALS
[ANNOTATE]$S=FOOTER $S=ANNOT LAYER S ANN02;;+
(if #dwgsc nil (setq #dwgsc (getvar "DIMSCALE")))
[ISO DIMS]^C^C^C(caload "/CA-ACAD/" "ISODIM") SNAP S I;;+
DIM DIMSE1 ON DIMSE2 ON DIMBLK ARROW-R STYLE DIM-R EXIT +
ISOPLANE RIGHT $S=ISOFOOTER $S=ISO $T1=TISO
[HATCH   ]^C^C^CLAYER M HAT01 C 3 ;$S=FOOTER $S=HATCH
[--------]
[SETTINGS]$P1=P1-SETVARS $P1=*
[TITLEBLK]^C^C^CSETVAR TEXTEVAL 1 LAYER S BORDER;;+
(caload "/CA-ACAD/" "TIMELIB") $S=TITLEFOOTER $S=TITLEBLK
[   3D   ]^C^C^C$S=3D $S=FOOTER LAYER S OBJ02;;
[TRANVIEW]$S=FOOTER $S=TRANVIEW +
(if #dwgsc nil (setq #dwgsc (getvar "DIMSCALE")))
[REPORT  ]$S=REPFOOTER $S=REPORT
[]
[]
[]
```

Save, exit and return to AutoCAD.

Command: **MENU**   Load the CA-MENU.MNU. Try it.

You can suppress the screen "echoing" of prompts and messages by menu macros. The AutoCAD system variable MENUECHO controls the echo. It is similar to AutoLISP's CMDECHO system variable. You can set MENUECHO to 0 (normal), 1, 2, 3, 6, or 7 to control the echo. Use the CA-SETUP.MNU item [INITIATE] as your example. [INITIATE] scrolls several lines on the screen. Try cleaning it up.

Finalizing the CA-SETUP.MNU

Command: **ED**   Edit your CA-SETUP.MNU file.

Add these two MENUECHO settings at the start and end of the [INITIATE] macro:

```
[]
[INITIATE]^c^c^cSETVAR MENUECHO 3;+
GRID !#dwgsc ^GSNAP (/ #dwgsc 8) AXIS (/ #dwgsc 2) AXIS OFF;+
SETVAR DIMSCALE !#dwgsc LIMITS 0,0 (list (* x #dwgsc) (* y #dwgsc));+
REGENAUTO ON ZOOM W 0,0 (getvar "LIMMAX") ZOOM .75X REGENAUTO OFF;+
STYLE STANDARD SIMPLEX (* 0.125 #dwgsc) ;;;;;+
STYLE STD1-16 SIMPLEX (* 0.0625 #dwgsc) ;;;;;+
STYLE STD3-32 SIMPLEX (* 0.09375 #dwgsc) ;;;;;+
STYLE STD3-16 SIMPLEX (* 0.1875 #dwgsc) ;;;;;+
STYLE STD1-4 SIMPLEX (* 0.25 #dwgsc) ;;;;;+
STYLE STD1-8 SIMPLEX (* 0.125 #dwgsc) ;;;;;+
VIEW S A LTSCALE (* 0.375 #dwgsc) INSERT !str1 0,0 !#dwgsc ;;+
MENU !#menu ^GAXIS ON SETVAR MENUECHO 0
```

Save and exit to AutoCAD.

Command: **MENU**   Load the CA-SETUP.MNU. Try it.

[INITIATE] should run "cleaner." If you are using Release 9, it suppresses the AutoLISP variables from the menu, but Version 2.6 still echoes them.

NOTE. You can toggle MENUECHO on and off for portions of a menu item. You set MENUECHO to a value from 1 to 3, and enclose the portion to echo in a pair of "^P"s. For example, to echo LTSCALE to the screen, use:

VIEW S A ^P**LTSCALE** (* 0.375 #dwgsc) ^PINSERT !str1 0,0 !#dwgsc

We recommend leaving MENUECHO off and using ^P where you want to echo menu items. This produces the cleanest interface.

We hope that you will take what you have learned with this book and create your own custom system. You probably will no longer have any use for the Standard AutoCAD Menu. But if you wish to integrate your customized menus with the ACAD.MNU, here is a minimum integration.

Integration with ACAD.MNU

Copy the original ACAD.MNU file into the \CA-ACAD directory.

Command: **ED**   Edit the copied ACAD.MNU.

Find the **S page below the ***SCREEN device name.

Add this line near the bottom of the **S page:

[CA-MENU   Loads the Customizing AutoCAD menu.]MENU CA-MENU

Save, exit to AutoCAD.

Command: **MENU**   Load the modified ACAD.MNU.

Select **[CA-MENU]**   And the CA-MENU should load.

Command: **ED**   Edit CA-MENU.

Add an [ACADMENU] item to the 19th label [] of the **ROOT page:

```
[]
[ACADMENU] MENU ACAD
```

Save and exit to AutoCAD

Command: **MENU**   Load the modified CA-MENU. [ACADMENU] should be at line 19.

Select one of the application menu pages. Its menu should cover [ACADMENU].

Select **[root]**   [ACADMENU] should reappear. Try several more page changes.

Select **[ACADMENU]**   The ACAD menu should load.

Command: **QUIT**

This is the minimum integration. If you want to integrate further, you can merge the two menus into one. If you do:

- Put your custom menu first, followed by the ACAD.MNU.
- Check for duplicate ***device and **page names.
- Remember, if ***device and **page names are duplicated, the first occurrence overrides.
- If you modify the ACAD.MNU code, you don't want to update pages scattered through ACAD.MNU with each new release of AutoCAD. Duplicate any pages you want to modify and place the modified copies in the ACAD.MNU file above the body of the menu. Then you can delete the body of the ACAD.MNU and append the new one when you get updates.

There is a problem integrating the ACAD.MNU Popups, Icons and Tablet menus into a custom menu. Many of these menu items flip pages in the screen menu. Remember the book's IsoDim menu? IsoDim and other menus which use GRTEXT, or menu label toggles, will scramble if the user picks a tablet, or other device item that pages the screen.

Even if you don't integrate the ***SCREEN device section of ACAD.MNU, these $S=pagename screen calls may cause problems if your menu duplicates any ACAD.MNU page names. There is a solution:

- Integrate only the Popups, Icons and Tablet devices.
- Copy the sections to integrate to a temporary file.
- Do a search-and-replace. Replace all: $S= with: $S=QZX and save it.
- Put your menu first and append the modified temporary file.

Now, if you select a tablet item that has a screen call, it will try to find a non-existent screen page and do nothing. For example, the ACAD.MNU ***TABLET4 contains the submenu **TEXT1 33, which includes the macro:

```
$S=X $S=TEXT ^C^CTEXT
```

This macro calls the **TEXT screen menu. The CA-MENU also has a **TEXT screen. The example search-and-replace changes the macro to:

```
$S=QZXX $S=QZXTEXT ^C^CTEXT
```

Since the page names **QZXX and **QZXTEXT are non-existent, the offending screen calls are harmless.

Let's see! We have covered ACAD.LSP setup, looked at redefining commands, looked at keyboard macros, talked a little about security, stripped comments out of an AutoLISP file, and pulled together CUSTOMIZING AutoCAD's menu system.

Authors' Farewell

This must be the end!

But don't stop here. You will find valuable charts and more information in the appendices. We hope that you proceed along the path to customizing AutoCAD, and that we have helped with a book chock full of ideas, information and techniques. We hope to have given you enough ideas to extend and expand your application with AutoCAD customization.

Good luck customizing!

APPENDIX A

Menu Macros and AutoLISP Functions

Menus and Macros

CHAPTER 4

TEST.MNU
Introductory menu macro routines to create Bubbles, Leaders, Piping Lines, and miscellaneous utilities.

CHAPTER 5

**BUTTONS
This is the default button menu for Customizing AutoCAD, providing a number of selection options.

**FLOWSYMB
A screen menu page showing symbol library use.

**OSNAP
A screen menu page containing Object Snap options.

**P1-SETVARS
A popup menu page providing a means to set system variables.

**POINTS
An icon menu providing a visual means to establish point shapes.

**PTFILTERS
A screen menu page containing XYZ point filter options.

**TRANSVIEW
A menu page containing transparent view selections.

CHAPTER 6

**TEXT
A simple screen menu page used to set and select different text sizes.

CA-MENU.MNU
This will become the final menu created by the Customizing AutoCAD exercises. It is structured to receive the chapter menu pages as they are created. See chapter 23, CAMASTER.MNU for the final CA-MENU.

CA-SETUP.MNU
This is Customizing AutoCAD's menu file containing routines to initialize a new drawing. **DWG is a menu page providing drawing size selection. **SCALE is a menu page used to select drawing scale and [INITIATE] routines. The [INITIATE] selection is a large macro that establishes LIMITS, GRID, SNAP, text STYLEs, border sheet insertion, and that loads a current application menu.

CHAPTER 7

**CA-LAYER
A screen menu page that performs layer operations based on entity selection. Functions are included to set the current layer to a selected entity's layer, and to isolate the layer of a selected entity by turning all other layers off.

**LISPSAVE
A screen menu page containing AutoLISP expressions to save and retrieve user input data.

CHAPTER 8

**SPECIALS
A screen menu page used to type special characters, like fractions.

CHAPTER 9

**HATCH
A screen menu page with a checkered plate and concrete pattern. It includes predefined hatch border tools.

CHAPTER 17

**3D

A screen menu to assist in creating 3D plans of walls. MAKEWALL creates a 3D wall by selecting a polyline indicating the center line of the wall. WINDOW and DOOR create 3D windows and doors within the walls.

CHAPTER 18

**ISO

A screen menu page designed for isometric dimensioning. This menu demonstrates how menus can control an application and provide information to a user. The ISODIM.LSP functions are required for this menu page.

**TISO

This is a tablet menu that works with the isometric screen menu, providing a visual selection of iso-planes. Selections made from the tablet are displayed on the screen menu page. The ISODIM.LSP functions are required for this tablet menu.

CHAPTER 19

**TITLEBLK

This menu page automates updating title block information.

CHAPTER 20

**PIPEFITS

This screen menu page is designed for a parametric piping application. It includes a toggle for material tagging and a label to indicate the current pipe diameter.

CHAPTER 22

**REPORT

This menu page extracts attribute tag information.
[PIPE OUT] and [FIT OUT] are macros that extract pipe length and fitting attributes from a drawing and call the Lotus 123 program.
[FTEXT IN] imports ASCII print file output from Lotus using the FTEXT AutoLISP program.
[REP STAT] and [PROTRAK] call the REPSTAT.LSP and PROTRAK.EXE programs.

CHAPTER 23

CA-MENU.MNU

This becomes the final menu created by the Customizing AutoCAD exercises. It is structured to receive chapter menu pages as they are created. CAMASTER.MNU contains the final CA-MENU.

AutoLISP Functions and Programs

CHAPTER 10

SLOT.LSP

C:SLOT draws a solid filled slot using a pline. The AutoLISP command requires the user to pick the center of the slot and supply the length and diameter of the slot.

CHAPTER 11

ANGTOC.LSP

ANGTOC formats AutoLISP angles for use by an AutoCAD command. This function is included in CA-ACAD.LSP.

ATEXT.LSP

C:ATEXT places a text string in an arc about a radius point. It places text towards or away from the radius point. The program accepts fixed or non-fixed height text styles.

GODIM.LSP

GODIM is a function that draws a dimension line with arrows on each end, places text above the middle of the dimension line, and places the text on a different layer than the dimension line.

UFUNS.LSP
UFUNS is a collection of functions that filter and format user input. These functions are included in CA-ACAD.LSP.
UANGLE improves AutoLISP's GETANGLE function. UANGLE formats, shows defaults and filters user input.
UDIST improves AutoLISP's GETDIST function. UDIST formats, show defaults and filters user input.
UINT formats a prompt, shows defaults, and filters input for the GETINT function.
UKWORD implements the GETKWORD function of AutoLISP, but formats prompts, shows defaults, and filters input.
UPOINT formats a prompt, shows defaults, and filters input for the GETPOINT function.
UREAL formats a prompt, shows defaults, and filters input for the GETREAL function.
USTR formats a string prompt, provides a default string and determines if spaces are allowed on input. It uses AutoLISP's GETSTRING.

CHAPTER 12

APLATE.LSP
C:APLATE is a modified area program that automatically subtracts the area of holes and cutouts in a plate.

BSCALE.LSP
C:BSCALE provides independent rescaling of the X, Y and Z factors of blocks previously inserted in a drawing.

CSCALE.LSP
C:CSCALE automatically changes the scale of a drawing. It resizes text and block entities, retaining their alignments.

DXF.LSP
DXF takes a dxf code and an entity data list and returns the data element of the association pair. This function is included in CA-ACAD.LSP.

MOFFSET.LSP
C:MOFFSET performs an incremental offset of a single entity, when given the number of offsets required and the offset distance.

RCLOUD.LSP
C:RCLOUD lets you sketch a crude revision cloud then reprocesses it into a neatly formatted revision cloud.

SSTOOLS.LSP
SSTOOLS is a collection of utilities to manipulate selection sets.
CATCH creates a selection set of all the entities from the MARKed position through the end of the database.
LASTN creates a selection set of a given number of entities starting from the end of the database.
MARK places a marker in the AutoCAD Database for CATCH to use as a starting point.
SSDIFF creates a new selection set by taking the difference between two selection sets.
SSINTER builds a selection set of entities common to two other sets.
SSUNION creates a new selection set by combining two selection sets.

CHAPTER 13

ETOS.LSP
ETOS function takes any AutoLISP expression and converts it to a string. This function is included in CA-ACAD.LSP.

LEGEND.LSP
This program creates a legend by scanning the block table and extracting all specially marked legend blocks. It builds the legend and places it in the drawing at a user specified location and format.
C:LEGEND collects legend blocks and formats them in a special list.
OUTLEG draws the legend.
LEDSWAP lets the user easily swap items in the legend.

PVAR.LSP
This program provides the function to create a personal variable system that permanently retains personal data in a drawing. The variables can be numeric points or strings.
C:PVAR is used to query, view or update the personal variables.
PUTVAR stores the variable as a linetype definition.
RETVAR retrieves the variable from the linetype definition.

PVCHK verifies the personal variable block is defined, or inserts it if it is undefined.
GETPVAL is used to get a new personal variable value from the user.

WLAYER.LSP
C:WLAYER scans a drawing's layer list and erases, skips or Wblocks the contents of each layer to a drawing file name using the first 8 characters of the layer name.

CHAPTER 14

ANSILIB.LSP
ANSILIB is a library for text screen formatting using the ANSI escape calls. These functions will clear the text screen, control cursor movements, control text color and effects such as blinking.

FILELIB.LSP
FILELIB is a collection of AutoLISP utilities that retrieve the current DOS directory path, verify a file's existence, format a file name, make a backup up file with an extension .BAK, and merge two files. Chapter 21 has additional utilities for path and file name handling.

PATTERN.LSP
C:PATTERN creates a hatch pattern definition from selected entities and appends the pattern to the ACAD.PAT file.

REFDWG.LSP
C:REFDWG displays a list of project drawings on the text screen, using ANSI library functions. The user indicates the reference drawings to list in the current drawing. After all the selections are made, the program places the text in the drawing at a user specified point. Functions supporting the C:REFDWG command are included.

CHAPTER 15

DDRAW.LSP
C:DDRAW is a drawing aid, automating drawing lines at preset angles. The program displays a compass icon to assist the user in selecting the angle for each line.

STRIP removes a specified association list from an entity data list. This function is included in CA-ACAD.LSP.

DVIEW.LSP
C:DVIEW is a command that uses GRDRAW to display a box around the views defined in a drawing. It uses GRTEXT to display the view name on the status line. Each view is displayed in a continuous loop until the user selects a view, or quits the program.

ETEXT.LSP
C:ETEXT is a text editing program demonstrating the use of GRREAD and the ANSI.SYS codes. The program lets you select a single line of text and edit it, adding or deleting characters with the cursor key.

XPOINT.LSP
C:XPOINT is a function that demonstrates alternative methods to mark key points on the graphic screen.

CHAPTER 16

CA-ACAD.LSP
This is Customizing AutoCAD's ACAD.LSP file. It contains a collection of AutoLISP functions and expressions that are used in the book's application programs. These include global variables to establish and control the editing environment. It includes the DXF.LSP, DDRAW.LSP, ETOS.LSP, ANGTOC.LSP functions, and the User Interface Functions from the UFUNS.LSP file. Other functions are:
CALOAD manages the loading of files.
MLAYER checks to make sure a layer is defined.
ERROR evaluates errors and displays the appropriate error message. It also evaluates a list of functions to restore the drawing environment at the point of error. A variety of functions support error handling to maintain the environment.
RESET uses the error function modules to restore the drawing environment after normal program completion.

LEADTEXT.LSP
This program draws a straight line leader with an unlimited number of segments. The program calculates the leader text

placement and justification. LTEXT lets the user add multiple lines of text to the leader. The program demonstrates the use of error handling.
C:S-LEAD draws a pline leader with as many segments as desired. The arrow head is incorporated as part of the pline.
LTEXT determines the text justification and initiates a DTEXT command to let the user place text.

CHAPTER 17

3DFILL.LSP
C:3DFILL fills a 3D area, defined by 4 points, with 3DFACEs. The program divides the fill area into a user supplied number of rows and columns.

3DTOOLS.LSP
3DTOOLS is a collection of functions to determine and manipulate points, lines and curves in 3D space.
C:PICKVPT lets the user select from predetermined views by entry of a key letter. The user also can specify angles in the xyz planes, or specify a target/eye view.
C:ZARC creates an arc in the z axis by representing the arc with 3DLINEs.
C:3DPOLAR provides a 3D equivalent to the AutoLISP polar function.

MELEV.LSP
This program draws an elevation view of a 3D wall plan, including window and door openings. The elevation can be created at the users specified view point.
C:MELEV draws an elevation plan. It prompts the user for 3D plan selection for drawing an elevation. Supporting functions evaluate the entities and draw the elevation.

CHAPTER 18

ISODIM.LSP
This file contains two isometric dimensioning AutoLISP commands. It includes initialization expressions to insure Blocks and Text Styles exist. These functions work in conjunction with the Isometric menu page.
C:DIMLINE function draws an isometric dimension line and text.
C:EXTLINE function draws an isometric extension line.

CHAPTER 19

AUTOBLK.LSP
AUTOBLK is a set of functions to automatically insert a block on a line, breaking the line to fit.
GETBLK gets the block name, searches and returns block and break dim data.
LINEBLK inserts the autobreaking block and breaks the entity.
C:BLINE draws a line and inserts autobreaking blocks.
C:BBLOCK inserts autobreaking blocks in an existing entity.

TIMELIB.LSP
This file contains the time and date library functions.
TODAY returns the current date as mm/dd/yy.
TIME returns the current time as hh:mm:ss.
YEAR returns the current year in form the 1988.

UPDATE.LSP
This program contains functions that find and increment drawing revision levels. It adds a new revision bar and prompts for revision notes and drawing reissue data.
C:UPDATE is the AutoLISP command to initiate the program.

CHAPTER 20

ATTFITS.LSP
This file contains functions to tag materials of the parametrc elbows and pipes.
ATTFIT tags elbows.
ATTPIPE tags pipes.

DBLINE.LSP
DBLINE is a parametric function that draws a section of pipe complete with flanges.

EL90-S.LSP
EL90-S is a parametric function that draws the side view of a pipe elbow.

EL90-FB.LSP
EL90-FB is a parametric function that draws the front or back view of a pipe elbow.

SETSIZE.LSP
SETSIZE function manages the pipe data retrieval process used in the parametric pipe routines.
GETSIZE is a data search and retrieval function for asterisk tagged "*name" comma delimited data file formats.

CHAPTER 21

BATCHSCR.LSP
This file contains functions and AutoLISP commands to automate batch building and processing.
GETFIL creates a list of user specified files.
GETSCR creates a list of script commands.
BATCH creates a BATCH.SCR file by assembling a list of files and commands.
C:MSCRIPT is a general purpose AutoLISP command for the BATCH function, handling the user interface and formatting.
C:CSLIDE demonstrates the use of the batch functions to create a set of slides.
C:SSLIDE demonstrates the use of the batch functions to show a set of slides.

FTEXT.LSP
C:FTEXT imports an ASCII file into AutoCAD as Text. It wordwraps text to a user specified maximum line length. Or it imports unformatted text for monospaced tables.

LBLOCK.LSP
C:LBLOCK manages libraries of symbols by Wblocking all defined or redefined library blocks in a master drawing to a designated library directory.
UPDSCR function creates UPDBLK.SCR. This script automatically redefines master library blocks from a list of drawings. UPDSCR uses the batch functions and demonstrates how AutoLISP functions can write other AutoLISP functions.

XINSERT.LSP
C:XINSERT extracts Blocks from another drawing file without leaving the current drawing. It works clearly, without adding anything else to the current drawing's database. See INSERT.TXT on the CA DISK for an example of its use.

CHAPTER 22

REPSTAT.LSP
C:REPSTAT extracts title block data and revision information from a drawing. The information is contained in two files and merged together to form a master PROJECT.DAT file. This file is read by the PRO_TRAK.EXE database program, a Clipper compiled dBASE III program.

EXTERNAL PROGRAMS
dBASE III, optional.
Lotus 123, required.

CHAPTER 23

KEY REDEFINITION
ACAD.KEY is an external text file to redefine keys for use with AutoCAD.
TOGGLE.KEY toggles key definitions off for standard use outside of AutoCAD.
KILL.KEY clears key definitions.

GPATH.LSP
GPATH takes a path name and a file name. It combines and returns the joined file specification.

LSPSTRIP.EXE
LSPSTRIP.EXE reduces an AutoLISP file's size by stripping comments and white space. It also provides a help display for AutoLISP programming.

MFLY.LSP
C:MFLY creates a macro on the fly. The function records a command sequence from the keyboard and replays the sequence each time FLY is typed.

UNDEFINE.LSP
C:END is a redefined END command. The new END command insures that layers and other setting are correct, Zooms All, and prompts the user to make a record slide.

EXTERNAL PROGRAMS
MENU and AutoLISP TOOLS available (on request) from Autodesk, Inc.
MNUCRYPT.EXE
PROTECT.EXE
KELV.EXE, also called the Kelvinator

APPENDIX B

Setup, Memory and Errors

Setup Problems with CONFIG.SYS

If your CONFIG.SYS settings do not run smoothly, your only indication may be that things don't work. If you get the error message:

`Bad or missing FILENAME`

DOS can't find the file as it is specified. Check your spelling, and provide a full path.

`Unrecognized command in CONFIG.SYS`

Means that you made a syntax error, or your version of DOS doesn't support the configuration command. Check your spelling.

Watch closely when you boot your system. These error messages flash by very quickly. If you suspect an error, temporarily rename your AUTOEXEC.BAT so that the system stops after loading CONFIG.SYS. You also can try to send the screen messages to the printer by hitting <CTRL-PRINTSCREEN> as soon as DOS starts reading the CONFIG.SYS file. Another <CTRL-PRINTSCREEN> turns the printer echo off.

Problems with AUTOEXEC.BAT

Errors in AUTOEXEC.BAT are harder to troubleshoot. There are many causes. Often, the system just doesn't behave as you think it should. Here are some troubleshooting tips:

- Isolate errors by temporarily editing your AUTOEXEC.BAT. You can disable a line with a leading colon ":", for example:
  `: NOW DOS WILL IGNORE THIS LINE!`

- Many AUTOEXEC.BAT files have echo to the screen turned off by the command ECHO OFF or @ECHO OFF. Disable echo off to see what they are doing. Put a leading ":" on the line.

- Echo to the printer. Hit <CTRL-PRINTSCREEN> while booting to see what is happening.

- Make sure PROMPT, PATH and other environment settings precede any TSR (memory resident) programs in the file.

- Check your PATH for completeness and syntax. Unsophisticated programs that require support or overlay files in addition to their .EXE or .COM files may not work, even if they are in the PATH. Directories do not need to be in the PATH, unless you want to execute files in them from other directories.

- APPEND (DOS 3.3 or later) works like PATH and lets programs find their support and overlay files in other directories. It uses about 5K of RAM. All files in an APPENDed directory are recognized by programs as if they were in the current directory. If you use APPEND, USE IT CAUTIOUSLY. If you modify an appended file, the modified file will be written to the current directory, NOT the APPENDed directory. Loading an AutoCAD .MNU file from an APPENDed directory creates a .MNX file in the current directory. AutoCAD searches an APPENDed directory before completing its normal directory search pattern, so appended support files will get loaded instead of those in the current directory.

- SET environment errors are often obscure. Type SET <RETURN> to see your current environment settings. If a setting is truncated or missing, you probably are out of environment space. Fix it in your CONFIG.SYS file. Do not use extraneous <SPACE>s in a SET statement:

SET ACADCFG=\CA-ACAD ....OK. Sets "ACADCFG" to "\CA-ACAD".
SET ACADCFG =\CA-ACAD ....Wrong. Sets "ACADCFG ".
SET ACADCFG= \CA-ACAD ....Wrong. Sets to " \CA-ACAD"

- If your AUTOEXEC.BAT doesn't seem to complete its execution, you may have tried to execute another .BAT file from your AUTOEXEC.BAT file. If you nest execution of .BAT files, the second one will take over and the first will not complete. There are two ways to nest .BATs. With DOS 3.0 and later, use:

  COMMAND /C NAME

where NAME is the name of the nested .BAT file. With DOS 3.3, use:

  CALL NAME

- If you are fighting for memory, insert temporary lines in the AUTOEXEC.BAT to check your available memory. Once you determine what uses how much, you can decide what to sacrifice. Use:

  CHKDSK
  PAUSE

at appropriate points. Reboot to see the effect. Remove the lines when you are done. We use an alternative FREEWARE program called MEM.COM. It reports available RAM, quickly.

- If you have unusual occurrences or lockups, and you use TSRs, suspect the TSRs as your problem source. Cause and effect may be hard to pin down. For example, there is a simple screen capture program that, if loaded, locks up our wordprocessor—even when inactive! Disable TSRs one at a time in your AUTOEXEC. Reboot and test.

These are the most common problems. See a good DOS book if you need more information.

Problems with DOS Environment Space

Running out of space to store DOS environment settings may give the error:

```
Out of environment space
```

An environment space problem also may show up in unusual ways, such as a program failing to execute, AutoLISP not having room to load, or a block Insertion not finding its block. This occurs because the PATH, AutoCAD settings limiting EXTended/EXPanded memory, and AutoCAD configuration, memory and support file settings are all environment settings.

To find out how much environment space you need:

- Type SET at the DOS prompt.
- Count the characters displayed, including spaces.
- Add the count of characters for all SET statements you wish to add. Include revisions to your AUTOEXEC.BAT and startup files, like CA.BAT.
- Add a safety margin of 10%.

For Customizing AutoCAD, you need about 240 bytes (characters). DOS defaults environment size to 160 bytes or less, depending on the DOS version. The space expands if you type in settings, but cannot expand during execution of a .BAT file, including your AUTOEXEC.BAT. Loading a TSR (memory resident) program or utility such as SIDEKICK, PROKEY, some RAMdisks and print buffers, the DOS PRINT or GRAPHICS commands, and other utilities, freezes the environment space to the current size.

Fortunately, you can easily expand the space with DOS 3.0 or later versions of DOS. Add this line to your CONFIG.SYS:

SHELL = C:\COMMAND.COM /P /E=nnn

Substitute your boot drive for "C:", if your boot drive isn't C. Do not use "nnn", replace nnn with an integer value:

- For DOS 3.0 and 3.1, nnn is the desired environment size divided by 16. If you want 512 bytes, use 32 since (512/16=32). The maximum for nnn is 62.
- For DOS 3.2 and 3.3, nnn is the actual size setting. For 512 bytes, use 512. The maximum for nnn is 32768.

If you must use DOS 2, your solutions are more difficult. You have few choices. You can modify your COMMAND.COM file to allocate a larger environment.

Using DOS 2

We do not recommend that you run AutoCAD with DOS 2. If you have reason to run with DOS 2 and you encounter environment space problems, we provide a troubleshooting technique to debug the DOS COMMAND.COM to set 512 bytes. Proceed at your own risk. Have someone familar with using DOS's DEBUG program help you if you are on unfamilar ground. Changing your

COMMAND.COM will make the change permanent and automatic, unless you overwrite the change later.

Debugging DOS 2 COMMAND.COM

This change is for PC DOS 2.0 and 2.1 only. It assumes your boot drive is C:. To set 512 bytes:

`C:\> COPY COMMAND.COM *.20`	Make a backup. Use a name like .20 for DOS 2.0, or .21 for DOS 2.1.
`C:\> COPY COMMAND.COM *.20X`	A copy to debug. Use .20X for DOS 2.0 or .21X for DOS 2.1. DEBUG.COM must be in a directory on the path.
`C:\> DEBUG COMMAND.20X`	Or .21X for DOS 2.1. You get a "-" prompt:
`-D ECE ECF`	Display the current value:

```
3585:0EC0                                              BB 0A
```

CAUTION: If you didn't get BB 0A as shown above, then Quit now!
Type Q <RETURN> to quit. You have a different version of DOS.

`-E ECF 20`	Replaces the value 0A with 20.
`-D ECF ECF`	Check it:

```
3585:0EC0                                              BB 20
```

If BB 20 was shown above, Type W to write it. Otherwise Quit instead.

`-W`	
`Writing 4500 bytes`	Confirms the write of xxx bytes.
`-Q`	Quits, but the W saved the change.
	If all went OK, test it:
`C:\> FORMAT A:/S`	Format a diskette with the DOS boot files.
`C:\> COPY COMMAND.20X A:*.COM`	Copy the modified file to drive A.

Reboot the system, with the diskette in drive A.

The system won't boot your CONFIG.SYS or AUTOEXEC.BAT, but it should boot up DOS correctly. If it appears OK, make a backup file of both the modified and original COMMAND.COM files. Then, copy the modified file to drive C: and reboot.

If you have any problems, reboot your system from a diskette with a copy of the original COMMAND.COM, copy the original to drive C:, and reboot again. Your system will be back in its former unchanged state.

Memory Settings and Problems

There are several DOS environment settings that deal with AutoCAD's memory usage. These settings are explained and shown in a chart of memory usage in the Release 9 AutoCAD Installation and Performance Guide (IPG). None of

AutoCAD's settings affect other programs, or tie up memory when AutoCAD is not running. The AutoCAD Status command shows several memory values:

```
Free RAM:        9784 bytes     Free disk: 1409024 bytes
I/O page space:  109K bytes     Extended I/O page space: 592K bytes
```

"Free RAM" is the currently unused portion of RAM for AutoCAD to work in. "I/O page space" is RAM used by AutoCAD to swap data in and out. An I/O page value of 60K or more is adequate for most use, and we've worked (slowly!) with values under 20K.

"SET ACADFREERAM=nn" reserves RAM for AutoCAD's working space. Too small a size can cause "Out of RAM" errors, slow down Spline fitting, and cause problems with AutoCAD's HIDE, TRIM and OFFSET commands. However, a large ACADFREERAM setting reduces I/O Page swap space, and slows down AutoCAD. The default is 14K, the maximum depends on the system, usually about 24-26K. A setting larger than the maximum is equivalent to maximum. There is no magic number. We recommend moving ACADFREERAM up and down if you get errors or want to try to increase I/O page space.

"Extended I/O page space" is another factor that affects I/O page space. EXTended memory is IBM AT-style memory above the 1Mbyte mark. It is commonly used as RAMdisk (VDISK), a print buffer (PRINT), and by some sophisticated programs like AutoCAD. AutoCAD uses EXTended memory as "Extended I/O page space" to swap its temporary files into EXTended RAM instead of slower disk space. EXPanded memory is Intel Above Board style memory above the 640K mark. It is also known as LIM and/or EMS memory. It is used for RAM disks and print buffers. It is designed for page swapping. AutoCAD uses it as "Extended I/O page space".

It may take some investigation to determine which type of memory you have. You often can configure Intel Above Boards and their imitators as EXPanded or EXTended memory. EXPanded memory is relatively clean, with established techniques for programs to share it. Unfortunately, EXTended memory lacks protection. Programs wishing to use it are not always able tell if it is in use by another program.

If you use EXTended memory and sometimes get unexplained crashes, you may have a memory conflict between programs. Crashes may appear random. Check to see what, if any, programs other than AutoCAD may be using your EXTended memory. In particular, check for DOS VDISK and PRINT. Check your CONFIG.SYS and AUTOEXEC.BAT. Figure out what addresses are in use so that you can set AutoCAD to avoid the conflict. If you have EXTended RAM, and it is used by any other program, like DOS VDISK (RAMdisk) or DOS PRINT, you can use "SET ACADXMEM=start,size" to avoid conflict. See your AutoCAD Installation and Performance Guide for details.

Even if you do not encounter crashes, you may want to examine your memory usage. To enable "Extended I/O page space", AutoCAD uses up some Free RAM. If you have 2MB or more of EXTended or EXPanded memory, you may need to actually restrict AutoCAD's use of it. Each 1Kbyte of "Extended I/O page space" reduces Free RAM and/or I/O page space by 16 bytes. You also may want to restrict your EXTended or EXPanded memory use, if you run Software Carousel or another multi-tasking setup.

"SET ACADLIMEM=value" (for Release 9 or later) configures EXPanded memory usage (LIM-EMS). This setting is probably not critical, since EMS avoids conflicts. If you must restrict it, or wish to reserve some of your LIM-EMS RAM for other programs to use via SHELL, refer to your AutoCAD Installation and Performance Guide.

Using CUSTOMIZING AutoCAD with a RAMDISK

Running AutoCAD from a RAM disk can be more efficient than using EXTended/EXPanded memory for I/O page space. If you want to run AutoCAD from a RAM disk, there are three things that you want to look at: AutoCAD's program files, temporary files, and the drawing itself.

AutoCAD locates its temporary files on the same drive as the drawing, unless you tell it to locate its files in another directory. In the configuration menu, under item 8. Configure operating parameters, you will find item 5. Placement of temporary files. You can set item 5 to the drive for a RAMDISK to locate your temporary files there. Item 5 defaults to <DRAWING>, for the drawing's directory. If you want to use a RAMDISK, we recommend allowing about 1.5 times the size of the largest drawing if you use a RAMDISK in association with extended I/O page space. If you use a RAMDISK without extended I/O page space, use 2+ times the size of the largest drawing.

Regardless of the temporary file setting, AutoCAD still locates the temporary file(s) that eventually becomes the new drawing file in the directory of the drawing file itself. This occurs because AutoCAD closes up the file(s) instead of copying them when the drawing is ended.

To speed drawing disk access, you can run the current drawing from your RAMDISK. If you do this, you need to insure that the finished drawing is copied or Saved to a real drive. Use a menu macro, or redefined END command for safety. If you have sufficient I/O page space, you will gain little from putting the drawing on RAMDISK. If you do put the drawing on your RAMDISK, allow more than twice the largest drawing size.

The easiest way to put the AutoCAD program on RAMDISK is to copy all files to the RAMdisk. It is a little wasteful of memory, but if you have plenty, it works. Make sure that you have set a real disk directory with SET ACADCFG=, ensuring any configuration changes are copied back to a real disk.

If you want to place AutoCAD on the RAMDISK, but need to make efficient use of RAMDISK space, copy selective AutoCAD program files to it. There is little if any advantage to putting support files or the ACAD.EXE file on a RAMDISK. Start with the overlay files. They are listed in the Software Installation section of your AutoCAD Installation and Performance Guide. The most important are: ACAD.OVL, ACAD0.OVL, and ACADVS.OVL. The ACAD2.OVL and ACAD3.OVL contain ADE2 and ADE3 commands, and the ACADL.OVL is AutoLISP. If you have to decide between them, you can refer to your AutoCAD User Reference Appendix on Commands. Decide whether you use ADE2 or ADE3 commands the most!

You need to make sure that the files copied to the RAMDISK are not also found in the \ACAD directory of your hard disk. Make an \ACAD\RAM directory to store the files. Let's assume that you have the following drives and directories:

```
C:                      Hard disk.
D:                      RAM disk.
C:\ACAD                 Std. ACAD support files and program files
                        except the .OVL files.
C:\ACAD\RAM             AutoCAD .OVL files to go on RAM disk.
C:\SUPFILES             Project specific support files.
C:\PROJECT              Working drawing project directory.
C:\CFGFILES             Specific configuration.
```

If D: is on your PATH, and you have copied the .OVL files to D:, you start AutoCAD with the following:

```
SET ACAD=SUPFILES
SET ACADCFG=CFGFILES
C:\ACAD\ACAD
```

This works fine. AutoCAD will find all needed files. The \ACAD\ACAD explicitly tells DOS to look in the \ACAD directory for the ACAD.EXE file. AutoCAD will also look there for .OVL files and remember it for support files. AutoCAD is a smart program, and it will also search the PATH to find its other needed support files. Since D: is on the PATH, it finds its .OVLs, and remembers D: as a possible source of support files.

If your path is wrong, or a file is in neither \ACAD or D:, you will get:

```
Can't find overlay file ACAD0.OVL.
Enter file name prefix (path name\ or X:) or '.' to quit=
```

Finding Support Files

When you ask AutoCAD to find a support file, like a menu file, it searches in a particular order. Given the above settings, you would get:

```
"STUFF.mnu": Can't open file
  in C:\PROJECT (current directory)     First the current directory.
  or C:\SUPFILES\                       Then the directory designated by SET ACAD=.
  or D:\                                Then the .OVL directory found on the PATH, if any.
  or C:\ACAD\                           Last the program directory, home of ACAD.EXE.
Enter another menu file name (or RETURN for none):
```

If you keep AutoCAD's search order in mind, it will help you avoid errors in finding the wrong support files. A common cause of finding the wrong support files is setting ACAD=somename in a startup BATch file. Make sure to SET ACAD= to clear it at the end of the BATch file. Clear your SET ACADCFG= settings.

Current Directory Errors

If you use SHELL to CD (Change Directories) from inside AutoCAD, you may get strange results. Parts of AutoCAD recognize the change and parts do not. New drawings will not default to the changed current directory, yet SAVE defaults to save files in the changed current directory. Subsequent attempts to load support files, such as .MNX files, can crash AutoCAD.

If you must CD on SHELL excursions, automate it with a BATch file that also changes back to the original directory.

SHELL Errors

Here are some common errors encountered in using SHELL:

`SHELL error swapping to disk`

is most likely caused by insufficient disk space. Remember that the temporary files used by AutoCAD easily can use up a megabyte of disk space.

`SHELL error: insufficient memory for command.`

can be caused by an ill-behaved program executed during a previous shell, or before entering AutoCAD. Some ill-behaved programs leave a dirty environment behind that causes AutoCAD to erroneously believe insufficient memory exists.

`Unable to load XYZABC: insufficient memory`
`Program too big to fit in memory`

If SHELL got this far, these are correct messages. You need to modify your ACAD.PGP to allocate more memory space.

`SHELL error in EXEC function (insufficient memory).`

can be caused by the default (24000) byte "SH" Shell memory allocation being too small to load DOS. Exactly how much memory you need to allocate depends on your versions of DOS and AutoCAD, and on what you have in your CONFIG.SYS file. Recall that DOS 3.2 and later DOS versions must have at least 25000 bytes allocated in the ACAD.PGP. Use 30000 to give a little cushion.

Common AutoLISP Errors

The AutoLISP Programmers Reference gives a complete listing of error messages. The following list gives a few hints of where and how to look for some other causes.

`error: invalid dotted pair`
`error: misplaced dot`

Look for a missing or an extra " double quote above the apparent error location.

Look for " inbedded in a string where it should be \".

Look for strings that exceed 132 characters. STRCAT two strings if you need to.

`n>`   prompts such as   `3>`

Look for the same " double quote errors as shown in the dot errors example.

Look for a missing) parenthesis.

If the error occurs while you are LOADing a .LSP file, look in the file.

`Unknown command`

May be caused by AutoLISP, if you have a COMMAND function containing a "". The ""tries to repeat the last command entered AT the ACAD prompt, NOT the last command sent to ACAD via the COMMAND function.

Miscellaneous Problems

If you run under a multitasking environment, like CAROUSEL or DESKVIEW, you may get an error claiming a file should be in a directory it never was in. For example, you may get:

`Can't find overlay file D:\ACAD\ACAD.OVL`
`Retry, Abort?`

Or any other .OVL or ACAD.EXE. Don't type an "A" until you give up. Try an "R" to retry. If that doesn't work, copy the file to the directory listed in the error message. Flip partitions. You may need to hit another "R" during the flip. Copy the file, flip back and "R" again.

`Expanded memory disabled`

When starting ACAD from DOS, this error message can be caused by a previously crashed AutoCAD. Sometimes a crashed AutoCAD does not fully clear its claim on Expanded memory. This causes the program to think none is available. Reboot to clear it.

Insufficient Files Errors

You may encounter file error messages that may be caused by AutoCAD's inability to open a file. This may be caused by AutoLISP's OPEN function leaving too many files open. If you repeatedly crash an AutoLISP routine which opens files, you may see the following:

```
Command: (load"test")
can't open "test.lsp" for input
"test"
```

If you know the variable names SETQed to the files, try to CLOSE them before you get the message. Otherwise, you have little choice but to QUIT AutoCAD and reboot the system to clean things up. This error can also show up as:

`Can't find overlay file D:\ACAD\ACAD.OVL`

Tracing and Curing Errors

You and your users are your best source of error diagnosis. When problems occur, log them so you can recognize patterns to cure them. Here are some tips and techniques:

- Use screen capture programs to document the textscreen.
- Dump the screen to the printer.

- Write down what you did in as much detail as possible, as far back as you can.
- Dump a copy of AutoCAD's STATUS screen to the printer.
- Dump a copy of the screen of the DOS command SET to check settings.

Avoidance is the best cure.

APPENDIX C

Reference Tables

This appendix provides two tables: an annotated Table of AutoCAD System Variables; and a Table of AutoCAD DXF Codes.

AutoCAD System Variables

Use the AutoCAD System Variables Table to determine AutoCAD drawing environment settings. The table presents all the variable settings available through the AutoCAD Setvar command or AutoLISP Setvar and Getvar functions. The system variable name and the default ACAD prototype drawing settings are shown. A brief description is given of each variable and the meaning of code flags. All values are saved with the drawing unless noted with **<CFG>** for ConFiGuration, or **<NS>** for Not Saved.

DXF Group Codes

Refer to the Table of DXF Group Codes when processing AutoCAD entities and reference tables. The DXF Group Codes apply to both data returned by AutoLISP and DXF (Drawing eXchange Format) output files created with the DXFOUT command. A DXF "Group" has two parts. The DXF code, an integer number from 0 to 74, identifies the data type. It is always followed by the data value. While identical numbers are used both with AutoLISP and DXF files, the data elements of a DXF group are not identical.

AutoLISP returns full point coordinates as one data value, such as (0.000000 1.000000). In AutoLISP, both the 10 and 20 DXF group codes of a point are combined under one 10 code. The DXF file lists each coordinate of the point separately. A portion of the DXF file listing the point would appear as:

```
   10
0.000000
   20
1.000000
```

DXF files do not save entity names. AutoLISP uses entity names to index the entities of a drawing. Entity names are included in the returned data. They are preceded by negative DXF code numbers, like -1 or -2. These are temporary entity names that are uniquely assigned at the start of each drawing edit.

Entity properties like elevation, thickness, linetype, and color are not stored in the DXF file or returned by AutoLISP unless they are explicitly set to a value other than the default. The default values are 0 for elevation and thickness and BYLAYER for linetype and color.

AutoCAD System Variables

SYSTEM VARIABLE	ACAD.DWG DEFAULTS VALUE	SETTING	COMMAND NAME	VARIABLE DESCRIPTION	
ACADPREFIX	"C:\ACAD\"			AutoCAD Directory Path <NS>	(read only)
ACADVER	"9.0"			AutoCAD Release Version	(read only)
AFLAGS	0		ATTDEF	Sum of: Invisible=1 Constant=2 Verify=4 Preset=8	
ANGBASE	0	EAST	UNITS	Direction of Angle 0	
ANGDIR	0	CCW	UNITS	Clockwise=1 Counter Clockwise=0	
APERTURE	10	10	APERTURE	Half of Aperture Height in Pixels <CFG>	
AREA	0.0000		AREA,LIST	Last Computed Area <NS>	(read only)
ATTDIA	0	PROMPTS	INSERT	DDATE Dialogue Box=1 Issue Attribute Prompts=0	
ATTMODE	1	ON	ATTDISP	Attribute Display Normal=1 ON=2 OFF=0	
ATTREQ	1	PROMPTS	INSERT	Use Prompts=1 Use Defaults=0	
AUNITS	0	DEC. DEG.	UNITS	Angular Units Dec.=0 Deg.=1 Grad=2 Rad= 3 Survey=4	
AUPREC	0	0	UNITS	Angular Units Decimal Places	
AXISMODE	0	OFF	AXIS	Axis ON=1 Axis OFF=0	
AXISUNIT	0.0000,0.0000		AXIS	Axis X,Y Increment	
BLIPMODE	1	ON	BLIPMODE	Blips=1 No Blips=0	
CDATE	19871204.142641529			Date.Time <NS>	(read only)
CECOLOR	"BYLAYER"		COLOR	Current Entity Color	(read only)
CELTYPE	"BYLAYER"		LINETYPE	Current Entity Linetype	(read only)
CHAMFERA	0.0000		CHAMFER	Chamfer Distance for A	
CHAMFERB	0.0000		CHAMFER	Chamfer Distance for B	
CLAYER	"0"		LAYER	Current Layer	(read only)
CMDECHO	1	ECHO	SETVAR	Command echo in AutoLISP Echo=1 No Echo=0 <NS>	
COORDS	0	OFF	[^D] [F6]	Update Display Picks=0 ON=1 Dist>Angle=2	
DATE	2447134.60555926			Julian Time <NS>	(read only)
DIMALT	0	ON	DIMALT	Use Alternate Units ON=0 OFF=1	
DIMALTD	2	0.00	DIMALTD	Decimal Precision of Alternate Units	
DIMALTF	25.4000		DIMALTF	Scale Factor for Alternate Units	
DIMAPOST	""	NONE	DIMAPOST	Suffix for Alternate Dimensions	(read only)
DIMASO	1		DIMASO	Associative=1 Line,Arrow,Text=0	
DIMASZ	0.1800		DIMASZ	Arrow Size (also controls text fit)=Value	
DIMBLK	""	NONE	DIMBLK	Name of Block to be Drawn instead of Arrow or Tick	(read only)
DIMCEN	0.0900	MARK	DIMCEN	Center Mark Size=Value Add Center Lines=Negative Value	
DIMDLE	0.0000	NONE	DIMDLE	Dimension Line Extension=Value	
DIMDLI	0.3800		DIMDLI	Increment between Continuing Dimension Lines	
DIMEXE	0.1800		DIMEXE	Extension Distance for Extension Lines=Value	
DIMEXO	0.0625		DIMEXO	Offset Distance for Extension Lines=Value	
DIMLFAC	1.0000	NORMAL	DIMLFAC	Overall Linear Distance Factor=Value	
DIMLIM	0	OFF	DIMLIM	(Tolerance Limits=Text) ON=1 OFF=0	
DIMPOST	""	NONE	DIMPOST	User Defined Dim. Suffix (eg: "mm")=Value	(read only)
DIMRND	0.0000	EXACT	DIMRND	Rounding Value for Linear Dimensions	
DIMSCALE	1.0000		DIMSCALE	Overall Dimensioning Scale Factor=Value	
DIMSE1	0	OFF	DIMSE1	Suppress Extension Line 1 Omit=1 Draw=0	
DIMSE2	0	OFF	DIMSE2	Suppress Extension Line 2 Omit=1 Draw=0	
DIMSHO	0	OFF	DIMSHO	Show Associative Dimension while Dragging	
DIMTAD	0	OFF	DIMTAD	Text Above Dim. Line ON=1 OFF(In Line)=0	
DIMTIH	1	ON	DIMTIH	Text Inside Horizontal ON=1 OFF(Aligned)=0	
DIMTM	0.0000	NONE	DIMTM	Minus Tolerance=Value	
DIMTOH	1	ON	DIMTOH	Text Outside Horizontal ON=1 OFF(Aligned)=0	
DIMTOL	0	OFF	DIMTOL	Append Tolerance ON=1 OFF=2	
DIMTP	0.0000	ARROWS	DIMTP	Plus Tolerance=Value	
DIMTSZ	0.0000		DIMTSZ	Tick Size=Value Draw Arrows=0	
DIMTXT	0.1800		DIMTXT	Text Size=Value	
DIMZIN	0		DIMZIN	Controls Leading Zero (See AutoCAD Manual)	
DISTANCE	0.0000		DIST	Last Computed Distance <NS>	(read only)
DRAGMODE	2	AUTO	DRAGMODE	OFF=0 Enabled=1 Auto=2	
DRAGP1	10		SETVAR	Drag Regen Rate <CFG>	
DRAGP2	25		SETVAR	Drag Input Rate <CFG>	
DWGNAME	"TEST"			Current Drawing Name	(read only)
DWGPREFIX	"C:\CA-ACAD\"			Directory Path of Current drawing <NS>	(read only)

AutoCAD System Variables

SYSTEM VARIABLE	ACAD.DWG DEFAULTS VALUE	SETTING	COMMAND NAME	VARIABLE DESCRIPTION	
ELEVATION	0.0000		ELEV	Current Default Elevation	
EXPERT	0	NORMAL	SETVAR	Suppress "Are you sure prompts" (See AutoCAD Manual)	
EXTMAX	-1.0000E+20,-1.0000E+20			Upper Right Drawing Extents X,Y	(read only)
EXTMIN	1.0000E+20,1.0000E+20			Lower Left Drawing Extents X,Y	(read only)
FILLETRAD	0.0000		FILLET	Current Fillet Radius	
FILLMODE	1		FILL	Fill ON=1 Fill OFF=0	
GRIDMODE	0	OFF	GRID	Grid ON=1 Grid OFF=0	
GRIDUNIT	0.0000,0.0000		GRID	X,Y Grid Increment	
HIGHLIGHT	1		SETVAR	Highlight Selection ON=1 OFF=0 <NS>	
INSBASE	0.0000,0.0000		BASE	Insert Base Point of Current Drawing X,Y	
LASTANGLE	0		ARC	Last Angle of the Last Arc <NS>	(read only)
LASTPOINT	0.0000,0.0000		"@"	Last Pickpoint X,Y <NS>	
LASTPT3D	0.0000,0.0000,0.0000		"@"	Last Pickpoint X,Y,Z <NS>	
LIMCHECK	0	OFF	LIMITS	Limits Error Check ON=1 OFF=0	
LIMMAX	12.0000,9.0000		LIMITS	Upper Right X,Y Limit	
LIMMIN	0.0000,0.0000		LIMITS	Lower Left X,Y Limit	
LTSCALE	1.0000		LTSCALE	Current Linetype Scale	
LUNITS	2	DEC.	UNITS	Liner Units Scientific=1 Dec.=2 Eng.=3 Arch=4 Frac.=5	
LUPREC	4	0.0000	UNITS	Unit Precision Decimal Places or Denominator	
MENUECHO	0	NORMAL	SETVAR	Normal=0 No Input=1 No Prompts=2 No Input or Prompts=3 <NS>	
MENUNAME	"ACAD"			Current Menu Name	(read only)
MIRRTEXT	1	YES	MIRROR	Retain Text Direction=0 Reflect Text=1	
ORTHOMODE	0	OFF	[^O] [F8]	Ortho ON=1 Ortho OFF=2	
OSMODE	0	NONE	OSNAP	Sum of: Endp=1 Midp=2 Cent=4 Node=8 Quad=16 Inte=32	
				Inse=64 Perp=128 Tang=256 Near= 512 Quick=1024	
PDMODE	0		POINT	SETVAR Controls Style of Points Drawn	
PDSIZE	0.0000		POINT	SETVAR Controls Size of Points	
PERIMETER	0.0000		AREA,LIST	Last Computed Perimeter <NS>	(read only)
PICKBOX	3		SETVAR	Half the Pickbox Size in Pixels <CFG>	
POPUPS	0			AUI Support=1 No Support=0 <NS>	(read only)
QTEXTMODE	0	OFF	QTEXT	Qtext ON=1 Qtext OFF=0	
REGENMODE	1	ON	REGENAUTO	Regenauto ON=1 Regenauto OFF=0	
SCREENSIZE	570.0000,410.0000			Size of Display in X,Y Pixels <NS>	(read only)
SKETCHINC	0.1000		SKETCH	Recording Increment for Sketch	
SKPOLY	0	LINE	SETVAR	Polylines=1 Sketch with Line=0	
SNAPANG	0		SNAP	Angle of SNAP/GRID Rotation	
SNAPBASE	0.0000,0.0000		SNAP	X,Y Base point of SNAP/GRID Rotation	
SNAPISOPAIR	0	LEFT	SNAP [^E]	Isoplane Left=0 Top=1 Right=2	
SNAPMODE	0	OFF	SNAP [^B] [F9]	Snap ON=1 Snap OFF=0	
SNAPSTYL	0	STD	SNAP	Isometric=1 Snap Standard=0	
SNAPUNIT	1.0000,1.0000		SNAP	Snap X,Y Increment	
SPLFRAME	0		PEDIT	Display Spline Frame ON=1 OFF=0	
SPLINESEGS	8		PEDIT	Number of Line Segments in each Spline Segment	
TDCREATE	2447134.60507674		TIME	Creation Time (Julian)	(read only)
TDINDWG	0.00436285		TIME	Total Editing Time	(read only)
TDUPDATE	2447134.60507674		TIME	Time of Last Save or Update	(read only)
TDUSRTIMER	0.00436667		TIME	User Set Elapsed Time	(read only)
TEMPPREFIX	""	DWG		Directory Location of AutoCAD's Temporary Files <NS>	(read only)
TEXTEVAL	0	TEXT	SETVAR	Evaluates Leading "(" and "!" Text=0 AutoLISP=1 <NS>	
TEXTSIZE	0.2000		TEXT	Current Text Height	
TEXTSTYLE	"STANDARD"		TEXT,STYLE	Current Text Style	(read only)
THICKNESS	0.0000		ELEV	Current 3D Extrusion Thickness	
TRACEWID	0.0500		TRACE	Current Width of Traces	
VIEWCTR	6.2518,4.5000		ZOOM,PAN,VIEW	X,Y Center Point of Current View	(read only)
VIEWSIZE	9.0000		ZOOM,PAN,VIEW	Height of Current View	(read only)
VPOINTX	0.0000		VPOINT	X Coordinate of VPOINT	(read only)
VPOINTY	0.0000		VPOINT	Y Coordinate of VPOINT	(read only)
VPOINTZ	1.0000		VPOINT	Z Coordinate of VPOINT	(read only)
VSMAX	12.5036,9.0000		ZOOM,PAN,VIEW	Current Upper Right of Virtual Screen X,Y <NS>	(read only)
VSMIN	0.0000,0.0000		ZOOM,PAN,VIEW	Current Lower Left of Virtual Screen X,Y <NS>	(read only)

ENTITY DXF GROUP CODES

Abbrev	Entity	Abbrev	Entity	Abbrev	Entity
LN =	LINE	PT =	POINT	CI =	CIRCLE
SH =	SHAPE	BK =	BLOCK	IN =	INSERT
DM =	DIMENSION	3L =	3DLINE	3F =	3DFACE
AR =	ARC	AD =	ATTDEF	SQ =	SEQEND
TR =	TRACE	AT =	ATTRIB	SD =	SOLID
TX =	TEXT	VT =	VERTEX	PL =	POLYLINE

ENTITIES

CODE	DESCRIPTION	LN	PT	CI	AR	TR	SD	SH	TX	BK	IN	AD	AT	PL	VT	DM	3L	3F	SQ
0	Primary Text Value								1										
1	Name: Shape, Block, Tag							2		2	2	2	2			1			
2	Prompt Strg											3				2			
6	Line Type Name	6	6	6	6	6	6	6	6	6	6	6	6	6	6	6	6	6	6
7	Text Style Name								7			7	7						
8	Layer Name	8	8	8	8	8	8	8	8	8	8	8	8	8	8	8	8	8	8
10	X - Start or Insert Point	10	10						10	10	10	10	10		10		10	10	
	X - Center Point			10	10														
	X - Corner Point					10	10									10			
11	X - Definition Point	11														11			
	X - End or Insert Point					11	11		11			11	11				11	11	
	X - Alignment Point																		
12	X - Middle Point Of Dim.					12	12									12		12	
	X - Corner Point																		
13	X - Corner Point					13	13											13	
14	X - Definition Point															14			
15	X - Definition Point															15			
16	X - Definition Point															16			
20	Y - Start or Insert Point	20	20						20	20	20	20	20	20	20		20	20	
	Y - Center Point			20	20														
	Y - Corner Point					20	20									20			
21	Y - Definition Point	21														21			
	Y - End or Insert Point					21	21		21			21	21				21	21	
	Y - Alignment Point																		
22	Y - Middle Point of Dim.					22	22									22		22	
	Y - Corner Point																		
23	Y - Corner Point					23	23											23	
24	Y - Definition Point															24			
25	Y - Definition Point															25			
26	Y - Definition Point															26			
30	Z - Corner Point																30	30	
31	Z - Corner Point																31	31	
32	Z - Corner Point																	32	
33	Z - Corner Point																	33	
38	Entity Elevation	38	38	38	38	38	38	38	38	38	38	38	38	38	38	38	38	38	38
39	Entity Thickness	39	39	39	39	39	39	39	39	39	39	39	39	39	39	39	39	39	39

ENTITIES

CODE	DESCRIPTION	LN	PT	CI	AR	TR	SD	TX	SH	BK	IN	AD	AR	PL	VT	DM	3L	3F	SQ
0																			
40	Radius			40	40			40	40			40	40	40	40	40			
	Height, Size or Width																		
	Leader Length																		
41	X Scale Factor or Width							41	41		41	41	41	41	41				
42	Y Scale Factor or Buldge										42				42				
43	Z Scale Factor										43								
44	Column Spacing										44								
45	Row Spacing										45								
50	Rotation Angle				50			50	50		50	50			50	50			
	Start Angle																		
	Curve Fit Tangent																		
51	End Angle				51			51	51			51							
	Obliquing Angle																		
62	Color	62	62	62	62	62	62	62	62	62	62	62	62	62	62	62	62	62	62
66	Entities follow flag										66			66					
70	Vertex or Polyline Flag									70	70	70		70	70	70			
	Attribute Flag																		
	Column Count																		
	Block Type Flag																		
71	Text Generation Flag							71			71	71							
	Row Count																		
72	Text Justification							72				72							
73	Field Length											73							

TABLE DXF GROUP CODES

LT = LINETYPE   LY = LAYER   ST = STYLE   VW = VIEW

CODE	DESCRIPTION	LT	LY	ST	VW
0	Table Name	2	2	2	2
2	Descriptive Text	3			
3	Font File Name			3	
4	Bigfont File Name			4	
6	Line Name		6		
10	X - View Center Point				10
11	X - View Direction				11
20	Y - View Center Point				20
21	Y - View Direction				21
31	Z - View Direction				31
40	View Height	40		40	40
	Pattern Length				
	Fixed Text Height				
41	View Width		41	41	41
42	Text Width			42	
49	Last Height Used	49			
	Dash length				
50	Obliquing Angle			50	
62	Color		62		
70	Number of Table Entries	70	70	70	70
71	Text Generation Flag			71	
72	Alignment Codes	72			
73	Number of dash items	73			

APPENDIX D

The Authors' Appendix

How Customizing AutoCAD Was Produced

We thought you might be interested in the hardware, software, tools and techniques that we used to put the book together. This appendix tells a little about how the book was produced, and also tells you about some of our favorite tools and sources of information for customization.

Our Hardware

Three of us used four computer systems to produce the book. The computers ran Intel 286 processors and coprocessors at 10Mhz. All computers had 2Mbytes or more of Intel Above Board Expanded memory. Three computers were used for development and writing. All three machines ran AutoCAD Version 2.6 or AutoCAD Release 9. The AutoCADs were provided by Autodesk, Inc. A fourth computer was used as the production printing machine, running Xerox Ventura Publisher Version 1.1. Ventura Publisher was provided by the Xerox Corporation.

We used Verticom M256E and H256E monitors and video cards. These were provided by Verticom, Inc. The Verticoms provided 640x480 graphics in 256 colors. The Verticom boards are IBM CGA and PGC compatible. They can run AutoShade. We also used a VMI Image Manager 1024 video card, provided by Vermont Micro Systems, Inc. We used a Verticom 2Page Display System on the production workstation. Pointing was done by Hitachi Tiger Tablets, and CALCOMP 2500 and 2300 digitizers. The CALCOMP digitizers were provided by CALCOMP. We used Logitech mice.

Final printing was done on a NEC LC 890 Postscript Page (laser) Printer. The NEC was provided with the help of KETIV Technologies. Drafts and illustrations were printed on an HP LaserJet II.

Our AUTOEXEC.BAT Files

Here is Rusty's annotated AUTOEXEC.BAT file. It shows a lot of our favorite tools.

```
c:\boot\MAXIT /CA /DB0              These install the extra MAXIT memory, from McGRAW Hill.
c:\boot\MAXIT /CA /DB0              It increases DOS memory to 736 Kbytes.
set ACADFREERAM=26                  26 also "codes" the ACAD version (29 for Release 9)
set LISPHEAP=34500
set LISPSTACK=10000
PROMPT $P$G                         Displays current directory at DOS prompt.
PATH d:\NORTON;d:\DOS;c:\BAT;c:\UTIL;d:\MOUSE;C:\;D:\;d:\WORD
d:\dos\FASTOPEN d:=84 c:=48         Reduces disk access with DOS 3.3.
c:\boot\SAFEPARK 15                 A utility parks the hard disk after 15 seconds.
:   c:\boot\vloadh c:\boot\h8vdi.ker   The ":" makes this Verticom H256 line inactive w/o deleting.
ECHO ^[[1;0;59p                     An ANSI.SYS key reassignment.
c:\boot\CTRLALT                     An invaluable cut and paste utility.
c:\boot\SPEEDFXR -X 2 28            Speeds up the screen text cursor.
c:\util\MAP > MAPMEM.$              Reads memory use map and redirects to filename MAPMEM.$.
ask "Run Carousel? ", yn            Uses Norton Utilities ASK to optionally run CAROUSEL.
if errorlevel 2 goto SETUP          If ASK got N (No=2), go to label :SETUP
if errorlevel 1 goto CAROUSEL       If ASK got Y (Yes=1).
:SETUP                              BATCH label allows optional execution of following.
c:\util\PUSHDIR                     Initializes utility, saves current dir on a stack.
CALL PUP   rem POP restores path    (DOS 3.3) Calls the PUP.BAT command, and returns here.
ask "Run JOT? ", yn                 Another optional load.
if errorlevel 2 goto CONTINUE       Skip JOT.
if errorlevel 1 CD\JOT
1\JOT                               Installs the JOT! abbreviation macro program.
CD\                                 Back to root directory.
:CONTINUE
COPY d:\a26\RAM E:                  Loads AutoCAD 2.6 to ramdisk.
COPY d:\acad\RAM F:                 Loads AutoCAD Release 9 to ramdisk.
D:
goto end                            Skip the rest. Don't load Carousel.
:CAROUSEL
C:\CAROUSEL\CAROUSEL                Load The Software Carousel. This was invaluable.
:END                                That's all!
```

Here is Joe's AUTOEXEC.BAT file.

```
path=c:\dos;c:\acad;c:\ws4;c:\norton;c:\
prompt $p$g
cls
```

Directories do not need to be in the PATH, unless you want to execute files in them from other directories. Any program.EXE, program.COM or batch.BAT file stored in a directory in the PATH can be executed from any directory. For example, UTIL, DOS and NORTON are in our PATH so that we can use them.

We use several TSR (RAM resident) utilities, which are always available in the background. They typically "popup" when you hit some "hotkey". We use typical TSRs, including PROKEY. We use CTRLALT for text screen capture and copying, PUSHDIR to save and restore current directories, JOT to create keyboard macro abbreviations, and HotShot for screen capture.

Our development environment used The Software Carousel, Norton's Editor, CTRLALT and Microsoft Word.

Our Text Editor

A good text editor is invaluable in customizing AutoCAD. Besides creating ASCII files, it helps if your editor is comfortable, compact, quick, and easy to use. Norton's Editor was our choice as a text editor. Norton's Editor provides all the following features.

- Quick loading, compact size. Quick saving and exit. Save to another file name. Ability to handle large files. Merge files.
- Compare text for differences.
- Line and column number counter display. Go to line number. Row mode, including block copy and move.
- Versatile cursor control. Versatile delete. Undelete text. Versatile block commands. Versatile search and replace.
- Find matching punctuation, like: ({[ ]}).

Perhaps the most valuable feature in Norton's Editor is the punctuation matching. It helps you find your way back out when you get "Lost In Stupid Parentheses" within AutoLISP.

Our Multitasking Interactive Environment

Some operating systems and operating system extensions let you load more than one program at a time. Concurrent ones, like Windows 386 and Deskview, can run programs in the background while you work. They require a 386 machine. Others, like The Software Carousel, from Softlogic Solutions, run on PC/XT/AT machines. They can flip back and forth between programs, but only one program is active at a time.

We found The Software Carousel invaluable in writing this book. It integrated our system. We set up five "partitions", each was like a separate computer. All five coexisted in the single system, but each could maintain its own settings and environment. The partitions were:

```
1          Used to run AutoCAD Release 9.
2          To run AutoCAD Version 2.6.
3          To run Norton's Editor and other utilities.
4          More utilities and housekeeping.
5          Microsoft Word.
```

CTRLALT ties the system together by extending across all partitions. CTRLALT is a memory resident popup Freeware program by Barry Simon and Richard Wilson. It has many features including:

- Popup ASCII Table.
- Popup Hex Table.
- Popup ANSI Table.
- Print.
- Cut and Paste.

The Cut and Paste feature made the tedious and error prone task of documenting the book's AutoCAD and AutoLISP routines easier. We could flip into AutoCAD, hit <CTRL-ALT-RETURN>, highlight part of the textscreen and capture it. This captured the screens exactly as you should see them.

Document Illustration, Formatting and Printing

Microsoft Word 3.1 was used to write the document. Xerox Ventura Publisher was used to format and print. Illustrations were done in AutoCAD. Screen images, both text and graphics were captured and cleaned up with HotShot, from SymSoft. SymSoft provided a beta version of their HotShot Plus.

Tools, Sources and Support

No one knows everything there is to know about AutoCAD. Knowing where to find AutoCAD information can help you solve your application problems. Here are other tools and resouces that we'd like to share.

Advanced AutoCAD Classes

Autodesk, Inc. has authorized a hundred or more AutoCAD Training Centers in the US and several foreign countries. These training centers are usually affiliated with colleges and technical institutes. Many centers have specialized and advanced courses on AutoLISP. Gold Hill Computers, the makers of Golden Common Lisp, is providing AutoLISP courses through selected Authorized AutoCAD Training Centers. You can get information about the centers by calling Autodesk, Inc. (800) 445-5415 or (415) 332-2344. In addition, many AutoCAD dealers also teach courses that meet or exceed the quality of the Authorized AutoCAD Training Centers.

A Brief Look at the CompuServe Autodesk Forum

Autodesk's forum on CompuServe may be the best source of support available today. The CompuServe Information Service, Compuserve, Inc., is available to anyone with a computer, modem, communications software, telephone and password.

You join CompuServe by getting a membership kit at a computer store, software store, or bookstore. When you follow the instructions and "Log on", the CompuServe main menu "TOP" shows:

A Brief Look at the CompuServe Autodesk Forum   D—5

```
CompuServe                    TOP

1  Subscriber Assistance
2  Find a Topic
3  Communications/Bulletin Bds.
4  News/Weather/Sports
5  Travel
6  The Electronic MALL/Shopping
7  Money Matters/Markets
8  Entertainment/Games
9  Home/Health/Family
10 Reference/Education
11 Computers/Technology
12 Business/Other Interests

Enter choice number !GO ADESK
```
If you shortcut with **GO ADESK** you get the Autodesk Top Menu

```
Autodesk

FUNCTIONS

1  (L)   Leave a Message
2  (R)   Read Messages
3  (CO)  Conference Mode
4  (DL)  Data Libraries
5  (B)   Bulletins
6  (MD)  Member Directory
7  (OP)  User Options
8  (IN)  Instructions

Enter choice !
```

Most forums are divided up into several parts:

Private, and weekly group 2-way interactive sessions.

Announcements such as new releases are promptly posted here.

Help!

You will be interested mainly in 1 (L) Leave a Message, 2 (R) Read Messages, and 4 (DL) Data Libraries. The messages and data libraries are broken up into several subject areas. If you are only interested in AutoLISP, you can set your topic user options to ignore everything else. If you select 4 (DL) Data Libraries, you will see a list of topics:

```
Data Libraries Available:

1  General Information
2  AutoCAD
3  AutoLISP
4  3rd Party Software
5  Developers Corner
6  AEC
7  CAD/camera
8  AutoSketch
9  New UPLOADS
10 Wishlist/Solutions
11 Engineer Works
12 News/Reviews
13 Education
14 International
15 Dealer Comm.
16 Shipping/Receiving

Enter Choice !3
```

Reserved for dealers.

If you select 3 AutoLISP, you get a typical DL:

```
DL 3 - AutoLISP

1 (DES)   Description of Data Library
2 (BRO)   Browse thru Files              Read short descriptions, filtered by keyword and/or date.
3 (DIR)   Directory of Files
4 (UPL)   Upload a New File              Contribute something you've created.
5 (DOW)   Download a File                Copy a file to your system.
6 (DL)    Change Data Library            Shortcut to another DL.
7 (T)     Return to Function Menu        Back to the previous menu.
8 (I)     Instructions                   Help is always available.
```

Browsing with 2 (BRO) shows short file descriptions. You can enter keywords to search for specific topics. Read it on-line, or copy it to your computer. There are many valuable AutoLISP routines that users have donated in DL 3. Upload time is free, and you can download them for just a few dollars per hour.

You can leave a message and get a reply in less than 24 hours. Some of the most knowledgeable Autodesk people contribute. Plus you have the benefit of the largest CAD users group on-line. Many users will have solved a problem you've encountered, and reply.

Here are some tips that we have learned. Set options to not stop between messages. Figure out how to make your communications software capture text to a disk file as it scrolls on the screen. Read the messages later, offline, and formulate replies to upload later. You can download the free communications package ATO from the IBMSW (IBM SoftWare) forum. This package is designed to automate using CompuServe forums.

We recommend using the CompuServe forum. It is the best support money can buy. It averages about $6/hour of connected time. It's well worth the time it takes to learn to use it.

Bulletin Boards

Many other local AutoCAD bulletin boards exist. You will find a listing, maintained by New Riders Publishing in the Autodesk Forum on CompuServe. GO ADESK and check the library for the newest upload of the list, or contact us at New Riders Publishing, Attn: Bulletin Board.

Users groups

There are numerous AutoCAD users groups. Some of the best overall support from these members. You may find members, or whole groups, who share a similar application to yours. New Riders also maintains a list of User's groups. GO ADESK and check for the newest upload of the list, or contact us at New Riders Publishing, Attn: Users Group.

Magazines

AutoCAD has two independent magazines.

CADalyst
Subscription Manager
314 E Holly, # 106
Bellingham, WA 98255
(604) 873-0811

CADalyst Publications LTD.
282-810 W. Broadway
Vancouver BC Canada V5Z 4C9

CADENCE
Circulation Dept.
POB 203550
Austin, TX 78720-3550
(512) 335-1731

Books

Of course you have all of New Riders' books including this one, but what about AutoLISP books. After you have worked your way through CUSTOMIZING AutoCAD, you may find that you want to learn more about LISP. Keep in mind that AutoLISP is its own dialect. Here are three books that we found useful.

- The "bible" of LISP is Winston and Horn's LISP (Addison-Wesley).

- David Touretsky's LISP A Gentle Introduction to Symbolic Computation (Harper & Row) is a good easy introduction to LISP.

- Tony Hasemer's Looking at LISP (Addison-Wesley) is another good book.

Commercial Utilities

There are many utility programs to ease dealing with DOS, the computer, and, of course, your proliferating files. The best source of current information and reviews are the magazines: PC Magazine, PC World, Info World, and PC Week.

Here are a few categories along with a few of our favorite programs:

Hard Disk Management and Backup

Use something more efficient than DOS BACKUP for archiving your files. We use Fastback, Fifth Generation Systems, for its speed and superb error recovery.

After you back up your hard disk, you should clean it up. Unfortunately, just deleting garbage isn't enough. DOS tends to break files into smaller and smaller chunks as time goes on. The average hard disk is probably operating at less than half of its optimum speed. The solution is weekly or monthly cleanups with an inexpensive utility, like Disk Optimizer from Softlogic Solutions. We us the Disk Optimizer. We also recommend MACE from Paul Mace Software, or the SpeedDisk utility on Norton's Utilities. These programs will rearrange files for faster access.

A unique program for problem hard disks is the Disk Technician, Prime Solutions Inc. It diagnoses and actually repairs flaky disks by maintaining a comparative record, and doing a low level format on individual problem tracks. It also includes the SAFEPARK utility which we use.

If you erase a file or even a directory accidentally, there is hope of recovery if you catch it quickly. Hope is slim if you do anything that writes to the disk before recovery. The best known utility, which we use, is the NU unerase utility of the Norton's Utilities. Others are: MACE, Paul Mace Software, or PC Tools, Central Point Software.

Keyboard Macros

You can use electronic shorthand programs, like Prokey from RoseSoft, to speed entry of repetitive text and commands and reduce errors. We used Prokey extensively in the past, and still use it for some tasks. Others are Newkey, FAB Software's low cost shareware, Keyworks from Alpha Software Corp. with an optional program language, or Metro from Lotus Development Corp.

Now we use an abbreviation program, JOT!, Beacon Software International. It expands abbreviations automatically while you type. If you type LI, it sends LINE to AutoCAD. Another is PRD Plus, Productivity Software Int'l. Some keyboard macros, like Prokey, have a similar feature, but they are not automatic and require a trigger key to expand the abbreviation.

User Interface Shells

If you hate DOS, or you are a system manager who needs to insulate users from DOS, there are numerous DOS Shells. They generally provide a point and shoot interface to DOS through menus. HOT, Executive Systems Inc., the Word Perfect Library, and many others are available for almost any taste. Avoid memory resident Shells, unless they use EMS. Otherwise AutoCAD will probably be left RAM hungry.

General Utility Packages

There are dozens of little utilities that you may find useful, even if you only manage a single system. Many useful utilities are available as Shareware or Freeware. There are several comprehensive utilities packages available commercially.

We use Norton Utilities, from Peter Norton Computing. The 4.0 version includes an easy point and shoot interface, and the Advanced Edition includes disk cleanup. PC Tools, Central Point Software, has a unique directory cut and past feature, which can move entire directory trees intact. But don't try it with CAROUSEL loaded!

Other typical features include Directory and File name Sorting, Text Searching groups of files, Finding Files no matter where they are on the disk, changing File Attributes, Directory name Listing, Disk Testing, and almost anything else that DOS forgot and we need.

Freeware and Shareware

Some of the best things in your computer's life are free, or nearly so. Most of the AutoCAD command line and textscreen sections in this book were "captured" from the AutoCAD textscreen by CTRLALT. This is freeware. We got it by downloading it from the IBMSW (IBM SoftWare) Forum on CompuServe. Here are some other popular utilities, with an asterisk indicating programs that we use.

ARC521.COM	an ARC utility to compress and archive files into libraries. Saves 30-65% for DWG files. *
BADSEC.ARC	finds bad sectors on hard or floppy disks.
BASERE.ARC	is a pop-up number base converter.
BC.EXE	is a Backup Companion, a simple backup utility. We use FastBack.
CARUTL.ARC	are Software Carousel utilities. *
COMSWT.ARC	switches COM1: and COM2:
CTLALT.ARC	are pop-up utilities, tables, cut/paste from screen into other programs. *
D.ARC	keeps track of files added/erased in hard disk subdirs. It makes cleanup easy. *
FDIR.EXE	Directory listing in user set alpha-sorted order. We use a similar Norton Utility, DS.
FILL.ARC	Copies group of files on minimum number of floppies, fills each full. *
GLOBAL.ARC	Execute a DOS command, program or .BAT file on a group of files or directories. Great for cleanup. *
GUDLUK.ARC	A side-by-side text file comparison utility.
HELP.ARC	Help facility for DOS commands, function, syntax, etc.
HIDE.ARC	Two programs to HiDe or UnHide directories. Good for security.
HOTKEY.ARC	Cursor speed-up and run-on stopper.
KEYBUF.COM	Extend your keyboard buffer. Get rid of the beeps and dropped character errors.
KEYLOC.ARC	Change shifting keys (Ctrl, alt, shift) into on/off toggles.
LOC.COM	Small fast File Finder (wildcards allowed). We use Norton's FF.
MOVE.ARC	Move files (or subdirectories) between directories. We made our own .BAT program.
NEWER.COM	Find files newer (or older) than other files. *
NEWKEY.ARC	Keyboard macro processor, improves macro file handling. Highly rated Shareware.
NOTEPAD.COM	Small, memory resident notepad, like Sidekick.
POPDIR.COM	Restore current directory prev. saved on stack with PUSHDIR.COM. Great in .BAT files. *
PWORD.ARC	Prevents booting without entering password (source incl).
QUERY.ARC	Create .COM file to display ASCII text. *
REMIND.ARC	Daily reminder utility (demo). Shareware.
SEARCH.ARC	PATH-like access to data, overlay, etc. files. We use DOS 3.3 APPEND.
SRCH82.ARC	Fast text search and replacement program.
VDL.ARC	Mass file delete with verification. Excellent for cleanup. *
WAITEX.ARC	Load/run programs at appointed time.
WHATIS.ARC	Add 60 character comments to files. We use Norton's FI.

To obtain any of these utilities, follow your CompuServe instructions.

There also is an organization called the PC Software Interest Group (PC-SIG) which maintains and distributes a large library of public domain software, freeware, and shareware. PC-SIG publishes annual catalogs and monthly updates. PC-SIG checks its software. They charge a distribution fee of about $6 per disk. This fee is in addition to any possible shareware fees. Their entire 490+ disk 9000 program library is available on a single CD-ROM for about $200! here is their address:

PC-SIG
1030 E. Duane Ave., Suite D
Sunnyvale, CA 94086
(408) 730-9291

Authors Last Word and Mail Box

We had a lot of fun and learned a lot doing the book. We hope that you enjoy using CUSTOMIZING AutoCAD and that you come back to the book again and again for customizing ideas. Borrow from our ideas. Modify and adapt the

AutoLISP routines so that they do what you want them to do. We only ask that you give us credit for what we provide, and that you pass the credit on.

If you want to correspond with us, we can be reached electronically on Compuserve. Use Compuserve's EMAIL feature to send us a message. Rusty Gesner's number is: 76310,10. Please leave your name, and telephone number, the version of AutoCAD that you are using, a description of the hardware, and your comments. Send us tips and new ideas that were inspired by the book.

We will respond to you on Compuserve or by phone as soon as possible. We try to respond within 48 hours. We scan our mail each day. Of course, you can reach us by phone or good old fashioned mail at the New Riders number and mail address in the back of the book.

Good luck with your customizing!

Index

👉 Indicates "must do" items, I-9
  ACAD.PGP, 2-1
  AUTOEXEC.BAT, 1-2
  CA-ACAD.LSP to ACAD.LSP, 16-2
  CA19SET.MNU to CA-SETUP.MNU, 19-6
  CA.BAT, 2-1
  CONFIG.SYS, 1-2
  Prototype drawing, 2-1, 6-10
  SYNPLANT.DWG, 19-23

💾 Do if you have the CA DISK, I-9

 Do if you DON'T have CA DISK, I-9

! Variable evaluator, 7-8
\> redirection symbol in DOS, 14-8
#ARGLIST global variable, 20-8, 20-11
#DWGSC global variable, 6-3, 6-8, 6-12
#ERROR global variable, 16-12
#MARK global variable, 12-14
$S code ; semicolon warning, 5-10
$ menu page codes, 5-10, 5-16
$S screen code, 5-10
<ESC> escape character, 14-20, 23-9
' QUOTE function, 11-3, 21-10
* multiply function, 7-9, 10-10
* Next menu paging, 5-8
;*comments, displaying, 23-16
ERROR CA function, 16-0, 16-12, 16-15, 23-18
ERROR standard function, 16-11
\+ addition function, 7-10
\- subtract function, 10-10
/ divide function, 10-10
/= not equal to function, 10-13
= equal to function, 10-13
\> greater than function, 10-13
< less than function, 10-13
\>= greater than or equal function, 10-13
<= less than or equal to function, 10-13
@ lastpoint, 4-11
\ backslash, 4-4
^M as return in Lisp/Macros, 7-6, 11-8
^P Menuecho, 23-21

1+ increment function, 10-19
1- decrement function, 21-6
1> Prompt, 7-8
2 1/2-D extrusions, 17-2
2-D integrating 3-D views, 17-24
3-D, 17-1
  3-D tools, Benefits, 17-1
  Angles, 17-7
  Arcs, 17-10
  AutoLISP tools, 17-5
  Calculating points, 17-5
  Distance, 17-5
  Entities, 17-2
  Formulas, 17-7
  In 2-D drawings, 17-24
  Mesh generator, 17-12, 17-13, 17-15
  Midpoint, 17-6
  Points, and tip 17-3, 17-26
  Polar function, 17-6, 17-7, 17-9
  Tips, 17-26, 17-27
  Viewpoint, 17-8
3D menu, 17-18
3DDIST CA function, 17-5
3Dface command, 17-12
3DFILL C:command, 17-13
3DLine command, 17-2
3DPOLAR C:command, 17-9
3DPOLAR CA function, 17-9, 17-10

A

ABS function, 14-15
ACAD.CFG, 1-8
ACAD.HDX, 20-7, 20-8
ACAD.HLP, 20-7
ACAD.LIN, 9-2, 13-18
ACAD.LSP, 2-7, 16-2, 16-6, 23-3, 23-5
  Chain loading files, 16-20
  Conflicts, 2-6, 2-7
  Finalizing file, 23-18
  Global variables, 16-6
  How to, tips, 16-2, 16-23
  Limitations, 16-3
  Loading, 16-15
ACAD.MNU standard menu, 23-21
ACAD.PAT, 9-5
ACAD.PGP, 2-2, 2-3, 2-5
ACAD??.OVL device overlays, 1-8
ADDERR CA function, 16-13
Addition + function, 7-10
ADI Drivers, 1-9
ADI Video, I-8

ALLOC function, 16-3
ALTITD (altitude) CA function, 17-11
AND function, 10-12, 10-13, 14-11
Angle, included, 17-11
ANGLE function, 11-19, 20-13
Angles
   3-D, 17-7
   Command input tip, 11-29
   Controlling, 15-17
   Snapping, 15-17
   To radians, 16-8
ANGTOC CA function, 11-14, 16-7
ANGTOS function and formats, 11-12
ANNOT menu, 6-15
ANSI.SYS, 1-12, 14-19, 14-21, 23-8
   Codes in AutoLISP, 14-19, 14-21
   Example codes, 15-13
   Format codes, 14-19, 14-21
   Format tips, 14-32, 14-33
   Key Redefinition, 23-9, 23-11
   Key Redefinition Tips, 23-11, 23-12
   Library, 14-21
   Screen positions, 14-21
APLATE C:command, 12-8
APPEND DOS command, caution, 14-9
APPEND function, 7-18, 10-18, 13-5
Appending
   Chapter menus, 6-18
   Files, 14-5, 14-11
Applications
   Documenting, 23-16, 23-17
   Menus, 6-11, 6-13, 6-15
   Presentation, 23-17
   User interface, 23-17
   Working with others, 16-19, 16-21
ARCLEN (arc length) CA function, 17-11
Arcs
   3-D, 17-10
   Bulge, 12-30
   Length, 17-11
Area system variable, 12-9
Argument list, Global, 20-8
Arguments, 7-9, 10-4
   Declaring, 10-6
ASCII, 8-5
   Codes in AutoLISP, 11-6, 11-7
   Extended codes, 23-9
   File Format, 1-10
   To numbers, 11-10
ASCII function, 11-6, 14-29
ASSOC function, 7-19, 12-17
Associations lists, Entities, 12-17
Associative Dimensioning Caution, 18-16
ATAN function, 17-7
ATEXT C:command, 11-15
ATOF function, 11-10, 11-13, 19-26, 20-10
ATOI function, 11-10
Atom, 10-3
ATOMLIST, 7-5, 7-7, 10-5, 16-5
ATOMLIST arithmetic functions
   * multiply, 7-9, 10-10
   + addition, 7-10
   - subtract, 10-10
   / divide, 10-10
   1+ increment, 10-19
   1- decrement, 21-6
   ABS, 14-15
   SQRT, 17-5
ATOMLIST basic functions
   ' QUOTE, 10-3, 11-3, 21-10
   APPEND, 7-18, 10-18, 13-5, 14-28, 15-18
   ASSOC, 7-19, 12-17
   BOUNDP, 11-3
   CAAR, CADAR, CADDR, CADR, CAR 7-17
   CAR, 12-17
   CDR, 7-17, 12-17
   COND, 10-11, 10-15
   CONS, 12-17, 12-27
   DEFUN, 10-5, 16-4
   EVAL, 11-17, 23-13
   FOREACH, 10-19, 14-24, 20-13
   IF, 10-11, 10-13
   LAMDA, 16-5
   LAST, 7-18
   LENGTH, 7-18
   LIST, 7-18, 13-26, 14-28
   LISTP, 11-3, 11-4
   LOAD, 10-11
   MAPCAR, 13-26, 17-5, 17-6, 20-13
   MEMBER, 14-29
   MIN, 17-22
   MINUSP, 11-4
   NTH, 7-17, 7-18, 14-29
   NUMBERP, 11-4, 13-27
   PROGN, 10-14
   QUOTE, 10-3, 11-3
   REPEAT, 10-16
   REVERSE, 7-18
   SET, 11-6, 20-13
   SETQ, 7-8, 7-10, 20-13
   SUBST, 12-26, 12-27
   SUBSTR, 13-27, 14-24, 21-7
   TYPE, 10-3, 11-2
   WHILE, 10-17, 10-18
   ZEROP, 11-4, 17-7
ATOMLIST conversion functions
   ANGTOS, 11-12
   ASCII, 11-6, 14-29
   ATOF, 11-10, 11-13, 19-26, 20-10
   ATOI, 11-10
   CHR, 11-6, 14-27
   FIX, 15-18
   ITOA, 11-10
   RTOS, 7-15, 11-11, 11-13, 14-15, 20-23
ATOMLIST device access functions
   GRCLEAR, 15-6
   GRDRAW, 15-0, 15-5, 15-8
   GRREAD, 15-12, 15-18
   GRTEXT, 15-0, 15-2, 15-8, 20-4
ATOMLIST entity access functions
   ENTDEL, 12-3, 12-4, 12-13
   ENTGET, 7-19, 12-16, 12-17, 13-4, 13-6
   ENTLAST, 7-19, 12-3, 12-13
   ENTMOD, 12-26, 12-27, 15-18
   ENTNEXT, 12-4, 12-28
   ENTSEL, 12-3, 12-17, 12-28, 13-9
   ENTUPD, 12-26, 12-31, 19-20

SSADD, 12-7, 12-14
SSDEL, 12-7
SSGET, SSGET "X" 12-6, 12-28, 19-20
SSLENGTH, 12-7, 12-9
SSMEMB, 12-7, 12-12
SSNAME, 12-7

ATOMLIST geometric functions
ANGLE, 11-19, 20-13
ATAN, 17-7
COS, 14-14, 17-9
DISTANCE, 11-19, 17-4
INTERS, 17-4, 17-22, 20-12
POLAR, 10-10, 13-8, 17-4
SIN, 14-14, 17-9

ATOMLIST I/O functions
CLOSE, 14-4, 14-6, 14-8
OPEN, 14-4, 14-5, 14-8
PRIN1, 11-10, 14-3
PRINC, 10-3, 11-10, 13-26, 14-3
PRINT, 11-10, 14-3
PROMPT, 10-3, 11-7, 11-9, 14-3
READ, 11-6, 13-16, 13-27
READ-CHAR, 14-6
READ-LINE, 14-6, 21-9
WRITE-CHAR, 14-3
WRITE-LINE, 14-3, 14-5, 14-15

ATOMLIST logical functions
AND, 10-12, 10-13, 14-11
NOT, 10-12
OR, 10-12 - 10-14, 14-11

ATOMLIST misc functions
ERROR, 16-11
ALLOC, 16-3
COMMAND, 10-7, 11-16
DISTANCE, 7-15
EXPAND, 16-3
GC, 16-3
GRAPHSCR, 11-8
OSNAP, 19-27
SETVAR, 7-5
TERPRI, 11-8
TEXTSCR, 11-8, 14-26
VER, I-6
VMON (Virtual Memory ON), 16-4

ATOMLIST relational functions
/= not equal to, 10-13
= equal to, 10-13
> greater than, 10-13
< less than, 10-13
>= greater than or equal to, 10-13
<= less than or equal to, 10-13
EQ, 10-13
EQUAL, 10-13

ATOMLIST string functions
STRCASE, 10-16, 11-5
STRCAT, 10-6, 11-5, 11-6, 11-22
STRLEN, 11-5, 21-7
SUBSTR, 11-5

ATOMLIST symbol table functions
TBLNEXT, 13-4, 13-6
TBLSEARCH, 7-20, 11-16, 13-2, 13-4

ATOMLIST user input functions
GETANGLE, 7-11, 7-13, 7-15
GETCORNER, 7-13
GETDIST, 7-11, 7-13, 7-15, 10-10

GETORIENT, 7-13
GETPOINT, 7-11, 7-13, 10-10
GETREAL, 7-10
GETSTRING, 7-10
INITGET, 11-22
Attdef command, 19-5
Attedit, Automating, 19-10, 19-11
Attext command, 22-5
ATTFIT CA function, 20-22
ATTPIPE CA function, 20-23
Attributes, overview, benefits, 19-1, 19-2
  "Lost", 19-28
  Accessing data, 19-17
  Application examples, 19-1
  Attedit, 19-10, 19-11
  ATTEXT Tips, 22-31
  ATTFIT, 20-21
  ATTPIPE, 20-21
  Attreq system variable, 19-17
  Block redefinition, 19-28
  Caution, 22-5
  Control, 19-2
  dBase, 22-31
  Defined, 19-1
  DXF codes, 19-17, 19-18
  Exporting to dBase, 22-20
  Extract template file, 22-2 - 22-5
  Extracting, 22-5
  In macros, 19-6, 19-7, 19-9
  Invisible, 20-21
  Lotus, 22-31
  Preset, 19-17
  Record order, 22-21, 22-23
  Tips, 19-29
  Variable, 19-17
  Variables, 19-4
AUto selection, 4-18
Autobreaking blocks, 19-24, 19-25, 19-27

AutoCAD Commands
3Dface, 17-12
Adding AutoLISP commands, 10-9
and AutoLISP Angles, 11-13
Attdef, 19-5
Attext, 22-5
Block, 3-2
Command: prompt, I-6
Dimasz, 2-10
Dimblk, 18-6
Dimscale, 2-10, 3-6
Dimse1, Dimse2, 18-8
Dimtxt, 2-10
Drag, 4-15
Dtext, 8-15
Dxbin, 17-24
End, 23-7
Exit, 2-5
Explode, 3-14, 12-14
External commands, PGP, 2-2
Files, 14-11
Help, 23-17
Hide, and problems B-5
Insert, Insert*, 3-4
List, 3-8
Load shapes, 3-15
Ltscale, 9-3

Minsert, 3-13
Multiple, 4-17
Offset, B-5, 12-5, 20-14
Osnap, Osnap QUIck, 4-13, 18-9
Pdmode, Pdsize, 5-23, 5-24
Pedit, 4-11
Pline, 4-11
Redefine, 23-6
Replacing with C:commands, 23-6
Resume, 21-16
Select, 4-12, 4-17, 12-12
Setvar, 7-3
Shape, 3-15
Shell, 2-4
Sketch, 12-29
Style, 8-2
Trim, B-5
Undefine use, caution, 23-6, 23-8
Vpoint, 17-8, 17-23
Vslide, 23-17
Wblock, 3-4
AutoCAD versions, converting, 21-28
AUTOEXEC.BAT, 1-12
ECHO, B-1
Examples, D-1, D-2
Problems, B-1
AutoLISP
\nnn codes, 11-7
^C canceling, 10-8
and AutoCAD commands, 10-7
Angles and AutoCAD commands, 11-13
Arguments, 10-4
ASCII Codes, 11-6
Atom, 10-3
ATOMLIST, 7-7
Backslash formatting, 21-17
Benefits of using, 10-1
Branching, 10-11
Character functions, 11-6
Conditionals, 10-12, 10-17
Data types, 11-2
Data types to strings, 13-17
Declaring arguments, 10-6
Defining functions, 10-4
DOS batch files, 14-12
DOS file handling, 14-11
DOS, integrating utilities, 14-1
Dotted pair, 12-17
Efficient programming, 16-5
Encryption, 23-14
Entity names, 12-2, 12-3, 12-5
Errors, B-8
Evaluation, 7-8, 10-3
Expressions, 7-8
Files, 10-9
Functions, tips, 10-6, 10-20, 10-23
Global variables, 10-5
Heap, 16-4
IF-THEN-ELSE, 10-13
Illegal characters, 10-4
Importing text, 21-5, 21-7
INTegers, 11-2
Intelligent programs, benefits, 12-1
Introduction, 7-1
LISP books, D-7

LIST, 7-16
Lists, 10-1, 10-2, 10-15, 11-2
Loading functions, 10-11, 16-8, 21-31, 23-15
Local variables, 10-5
Logical operators, 10-12
Looping, 10-16
Macros and AutoLISP, 7-1
Macro control codes, 11-9
Maximum string length, 11-5
Memory, program vs. data 16-4
NIL, NON-NIL, 10-8, 10-11, 10-12
PI constants, 10-10
Predicates, 11-2
Program vs. data, 10-1 - 10-3
PROTECT.EXE, 23-14
REALs, 11-2
Relational operators, 10-13
Reserved characters, 10-4
Resetting environment, 16-13
Returns, 10-3
Selection sets limitations, 12-10
SET vs. SETQ, 11-6
Stack, 16-4
Standard subroutines, benefits, 16-1
STRings, 11-2, 11-5
SYMbols, 11-2, 11-3, 21-10
Symbols, creating with SET, 11-6
T, (True) 10-12
Upper/lower case, 10-14
Variables, 7-6, 16-4
Virtual memory, 16-4
vs. Scripts, 21-3
WHILE-NOT-OR-AND, 14-11
XLISP, 10-1
AutoSHADE, Tips, 17-15, 17-26
AUXILIARY menu device, 5-11
Axis, Isometric, 18-8

B

Backslash in macros, 4-4, 4-6, 5-17
Backslash path formatting, 21-17
BACKUP CA function, 14-11
BANGLE CA function, 17-7, 17-8
Batch Processing, and benefits, 21-1, 21-8
Batch processing tips, 21-31
BATCH.SCR, 21-11
BATSCR CA function, 21-11
Bigfonts, 8-2
Base offset, 8-11
Escape characters, 8-10
Integration, 8-14
Text, 8-9
Bill of Materials, 20-2, 22-18
Binding Variables, 7-7
BLINE C:command, 19-27
Block command, 3-2, 3-3
Block def, 3-3
Block definition, 3-3
Symbol table, 13-2

C X—5

Block Insert or reference, 3-4
Blocks, 3-0, 3-1
  As temporary markers, 15-17
  Attdefs, 19-18
  Autoblocking, 19-24, 19-25, 19-27
  Colors, 3-7
  Conventions, 3-10
  DXF, 21-27
  Efficiency, 3-13 - 3-15
  Exploding, 3-14
  Extracting, 21-22
  Importing, 21-22
  Layers, 3-7
  Linetypes, 3-7
  Management, 3-1, 3-7, 3-16, 3-17, 21-18
  Minsert, 3-11
  Nested, 3-11
  Organization of, benefits, 3-1
  Parametric, 19-24, 19-25, 19-27
  Purging, 3-12
  Redefinition, 19-28, 21-18
  Scale, modifing, 12-27
  Storing data in, 19-24, 19-25, 19-27
  Subdirectories, 3-1, 3-7, 3-17
  Table data, 13-4
  TBLSEARCH function, 13-4
  Tips, 13-17, 19-28
  Unnamed, 9-5
  Updating, 21-18
  vs. Shapes, 3-15
BOLD CA function, 14-22, 14-27
Bonus Disk, I-8, 21-29
Bootup, 1-11
BOUNDP function, 11-3
BOX selection, 4-18
BSCALE C:command, 12-27
BSLASH (backslash) CA function, 21-17
Buffers, 1-11
Bulge, 12-30
  Formula, 17-11
  Polyline, 12-30
Bulletin boards, D-6
BUTTONS menu device, 5-11, 5-14
Buttons Menu toggles, 5-18
Byblock, Bylayer, 3-7, 12-23

C

CA Bonus Disk, I-8, 21-29
CA DISK, I-8, I-9
  How to use menus, 4-2, 4-3
  Installation and use, 1-9
CA-ACAD directory, 1-8
CA-ACAD.PAT, 9-16
CA-LAYER menu, 7-21
CA-LOAD CA function, 23-18
CA-MENU.MNU, 23-4, 23-19
CA-PROTO.DWG, 2-9
CA.BAT, 2-7
CAAR, CADAR, CADDR functions, 7-17
CADR function, 7-17

CALC-INC CA function, 17-13
CALCELV CA function, 17-21
CALOAD CA function, 16-8
CAR function, 12-17
CATCH CA function, 12-15
CDF Files, 14-23, 20-5
CDR function, 7-17, 12-17
CENTER CA function, 14-22, 14-27
Centerline Character, 8-4, 8-5, 8-7
Change command, 3-8
Character functions, AutoLISP, 11-6
Characters, expanding Control, 14-4
CHECKPlate hatch, 9-6
CHKDSK, 2-13
CHR function, 11-6, 14-27
CLEAN memory mgmt function, 16-5
Clipper database program, 22-24
CLOSE function, 14-4, 14-6, 14-8
CLS CA function, 14-22, 14-27
Cmdecho system variable, 23-20
Colors
  0, 3-8
  Byblock, 3-7
  Bylayer, 3-7
  Conventions, 3-10
  Entity, 3-7
  Explicit, 3-7
  Pens, 2-11
  Setup, 2-10
  Table, 2-11
Comma delimited files, 14-23, 20-5
COMMAND function, 10-7, 11-16
  ^C canceling, 10-8
  Format, 10-7
  NIL, 10-8
  PAUSE in, 18-10
Command pipeline, 10-7
Command redefinition, 23-6, 23-7
COMMAND.COM, 1-11
Commands
  See AutoCAD commands
  in scripts, 21-16
  Defining in AutoLISP, 10-2
Compile shapes, 8-7
Complex entities, 12-28
CompuServe, D-8
CompuServe Autodesk forum, D-4 - D-6
CONCREte Hatch, 9-8
COND function, 10-11, 10-15, 12-24
Conditional tests, 10-11, 10-17
CONFIG.SYS, 1-11
  Problems, B-1
Configuration
  Files, 1-8
  Prototype drawings, 2-12
  Tablet menu, 5-12
  Temporary files, B-6
CONS function, 12-17, 12-27
CONsole device, 14-3
Contours, tip 12-34
Control

characters, 11-7
Codes, 11-7
Debug flag in LISP, 16-6
Development, 16-6
Directories, 23-4
Dwg environment, 6-10, 6-21, 16-3, 23-5
ENDing, 23-7
Environment, 6-10
Error, 16-11, 16-13, 16-15, 16-17
Existing drawings, 23-5
Files, 23-2, 23-3
Input, 15-1, 15-12
Layers, 16-9
Memory, 16-3
Menu pages, 6-10
Osnap, 18-9
Reset, 23-6, 23-7
Standardization, 16-1
System, 23-2
System management benefits, 23-1
User input with INITGET, 11-22, 11-23

Conventions
Blocks, 3-10
Colors, 3-10
Layers, 3-10
Linetypes, 3-10
Scale, 3-7

Coordinates
Continuous tracking, 15-16, 15-17, 15-19
Display, 15-3

CORNERS CA function, 1-3
COS function, 17-9
CPATH CA function, 14-8
CSCALE C:command, 12-23
CSLIDE C:command, 21-14

Customizing AutoCAD DISK Files
See also Files
3DFILL.LSP, 17-13
3DTOOLS.LSP, 17-5
ACAD.KEY, 23-9
ACAD.PAT, 9-6, 9-8
ACAD.PGP, 2-5
ANGTOC.LSP, 11-13
ANSI.HLP, 14-19
ANSILIB.LSP, 14-21, 15-13
APLATE.LSP, 12-8
ARROW.DWG, 3-5
ATEXT.LSP, 11-15
ATTFIT.DWG, 20-21
ATTFIT.TXT, 22-3
ATTFITS.LSP, 20-22
ATTPIPE.DWG, 20-21
ATTPIPE.TXT, 22-3
ATTSHT-D.DWG, 19-6
AUTOBLK.LSP, 19-25
BATCHSCR.LSP, 21-9
BSCALE.LSP, 12-27
CA-ACAD.LSP, 16-6
CA-BLOCK.DWG, 1-9
CA-BLOCK.SCR, 1-9
CA-MENU.MNU, 6-11, 23-19
CA-SETUP.MNU, 6-3
CA-SIMPX.SHP, 8-11
CA.BAT, 2-7
CA.SHP, 3-15

CA18PIPE.DWG, 18-11
CADATA.DBF, 22-24
CAMASTER.MNU, 23-19
CATEMP.DBF, 22-24
COMPASS.DWG, 15-17
CONFIG.CA, 1-11
CONVDXF.EXE, 21-29
CONVDXF.LSP, 21-28
CSCALE.LSP, 12-23
DBLINE.LSP, 20-19
DDRAW.LSP, 15-17
DVIEW.LSP, 15-7
DXF.LSP, 12-22
EL90-FB.LSP, 20-16
EL90-S.LSP, 20-12
ENTITY.DWG, 12-16, 21-24
ETEXT.LSP, 15-14
ETOS.LSP, 13-17
FILELIB.LSP, 14-8, 21-17
FLOW????.DWG, 5-5
FTEXT.LSP, 21-6
GODIM.LSP, 11-19
GPATH.LSP, 23-3
HATCORN.DWG, 9-15
HATEDGE.DWG, 9-15
HATPOND.DWG, 9-15
HATSQR.DWG, 9-15
ISO-INIT.DWG, 18-3
ISO-TAB.DWG, 18-12
ISODIM.LSP, 18-9
KILL.KEY, 23-9
LBLOCK.LSP, 21-18
LEADTEXT.LSP, 16-15
LEGEND.DWG, 13-3
LEGEND.LSP, 13-5
LSPSTRIP.EXE, 23-15
MELEV.DWG, 17-18
MELEV.LSP, 17-19
MFLY.LSP, 23-13
MOFFSET.LSP, 12-5
MOREACAD.LSP, 16-20
PATTERN.LSP, 14-14
PERVARS.DWG, 13-19
PIPEFIT.DAT, 20-6
PIPEFIT.HLP, 20-7
PIPEMATL.DWG, 20-25, 22-3
PIPING.WK1, 22-14, 22-15
POINT.SLB, 5-23
PRO_TRAK.EXE, 22-24
PVAR.LIN, 13-18
PVAR.LSP, 13-22
RCLOUD.LSP, 12-30
REFDWG.LSP, 14-27
REFDWG.TXT, 14-26
REPSTAT.LSP, 22-21
REVBLOCK.DWG, 19-15, 19-16
REVBLOCK.TXT, 22-20
SAMPTEXT.TXT, 21-5
SCR-MAP.DWG, 14-20
SETSIZE.LSP, 20-9
SHEET-B.DWG, 6-5
SHEET-D.DWG, 6-5
SLOT.LSP, 10-10
SSTOOLS.LSP, 12-11
SYNPLANT.DWG, 19-8
TESTACAD.LSP, 16-20

TIMELIB.LSP, 19-14
TITLEBLK.TXT, 22-20
TOGGLE.KEY, 23-9
UDIST.LSP, 11-21
UFUNS.LSP, 11-25
UNDEFINE.LSP, 23-7
UPDATE.LSP, 19-20
VALVE-B.DWG, 19-24
VALVE.DWG, 4-14
WLAYER.LSP, 13-12
XINSERT.LSP, 21-22
XINSERT.TXT, 21-22
XPOINT.LSP, 15-10

Customizing AutoCAD Functions
See annotated listing in Appendix A
ERROR, 16-12, 16-15, 23-18
3DDIST, 17-5
3DPOLAR, 17-9, 17-10
ADDERR, 16-13
ALTITD (altitude), 17-11
ANGTOC, 11-14, 16-7
APLATE, 12-8
ARCLEN (arc length), 17-11
ATTFIT, 20-22
ATTPIPE, 20-23
BACKUP, 14-11
BANGLE (Inclined angle), 17-7, 17-8
BATSCR, 21-11
BOLD, 14-22, 14-27
BSLASH (Back SLASH), 21-17
C:3DFILL, 17-13
C:3DPOLAR, 17-9
C:ATEXT, 11-15
C:BLINE, 19-27
C:BSCALE, 12-27
C:CSLIDE, 21-14
C:DDRAW, 15-18
C:DIMLINE, 18-9
C:DVIEW (Dynamic VIEW), 15-7, 15-8
C:END, 23-7
C:ETEXT, 15-14
C:EXTLINE, 18-10
C:FLY, 23-13
C:FTEXT, 21-6
C:LBLOCK, 21-1
C:LEGEND, 13-6
C:MARK, 12-14
C:MELEV, 17-20
C:MFLY, 23-13
C:MSCRIPT, 21-13
C:PATTERN, 14-14
C:PICKVPT, 17-8
C:PVAR, 13-26
C:RCLOUD, 12-30
C:REFDWG, 14-27
C:REPSTAT, 22-21
C:S-LEAD (Straight LEADer), 16-16
C:SH-BOX, 1-5
C:SLOT, 10-10
C:SSLIDE, 21-14
C:UPDATE, 19-20
C:WLAYER, 13-12
C:XINSERT, 21-22
C:XPOINT, 15-10
C:ZARC (vertical arc), 17-10
CA-LOAD, 23-18

CALC-INC, 17-13
CALCELV, 17-21
CALOAD, 16-8
CATCH, 12-15
CENTER, 14-22, 14-27
CLS, 14-22, 14-27
CORNERS, 1-3
CPATH, 14-8
CSCALE, 12-23
DBLINE, 20-19
DELERR, 16-13
DOWNROW, 14-22
DTOR, 16-8
DXF, 12-22, 16-7
EL90-FB, 20-16
EL90-S, 20-12
ETOS, 13-17, 16-7
FFNAME, 14-27, 21-20
FFNAME (Format File NAME), 14-10
FSLASH (Forward SLASH), 21-17
FTEXT, 22-19
GETBLK, 19-26
GETFIL, 21-9
GETPVAL (GET Pvar VALue), 13-24
GETSCR, 21-10
GETSIZE, 20-10
GODIM, 11-19
GOTO, 14-22, 14-27
GPATH, 23-3
INCANG (included angle), 17-11, 20-16
INITERR, 16-13
L-TEXT (Leader TEXT), 16-17
LASTN, 12-14
LBLOCK, 21-19
LEDSWAP, 13-9
LINEBLK, 19-27
MERGEF, 14-11
MERGEV, 14-12, 14-16
MID3D, 17-6
MIDOF, 16-8
MLAYER, 16-9
MOFFSET, 12-5
NORMAL, 14-22, 14-27
OUTLEDG, 13-7
PUTVAR (PUT VARiable), 13-23
PVAR, 13-22
PVCHK (PVar CHecK), 13-22
RADIUS, 17-11
REFOUT, 14-30
REFSEL, 14-29
RESET, 16-13
RESTPOS, 14-22, 14-28
RETVAR (RETurn VARiable), 13-23
REVALL, 19-22
REVTIME, 19-23
SAVEPOS, 14-22, 14-28
SETSIZE, 20-9
SSDIFF, 12-12
SSINTER, 12-12
SSUNION, 12-11
STRIP, 15-17, 16-7
TIME, 19-14
TODAY, 19-14, 19-21
UANGLE, 11-26, 16-10
UDIST, 11-21, 12-5, 16-10
UINT, 11-26, 12-5, 13-25, 16-10

UKWORD, 11-27, 13-25, 16-10
UNDEF, 23-7
UPD, 19-23
UPDSCR, 21-20
UPOINT, 11-26, 12-5, 16-10
UREAL, 11-26, 16-10
USTR, 11-27, 16-10
VFFILE (VeriFy FILE), 14-9, 14-16
VFPATH, 21-17, 21-19
YEAR, 19-14

D

Data
   Attributes, 19-17
   DXF codes, 21-24
   External data storage, benefits, 14-1
   External files, 20-5, 20-8, 20-9
   Getting data, 11-1
   Handling external, 14-23, 14-25
   Import/export, tips 22-1, 22-18, 22-31
   Importing to dBase, 22-26, 22-27
   Importing to Lotus, 22-6, 22-7, 22-9,
      22-11, 22-13, 22-15, 22-17
   Parametric, 20-5
   Reporting, 22-28
Data files
   CDF comma delimited format, 20-5
   Large, 20-26
   Tips, 20-26
Data types, 13-24
   Determining, 11-2, 11-3
   How to use, 11-1
Date stamping and calculating, 19-13
dBASE
   Conditionals, 22-26
   CREATE, 22-25
   Data from AutoCAD, 22-21
   Importing data, 22-26, 22-27
   Looping, 22-26
   Preparing input, 22-21, 22-23
   Procedures, 22-24
   Programming, 22-24
   Prompting, 22-25
   Structure, creating, 22-25
   Tips, 22-31
   Transferring data, 22-27
   Using with AutoCAD, 22-19
DBLINE CA function, 20-19
DDRAW C:command, 15-18
DEBUG flag control, 16-6, 16-8, 16-10,
      16-12, 16-14, 16-19, 23-18
Decimal point, 7-12
DEFUN function, 10-5, 16-4
DELERR CA function, 16-13
Device access, 14-1
   GRTEXT, 15-2
   Alternatives, 15-9, 15-11
   GRDRAW, 15-5
   GRREAD, 15-12
Devices, 14-1
   Access modes, 14-5

CONsole, 14-3
I/O, 14-2
NUL, 14-7
Printer, 14-6
Table of, 14-4
Digitizer
   Input control, 15-13
   Tracking, 15-16
Dimasz, 2-10
Dimblk, 18-6
Dimensioning, custom benefits, 18-1
   Associative caution, 18-16
   AutoLISP, 11-18
   Customizing, 18-1
   Macro, 7-14
   Tip on fitting, 18-16
DIMLINE C:command, 18-9
Dimscale, 2-10, 3-6, 6-8
Dimse1, Dimse2, 18-8
Dimtxt, 2-10
Dimtypes
   Iso-dimensioning, 18-3
   Iso-dimensioning illustrated, 18-4
Dimzin and RTOS, 11-11
Directories
   CA-ACAD, 1-18
   Control, 14-33, 23-2, 23-3
   DOS Subst control, 23-4
Disabling functions, 13-6
Disk (CA DISK)
   Customizing AutoCAD disk set, I-8
   CA Bonus Disk, I-8, 21-29
Distance
   3-D, 17-5
   Formatting, 11-11
DISTANCE function, 7-15, 11-19, 17-4
Divide / function, 10-10
Documentation, 23-16, 23-17
   Benefits, 23-1
   Displaying .LSP, 23-16
   Menu, 6-5
[DOORS] macro, 17-19
DOS, I-7
   Append command, caution, 14-9, B-2
   AutoCAD environment, 2-1
   AutoLISP file handling, 14-11, 21-16
   Commands, adding to AutoCAD, 2-1
   Debug command, B-3
   Directories, 1-7
   Environ. settings, problems, B-1, B-3
   Path, 1-8, 1-12, 23-3, B-2, D-2
   Print command caution, B-5
   Prompt, 1-12, 2-5
   Redirection symbol, 14-8
   Replaceable parameters, 2-9, 22-14
   Tips, 1-13
   Vdisk command caution, B-5
   Version, 1-13
   Version 2, use and caution, B-3
Dot errors, B-8
Dotted pair, 12-17
DOWNROW CA function, 14-22
Drag command, 4-15

Drawing database access, benefits, 12-1
Drawing exchange format DXF, 7-19
Drawing file formats, 21-22
Drawing revision system, 19-13
Drawings
    Control, 23-5
    Prototype, 23-5
Dtext command, 8-15
DTOR CA function, 16-8
DUP PGP command, 2-5
DVIEW C:command, 15-7, 15-8
DWGS PGP command, 2-5
DXB
    Configuration, 17-24
    Files, 17-24
    Tip, 17-25
Dxbin command, 17-24
DXF
    -1 entity code, 12-19
    -2 subentity code, 13-4
    Attributes, 19-17, 19-18
    Block Reference, 21-27
    Codes, 7-19, C-4
    Entities section, 21-27
    Entity codes, Table 12-20, C-4
    Extraction function, 12-22
    File format, 21-25
    Group codes, 12-18, 12-19, 12-21
    Group codes and data, 21-24
    Header information, 21-25
    Importation, 21-4
    Processing files, benefits, 21-1, 21-29
    Spelling checker, 21-31
    Tables section, 21-26
    vs. System variables, 21-25
    When to use, tips, 21-8, 21-23
DXF CA function, 12-22, 16-7
DXFTEST.DXF, 21-24

E

ECHO, B-1
ED PGP command, 2-5
Editing commands vs. ENTMOD, 15-18
EDLIN, 1-10
EL90-FB CA function, 20-16
EL90-S CA function, 20-12
Elevation, 12-23, 17-2
    Tips, 17-26
Elevations, Automatic, 17-16 - 17-23
Ellipse, 18-16
Encryption, 23-14
END C:command, 23-7
End command, 23-7
End-of-file character <^Z>, 14-11, 14-16
ENDBLK, 21-25
ENTDEL function, 12-3, 12-4, 12-13
    vs. Erase, 12-4, 15-10
ENTGET function, 7-19, 12-16, 13-6
Entities
    3-D, 17-2
    Accessing data, benefits, 12-1, 12-16
    Association lists, 12-17
    Byblock, 12-23
    Complex, 12-28
    Database order, 12-13, 12-15
    Default properties, 12-22
    DXF codes, 12-18, 12-19, 12-21, 21-27
    DXF extraction function, 12-22
    Elevation, Thickness, 12-23
    in Block definitions, 13-2
    Inserts, 19-17
    Modifing, 12-2, 12-26, 12-27
    Order of, 13-8
    Polylines, Modifing, 12-31
    Selection by point, 20-14
    Selection, see Selection Sets
    SEQEND, 19-17
    Subentities, 12-28, 13-4
    Updating, 12-26, 12-27
Entity Color, Linetype, 3-7
Entity data, 7-19
    Appending, 15-17
    Attdefs in Blocks, 19-18
    Attributes in Inserts, 19-18
    How to use, 12-1
    Removing sublists, 15-17
Entity names, 7-19
    AutoLISP, 12-2, 12-3, 12-5
    DXF -1 code, 12-19
    DXF -2 subentity code, 13-4
ENTLAST function, 12-3, 12-13
ENTMOD function, 12-26, 12-27, 15-18
    vs. Editing commands, 15-18
ENTNEXT function, 12-4, 12-28
ENTSEL funct, 12-3, 12-17, 12-28, 13-9
    Entity selection by point, 20-14
ENTUPD function, 12-26, 12-31, 19-20
Environment
    Control drawing, 16-3
    Initialization, tip 18-17
    Resetting, 16-13
EQ function, 10-13
EQUAL function, 10-13
Equal to = function, 10-13
Erase command vs. ENTDEL, 15-10
Errors
    AUTOEXEC.BAT, B-1
    AutoLISP, B-8
    CONFIG.SYS, B-1
    Control of, 16-11, 16-13, 16-15, 16-17
    Directory, B-7
    DOS, B-1, B-3
    dotted pair, misplaced dot, B-8
    Environment settings, B-2, B-3
    Free RAM, B-5
    Insufficient files, B-9
    LOADing LISP files, B-9
    Memory, B-3 - B-5, B-9
    Opening files, B-9
    Overlays, B-9
    Recovery in AutoLISP, 16-0
    SHELL, B-8
    Tracing and curing, B-9, B-11

Unknown command, B-9
ETEXT C:command, 15-14
ETOS CA function, 13-17, 13-26, 16-7
EVAL function, 11-17, 23-13
Evaluation, LISP, 7-8, 7-10, 10-3, 10-5
EXAMPLES menu, 6-12
Exercises
   See listing by chapters, following the general index
   How shown, I-6
   Syntax, I-6, I-9
Exit command, 2-5
EXPAND function, 16-3
Expanded ASCII codes in AutoLISP, 11-7
EXPanded memory, B-5
EXPanded memory disabled error, B-9
Expanding control characters, 14-4
Expert System Variable, 13-23
Explode command, 3-14, 12-14
   Order of entities, 13-8
   Explode tip, 12-34
Expressions AutoLISP, 7-7, 7-8
EXTended memory disabled error, B-9
Extended ASCII codes in DOS, 23-9
Extended I/O page space, B-5
Extended I/O page space, too much, B-6
External commands, 2-2
EXTLINE C:command, 18-10

F

FFNAME CA function, 14-10, 14-27, 21-20
Figures
   See listing following general index
File access tips, 14-32
Files, 14-4
   See also Customizing AutoCAD DISK Files
   .DWG format, 21-22
   .DXB, 17-24
   .LSB, 23-16
   .LSC, 23-14
   .LSP, 10-9
   .MNX, 23-14
   .SHP, 8-7
   .SHX, 8-7
   .SLB, 5-23
   ACAD.CFG, 1-8
   ACAD.LIN, 9-2
   ACAD.LSP, 16-6
   ACAD.PAT, 9-5
   ACAD.PGP, 2-3, 2-5
   ACAD??.OVL device overlays, 1-8
   Appending, 14-11
   AutoCAD's Temporary, B-6
   AutoLISP, 10-9
   BATCH.SCR, 21-11
   CA-ACAD.PAT, 9-16
   CDF comma delimited, 14-23, 20-5
   CONFIG.SYS, 1-12
   Configuration, 1-8
   Data formats, 20-5
   Directory control, 23-2, 23-3
   Drivers, 1-8
   DXF format, processing, 21-25, 21-29
   Encryption, 23-14
   End-of-file character, 14-11
   Error finding, B-9
   Error opening, B-9
   Error, insufficient, B-9
   External, 20-5
   Finding support files, B-7
   Formats, 14-23, 14-25
   Handles, functions for, 14-4, 21-16
   Importing Data, tips 22-31
   Library files, tips, 18-17
   Linetype, 9-2, 13-18
   LIST format, 14-25
   Manually Importing to 123, 22-6
   Organization, 23-2, 23-3
   Parametric, 20-5
   PGP, 2-3, 2-5
   Read-only protection, 23-14
   Reducing .LSP size, 23-15
   Required, 1-8
   SDF Standard Data Format, 14-24
   Security, 23-14
   SSLIDE.SCR, 21-14
   Stripping .LSP files, 23-15
   Support, 1-8
   Templates, Attribute Extraction, 22-3
   TEST.DXF, 21-4
   Testing existence, 14-7, 14-9
   Text fonts, 8-1
Files command, 14-11
Fills, irregular, 9-1
FITTINGS.TXT, 22-5
FIX function, 15-18
Flags, 7-9, 20-12
FLOWLINE menu, 6-15
FLOWSYMB menu, 5-9
FLY C:command, 23-13
Fonts, Text, 8-1
   Definition codes
   Table, 8-6
FOOTER menu, 6-12
FOREACH function, 10-19, 14-24, 20-13
Fractions, 8-2
   Special text characters, 8-8, 8-11
Free RAM, B-5
Freeware, D-8
FSLASH (fwdslash) CA function, 21-17
FTEXT C:command, 21-6
FTEXT CA function, 22-19
Functions, 10-2
   Arguments, 10-4
   Defined, 7-9
   Disabling, 13-6
   Executing commands while loading, 18-10
   Loading, 10-11, 16-8
   Recursive, 13-24
   Self-loading, 23-18
   Testing, 13-6
   User defined, 10-6

G

GC function, 16-3
GET functions, 7-10
    Base point, 7-13
    Menus, 7-13
    Pausing Problems, 7-15
    UDIST, 11-0
GETANGLE function, 7-11, 7-13, 7-15
    UANGLE, 11-25
GETBLK CA function, 19-26
GETCORNER function, 7-13
GETDIST function, 7-11, 7-13, 7-15, 10-10
    UDIST, 11-21
GETFIL CA function, 21-9
GETINT function
    UINT, 11-25
GETKWORD function
    UKWORD, 11-25
GETORIENT function, 7-13
GETPOINT function, 7-11, 7-13, 10-10
    UPOINT, 11-25
GETPVAL (Pvar) CA function, 13-24
GETREAL function, 7-10
    UREAL, 11-25
GETSCR CA function, 21-10
GETSIZE CA function, 20-10
GETSTRING function, 7-10
    USTR, 11-25
GETVAR function, 7-5
Global variables, 10-5, 16-3
GODIM CA function, 11-19
GOTO CA function, 14-22, 14-27
GPATH CA function, 23-3
Graphics screen, writing to, 15-1
GRAPHSCR function, 11-8
GRCLEAR function, 15-6
GRDRAW function, 15-0, 15-5, 15-8
    Blanking lines, 15-6
    Highlighting lines, 15-5
    Reversing colors, 15-6
    Tips, 15-21
Greater than > function, 10-13
Greater than or equal to >=, 10-13
Grid, Isometric, 18-8
GRREAD function, 15-12, 15-18
    Cautions, 15-16
    Devices, 15-13
    Table of codes, 15-13
    Tips, 15-21
GRTEXT function, 15-0, 15-2, 15-8, 20-4
    Boxes, 15-2
    Clearing, 15-4
    Highlighting, 15-4
    Updating boxes, 15-4
    Tips, 15-21

H

Hard Disk
    Mgmt, backup, speed tips D-7
HATCH menu, 9-15
Hatches, 9-1
    * Hatches, 9-5
    and Blocks, 9-5
    and Fit polylines, 14-18
    Automatic generation, 14-13, 14-15, 14-17
    Base point, 9-5
    CHECKPlate, 9-6
    CONCREte, 9-8, 9-14
    Custom, benefits, 14-13, 14-15, 14-17
    Defining, 9-5
    Development box, 9-8
    Efficiency, tips 9-17, 9-19
    Exploding, 9-5
    File conflict, ACAD.PAT, 9-16
    HATCORN, 9-15
    HATEDGE, 9-15
    HATPOND, 9-15
    HATSQR, 9-15
    Irregular, 9-15
    Line families, 9-5
    Partial, 9-15
    Snap, 9-5
    X,Y offset and X,Y origin, 9-5
HEAP, 2-8, 16-4
Help
    Customizing AutoCAD's, 20-7
    Displaying .LSP files, 23-16
    Index, HELP.HDX, 20-7, 20-8
    Slides as, 23-17
Help command, 23-17
Hide command, B-5
Highlight system variable, 12-12
Highlighting
    Graphics, GRDRAW, 15-5
    Text, GRTEXT, 15-4

I

I/O Devices, 14-2
ICON menu and device, 5-11, 5-23, 5-25
IF function, 10-11, 10-13
IF-THEN-ELSE, 10-13
Illegal characters, 10-4
Illegal variables, 7-6
INCANG (included angle) CA function, 17-11, 20-16
INITERR CA function, 16-13
INITGET function, 11-22
    3-D points, 17-4
    Control bit table, 11-22
    Keywords, 11-23
[INITIATE] macro, 6-8, 19-7
Input
    Controlling, benefits, 15-1, 15-12
    Filtering, 15-12, 16-23
    GRREAD control, 15-12
    Parametric, 20-9
Insert command, 3-4

Insert* command, 3-4
Inserts, 3-0, 3-8, 12-25, 12-28, 17-2
    INS key in menus, 4-21
    Attributes in, 19-18
INTeger variables, 7-4, 11-2
    To ASCII, 11-10
Integrating menus
    3D menu, 17-25
    HATCH menu, 9-16
    ISO DIMS menu, 18-15
    PIPEFITS menu, 20-25
    SPECIALS, 8-16
    TEXT menu, 6-18, 11-28
    TITLBLK menu, 19-29
INTERS function, 17-4, 17-22, 20-12
Invalid Characters, 16-21
Invalid dotted pair, 7-12
Invalid variables, 7-7
ISO menu, 18-6
Iso-dimensioning
    Associative caution, 18-16
    Dimtypes illustrated, 18-4
    Dimtypes table, 18-3
    Overview, benefits, 18-1, 18-2
    Symbols, Text styles, 18-3
    Tablet menu, 18-12, 18-13
Isometric
    Axis, 18-8
    Grid, 18-8
    Snap, 18-8
Isoplanes, 18-3
Iteration, 10-18
ITOA function, 11-10

K

Kelvinator, 23-14
Keyboard
    ALT codes, 8-7
    ANSI.SYS redefinition, 23-9, 23-11
    Control, 23-8
    Input control, 15-13
    Macro programs, tips, 5-27, 23-8
    Macros, D-8
Keywords
    Association list, 12-17
    INITGET, 11-23

L

L-TEXT (LeaderTEXT) CA function, 16-17
LAMDA function, 16-5
LASTN CA function, 12-14
Layers
    0, 3-7
    Color, 3-7
    Conventions, 2-10, 3-10
    Filtering names, 2-11, 13-12
    Linetype, 3-7

Macros, 4-14
Management, 16-9
Pen assignment, 2-11
Properties, 3-7
Protected, 13-20
Reference, 6-5
Setup, 2-10
Symbol table access, 7-20, 13-2, 13-11
Table data, 13-12
Wildcards, 2-11
LBLOCK C:command, 21-1, 21-19
LEDSWAP CA function, 13-9
LEGEND Attributes, 13-5
LEGEND C:command, 13-6
LEGEND CA function, 13-0
LENGTH function, 7-18
Less than < function, 10-13
Less than or equal to <=, 10-13
LINEBLK CA function, 19-27
Lines
    Blanking, 15-6
    GRDRAW, 15-5
    Temporary, 15-5
Linetypes, 9-1, 9-2
    Alignment code, 9-3
    Byblock, Bylayer, 3-7
    Conventions, 3-10
    Creating, 9-2
    Definition, 9-3
    Entity, 3-7
    Explicit, 3-7
    Files, 13-18
    How stored, 13-23
    Scale, Plot scale, 9-3
    Setup, 2-10
    Table of the book's, 2-11
    Table access, data, 13-13, 13-15
LISP, Defined, vs AutoLISP, 10-1
LISPSAVE menu, 7-12
LIST, AutoLISP data, 7-16, 10-2
List command, 3-8
LIST function, 13-26, 14-28
LISTP function, 11-3, 11-4
Lists, 10-1, 10-15, 11-2
    As COMMAND input, 11-16
    Processing, 10-19
    Long list tips, 11-29, 16-5
Load command, 3-15
LOAD function, 10-11
Loading
    ACAD.LSP, 16-15
    Functions, 16-8
    Scripts to load programs, 21-31
LOADing functions faster, 23-15
Local variables, 10-5
Logical drives, 23-4
Logical operators in AutoLISP, 10-12
Looping structures, 10-16
Lost attributes, 19-28
Lost clusters, 2-13
Lotus, 22-6, 22-14
    Automating, 22-14

Converting data, 22-10
Importing data, 22-6, 22-7, 22-9, 22-11,
  22-13, 22-15, 22-17
Integration 22-2
Macros in, 22-16
Output, 22-18
Parsing data, 22-8
Program flow, 22-0
Tips, 22-31
LSP PGP command, 2-5
LSPSTRIP caution, 23-16
LSPSTRIP.EXE, 23-15
Ltscale command, 9-3
Ltypes Symbol table, 13-2

M

Macros
  * repeating, 4-16
  + continuation character, 4-13
  ; Semicolons as return, 4-6
  <SPACE> character, 4-7
  ^C cancel, 4-9
  [labels] square brackets, 4-7
  Attribute input, 19-9
  AutoLISP tips, 7-23
  Backslashes / in, 4-4, 5-17
  [CA DIMS], 7-14
  [DOOR], 17-19
  DOS path names, 4-8, 4-9, 4-11
  Keyboard, D-8, 23-8
  [INITIATE], 6-8, 19-7
  Labels, 4-7
  Layers, 4-14
  Long, 4-12, 4-13, 21-4
  [MAKEWALL], 17-18
  On-the-fly, 23-12
  Osnap QUIck, 4-13
  Overview, benefits, 4-1, 4-2
  Previous selection, 4-13
  Repeating, 4-15, 4-16
  RETURN, 4-6, 4-7
  Select, 4-17
  Showing status, 18-6
  Special characters, 4-4 - 4-6
  Suspended commands, 5-17
  Syntax, 4-4, 4-5
  Table, 4-6
  Text in, 4-7
  Tips, 4-20, 4-21
  Toggle keys, 4-9
  Transparent commands, 5-19
  [WINDOW], 17-18
Magazines, D-6
MAIN.MNU, 6-13
Management
  Blocks, 21-18
  Drawing status, 22-24
  Hard disk, D-7
  How to manage system, 23-1
  Memory, 16-3
  Multitasking, D-3
  Project tracking, 19-3

RAMdisk and speed, B-6
System Organization, benefits, 1-1
Time, 22-29
MAPCAR function, 13-26, 17-5, 20-13
  Set variable lists, 20-13
MARK C:command, 12-14
Material tagging, 20-21, 20-23, 22-3
MELEV C:command, 17-20
MEMBER function, 14-29
Memory
  AutoLISP, 16-4
  Errors, B-9
  Expanded, B-5, 16-3
  Extended, 16-3
  Heap, 16-4
  LIM-EMS, B-5
  Management, B-2, D-1, 16-1, 16-3
  Paging virtual, 16-4
  Problems, B-4 - B-6
  Resident TSR programs, B-1, D-2
  Stack, 16-4
  Tips, 11-29, 16-5
  Virtual, 16-3
Menu, 6-1
  Applications, 6-11, 6-13, 6-15
  AutoLISP Configuration, I-6
  CA DISK, I-9
  Devices, and tips, 5-11, 5-26
  Icons, I-8
  Organization, standardization, 5-1, 6-1
Page names, 5-9
  Popups, I-8
  Standard AutoCAD, 4-1
  Toggling, 18-4
Menuecho system variable, 23-20
Menus
  See also Macros
  See annotated listing in Appendix A
  $ page code, 5-10
  $I page code, 5-23, 5-25
  $P page codes, 5-21
  $S screen code, 5-10
  * Next paging, 5-8
  ** page names, 5-9
  *** device names, 5-11
  ***BUTTONS, 5-14
  ***ICON device name, 5-23, 5-25
  ***POP device names, 5-21
  ***SCREEN, 5-13
  ***TABLET1, 5-13
  **3D, 17-18
  **ANNOT, 6-15
  **CA-LAYER, 7-21
  **EXAMPLES, 6-12
  **FLOWLINE, 6-15
  **FLOWSYMB, 5-9
  **FOOTER, 6-12
  **ISO, 18-6
  **LISPSAVE, 7-12
  **P1-SETVAR popup, 5-22
  **PIPEFITS, 20-4
  **POINTS, 5-24
  **PTFILTERS, 5-18
  **REPORT, 22-14, 23-19
  **ROOT, 6-11, 23-20

**SPECIAL, 8-14
**TEXT, 6-18
**TISO, 18-14
**TITLEBLK, 19-11, 22-23
**TRANSVIEW, 5-20
Analyzing needs, 6-1
Appending, 6-18
AutoLISP in, 7-8
AUXILIARY device, 5-11
Box numbers, 15-2
BUTTONS device, 5-0, 5-11
Button toggles, 5-0, 5-18
CA-MENU completed, 23-0
  Control of, 23-22
  Control points in, 6-10
  Devices, 5-11
  Documentation, 6-5, 23-16
  Drag in, 4-15
  Dummy pages, 23-16
  Dynamic labels, 15-2
  Echoing prompts, 23-20
  Encrypted, 16-22
  Finalizing the book's, 23-19
  Flat structure, 5-3
  Format, 4-19
  GET functions in, 7-13
  Highlighting boxes, 15-4
  How to use with the book, 4-3
  ICON, 5-11, 5-23, 5-25
  Insert INS key, 4-21
  Integrating ACAD.MNU, 23-21
  Integration tips and problems, 23-22
  Layout, 5-3
  Long macros, 4-12, 4-13, 21-4
  Next paging, 5-8
  Pages, 5-1
  Page access control, 6-10
  Page switching, 5-16, 6-10
  Parametrics, 20-2, 20-3
  POPUP, 5-11, 5-21
  Random structure, 5-3
  SCREEN, 5-11
  Security, 23-14
  Sequential structure, 5-3
  Setup, 6-0, 6-2, 23-5
  Single page, 5-3
  Structures, 5-1, 5-3, 5-11
  Structures, 5-3
  Suppressing prompts, 23-20
  TABLET, 5-11
  Template, 6-3
  TEST.MNU, 6-17
  Tips, 6-19
  Toggle labels, 20-4
  Toggling pages, 5-0, 5-18
  Tree structure, 5-3
MERGEF CA function, 14-11
MERGEV CA function, 14-12, 14-16
Mesh generator, 17-0
  3-D, 17-12, 17-13, 17-15
MFLY C:command, 23-13
MID3D CA function, 17-6
MIDOF CA function, 16-8
Midpoint, 16-8
  3-D, 17-6

MIN function, 17-22
Minsert command, 3-13
MINUSP function, 11-4
MLAYER CA function, 16-9
MNU PGP command, 2-5
MOFFSET C:command, 12-5
Move command vs. ENTMOD, 15-18
MSCRIPT C:command, 21-13
Multiple command, 4-17
Multiply * function, 7-9, 10-10
Multitasking environment, D-3

N

Next in menus, 5-8
NIL, NON-NIL, 10-8, 10-11, 10-12, 10-15
NORMAL CA function, 14-22, 14-27
Not equal to /= function, 10-13
NOT function, 10-12
NTH function, 7-17, 14-29
NUL device, 14-7
NUMBERP function, 11-4, 13-27
Numbers to strings, 11-10

O

Object selection, 5-14, 7-19, 12-4
  See also Selection Sets
  Last, 12-4
Offset command, B-5, 12-5, 20-14
OPEN function, 14-4, 14-5, 14-8
OR function, 10-12, 10-13, 10-14, 14-11
Organization
  AutoCAD System, benefits, 2-0, 2-1
  CA System, 1-0
  Files, 23-2, 23-3
  Hatches, 9-0
  Linetypes, 9-0
  Menu, 5-1
  Of the book, I-2
Osnap, 5-14, 12-4
  Caution, 20-12
  Control, 18-9
  QUIck, 4-13
Osnap command, 4-13, 18-9
OSNAP function, 19-27
OUTLEDG CA function, 13-7
Overlay files, Errors, B-9

P

P1-SETVARS popup menu, 5-22
Page, Menu, 5-1
Parametrics
  Attribute controlled, 19-24, 19-25, 19-27
  Autobreaking blocks, 19-24, 19-25, 19-27

Calculating points, 20-16
Calculations, 20-11
Data retrieval, 20-10
Door, 17-19
Drawing components, 20-11
Input, 20-9
Menus, 20-3
Multiple views, 20-15, 20-17
Osnap caution, 20-12
Overview, benefits, 20-1
Planning, 20-11
Program flow, 20-0
Tips, 20-26, 20-27
Window, 17-19
Parenthesis, Syntax, 7-7, 7-8

Parsing, 14-23, 14-24
Parts, 3-2
Scale, 3-4, 3-5
Unit blocks, 3-2
Path, 1-8, B-2, D-2, 23-3
AutoLISP variables to, 23-3
Control, 16-21
DOS, 1-12
In macros, 4-8, 4-9, 4-11
Testing, 14-7, 14-9
Verifying from AutoLISP, 21-17
PATTERN C:command, 14-0, 14-14
Patterns
Automatic hatches, 14-13, 14-15, 14-17
Custom, 14-13, 14-15, 14-17
Defined, 9-1
Dot-dash, 9-1
Linetype, 9-2
Mesh, 17-12, 17-13, 17-15
PAUSE in COMMAND function, 18-10
Pdmode, Pdsize, 5-23, 5-24
Pedit, 4-11
Pens
Colors, 2-11
Logical, 9-2
Table of book's, 2-11
Personal variables, Tips, 13-1, 13-29
PGP commands and file, 2-2
PI constants, 10-10
PICKVPT C:command, 17-8
Pipe fittings
Elbow, 20-12
Flanges, 20-19
Parametric, 20-1
Views, 20-6
PIPEFITS menu, 20-4
Pipeline, the COMMAND, 10-7
Pline, Width in macros, 4-11
Pline command, 4-11
Plot automation tips, 13-29
Plotting batch scripts, 21-16
Points, 17-2
3-D format, 17-3
Calculating, 20-16
Calculating 3-D, 17-5
Display size, 15-9
Processing lists of, 15-9

POINTS icon menu, 5-24
POLAR function, 10-10, 13-8, 17-4
3-D, 17-6, 17-7, 17-9
Polylines, 12-28
Modifing, 12-31
POPUP menu device, 5-11
Popup menus, 5-21
Popups
Opening, 5-21
Pulling down, 5-21
Predicates, 11-2
Alternatives, 11-4
Prerequisites, I-7
CONFIG.SYS, 1-11
Directories, 1-7
DOS, I-7
For exercises, I-9
Hardware, I-8
Memory, I-8
Version, I-6, I-8
PRIN1 function, 11-10, 14-3
PRINC function, 10-3, 11-10, 13-26, 14-3
Clean finish, 11-10
PRINT function, 11-10, 14-3
Printing strings, 11-9
PRO_TRAK, Using, 22-30
PROCEDUREs, dBASE, 22-24
PROGN function, 10-14
ProGram Parameter PGP file, 2-2
Project management tracking, 19-3
PROJECT.DAT, 22-21, 22-28
PROMPT function, 10-3, 11-7, 11-9, 14-3
Prompt, DOS, 1-12
in SHELL, 2-5
PROTECT.EXE, 23-14
Prototype drawing, 2-1, 2-9
Configuring, 2-12
PTFILTERS menu, 5-18
PUTVAR (Pvar) CA function, 13-23
PVAR C:command, 13-22, 13-26
PVARs, 13-18
PVCHK (PVar CHecK) CA function, 13-22

Q

QUOTE ' function, 10-3, 11-3

R

RADIUS CA function, 17-11
RAMdisk, B-5, B-6
Using with AutoCAD, B-6
RCLOUD C:command, 12-30
READ function, 11-6, 13-16, 13-27
Read only variables, 7-4
READ-CHAR function, 14-6
READ-LINE function, 14-6, 21-9

Read-only files, 23-14
Reading devices, 14-1
Real variables, 7-4
REALs, 11-2
   To strings, 11-11
Recursive functions, 13-24, 16-5
Redefine command, 23-6
Redefining commands, 23-6, 23-7
REFDWG C:command, 14-27
Reference Drawing Schedule, 14-26
REFOUT CA function, 14-30
REFSEL CA function, 14-29
Relational operators in AutoLISP, 10-13
REN PGP command, 2-5
REPEAT function, 10-16
Repeating Macros, 4-16
Replaceable parameters, 2-9
REPORT menu, 22-14, 23-19
REPSTAT C:command, 22-21
Reset controls, 18-17, 23-6, 23-7
Reset C:command, 23-6
RESET CA function, 16-13
RESTPOS CA function, 14-22, 14-28
Resume command, 21-16
RETURN
   ^M in Lisp/Macros, 11-8
   ASCII codes, 11-7
   In macros, 4-6
RETVAR (Pvar) CA function, 13-23
REVALL CA function, 19-22
REVBLOCK.DWG attributes, 19-15
REVERSE function, 7-18
Revisions
   Automating, 19-13
   Clouds, 12-29
   Level, 19-21
   Outputing status, 22-23
   System for, 19-13
   Tracking, 19-15, 22-24
   UPDATE C:command, 19-19, 19-21, 19-23
REVTIME CA function, 19-23
ROOT menu, 6-11, 23-20
Rotation angle, 11-12
RTOS function, 7-15, 11-11, 11-13, 14-15
   0 control, 11-11
   Dimzin, 11-11
   Formats, 11-11
Rubberbanding, 7-13
RUN123.BAT, 22-14

S

S-LEAD (str leader) C:command, 16-16
SSADD function, 12-7
SAVEPOS CA function, 14-22, 14-28
Scale, 2-9, 3-1, 6-7
   Convention, 3-7
   Linetype, 9-3
   Parts, 3-4, 3-5

   Plot, 3-4
   Symbols, 3-4
Schedules, in drawings, 14-26
SCREEN menu device, 5-11, 5-13
Screens
   ANSI format codes, 14-19, 14-21
   Dynamic labeling, 15-2, 15-3
   How shown in book, I-7
   Size calculations, 15-7
Screensize system variables, 15-7
Scripts, 21-3
   And AutoLISP, 21-18, 21-19, 21-21
   And plotting, 21-16
   Automate building, 21-8, 21-11, 21-13
   Environment, 21-16
   Repeating, 21-11
   Resuming, 21-16
   Rscript, 21-11
   Stopping, 21-11, 21-16
   vs. menus, AutoLISP, external prog, 21-3
   When to use, benefits, 21-1, 21-8
SDF Standard Data Files, 14-24
Security, 23-14
Select command, 4-12, 4-17, 12-12
Selection Sets, 4-17, 12-1
   See also Object Selection
   Adding, 12-11
   AUto, 4-18
   AutoLISP manipulation, 12-8
   BOX, 4-18
   Entities, 12-6, 12-7
   Intersection of, 12-11
   Modes, 4-18
   SIngle, 4-18
   Subtracting, 12-11
   Tips, 12-34
Self-loading functions, 23-18
SEQEND, 19-17, 21-25
SET
   DOS command, B-3
   Environment settings, B-2, 2-7, 2-8
   Environment variables, 2-12, 2-13
SET ACAD=, 2-8
SET ACADCFG=, 2-8
SET ACADFREERAM=, B-5, 2-8
SET ACADLIMEM=, B-6
SET ACADXMEM=, B-5
SET function, 11-6, 20-13
SET LISPHEAP=, 2-8
SET LISPSTACK=, 2-8
SET vs. SETQ, 11-6, 20-13
SETQ function, 7-8, 7-10, 20-13
SETSIZE CA function, 20-9
Setup
   AutoCAD's DOS enviornment, 2-1
   DOS, 1-1
   Drawing environment, 23-5
   Existing, 1-8
   Our setup for the book, D-1
   Subdirectory, 1-7
   Tips, D-1, 6-19
Setup menu, 6-2
Setvar tranparent commands, 7-4

Setvar command, 7-3
SETVAR function, 7-5
SH-BOX C:command, 1-5
SH PGP command, 2-5
Shape command, 3-15
Shapes, 3-0, 3-15
Shareware, D-8
Shell, 2-2, 14-12
   AutoCAD, 2-2
   CONFIG.SYS, 1-12
   Directory control, 22-14
   Errors, B-8
   Exit, 2-5
SHELL PGP command, 2-4
   and AutoLISP, 14-8
SHMAX PGP command, 2-5
SHOW PGP command, 2-6
SIMPLEX.SHP, 8-5
SIMPX.SHP, 8-5
SIN function, 14-14, 17-9
SIngle selection, 4-18
Sketch command, 12-29
Skpoly system variable, 12-29
Slide libraries, 5-23, 23-17
SLIDELIB program, 5-23, 21-16
Slides
   And scripts, 21-8
   Automating creation, 21-14
   Automating display, 21-14
   Libraries, 21-15
   Starting from DOS, 21-15
SLOT C:command, 10-10
Snap
   Isometric, 18-8
   rotated, 15-18
Solid fills, tips, 9-17
SPACE character, 4-6
   ASCII codes, 11-7
Special menu characters, Table, 4-6
SPECIALS menu, 8-14
Spelling checker with DXF, 21-31
Spreadsheets, 22-6, 22-7, 22-9, 22-11,
   22-13, 22-15, 22-17
SQRT function, 17-5
SSADD function, 12-14
SSDEL function, 12-7
SSDIFF CA function, 12-12
SSGET function, 12-6, 12-28
   SSGET "X" filtering, 12-10, 13-12 19-20
   SSGET "X" tips, 12-34
   Table of formats, 12-10
SSINTER CA function, 12-12
SSLENGTH function, 12-7, 12-9
SSLIDE C:command, 21-14
SSLIDE.SCR, 21-14
SSMEMB function, 12-7, 12-12
SSNAME function, 12-7
SSUNION CA function, 12-11
STACK, 2-8, 16-4
Standard Data Format, SDF, 14-24

Status line, 15-3
STRCASE function, 10-16, 11-5
STRCAT function, 10-6, 11-5, 11-6, 11-22
STRings, 11-1, 11-2, 11-5
   Case, 11-5
   Converting data to, 13-17
   Converting to other data types, 13-16
   Displaying, 11-9
   Formatting, 11-7
   Length limit, 11-5, 21-8
   Length tips, 11-28, 11-29, 16-5
   Printing, 11-9
   Prompts, 11-20, 11-21
   To numbers, 11-10
   Variables, 7-4
STRIP CA function, 15-17, 16-7
STRLEN function, 11-5, 21-7
Style command, 8-2
Styles
   Changing, 8-4
   Conventions, 2-10
   Symbol table access, 7-20, 13-2, 13-11
   Table data, 13-11
   Text, 8-2, 8-3
Subdirectories, see Directories
Subentities, 12-28, 13-4
   Entity name -1 DXF code, 12-19
   Subentity name -2 DXF code, 13-4
Subscript, Text, 8-11, 8-15
Subst DOS command, 23-4
SUBST function, 12-26, 12-27
SUBSTR funct, 11-5, 13-27, 14-24, 21-7
Subtract - function, 10-10
Superscript, Text, 8-11, 8-15
Support
   Bulletin boards, D-6
   Classes, D-4
   CompuServe Autodesk forum, D-4, D-5
   For this book, D-10
   Magazines, D-6
   New Riders, D-10
   Sources, D-4
   Users groups, D-6
Support files, 1-8, B-7
Suspended commands, 5-17
Symbol Tables, 13-2
   Block definition, 13-2
   How to access, benefits, 13-1
   Layers, 7-20, 13-2
   Ltypes, 13-2
   Styles, 7-20, 13-2
   Views, 13-2
Symbols, 3-1, 3-2, 11-2, 11-3
   AutoLISP, 21-10
   Creating with SET, 11-6
   In lines, 19-24
   Iso-dimensioning, 18-3
   Scale, 3-4
SYNPLANT.DWG, 22-20
Syntax, 1-2, I-6, I-9
   .LSP files, 10-9
   Attribute extract template, 22-4
   AutoLISP, 1-2, 7-9

Layers, 2-11
Macros, 4-4, 4-5
Parenthesis in LISP, 7-7
System
   AutoCAD diagram, 2-0
   Global control, 23-8
   Organization of book's, 1-1
System management, 23-1
System variables, 7-3, 16-3, C-2
   Cmdecho, 23-20
   Controlling in Lisp, 11-17
   Expert, 13-23
   Menu, in 5-22
   Menuecho, 23-20
   Personal user variables, 13-18, 13-19
   Screensize, 15-7
   Table of, annotated, C-2
   Tdcreate, Tdupdat, 23-6
   User, 13-13, 19-23
   vs. DXF, 21-25

T

T (True) NON-NIL, 10-12
Table access, see Symbol Tables
Tables
   $ menu page codes, 5-16
   Angle formats, 11-12
   ANGTOS units, 11-12
   Bigfont character map, 8-10
   Color/Pen/Linetype, 2-11
   Device names, 14-4
   Drawing factors, 6-7
   Drawing reference information, 21-26
   DXF entity codes, 12-20, 12-21
   Entity properties, default, 12-22
   Font definition codes, 8-6
   GRREAD codes, 15-13
   INITGET control bits, 11-22
   Iso-dimensioning, 18-3
   Layers, 13-11
   Linetype/Color/Pen, 2-11
   Linetypes, 13-13, 13-15
   Macro special characters, 4-6
   Modifing data, 13-15
   Pen/Linetype/Color, 2-11
   REVBLOCK attributes, 19-15
   RTOS formats, 11-11
   Scale factors, 6-7
   Special characters in macros, 4-6
   SSGET formats, 12-10
   Styles, 13-11
   Title block attributes, 19-4
   Views, 13-11
Tablet
   Iso-dimensioning, 18-12, 18-13
   Menu configuration, 5-12
TABLET menu device, 5-11
TABLET1 menu, 5-13
TBLNEXT function, 13-4, 13-6
TBLSEARCH function, 7-20, 13-2, 13-4
Tdcreate, Tdupdat system variables, 23-6

Templates, Attext, 22-2, 22-3, 22-5
TERPRI function, 11-8
TEST.DXF, 21-4
TEST.LSP, 6-12
TEST.MNU, 6-12, 6-17
Testing
   Files' existence, 14-7, 14-9
   Functions, 13-6
   Path, 14-7, 14-9
Text
   %% escape codes, 4-7, 8-7
   Alignment codes, 12-24
   AutoLISP evaluation, 7-5
   Bigfonts, 8-9
   Custom fonts, benefits, 8-1
   Curved, 11-15
   Fixed height, 2-10
   Fonts, 8-1
   Formatting with AutoLISP, 21-6
   Fractions, 8-11
   Height, 8-3
   Height table, 6-7
   Importing with AutoLISP, 21-5, 21-7
   Importing with DXF, 21-4
   In menus, 4-7
   Iso-dimensioning, 18-3
   Line editor for AutoCAD, 15-13, 15-15
   Special characters, 8-1
   Tips, 8-17, 18-16
   Styles, 2-10, 8-2, 8-3
   Subscript, Superscript, 8-11
   Writing to graphic screen, 15-2
Text Editors, 1-13, 6-12
   For AutoCAD text, 15-13, 15-15
   Integrating, 2-4
   Selecting, Tips 1-10, 1-13, D-3
TEXT menu, 6-18
Texteval, 7-5, 19-7
TEXTSCR function, 11-8, 14-26
Thickness, 12-23, 17-2
Time, logging and reporting, 22-29
TIME CA function, 19-14
Time stamping and automating, 19-13
Time system variables, 23-6
TISO menu, 18-14
Title block
   Attributes table, 19-4
   ATTSHT-D, 19-6
   Maintenance, 19-10
   System, 19-3, 19-5
   System: Illustrated, 19-0
TITLEBLK menu, 19-11, 22-23
TODAY CA function, 19-14, 19-21
Toggles, Menu label, 20-4
Translation, CAD drawings, 21-29
Transparent commands
   Macros, 5-19
   Setvars, 7-4
TRANSVIEW menu, 5-20
Trim command, B-5
TSR, memory resident programs, 1-12, B-1 - B-3, D-2

Tutorials, Tips for creating, 15-21
TYPE function, 10-3, 11-2
TYPE PGP command, 2-3

U

UANGLE CA function, 11-26, 16-10
UDIST CA funct, 11-0, 11-21, 12-5, 16-10
UINT CA funct, 11-26, 12-5, 13-25, 16-10
UKWORD CA funct, 11-27, 13-25, 16-10
UNDEF CA function, 23-7
Undefine command, 23-6
Undefine command caution, 23-8
Unit Parts, 3-2
Units, Angular, 11-12
UPD CA function, 19-23
UPDATE C:command, 19-20
UPDSCR CA function, 21-20
UPOINT CA function, 11-26, 12-5, 16-10
UREAL CA function, 11-26, 16-10
User input control, tips, 11-23, 11-29
User interface, 23-17, 23-21
    DOS Shells, D-8
    Examples, 12-5
    Informing users, benefits, 15-1
User interface functions, 16-10
    UANGLE, 11-25
    UDIST, 11-21
    UINT, 11-25
    UKWORD, 11-25
    UPOINT, 11-25
    UREAL, 11-25
    USTR, 11-25
User macros on-the-fly, 23-12
User variables, 13-13, 13-18
USTR CA function, 11-27, 16-10
Utilities, software, D-7
Utilities, Freeware, D-8
Utilities, Shareware, D-8

V

Variables
    AutoLISP, 7-2, 7-6
    Binding, 7-7
    Global, 10-5, 16-3, 16-6
    Illegal, 7-6
    Integer, 7-4
    Invalid, 7-7
    Local, 10-5
    Names, 7-7
    Personal, 13-18
    Read only, 7-4
    Real, 7-4
    String, 7-4
    System, 7-3
    Types, 7-3, 7-5
    User, 13-18, 19-23

Vector direction diagram, 8-6
Vectors, in Shapes 8-1
VER function, I-6
Versions, required, I-6, I-8
    Converting AutoCAD, 21-28
Vertexes, 12-29
VFFILE (VeriFy FILE) CA function, 14-9
VFFILE CA function, 14-16
VFPATH CA function, 21-17, 21-19
Video, ADI, I-8
Views
    Dynamic selection, 15-7
    Symbol table, 13-2
    Table data, 13-11
Virtual memory, 16-3
VMON function
    (Virtual Memory ON), 16-4
Vpoint, entity selection tip, 17-23
Vpoint command, 17-8, 17-23
Vslide command, 23-17

W

[WALLS] macro, 17-18
Wblock command, 3-3, 3-4
WHILE function, 10-17, 10-18
WHILE-NOT-EQUAL, 14-23
[WINDOWS] macro, 17-18
WLAYER C:command, 13-12
WRITE-CHAR function, 14-3
WRITE-LINE function, 14-3, 14-5, 14-15
Writing, to printer, 14-6
Writing to files, 14-5

X

XINSERT C:command, 21-22
XLISP, 10-1
XPOINT C:command, 15-10

Y

YEAR CA function, 19-14

Z

Z coordinate, 17-3
ZARC (vertical arc) C:command, 17-10
ZEROP function, 11-4, 17-7

EXERCISES

Introduction Exercises
AutoLISP, Checking Your Version, I-6
Testing for POPUPS and ICONS, I-8

Chapter 1, Getting Started Exercises
AutoLISP Automation with SH-BOX, 1-3
AutoLISP, Sample ZL command, 1-2
CONFIG.SYS and AUTOEXEC.BAT, 1-11
Copying Files to the CA-ACAD Dir, 1-8
Entity Access, a peek at , 1-6
Installing the CA DISK, 1-9
Reboot, 1-13
Subdirectory Setup, 1-7
Text Editor Test, 1-10

Chapter 2, AutoCAD System Exercises
CA.BAT, Creating , 2-7
Cleaning the Slate, 2-9
Config. CA-PROTO as Default, 2-12
Dummy ACAD.LSP File, 2-7
Text Editor, Integrating, 2-4

Chapter 3, Scaling and Block Exercises
Minserted RACKs, 3-13
Nesting Blocks, 3-11
Shape Performance Test, 3-15
Symbol Scale and Plot Scale, 3-5

Chapter 4, Menu Macro Exercises
[BUBBLE] Macro Using SELECT, 4-12
[LEADER] Macro, 4-9
Building Selection Sets in Advance, 4-17
Menu format improving, 4-19
Menu Label, making, 4-8
Multi-segment [<-LEADER] Macro, 4-11
Piping Macro with Layer Setup, 4-14
Repeating Commands and Macros, 4-16
Single Entity [LEADER->], 4-10
Writing a Simple Menu Item, 4-3

Chapter 5, Anatomy Of Menu Exercises
***BUTTONS Menu, 5-14
Adding OSNAPs to Screen Menu, 5-15
Adding Symbols to a Menu, 5-6
Icon Screen of PDMODEs, 5-24
Labeling and Switching Pages, 5-9
Point Filter Menu Page, 5-18
SETVARs on a PopUp Menu, 5-22
Slide Libraries, 5-23
Tablet, Assigning Menu Pages, 5-12
Testing Menu Independence, 5-17
Transparent Commands Macros, 5-20

Chapter 6, Menu System Exercises
Border Drawings, Making 6-6
Border Sheet Page, 6-5
CA-SETUP.MNU, 6-3
Creating the **TEXT Screen, 6-17
Development Menu Template, 6-13
Drawing Scale Menu, 6-7
Main Application Screen, 6-11
Merging Menu Files, 6-14
Testing the CA-MENU Test Menu, 6-16
Testing Setup Menu and Updating
   CA-PROTO.DWG, 6-9

Chapter 7, Little More LISP Exercises
Accessing System Variables, 7-5
Atomlist, Looking at 7-7
AutoCAD's System Variables
   SETVAR, 7-3
Changing Settings with 'SETVAR, 7-4
Dimensioning Macro [CA DIMS], 7-14
Editing with Entity Data, 7-19
GETDIST, GETANGLE, and
   GETPOINT, 7-11
Getting Input with GET, 7-10
Integrating Entity Access, Menus, 7-20
Math Functions in AutoLISP, 7-9
Menu to Save AutoLISP Data, 7-12
Pausing for Get Functions, 7-16
Quick Look at Table Access, 7-20
Working with LISP Lists, 7-16

Chapter 8, Text Styles, Fonts Exercises
Changing Text Size Using STYLE, 8-3
Integrating Bigfonts, 8-14
Integrating Spec. Char.Bigfonts, 8-11
SIMPX.SHP, Making, 8-5
Special Fractions, 8-8

Chapter 9, Linetypes, Hatches, Exercises
Adjusting Line Scale, 9-4
DASH3DOT Linetype, 9-2
Concrete Hatch Pattern, 9-8
Checkered Plate Pattern, 9-6
Hatch Menu, 9-15

Chapter 10, AutoLISP: Theory Exercises
C:SLOT.LSP Making a Function, 10-10
COMMAND Function, 10-8
DEFUN of VALDAT, 10-5
FOREACH with [MAKE PTS], 10-18
Logical Operations GETPOINT, 10-12
Making VALDAT a Command, 10-9
Program Iteration Using WHILE, 10-17
REPEAT Animation with AutoCAD, 10-16
Testing the Slot Function, 10-11
Transferring Data with VALDAT, 10-6
Trying IF with [ASK ME ?], 10-13
Using COND with a , 10-15
Using WHILE with [MAKE PTS], 10-17
VALDAT Program or Is It Data?, 10-3

Chapter 11, AutoLISP Data Exercises
ANGTOS, 11-13
ATEXT an Arc Text Command, 11-15
CHR and ASCII Functions, 11-6
Creating ANGTOC, 11-13
Expanded ASCII Codes, 11-8
GODIM Custom Dim. Command, 11-19
Making More UFUNS, 11-25
Menu Control Codes, 11-9
More Predicate Tests, 11-4
Processing Data with BOUNDP, 11-3
PROMPT, PRINT, PRIN1, PRINC, 11-10

RTOS, 11-11
String Conv. and Units Handling, 11-11
String Functions, 11-5
TEXTSCR, GRAPHSCR, TERPRI, 11-8
User Interface Function UDIST, 11-21

Chapter 12, Drawing Database Exercises
Automatic Change of Scale, 12-23
Creating Selection Sets, 12-7
Creating SSTOOLS, 12-11
DXF Extraction Function, 12-22
Examining Entities, 12-16
LASTN, Making the Function, 12-14
Modifying Data with ENTMOD, 12-26
Multiple Offset Command, 12-5
Plate Area Calculator, 12-8
RCLOUD, Making, 12-29
Selecting an Entity, 12-3
Using DXF Codes, 12-18
Using SSGET, 12-10
Variable Rescaling of Blocks, 12-27

Chapter 13, AutoLISP Table Access Exercises
Adding a SWAP Utility LEDSWAP, 13-9
Creating ETOS, 13-17
Enhancing the PVAR's Function, 13-18
Initializing a Legend Generator, 13-3
Legend Generator Program, 13-5
Outputting the Legend, 13-7
Personal Variables PVAR, 13-22
PVARs Using Linetypes, 13-14
Table Access, Views, Layers, Styles, 13-11
WLAYER Wblocking w/SSGET X, 13-12

Chapter 14, AutoLISP I/O Exercises
ANSI Library of Screen Control Functions, 14-21
Handling File Formats, 14-24
Making Functions to Check Files, 14-8
Opening and Closing Files, 14-5
PATTERN Auto Hatch Generator, 14-14
Reading, Writing Data to CON, 14-3
REFDWG, Creating, 14-27
Reference Drawing Data File, 14-26
Sample Text Screen Formatting, 14-19

Chapter 15, AutoLISP Device Exercises
DDRAW An AutoRotate Snap, 15-17
Dynamic VIEW Command, 15-7
GRDRAW to Draw Vectors, 15-6
GRREAD Accessing Data, 15-12
GRTEXT to Place Text on Screen, 15-3
Making ETEXT, 15-14
X Marks the Spot XPOINT, 15-10

Chapter 16, AutoLISP Toolkit Exercises
ACAD.LSP Toolkit, 16-6
Adding Error Handling to Toolkit, 16-11
Chain Loading ACAD.LSP Files, 16-20
Creating *ERROR* Trap Routines, 16-12
Leader/Text, Error Recovery, 16-15
Look at MEM, 16-4

Chapter 17, 3-D Manipulations Exercises
3D Curve Functions, 17-11
3D Distance Function 3DDIST, 17-5
3D Polar Function BANGLE, 17-7
DXB 2-D Dwg. From 3-D View, 17-24
Examining a 3-D Entity, 17-3
Mesh Generator 3DFILL, 17-13
More 3D Tools, 17-9
MELEV Elevation Program, 17-18
Pick and View Function PICKVPT, 17-8

Chapter 18, ISO Dim. System Exercises
ISODIM.LSP, 18-9
ISODIM Tablet Menu, 18-13
Isometric Dim. Screen Menu, 18-5
Isometric Symbols, 18-3
Testing the ISODIM System, 18-11

Chapter 19, Attrib. Data Tools Exercise
Accessing Attribute Data, 19-17
AUTOBLK.LSP Functions, 19-25
AutoBreak Block, 19-25
Automating ATTEDIT, 19-11
CA-SETUP.MNU and ATTSHT-D, 19-7
Creating UPDATE, 19-20
Revision Box Attributes REVBLOCK, 19-16
Time and Date Stamping, 19-14
Title Block with Attributes, 19-4
Updating SYNPLANT, 19-23

Chapter 20, Parametrics Exercises
DBLINE.LSP Pipes, 20-19
EL90-FB for Front/Back Elbows, 20-16
GETSIZE, 20-10
SETSIZE.LSP, 20-9
Fittings, 20-24
PIPEFITS Menu, 20-4
Material Tagging, 20-21
Pipe Fitting Data Library, 20-6
PIPEFIT Help Screen, 20-7
The EL90-S.LSP Function, 20-12

Chapter 21, AutoLISP, Script and DXF Batch Exercises
Block Library Update Program LBLOCK, 21-18
C:MSCRIPT Putting Pieces Together, 21-13
DXF File Format, 21-23
DXF Look Backwards, 21-28
AutoLISP Text Importation, 21-5
BATSCR Writing the Script, 21-11
Creating C:CSLIDE, 21-14
File Handling Functions, 21-16
GETFIL Getting the File List, 21-9
GETSCR Get Script Commands, 21-10
Importing a DXF Text File, 21-4

Chapter 22, LOTUS and dBASE Exercise
Automating the Lotus Link, 22-13
Generating Revision Records, 22-23
Importing .PRN File in AutoCAD, 22-19
REPSTAT Report Status Program, 22-21

Testing the Attribute Files, 22-20
PRO_TRAK, Using, 22-30

Chapter 23, System Controls Exercises
ANSI.SYS Key Redefinition, 23-9
Finalizing the CA-MENU, 23-19
Finalizing the CA-SETUP.MNU, 23-20
GPATH Dir. Path Management, 23-3
Integration with ACAD.MNU, 23-21
LSPSTRIP.EXE, Using, 23-15
Macro-on-the-Fly Command MFLY, 23-13
Self Loading Functions, 23-18
Undefine and Redefine Commands, 23-7

FIGURES

25 x 80 Screen Position Chart, 14-21
3-D Image Using 3DFILL, 17-15
[CA DIMS] Macro, 7-15
Accessing the Drawing Database, 12-0
Alpha and Beta Angles, 17-6
Arc Text Routine, 11-18
Array of Valves, 3-16
Arrow and Leader Comparison, 3-6
ATEXT.LSP Diagram, 11-14
AutoBreaking VALVE-B Block, 19-25
AutoCAD Graphics Screen Boxes, 15-3
AutoCAD System Organization, 2-0
AutoCAD's Entities, 12-16
AutoLISP Automation with SH-BOX, 1-3
AutoLISP LOCALS, GLOBALS and the ATOMLIST, 10-0, 10-19
Automatic Hatch Generator, 14-0
Automatic Mesh Generator, 17-0
Blank Template Page, 6-3
BLINE command, 19-24
Block and Wblock Commands, 3-3
Blocks, Files, Inserts and Shapes, 3-0
Border Drawings, 6-6
Border with Title Block, 19-3
Buttons and Screen Toggling, 5-0
BYBLOCK vs. BYLAYER, 3-10
BYLAYER, Explicit and BYBLOCK, 3-9
CA-ACAD directory, 1-8
CA-MENU Menu, 6-14
Calculated Points Fitting Sideview, 20-13
CenterLine Diagram, 8-7
Chain Loading, 16-20
Chapter 6 Menu, 6-19
Chapter 7 Menu, 7-23
Chapter 8 Menu, 8-16
Chapter 9 Menu, 9-17
Chapter 11 Menu, 11-28
Chapter 12 Menu, 12-33
Chapter 13 Menu, 13-29
Chapter 14 Menu, 14-32
Chapter 15 Menu, 15-20
Chapter 16 Menu, 16-22
Chapter 17 Menu, 17-26

Chapter 20 Menu, 20-26
Chapter 21 Menu, 21-30
Char Ascender, Descender, Baseline, 8-5
Checkered Plate Pattern, 9-7
Circle with Text, 4-5
Complex Entities, 12-28
Concrete Dots and Lines, 9-12
Concrete with Triangle Added, 9-13
COND Structure, 10-14
Creating A Database Structure, 22-26
Custom Title Block Program, 19-0
Custom UDIST Function, 11-0
DASH3DOT Linetype, 9-2
Data After Importing, 22-17
DDRAW Completed, 15-20
DDRAW Routine, 15-19
Development Boxes, 9-9
Diagram for the MELEV Program, 17-17
Diagram of C:MSCRIPT, 21-13
Door Block, 17-17
Double Line Pipe Points, 20-18
Drawing Compass, 15-16
Drawing Database Order, 12-2
Drawing Factors, 6-7
Drawing Revision Blocks, 19-19
Dxf Chart, 12-20
Elbow with Tag, 20-25
Entity Data Retrieval, 12-6, 12-16
Error Recovery System, 16-0
ETEXT Line Editor, 15-16
Example Chapter Menu, I-5
Final Hatch Pattern, 14-17, 14-19
Final Pattern, 9-14
First Dash Line Family, 9-11
First Dot Line Family, 9-10
Fittings in Three Views, 20-18
Five-sided Aggregate, 9-12
FOREACH Structure, 10-18
Formatted Text from FTEXT.LSP, 21-8
Formulas and Triangles, 17-7
Fractions, Subscripts, Superscripts, 8-15
Front and Back Views, 20-16
Generated Elevation, 17-23
GODIM Dimensioning Command, 11-18
GRTEXT and GRDRAW Functions, 15-0
Hatch Pattern Components, 9-5
Heap and Stack Use, 16-4
Help Slide, 23-18
Highlighted and Unhighlighted Menu Boxes, 15-5
Icons Screen Called from a Macro, 5-26
IF Branch, 10-13
Import Macro, 22-15
Imported File, 22-8
Imported Text in Drawing, 22-19
Importing Procedure, 22-7
Importing Project Records, 22-27
Initial Hatch Pattern, 14-16

FIGURES X—23

Inserted Valves, 5-7
Irregular Boundaries, 9-15
ISO Dimensioning System, 18-0
Isometric Arrows Diagram, 18-3
Isometric Menu, 18-5, 18-16
Isometric Piping Drawing, 18-12
Isometric Reference Drawing, 18-4
Isometric TABLET1 Template, 18-13
LBLOCK Block Lib Update Prog, 21-1
Leader Text using Error Recovery, 16-18
LEADER Macro, 4-9
LEGEND Program, 13-0
Legend Symbol Blocks Library, 13-2
Linetype and Hatch Organization, 9-0
Linetype Pattern Adjusting, 9-4
Marked Reference Drawing, 14-30
Menu Device Names, 5-12
Menu Macro Using AutoLISP, 7-0
Minserted Rack, 3-14
Multi-segment Pline Leader, 4-12
Multiple Offset Command, 12-6
Original ARROW at 96:1, 3-6
Output Legend, 13-8
Output Range, 22-11
Parametric Elbow Side View, 20-15
Parametric Elbow Fittings, 20-6
Parsing Input Data, 22-9
PGP Going Outside AutoCAD, 2-2
Pipe Fitting Material Spreadsheet, 22-12
Pipe Flow Valve, 4-14
Pipe Lines and Valves, 4-16
Pipe Segments Between Flanges, 20-20
PIPEFITS Menu, 20-3
Plate with Holes and Cutouts, 12-4
Point Filters Screen, 5-19
Pointing Arrow, 3-4
Pointing Arrow Block, 4-8
Populated Drawing, 20-23
PopUp Menu Locations, 5-21
Popup Menu Selection, 5-22
Prettier Screen Menu NEXT Paging, 5-9
Print Out Range, 22-18
PRO_TRAK Main Menu, 22-24, 22-31
Process Flow Diagram, 5-4
PROGN Structure, 10-14
Program Flow Passing Data to Lotus, 22-0
Program Flow, Parametric System, 20-0
Project Log Time, 22-30
Proposed Hatch Pattern, 9-8
Prototype Layers, 2-11
PVAR Program, 13-21
Raw Labels, Numeric Data in Lotus, 22-11
Reference Drawing as Text, 14-31
REPEAT Structure, 10-15
Revision Block with Attributes, 19-15
Revision Cloud, 12-32

Revision History of Drawing, 22-28
Sample Entities to Examine, 21-23
Sample Screen Shot, I-7
Scaling of Parts and Symbols, 3-2
Screen Menu Labels, 4-8
Screen Menu Macros, 4-0
Screen Menu with NEXT Paging, 5-8
Selecting the Input Columns, 22-10
SETUP Menu System, 6-0
Shaded Image in AutoShade, 17-16
Shadow Box, 1-5
Shadow Box with Text, 1-6
Single Entity Leader, 4-11
Slanted Style, 8-4
Small Valve, 5-5
Solid Slot Hole, 10-10
Some Standard Text, 8-3
Special 1/8 Fraction, 8-8
SSUNION, SSDIFF and SSINTER, 12-11
Standard Text Screen, 6-17
Switch Board Terminal, 3-10
Symbols, Files and Labels, 5-5
System Organization for CA, 1-0
Tablet Menu, 5-12
Tally of Pipe Length and Bolts, 22-17
Ten Button Cursor, 5-14
Terminals, 3-11
TEST.MNU Item and Page Labels, 5-11
TEST.MNU with CA-SETUP Menu, 6-9
Text Alignment Points, 12-23
Text Style and Font Organization, 8-0
Three Dimensional Axis System, 17-5
Title Block Menu, 19-11
Title Block with Attributes, 19-5
Title Block with Intelligent Data, 19-8
Translator Exerciser, 21-29
Trigonometry of a Curve, 17-11
Two RACK Inserts, 3-12
Types of Menus, 5-2
Typical Valve, 3-15
Variables and Fittings, 20-11
Vector Directions, 8-6
WHILE Program Loop, 10-16
Worksheet Display, 22-6
XPOINT Completed, 15-12
XPOINT in Progress, 15-11

THE NEW RIDERS' LIBRARY

COMPUTER AIDED DRAFTING AND DESIGN

CUSTOMIZING AutoCAD

A Complete Guide to Integrating AutoLISP Menus, Macros and More.
576 Pages, 180 Illustrations
ISBN 0-934035-18-0, $34.95

J. Smith and R. Gesner

This first comprehensive, personal guide to the customization of AutoCAD. *CUSTOMIZING AutoCAD* offers a true systems approach, giving users the concepts and techniques needed to create their own personal application programs. It features "inside" information unavailable elsewhere. It shows how to use AutoLISP to take control of the AutoCAD drawing editor; enhance it and have personal commands on-line at all times; use AutoLISP to access everything in an AutoCAD drawing database; develop advanced AutoLISP tools to build upon AutoCAD's 3-D commands, and more. It covers key database topics.

CUSTOMIZING AutoCAD DISK SET

Floppy Macro Disk Set (2 disks), $28.90

In order to reduce keyboard entry time and save time debugging AutoLISP routines, we offer a two-disk set for use with *CUSTOMIZING AutoCAD*. The disk set includes an AutoLISP Library plus all the supporting files. *CUSTOMIZING AutoCAD* supports the latest AutoCAD release. Additional AutoLISP routines and programs are available on a bonus disk.

INSIDE AutoCAD

A Teaching Guide to the AutoCAD Microcomputer Design and Drafting Program.
320 Pages, 325 Illustrations
ISBN 0-934035-08-3, Rev. Ed. $34.95

D. Raker and H. Rice.

The New Riders' Library of CAD books centers around our book *INSIDE AutoCAD*. It is the true reading reference for the AutoCAD program no matter what version you use. *INSIDE AutoCAD* provides you with the information you need. It can be used as a command syntax guide or as a guide to creating 3-D drawings and using advanced AutoCAD editing commands. *INSIDE AutoCAD* has sold over 200,000 copies and is the most widely used book on CAD across the United States. *INSIDE AutoCAD* is the key you need to learn AutoCAD.
It supports the latest AutoCAD release.

WORKING OUT WITH AutoCAD

An Step-by-Step Guide to Building Professional CAD Skills.
304 Pages, 300 Illustrations
ISBN 0-934035-10-5, $29.95

M. Lubow

The ideal advanced AutoCAD guide is Martha Lubow's *WORKING OUT WITH AutoCAD*. *WORKING OUT WITH AutoCAD* contains seven exercise sets and 23 structured exercises. It covers advanced setup techniques, including advanced use of MS-DOS. Integrated into the drawing exercises are 200 menu items and over 50 AutoLISP macros. Using these macros will help you learn AutoCAD drawing customization techniques.

WORKOUT DISK

Floppy Macro Disk, $14.95

In order to reduce keyboard entry time, we offer an inexpensive floppy disk for use with *WORKING OUT WITH AutoCAD*. The disk contains all the customization macros found in the book, plus special bonus AutoLISP routines. The disk will have you up and learning today!

New Riders' Library

STEPPING INTO CAD
A Complete Guide to Using AutoCAD for Technical Drafting
320 Pages, 140 Illustrations
ISBN 0-934035-05-9, Rev. Ed. $29.95

M. Merickel

STEPPING INTO CAD, The Professional Edition, is the latest version of this best selling drafting guide. It is a self-paced technical drafting guide, containing seven steps and 42 structured CAD exercises to teach Computer Aided Drafting Skills. It has been revised to cover the latest AutoCAD version. It includes extensive appendices and advanced Drafting ANSI Y14.5 Tablet and Screen menus. *STEPPING INTO CAD* is used by Autodesk, Inc. in their in-service CAD instructor training program.

DRAFTING DISK
Floppy Macro Disk, $14.95

In order to reduce keyboard entry time, we offer an inexpensive floppy disk for use with *STEPPING INTO CAD*. The Drafting Disk includes both English and Metric Drafting menus plus drafting macros and AutoLISP routines. The drafting diskette is available for use with the latest edition.

STEPPING INTO CAD has a companion INSTRUCTOR'S GUIDE. It provides a model CAD curriculum, insights into student drawing management, and aids for evaluating *STEPPING INTO CAD* drawings.

THE MICROCOMPUTER CAD MANUAL
Everything You Want to Know About CAD on a Micro.
350 Pages, 150 Illustrations
ISBN 0-934035-04-0, $27.95

H. Rice and D. Raker

Because the field of Micro-CAD is changing so fast, you need to have all the hardware and software information in one convenient place. You have it in *THE MICROCOMPUTER CAD MANUAL. THE MICRO-CAD MANUAL* is designed to give you everything you need to know in order to understand, evaluate, buy and use microcomputer-based systems. The *MANUAL* shows how to put together an evaluation bid list, including workstation integration, training and support. The *MANUAL* also includes an extensive list of CAD software and hardware resource references. *THE MICROCOMPUTER CAD MANUAL* is the book to buy, before you buy.

OTHER GOODIES

COOKIES — Low Cost Software

START YOUR COMPUTER DAY OFF RIGHT WITH *COOKIES!* — *COOKIES* is 100% IBM compatible.

IBM compatible Disk and booklet $6.95

Tired and listless in the morning when you boot your computer? *COOKIES* is a program that will give you a fortune cookie every time it is run. Install it, and get a cookie every day before you start work.

COOKIES also can tell you one-liners, helpful tips or whatever you decide! If the cookies get stale, just change the *COOKIE* file and mix up a batch of your own. Of course, if you don't have the time to mix up your own, check with us and see what we'er "cooking" up.

ORIGINAL COVER DESIGN POSTER ART
Only $9.95

You ask for it! We've got it. So many of you have asked about the cover for our *MICROCOMPUTER CAD MANUAL* that we have made it available as a poster. This 20" x 24" poster (with cover text removed) is a real work of art. Printed on glossy poster paper stock, suitable for framing, this just might be the idea gift to brighten up any office, den or computer user's face.

Desktop Publishing

Inside Xerox Ventura Publisher

A Guide to Professional-Quality Desktop Publishing on the IBM PC.
328 Pages, 130 Illustrations
ISBN 0-934035-13-X, $19.95

J. Cavuoto and Jesse Berst

Written by two of the industry's leading experts, *INSIDE XEROX VENTURA PUBLISHER* takes readers from beginning commands through advanced concepts in one easy-to-understand book. The book was produced with the cooperation and technical review of Xerox Corporation and Ventura Software, and fully covers the new, more powerful Ventura Version 1.1.

INSIDE XEROX VENTURA PUBLISHER includes inside information unavailable from any other source — dozens of undocumented short-cuts, advanced functions, and tips on how the use CAD images in your text. This invaluable how-to manual provides the information you need to unlock the inner secrets of *XEROX VENTURA PUBLISHER*.

Publishing Power with Ventura

The Complete Teaching Guide to Xerox Ventura Publisher.
576 Pages, 200 Illustrations
ISBN 0-934035-19-9, $24.95

M. Lubow and J. Berst

A comprehensive, hands-on tutorial for desktop publishing which provides step-by-step techniques for creating newsletters, marketing materials, corporate reports, directories, technical documents and more. You will learn how to use Ventura to integrate graphics with text from word processing software and how to master Ventura's style sheets to get better documents faster. *PUBLISHING POWER WITH VENTURA* is the ideal book to help both the beginner and the advanced user get maximum power and performance from the Xerox Ventura program.

Publishing Power Disk

IBM-compatible floppy disk, $14.95

This companion disk to *PUBLISHING POWER WITH VENTURA* contains every text file and style sheet used in the book. Beginners save time and effort because they don't have to type in text files to follow along with the book's examples. Advanced users also save time because the style sheets contain dozens of preformatted tags. Users are spared the effort of redefining basic tags, margins, columns, etc., so they can concentrate on advanced skills and techniques.

ORDER FORM

Yes, please send me the productivity-boosting materials I have checked below. Make check payable to New Riders Publishing.

☐ Check enclosed.

☐ Charge to my credit card:

☐ VISA ☐ MASTER CARD

Card # _____

Expiration: _____

Signature: _____

Name: _____

Company: _____

Address: _____

City _____

State: _____ Zip: _____

Phone: _____

The easiest way to order is to pick-up the phone and call (818) 9⬚-5392 between 9:00 AM and 5:00 PM PST. Please have your cre⬚ card readily available and your order can be placed in a snap!

Quantity	Description of Item	Unit Cost	Total Cost
	CUSTOMIZING AutoCAD	$34.95	
	CUSTOMIZING AutoCAD DISK SET	$29.90	
	INSIDE AutoCAD	$34.95	
	STEPPING INTO CAD PROFESSIONAL EDITION	$29.95	
	STEPPING INTO CAD DRAFTING MACRO DISK	$14.95	
	WORKING OUT WITH AutoCAD	$29.95	
	WORKING OUT WITH AutoCAD MACRO DISK	$14.95	
	THE MICROCOMPUTER CAD MANUAL	$27.95	
	COOKIES [Put the fun back into your computer!]	$6.95	
	COVER POSTER MICROCOMPUTER CAD MANUAL	$9.95	
	INSIDE XEROX VENTURA PUBLISHER	$19.95	
	PUBLISHING POWER WITH VENTURA	$24.95	
	Shipping and Handling: see information below.		
	Sales Tax: Californians please add 6.5% sales tax.		
	TOTAL:		

Send to:

New Riders Publishing
P.O. Box 4846
Thousand Oaks CA 91360
(818) 991-5392

Shipping and Handling: $3.50 for the first book and $1.00 for each additional book. Floppy disk add $1.50 for shipping and handling. Add $15.00 per book for overseas shipping and handling. If you have to have it NOW! We can ship product to you in 24 to 48 hours. For an additional $5.00 RUSH CHARGE for processing plus the actual cost of air freight you'll be able to recieve your item over night or in 2 days.

JDK-1

BONUS DISK GIFT CERTIFICATE

Name: _____

Company: _____

Address: _____

City _____

State: _____ Zip: _____

CUSTOMIZING AutoCAD has a companion two-disk set which includes all of the AutoLISP routines found in the book. It can be ordered on the other side of this form.

This certificate is good for the free BONUS disk, just return this card (along with $2.00 shipping and handling) and we will send you your CUSTOMIZING AutoCAD BONUS DISK. This disk contains many new AutoLISP routines not found anywhere else, along with ready-to-run versions of the dBASE program found in the book.

Quantity	Description of Item	Unit Cost	Total Cost
1	CUSTOMIZING AutoCAD BONUS DISK	FREE	FREE
	Shipping and Handling:		$2.00
	TOTAL:		$2.00

To order: Fill in the ORDER FORM, fold, and mail

NO POSTAGE NECESSARY IF MAILED IN THE UNITED STATES

BUSINESS REPLY MAIL
FIRST CLASS PERMIT NO. 53   THOUSAND OAKS, CA

POSTAGE WILL BE PAID BY

NEW RIDERS PUBLISHING
P.O. Box 4846
Thousand Oaks, CA 91360